Advanced Power MOSFET Concepts

B. Jayant Baliga

# Advanced Power MOSFET Concepts

Springer

B. Jayant Baliga
Department of Electrical and Computer
Engineering
North Carolina State University
Raleigh NC 27695
USA

ISBN 978-1-4419-5916-4 e-ISBN 978-1-4419-5917-1
DOI 10.1007/978-1-4419-5917-1
Springer New York Heidelberg Dordrecht London

Library of Congress Control Number: 2010929769



Printed on acid-free paper

Springer is part of Springer Science+Business Media (www.springer.com)

# Dedication

*The author would like to dedicate this book to his wife, Pratima, for her unwavering support throughout his career devoted to the enhancement of the performance and understanding of power semiconductor devices.*

# Preface

With increased awareness of the adverse impact on the environment resulting from carbon emissions into the atmosphere, there is a growing demand for improving the efficiency of power electronic systems. Power semiconductor devices are recognized as a key component of all power electronic systems. It is estimated that at least 50 percent of the electricity used in the world is controlled by power devices. With the wide spread use of electronics in the consumer, industrial, medical, and transportation sectors, power devices have a major impact on the economy because they determine the cost and efficiency of systems. After the initial replacement of vacuum tubes by solid state devices in the 1950s, semiconductor power devices have taken a dominant role with silicon serving as the base material. These developments have been referred to as the *Second Electronic Revolution*.

In the 1970s, the power MOSFET product was first introduced by International Rectifier Corporation. Although initially hailed as a replacement for all bipolar power devices due to its high input impedance and fast switching speed, the silicon power MOSFET has successfully cornered the market for low voltage (<100 V) and high switching speed (>100 kHz) applications but failed to make serious inroads in the high voltage arena. This is because the on-state resistance of silicon power MOSFETs increases very rapidly with increase in the breakdown voltage. The resulting high conduction loss, even when using larger more expensive die, degrades the overall system efficiency.

The large on-state voltage drop for high voltage silicon power MOSFETs and the large drive current needed for silicon power bipolar transistors encouraged the development of the insulated gate bipolar transistor (IGBT) [1]. First commercialized in the early 1980s, the IGBT has become the dominant device used in all medium and high power electronic systems in the consumer, industrial, transportation, and military systems, and even found applications in the medical sector. The US Department of Energy has estimated that the implementation of IGBT-based variable speed drives for controlling motors is producing an energy savings of over 2 quadrillion btus per year, which is equivalent to 70 Giga-Watts of power. This energy savings eliminates the need for generating electricity from 70 coal-fired power-plants resulting in reducing carbon dioxide emissions by over one-Trillion pounds each year.

With on-going investments in renewable energy sources such as wind and solar power that utilize power semiconductor device in inverters, it is anticipated that there will be an increasing need for technologists trained in the discipline of designing and manufacturing power semiconductor devices. My recently published textbook [2] provides a comprehensive analysis of the basic power rectifier and transistor structures. This textbook has been complemented with a monograph on "Advanced Power Rectifier Concepts" to familiarize students and engineering professionals with structures that exhibit improved performance attributes.

This monograph introduces the reader to advanced power MOSFET concepts that enable improvement of performance of these transistor structures. For the convenience of readers, analysis of the basic transistor structures, with the same voltage ratings as the novel device structures, have been included in the monograph to enable comparison of the performance. As in the case of the textbook, analytical expressions that describe the behavior of the advanced power MOSFET structures have been rigorously derived using the fundamental semiconductor Poisson's, continuity, and conduction equations in this monograph. The electrical characteristics of all the power MOSFETs discussed in this book can be computed using these analytical solutions as shown by typical examples provided in each section. In order to corroborate the validity of these analytical formulations, I have included the results of two-dimensional numerical simulations in each section of the book. The simulation results are also used to further elucidate the physics and point out two-dimensional effects whenever relevant. Due to increasing interest in the utilization of wide band-gap semiconductors for power devices, the book includes the analysis of silicon carbide structures.

In the first chapter, a broad introduction to potential applications for power devices is provided. The electrical characteristics for ideal power MOSFETs are then defined and compared with those for typical devices. The second and third chapters provide analyses of the planar DMOSFET structure and the trench-gate UMOSFET structure with 30-V blocking capability, which can be used as a benchmark for understanding the improvements achieved using the advanced device concepts. The analysis includes the on-resistance, the input capacitance, the gate charge, and the output characteristics.

The next four chapters are devoted to various advanced power MOSFET structures that allow improvement in the performance of devices with 30-V blocking capability. The fourth chapter discusses on the "Shielded Channel Planar Power MOSFET" structure, which allows a significant reduction in the gate charge while achieving a specific on-resistance close to that of the UMOSFET structure. The fifth chapter discusses the power CC-MOSFET structure, which utilizes the two-dimensional charge coupling effect to reduce the specific on-resistance by an order of magnitude. This structure is favorable for use as a synchronous rectifier in the sync-buck circuit topology used in voltage regulator modules for providing power to microprocessors in computers.

The next two chapters are devoted to high-voltage silicon device structures that utilize the charge-coupling concept to reduce the resistance of the drift region. In chapter six, the charge-coupling phenomenon is accomplished by using a graded doping profile in conjunction with an electrode embedded in an oxide coated trench to create the power GD-MOSFET structure. In chapter seven, the charge-coupling phenomenon is accomplished with adjacent p-type and n-type layers in the drift region to create the power SJ-MOSFET structure.

Chapter eight provides a detailed discussion of the body-diode within the various silicon power MOSFET structures. The body-diode can be used in place of the fly-back rectifier utilized in the H-bride circuit commonly used for motor control applications. It is demonstrated in this chapter that the judicious utilization of a Schottky contact within the power MOSFET cell structure can greatly improve the reverse recovery behavior of the body-diode.

Improvement in the performance of high voltage power MOSFET structures can also be achieved by replacing silicon with silicon carbide as the base material [3]. The much larger breakdown field strength for 4H-SiC allows increasing the doping concentration in the drift region by a factor of 200-times while shrinking the thickness of the drift region by one-order of magnitude. However, the silicon power MOSFET structure must be modified to shield the gate oxide from the much larger electric fields prevalent in silicon carbide to avoid rupture. In addition, the base region must be shielded to avoid reach-through breakdown. The on-resistance of these devices becomes limited by the channel resistance.

The final chapter provides a comparison of all the power MOSFET structures discussed in this book. The devices are first compared for the 30-V rating suitable for VRM applications and then with the 600-V rating suitable for motor control applications. In addition, the performance of all the devices is compared over a wide range of blocking voltages to provide a broader view.

I am hopeful that this monograph will be useful for researchers in academia and to product designers in the industry. It can also be used for the teaching of courses on solid state devices as a supplement to my textbook [2].

December, 2009 B. Jayant Baliga
Raleigh, NC

## References


1. B.J. Baliga, "How the Super-Transistor Works", Scientific American Magazine, Special Issue on 'The Solid-State-Century', pp. 34–41, January 22, 1988.
2. B.J. Baliga, "Fundamentals of Power Semiconductor Devices", Springer Scientific, New York, 2008.
3. B.J. Baliga, "Silicon Carbide Power Devices", World Scientific Press, Singapore, 2006.

# Contents

**1 Introduction** .......... 1
1.1 Ideal Power Switching Waveforms .......... 2
1.2 Ideal and Typical Power MOSFET Characteristics .......... 3
1.3 Typical Power MOSFET Structures .......... 5
1.4 Ideal Drift Region for Unipolar Power Devices .......... 6
1.5 Charge-Coupled Structures: Ideal Specific On-Resistance .......... 8
1.6 Revised Breakdown Models for Silicon .......... 12
1.7 Typical Power MOSFET Applications .......... 18
1.7.1 DC-DC Sync-Buck Converter .......... 18
1.7.2 Variable-Frequency Motor Drive .......... 19
1.8 Summary .......... 21
References .......... 21

**2 D-MOSFET Structure** .......... 23
2.1 The D-MOSFET Structure .......... 23
2.2 Power D-MOSFET On-Resistance .......... 25
2.2.1 Channel Resistance .......... 28
2.2.2 Accumulation Resistance .......... 29
2.2.3 JFET Resistance .......... 29
2.2.4 Drift Region Resistance .......... 31
2.2.5 $N^+$ Substrate Resistance .......... 32
2.2.6 Drain and Source Contact Resistance .......... 32
2.2.7 Total On-Resistance .......... 33
2.3 Blocking Voltage .......... 37
2.3.1 Impact of Edge Termination .......... 38
2.3.2 Impact of Graded Doping Profile .......... 39
2.4 Output Characteristics .......... 44
2.4.1 Simulation Example .......... 45
2.5 Device Capacitances .......... 46
2.5.1 Simulation Example .......... 49

2.6 Gate Charge ... 51
2.6.1 Simulation Example ... 52
2.7 Device Figures of Merit ... 54
2.8 Discussion ... 56
References ... 61

**3 U-MOSFET Structure** ... 63
3.1 The U-MOSFET Structure ... 63
3.2 Power U-MOSFET On-Resistance ... 66
3.2.1 Channel Resistance ... 67
3.2.2 Accumulation Resistance ... 68
3.2.3 Drift Region Resistance ... 68
3.2.4 Total On-Resistance ... 69
3.3 Blocking Voltage ... 73
3.3.1 Impact of Edge Termination ... 74
3.3.2 Impact of Graded Doping Profile ... 74
3.4 Output Characteristics ... 80
3.4.1 Simulation Example ... 80
3.5 Device Capacitances ... 81
3.5.1 Simulation Example ... 84
3.6 Gate Charge ... 86
3.6.1 Simulation Example ... 88
3.7 Device Figures of Merit ... 90
3.8 Thick Trench Bottom Oxide Structure ... 92
3.8.1 On-Resistance ... 92
3.8.2 Reverse Transfer Capacitance ... 92
3.8.3 Gate Charge ... 94
3.8.4 Device Figures-of-Merit ... 95
3.9 High Voltage Devices ... 101
3.9.1 Simulation Results ... 101
3.10 Inductive Load Turn-Off Characteristics ... 106
3.10.1 Simulation Results ... 111
3.11 Discussion ... 112
References ... 117

**4 SC-MOSFET Structure** ... 119
4.1 The SC-MOSFET Structure ... 120
4.2 Power SC-MOSFET On-Resistance ... 122
4.2.1 Channel Resistance ... 124
4.2.2 Accumulation Resistance ... 124
4.2.3 JFET Resistance ... 125
4.2.4 Drift Region Resistance ... 126
4.2.5 Total On-Resistance ... 127

4.3 Blocking Voltage ........ 133
4.3.1 Impact of Edge Termination ........ 133
4.4 Output Characteristics ........ 138
4.4.1 Simulation Example ........ 139
4.5 Device Capacitances ........ 139
4.5.1 Simulation Example ........ 144
4.6 Gate Charge ........ 147
4.6.1 Simulation Example ........ 149
4.7 Device Figures of Merit ........ 151
4.8 Discussion ........ 153
References ........ 158

**5 CC-MOSFET Structure** ........ 159
5.1 The CC-MOSFET Structure ........ 160
5.2 Charge-Coupling Physics and Blocking Voltage ........ 162
5.2.1 Simulation Results ........ 173
5.3 Power CC-MOSFET On-Resistance ........ 186
5.3.1 Channel Resistance ........ 187
5.3.2 Accumulation Resistance for Current Spreading Region ........ 188
5.3.3 Drift Region Resistance ........ 189
5.3.4 Total On-Resistance ........ 190
5.4 Output Characteristics ........ 195
5.4.1 Simulation Example ........ 195
5.5 Device Capacitances ........ 196
5.5.1 Simulation Example ........ 206
5.6 Gate Charge ........ 209
5.6.1 Simulation Example ........ 214
5.7 Device Figures of Merit ........ 217
5.8 Edge Termination ........ 220
5.8.1 Simulation Example ........ 222
5.9 High Voltage Devices ........ 224
5.9.1 Simulation Results ........ 224
5.10 Process Sensitivity Analysis ........ 232
5.11 Discussion ........ 235
References ........ 239

**6 GD-MOSFET Structure** ........ 241
6.1 The GD-MOSFET Structure ........ 242
6.2 Charge-Coupling Physics and Blocking Voltage ........ 244
6.2.1 Simulation Results ........ 250
6.3 Power GD-MOSFET On-Resistance ........ 263
6.3.1 Channel Resistance ........ 265
6.3.2 Accumulation Resistance for Current Spreading Region ........ 266

6.3.3 Drift Region Resistance ........ 266
6.3.4 Total On-Resistance ........ 268
6.4 Output Characteristics ........ 271
6.4.1 Simulation Example ........ 272
6.5 Device Capacitances ........ 273
6.5.1 Simulation Example ........ 276
6.6 Gate Charge ........ 278
6.6.1 Simulation Example ........ 280
6.7 Device Figures of Merit ........ 283
6.8 Edge Termination ........ 285
6.9 High Voltage Devices ........ 285
6.9.1 Simulation Results ........ 286
6.10 Process Sensitivity Analysis ........ 306
6.11 Inductive Load Turn-Off Characteristics ........ 310
6.11.1 Simulation Results ........ 315
6.12 Discussion ........ 317
References ........ 322

**7 SJ-MOSFET Structure** ........ 323
7.1 The SJ–MOSFET Structure ........ 324
7.2 Charge–Coupling Physics ........ 326
7.2.1 Simulation Results ........ 329
7.3 Power SJ-MOSFET On-Resistance ........ 346
7.3.1 Channel Resistance ........ 350
7.3.2 Accumulation Resistance for Current Spreading Region ........ 351
7.3.3 Drift Region Resistance ........ 351
7.3.4 Total On-Resistance ........ 353
7.4 Output Characteristics ........ 357
7.4.1 Simulation Example ........ 357
7.5 Device Capacitances ........ 357
7.5.1 Simulation Example ........ 364
7.6 Gate Charge ........ 367
7.6.1 Simulation Example ........ 369
7.7 Device Figures of Merit ........ 371
7.8 Edge Termination ........ 373
7.8.1 Simulation Example ........ 374
7.9 High Voltage Devices ........ 378
7.9.1 Simulation Results ........ 378
7.10 Process Sensitivity Analysis ........ 381
7.11 Inductive Load Turn-Off Characteristics ........ 385
7.11.1 Simulation Results ........ 389
7.12 Discussion ........ 391
References ........ 396

**8 Integral Diode** .... 399
8.1 Power MOSFET Body Diode .... 400
8.2 Computer Power Supplies .... 400
8.2.1 Power U–MOSFET Structure .... 402
8.2.2 Power CC–MOSFET Structure .... 407
8.2.3 Power JBSFET Structure .... 412
8.3 Motor Control Application .... 424
8.3.1 Power U–MOSFET Structure .... 425
8.3.2 Power JBSFET Structure .... 430
8.3.3 Power GD–MOSFET Structure .... 437
8.3.4 Power GD–JBSFET Structure .... 443
8.3.5 Power SJ–MOSFET Structure .... 461
8.3.6 Power SJ–JBSFET Structure .... 467
8.4 Discussion .... 472
8.4.1 Low–Voltage Devices .... 473
8.4.2 High–Voltage Devices .... 474
References .... 476

**9 SiC Planar MOSFET Structures** .... 477
9.1 Shielded Planar Inversion-Mode MOSFET Structure .... 478
9.1.1 Blocking Mode .... 479
9.1.2 Threshold Voltage .... 484
9.1.3 On–State Resistance .... 486
9.1.4 Capacitances .... 493
9.1.5 Gate Charge .... 496
9.1.6 Device Figures of Merit .... 498
9.1.7 Inductive Load Turn-Off Characteristics .... 499
9.1.8 Body-Diode Characteristics .... 503
9.2 Shielded Planar ACCUFET Structure .... 504
9.2.1 Blocking Mode .... 505
9.2.2 Threshold Voltage .... 510
9.2.3 On–State Resistance .... 512
9.2.4 Capacitances .... 517
9.2.5 Gate Charge .... 519
9.2.6 Device Figures of Merit .... 522
9.2.7 Inductive Load Turn-Off Characteristics .... 523
9.2.8 Body–Diode Characteristics .... 527
9.3 Discussion .... 531
References .... 533

**10 Synopsis** .... 535
10.1 Computer Power Supplies .... 536
10.1.1 Inadvertent Turn–On Suppression .... 537
10.1.2 Device Active Area .... 539

10.1.3 Switching Power Losses ..... 540
10.1.4 Input Capacitance ..... 540
10.1.5 Device Comparison ..... 540
10.2 High Voltage Motor Control ..... 543
10.3 Device Comparison ..... 549
10.4 Summary ..... 552
References ..... 552

**About the Author** ..... 553

**Index** ..... 557

# Chapter 1
# Introduction

Power devices are required for applications that operate over a broad spectrum of power levels as shown in Fig. 1.1 [1]. Based up on this figure, the applications can be broken down into several categories. The first category is applications that require low operating current (typically less than 1 A) levels. These applications, such as display drives, usually require a large number of transistors that must be capable of blocking up to 300 V. The small size of the low-current transistors allows their integration on a single chip with control circuits to provide a cost-effective solution.

The second category is applications where the operating voltage of the power circuit is relatively small (<100 V). Typical examples are automotive electronics and power supplies used in desktop computers and laptops. Silicon power MOSFET structures offer the best performance for these applications because of their low on-resistance and fast switching speed. This monograph describes a variety of power MOSFET structures that enable enhancement in their operating characteristics.

The third category is applications with high operating voltages (above 200 V). Typical examples are lamp ballasts, consumer appliances that utilize motors, and electric vehicle drives. The on-resistance of conventional silicon power MOSFET structures is too large to serve these applications. Consequently, these applications utilize silicon insulated gate bipolar transistors (IGBTs). The silicon IGBT combines the physics of the MOSFET structure with the physics of the bipolar transistor structure. Although the silicon IGBT has become ubiquitous for high voltage power electronic applications, silicon power MOSFETs that utilize two-dimensional charge coupling can be competitive with silicon IGBTs as shown in this monograph. In addition, power MOSFET structures built using silicon carbide as the base material have been shown to exhibit very promising characteristics for applications that require blocking voltages of up to 5,000 V [2]. Consequently, this monograph includes the discussion of the power MOSFET structures that are specially configured to obtain a high performance from silicon carbide.

B.J. Baliga, *Advanced Power MOSFET Concepts*,
DOI 10.1007/978-1-4419-5917-1_1, 

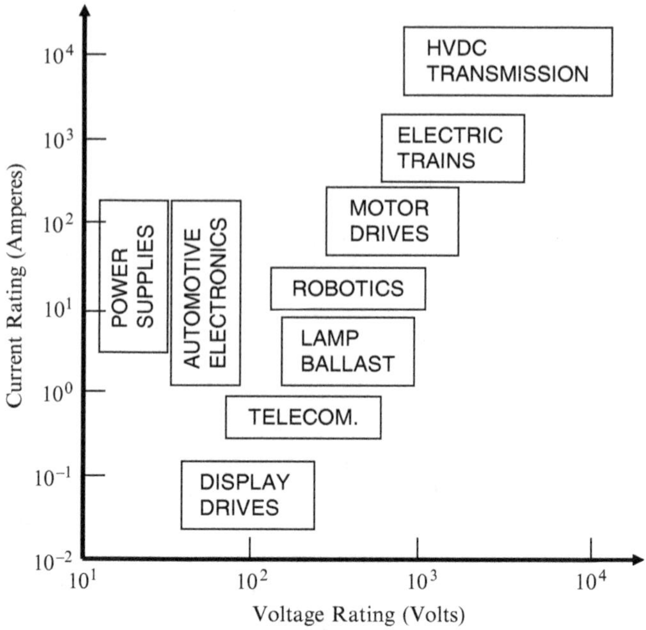


**Fig. 1.1** Applications for power devices

## 1.1 Ideal Power Switching Waveforms

An ideal power device must be capable of controlling the flow of power to loads with zero power dissipation. The loads encountered in systems may be inductive in nature (such as motors and solenoids), resistive in nature (such as heaters and lamp filaments), or capacitive in nature (such as transducers and LCD displays). Most often, the power delivered to a load is controlled by turning-on a power switch on a periodic basis to generate pulses of current that can be regulated by a control circuit. The ideal waveforms for the power delivered through a power switch are shown in Fig. 1.2. During each switching cycle, the switch remains on for a time up to $t_{ON}$ and maintains an off-state for the remainder of the period T. This produces pulses of current that flow through the circuit as controlled by the turning-on of power switches. For an ideal power switch, the voltage drop during the on-state is zero resulting in no power dissipation. Similarly, during the off-state, the (leakage) current in the ideal power switch is zero resulting in no power dissipation. In addition, it is assumed that the ideal power switch makes the transition between the on-state and off-state instantaneously resulting in no power loss as well.

Typical power MOSFET structures exhibit a finite voltage drop in the on-state and leakage current flow in the off-state. In addition, the power MOSFET structure exhibits power losses during the turn-on and turn-off transients. These power losses are related to their large terminal capacitances which must be charged

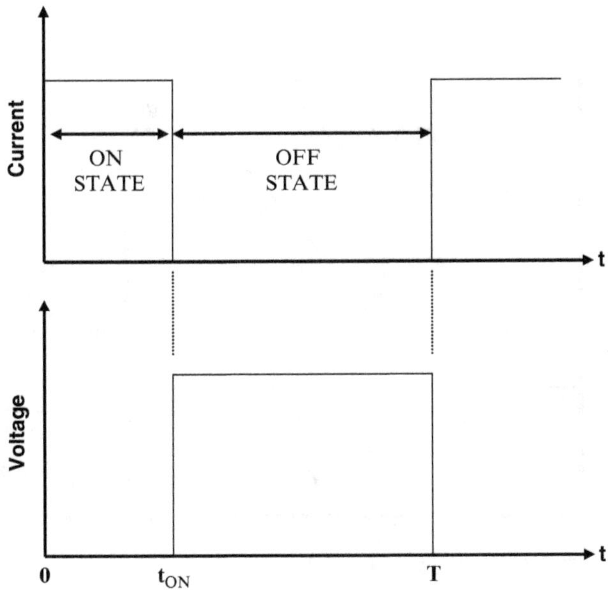


**Fig. 1.2** Ideal switching waveforms for power delivery

and discharged during each operating cycle. The novel power MOSFET structures that are discussed in this monograph were created to reduce the on-resistance and capacitance.

## 1.2 Ideal and Typical Power MOSFET Characteristics

The *i–v* characteristics of an ideal power switch are illustrated in Fig. 1.3. The ideal transistor conducts current in the on-state with zero voltage drop and blocks voltage in the off-state with zero leakage current. In addition, the ideal device can operate with a high current and voltage in the active region with the saturated forward current in this mode controlled by the applied gate bias. The spacing between the characteristics in the active region is uniform for an ideal transistor indicating a gain (transconductance) that is independent of the forward current and voltage.

The *i–v* characteristics of a typical power MOSFET structure are illustrated in Fig. 1.4. This device exhibits a finite resistance when carrying current in the on-state as well as a finite leakage current while operating in the off-state (not shown in the figure because its value is much lower than the on-state current levels). The breakdown voltage of a typical transistor is also finite as indicated in the figure with 'BV'. The typical transistor can operate with a high current and voltage in the active region. This current is determined by a gate voltage for a MOSFET as indicated in the figure. It is desirable to have gate voltage controlled characteristics because the

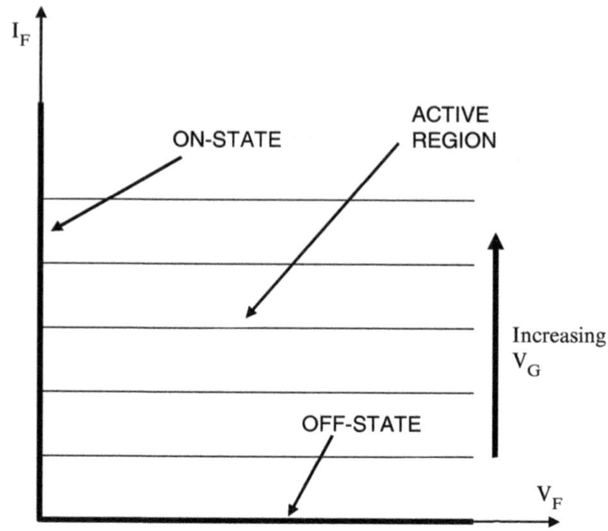


**Fig. 1.3** Characteristics of an ideal power switch

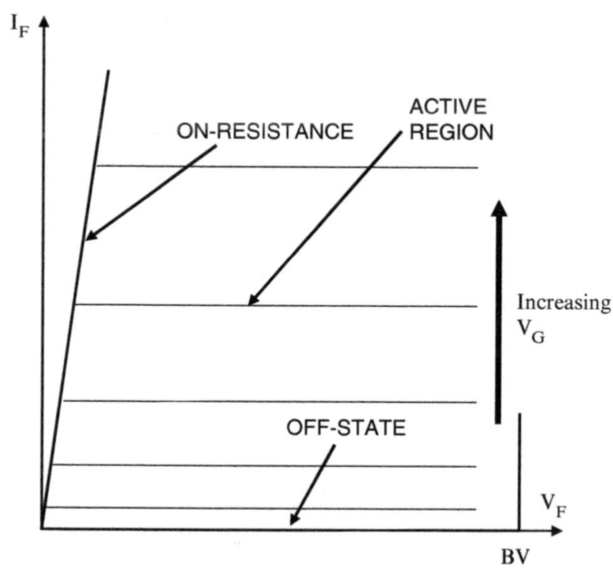


**Fig. 1.4** Characteristics of a typical power MOSFET structure

drive circuit can be integrated to reduce its cost. The spacing between the characteristics in the active region is non-uniform for a typical MOSFET with a square-law behavior for devices operating with channel pinch-off in the current saturation mode. Recently, devices operating under a new super-linear mode have been proposed and demonstrated for wireless base-station applications [3]. These devices exhibit an equal spacing between the saturated drain current characteristics as the gate voltage is increased. This is an ideal behavior when the transistor is used for the amplification of audio, video or cellular signals because it eliminates signal distortion that occurs with the characteristics shown in Fig. 1.4.

## 1.3 Typical Power MOSFET Structures

The most commonly used unipolar power transistor is the silicon power Metal-Oxide-Semiconductor Field-Effect-Transistor or MOSFET. Although other structures, such as JFETs or SITs have been explored [4], they have not been popular for power electronic applications because of their normally-on behavior. The commercially available silicon power MOSFETs are based upon the structures shown in Fig. 1.5. The D-MOSFET was first commercially introduced in the 1970s and contains a 'planar-gate' structure. The P-base region and the $N^+$ source regions are self-aligned to the edge of the polysilicon gate electrode by using ion-implantation of boron and phosphorus followed by their respective drive-in thermal cycles. The n-type channel is defined by the difference in the lateral extension of the junctions under the gate electrode. The device supports positive voltage applied to the drain across the P-base/N-drift region junction. The voltage blocking capability is determined by the doping and thickness of the N-drift region. Although low voltage silicon power MOSFETs have small on-resistances, the drift region resistance increases rapidly with

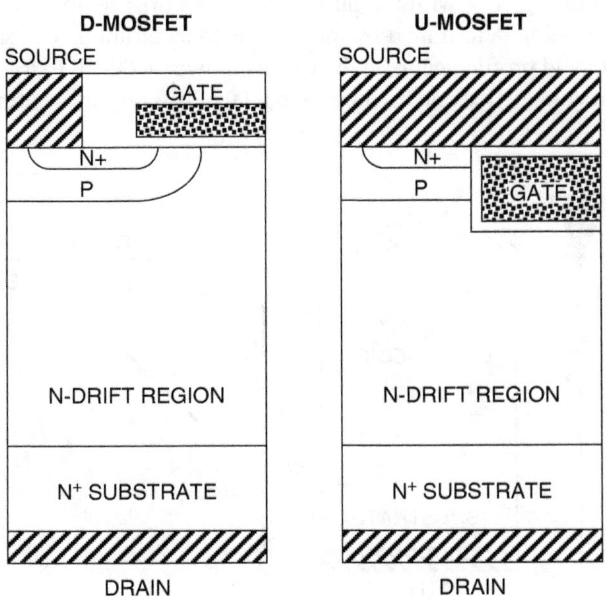


**Fig. 1.5** Typical silicon power MOSFET structures

increasing blocking voltage limiting the performance of silicon power D-MOSFETs to below 200 V.

The silicon U-MOSFET structure became commercially available in the 1990s. It has a gate structure embedded within a trench etched into the silicon surface. The N-type channel is formed on the side-wall of the trench at the surface of the P-base region. The channel length is determined by the difference in vertical extension of the P-base and $N^+$ source regions as controlled by the ion-implant energies and drive times for the dopants. The silicon U-MOSFET structure was developed to reduce the on-state resistance by elimination of the JFET component within the D-MOSFET structure.

## 1.4 Ideal Drift Region for Unipolar Power Devices

The power MOSFET structures discussed above contain a drift region which is designed to support the blocking voltage. The properties (doping concentration and thickness) of the *ideal drift region* can be analyzed by assuming an abrupt junction profile with high doping concentration on one side and a low uniform doping concentration on the other side, while neglecting any junction curvature effects by assuming a parallel-plane configuration. The resistance of the ideal drift region can then be related to the basic properties of the semiconductor material [5].

A Schottky rectifier structure is illustrated in Fig. 1.6. The solution of Poisson's equation in the voltage blocking mode leads to a triangular electric field distribution, as shown in Fig. 1.6, within a uniformly doped drift region with the slope of the field profile being determined by the doping concentration. The same behavior for the electric field profile occurs in the vertical power MOSFET structures as well. The maximum voltage that can be supported by the drift region is determined by the

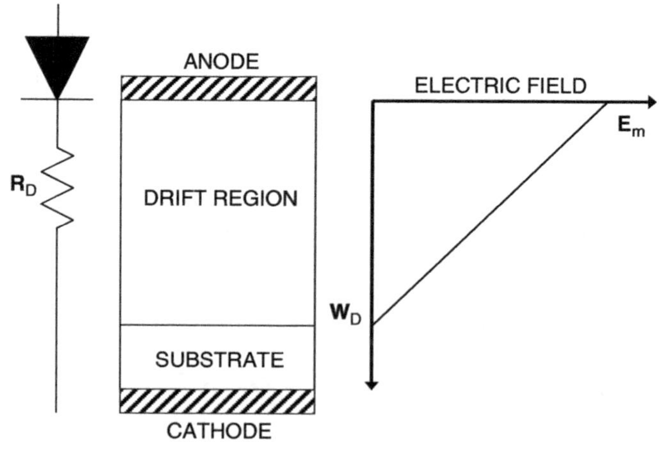


**Fig. 1.6** The ideal drift region and its electric field distribution

maximum electric field ($E_m$) reaching the critical electric field ($E_c$) for breakdown for the semiconductor material. The critical electric field for breakdown and the doping concentration then determine the maximum depletion width ($W_D$).

The specific resistance (resistance per unit area) of the ideal drift region is given by:

$$R_{on.sp} = \left(\frac{W_D}{q\mu_n N_D}\right) \quad (1.1)$$

where $N_D$ is the doping concentration of the drift region. Since this resistance was initially considered to be the lowest value achievable with silicon devices, it has historically been referred to as the *ideal specific on-resistance of the drift region.* More recent introduction of the charge-coupling concept, described later in this chapter, has enabled reducing the drift region resistance of silicon devices to below the values predicted by this equation. The depletion width under breakdown conditions is given by:

$$W_D = \frac{2BV}{E_C} \quad (1.2)$$

where BV is the desired breakdown voltage. The doping concentration in the drift region required to obtain this breakdown voltage is given by:

$$N_D = \frac{\varepsilon_S E_C^2}{2qBV} \quad (1.3)$$

Combining these relationships, the specific resistance of the ideal drift region is obtained:

$$R_{on-ideal} = \frac{4BV^2}{\varepsilon_S \mu_n E_C^3} \quad (1.4)$$

The denominator of (1.4) ( $\varepsilon_S\ \mu_n E_C^3$) is commonly referred to as *Baliga's Figure of Merit for Power Devices*. It is an indicator of the impact of the semiconductor material properties on the resistance of the drift region. The dependence of the drift region resistance on the mobility (assumed to be for electrons here because in general they have higher mobility values than for holes) of the carriers favors semiconductors such as Gallium Arsenide. However, the much stronger (cubic) dependence of the on-resistance on the critical electric field for breakdown favors wide band gap semiconductors such as silicon carbide [2]. The critical electric field for breakdown is determined by the impact ionization coefficients for holes and electrons in semiconductors.

As an example, the change in the specific on-resistance for the drift region with critical electric field and mobility is shown in Fig. 1.7 for the case of a breakdown

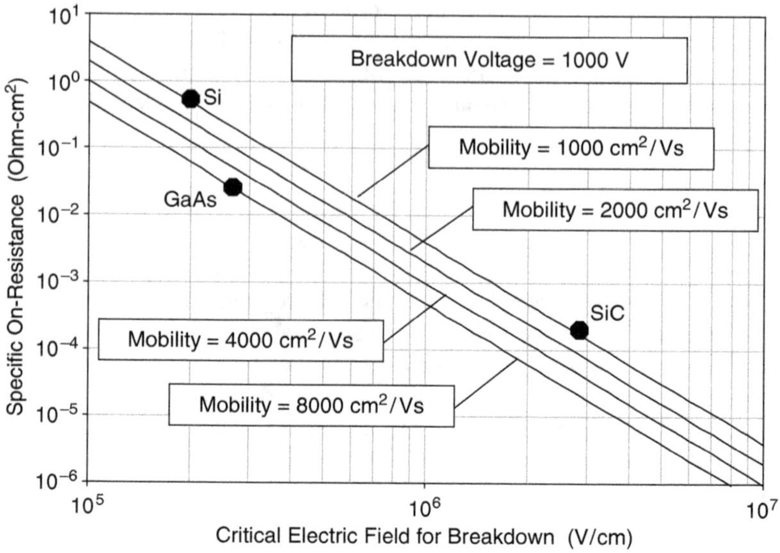


**Fig. 1.7** Specific on-resistance of the *ideal drift region*

voltage of 1,000 V. The location of the properties for silicon, gallium arsenide, and silicon carbide are shown in the figure by the points. The improvement in drift region resistance for GaAs in comparison with silicon is largely due to its much greater mobility for electrons. The improvement in drift region resistance for SiC in comparison with silicon is largely due to its much larger critical electric field for breakdown. Based upon these considerations, excellent high voltage Schottky rectifiers were developed from GaAs in the 1980s [6] and from silicon carbide in the 1990s [7]. Interest in the development of power devices from wide-band-gap semiconductors, including silicon carbide and gallium nitride, continues to grow.

## 1.5 Charge-Coupled Structures: Ideal Specific On-Resistance

The depletion region extends in one-dimension from a junction or Schottky contact during the blocking mode for the conventional structure discussed in the previous section. In the charge coupled structure, the voltage blocking capability is enhanced by the extension of depletion layers in two-dimensions. This effect is created by the formation of a horizontal Schottky contact on the top surface as illustrated in Fig. 1.18 which promotes the extension of a depletion region along the vertical or y-direction. Concurrently, the presence of the vertical P-N junction created by the alternate N and P-type regions promotes the extension of a depletion region along the horizontal or x-direction. These depletion regions conspire to produce a two-dimensional charge coupling in the N-drift region which alters the electric field profile. A similar phenomenon can be induced in the case of power MOSFET structures under the P-base region as discussed later in this monograph.

The optimization of the charge coupled structure requires proper choice of the doping concentration and thickness of the N and P-type regions. It has been found that the highest breakdown voltage occurs when the charge in these regions is given by:

$$Q_{optimum} = 2qN_D W_N = \varepsilon_S E_C \tag{1.5}$$

where q is the charge of an electron ($1.6 \times 10^{-19}$ C), $N_D$ is the doping concentration of the N-Type drift region, $W_N$ is the width of the N-type drift region as shown in Fig. 1.8, $\varepsilon_S$ is the dielectric constant of the semiconductor, and $E_C$ is the critical electric field for breakdown in the semiconductor. For silicon, the optimum charge is found to be $3.11 \times 10^{-7}$ C/cm$^2$ based upon a critical electric field of $3 \times 10^5$ V/cm. The optimum charge is often represented as a dopant density per unit area, in which case it takes a value of about $2 \times 10^{12}$/cm$^2$ for silicon. A slightly lower value for the doping concentration in the drift region may be warranted as discussed later in this section of the chapter.

The specific on-resistance for the drift region in the charge coupled structures is given by:

$$R_{D,sp} = \rho_D t\left(\frac{p}{W_N}\right) \tag{1.6}$$

where $\rho_D$ is the resistivity of the N-type drift region, t is the trench depth and p is the cell pitch. Here, the uniform electric field is assumed to be produced only along the

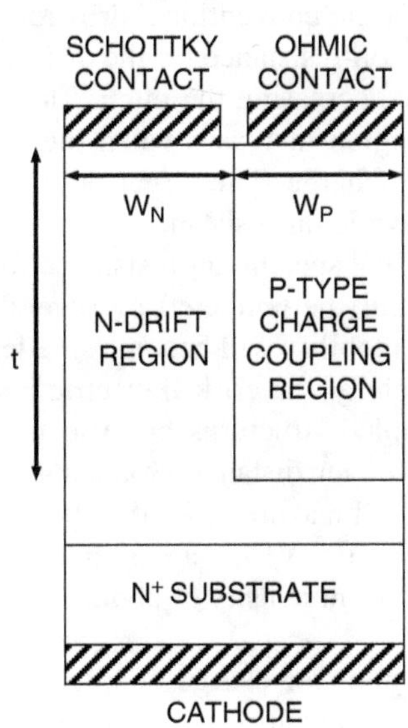


**Fig. 1.8** Basic charge coupled Schottky diode structure

trench where the charge coupling occurs and the resistance of the remaining portion of the N-drift region is neglected. Using the relationship between the resistivity and the doping concentration, this equation can be written as:

$$R_{D,sp} = \frac{tp}{q\ \mu_N N_D W_N} \tag{1.7}$$

Combining this expression with (1.5):

$$R_{D,sp} = \frac{2tp}{\mu_N Q_{optimum}} \tag{1.8}$$

If the electric field along the trench, at the on-set of breakdown in the charge coupled device structure, is assumed to be uniform at a value equal to the critical electric field of the semiconductor:

$$t = \frac{BV}{E_C} \tag{1.9}$$

Using this expression, as well as the second part of (1.5), in (1.8) yields:

$$R_{D,sp} = \frac{2BVp}{\mu_N \varepsilon_S E_C^2} \tag{1.10}$$

This is a fundamental expression for the *ideal specific on-resistance of vertical charge coupled devices*. By comparison of this expression with that for the one dimensional case (see 1.4), it can be observed that the specific on-resistance for the charge coupled devices increases linearly with the breakdown voltage unlike the more rapid quadratic rate for the conventional drift region. In addition, it is worth pointing out that the specific on-resistance for the drift region in the charge coupled structure can be reduced by decreasing the pitch. This occurs because the doping concentration in the drift region increases when the pitch is reduced in order to maintain the same optimum charge. The larger doping concentration reduces the resistivity and hence the specific on-resistance.

However, the analysis of the specific on-resistance for the drift region in charge coupled device structures must be tempered by several considerations. Firstly, it must be recognized that the mobility will become smaller when the doping concentration becomes larger. Secondly, the critical electric field for breakdown becomes smaller for the charge coupled structures because the high electric field in the drift region extends over a larger distance producing enhanced impact ionization. If a critical electric field for breakdown in the drift region for charge coupled structures is reduced to $2 \times 10^5$ V/cm, an optimum charge of $2.07 \times 10^{-7}$ C/cm$^2$, with a corresponding dopant density of about $1.3 \times 10^{12}$/cm$^2$, is more appropriate for silicon.

In designing the drift region for charge coupled structures, it is important to recognize that, unlike in the conventional one-dimensional case, the doping concentration of the drift region is dictated by the cell pitch and not the breakdown

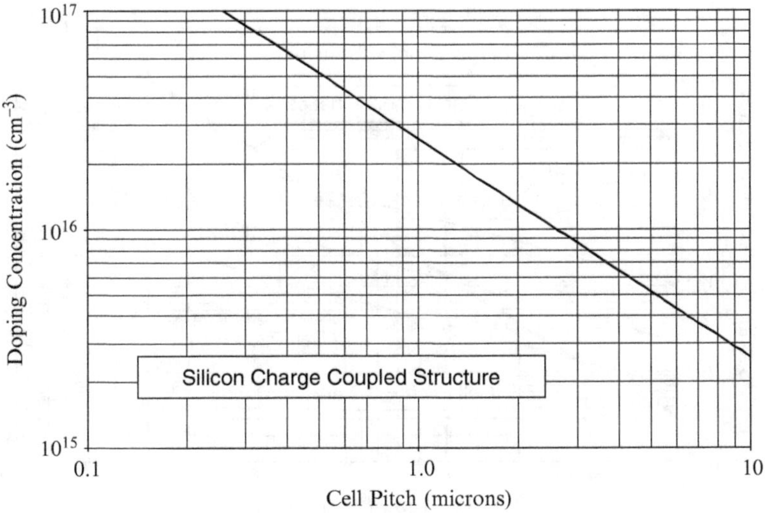


**Fig. 1.9** Drift region doping concentration for the charge coupled structure

voltage. The breakdown voltage in the charge coupled structure is determined solely by the depth of the trench used to provide the charge coupling effect and is independent of the doping concentration of the N-drift region. In the case of silicon charge coupled devices, the doping concentration for the N-type drift region is provided in Fig. 1.9 for the case of equal widths for the N-type and P-type charge coupling regions. For a typical cell pitch of 1 μm, the doping concentration in the N-type drift region is about $2.5 \times 10^{16}/\text{cm}^3$ when a critical electric field of $2 \times 10^5$ V/cm is assumed.

It is interesting to compare the ideal specific on-resistance for the drift region in the silicon charge coupled structures to that for the one-dimensional parallel-plane case. This comparison is done in Fig. 1.10 using three values for the cell pitch in the case of the charge coupled structures. The doping concentration in the N-drift region increases when the cell pitch is reduced from 5 to 0.2 μm, as already shown in Fig. 1.9, leading to a decrease in the specific on-resistance. The resulting reduction of the mobility with increasing doping concentration was included during the calculation of the specific on-resistance in Fig. 1.10. There is a cross-over in the specific on-resistance for the two types of structures. For the cell pitch of 1 μm, the cross-over occurs at a breakdown voltage of about 50 V. The cross-over moves to a breakdown voltage of about 130 V when the cell pitch is increased to 5 μm, and to about 20 V if a smaller cell pitch of 0.2 μm is used. Consequently, the charge coupled structure is more attractive for reducing the specific on-resistance when the cell pitch is smaller. This entails a more complex process technology with higher attendant costs.

As a particular example, consider the case of silicon devices designed to support 200 V. In the case of the conventional structure with a one-dimensional junction,

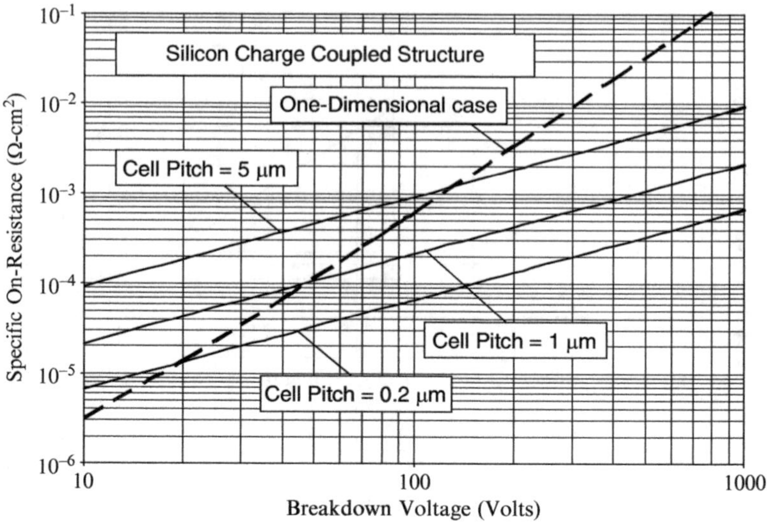


**Fig. 1.10** Ideal specific on-resistance for the charge coupled structure. (*solid lines*: charge coupled structures; *dashed line*: one dimensional case)

the specific on-resistance of the drift region is found to be 3.4 mΩ cm$^2$ if a critical electric field for breakdown of 3 × 10$^5$ V/cm is used. In contrast, the specific on-resistance for the drift region of the charge coupled structure with a cell pitch of 1 μm is found to be only 0.43 mΩ cm$^2$ if a critical electric field for breakdown of 2 × 10$^5$ V/cm is used. In this calculation, a bulk mobility of 1,120 cm$^2$/V s was used corresponding to a doping concentration of 2.6 × 10$^{16}$/cm$^3$ in the N-type portion of the drift region. In this example, the drift region for the charge coupled structure would have a thickness of 10 μm when compared with 12.5 μm needed in the conventional structure.

## 1.6 Revised Breakdown Models for Silicon

In the textbook [1], the breakdown voltage for silicon devices was analyzed by using the Fulop's power law relating the impact ionization coefficient to the electric field. The Fulop's power law [8] for impact ionization in silicon is given by:

$$\alpha_B(\mathrm{Si}) = 1.80 \times 10^{-35} E^7 \tag{1.11}$$

The values for the impact ionization coefficient obtained by using this equation are compared with the impact ionization coefficients measured for electrons and holes in silicon [9] as represented by Chynoweth's equation in Fig. 1.11. It can be observed that Fulop's power law falls between that for electrons and holes and

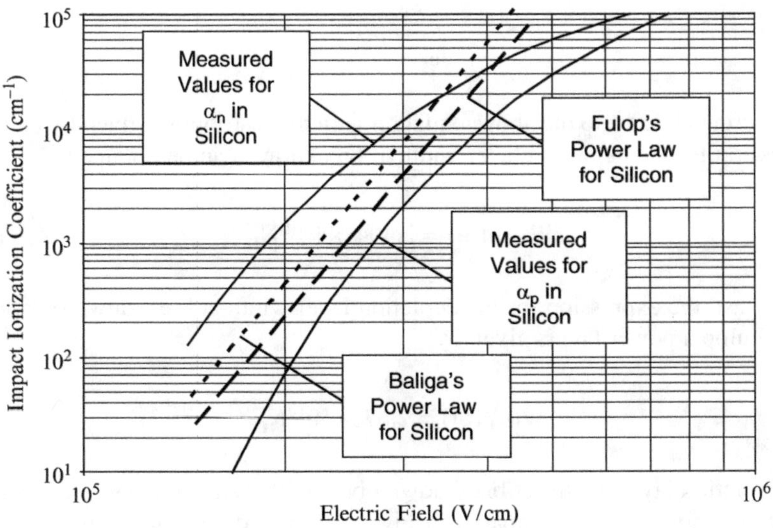


**Fig. 1.11** Impact ionization coefficients for silicon

consequently underestimates the values for the impact ionization coefficients for electrons. This results in the prediction of larger breakdown voltages than in actual devices when performing the analytical calculations as pointed out in the textbook.

A better prediction of breakdown in silicon devices using analytical models can be achieved by improving the match between the power law and the measured data for impact ionization coefficients for electrons and holes in silicon. The proposed Baliga's power law for impact ionization in silicon is given by:

$$\alpha_B(\mathrm{Si}) = 3.507 \times 10^{-35} E^7 \tag{1.12}$$

From Fig. 1.11, it can be observed that this equation provides a larger value for the impact ionization coefficients which will result in reducing the breakdown voltage.

In the case of one-dimensional parallel-plane junctions discussed in Chap. 3 of the textbook, the electric field takes a triangular distribution in the lightly doped side of the P-N junction given by:

$$E(x) = -\frac{qN_D}{\varepsilon_S}(W_D - x) \tag{1.13}$$

where $W_D$ is the depletion layer width, and $N_D$ is the doping concentration on the lightly doped side of the junction. The breakdown voltage in this case is determined by the ionization integral becoming equal to unity:

$$\int_0^{W_D} \alpha dx = 1 \tag{1.14}$$

Substituting (1.12) into the above equation with the distribution given by (1.13), an expression for the depletion layer width at breakdown can be obtained:

$$W_{PP,B}(Si) = 2.404 \times 10^{10} N_D^{-7/8} \tag{1.15}$$

In contrast, the expression for the depletion layer width at breakdown obtained by using Fulop's power law is given by:

$$W_{PP,F}(Si) = 2.67 \times 10^{10} N_D^{-7/8} \tag{1.16}$$

The depletion layer widths at breakdown obtained for silicon devices by using the above equations can be compared in Fig. 1.12. The depletion layer widths computed using Baliga's power law are 11% smaller than those predicted by Fulop's power law.

The maximum electric field located at the P-N junction for the one-dimensional parallel-plane case is given by:

$$E_M = \frac{qN_D}{\varepsilon_S} W_D \tag{1.17}$$

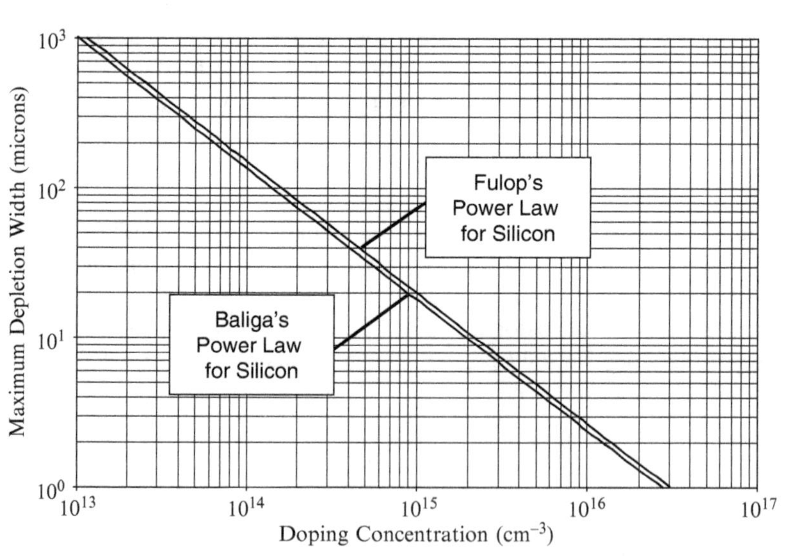


**Fig. 1.12** Depletion layer width at breakdown in silicon for the one-dimensional parallel-plane junction

The critical electric field for breakdown of the one-dimensional parallel-plane junction can be obtained by substituting the depletion layer width at breakdown into this equation. In the case of Baliga's power law, the critical electric field for breakdown of the one-dimensional parallel-plane junction is given by:

$$E_{C,1D,B}(Si) = 3,700\ N_D^{1/8} \quad (1.18)$$

In contrast, in the case of Fulop's power law, the critical electric field for breakdown of the one-dimensional parallel-plane junction is given by:

$$E_{C,1D,F}(Si) = 4,010\ N_D^{1/8} \quad (1.19)$$

The critical electric fields for breakdown of the one-dimensional parallel-plane junction obtained for silicon devices by using the above equations can be compared in Fig. 1.13. The critical electric fields for breakdown of the one-dimensional parallel-plane junction computed using Baliga's power law are 8.4% smaller than those predicted by Fulop's power law.

The one-dimensional parallel-plane breakdown voltage for abrupt P-N junctions in silicon can be computed using the critical electric field and the depletion layer width at breakdown:

$$BV_{PP} = \frac{1}{2} E_{C,1D} W_{PP} \quad (1.20)$$

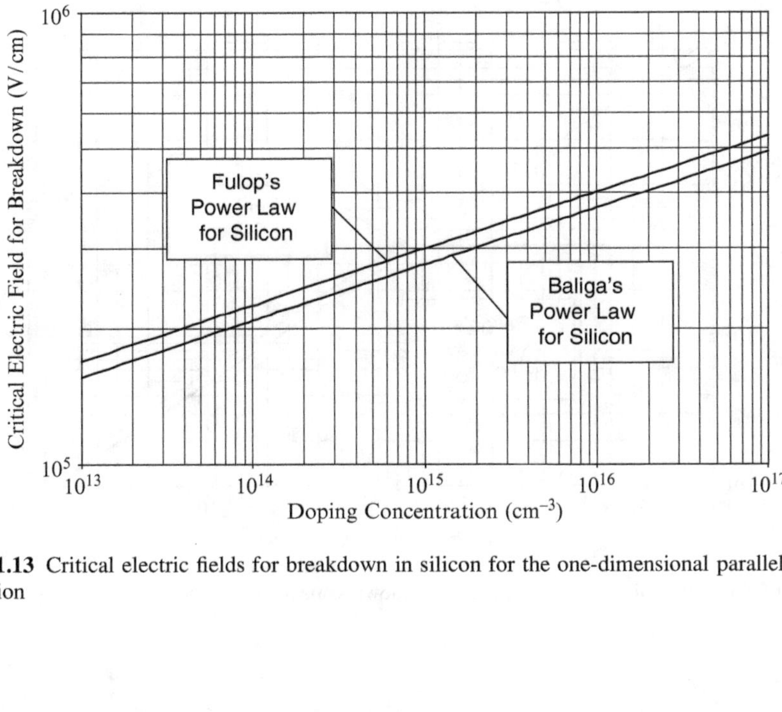


**Fig. 1.13** Critical electric fields for breakdown in silicon for the one-dimensional parallel-plane junction

Using the equations derived above for the critical electric field and the depletion layer width at breakdown with Baliga's power law for the impact ionization coefficients, the breakdown voltage of the one-dimensional parallel-plane junction is given by:

$$BV_{PP,B}(Si) = 4.45 \times 10^{13} N_D^{-3/4} \quad (1.21)$$

In contrast, in the case of Fulop's power law, the breakdown voltage of the one-dimensional parallel-plane junction is given by:

$$BV_{PP,F}(Si) = 5.34 \times 10^{13} N_D^{-3/4} \quad (1.22)$$

The breakdown voltage of the one-dimensional parallel-plane junction obtained for silicon devices by using the above equations can be compared in Fig. 1.14. The breakdown voltages for the one-dimensional parallel-plane junction computed using Baliga's power law are 20% smaller than those predicted by Fulop's power law.

The ideal specific on-resistance is defined as the resistance per unit area for the drift region with the doping concentration and thickness corresponding to each breakdown voltage. This resistance can be computed using:

$$R_{on,sp}(Ideal) = \frac{W_{PP}}{q\,\mu_n N_D} \quad (1.23)$$

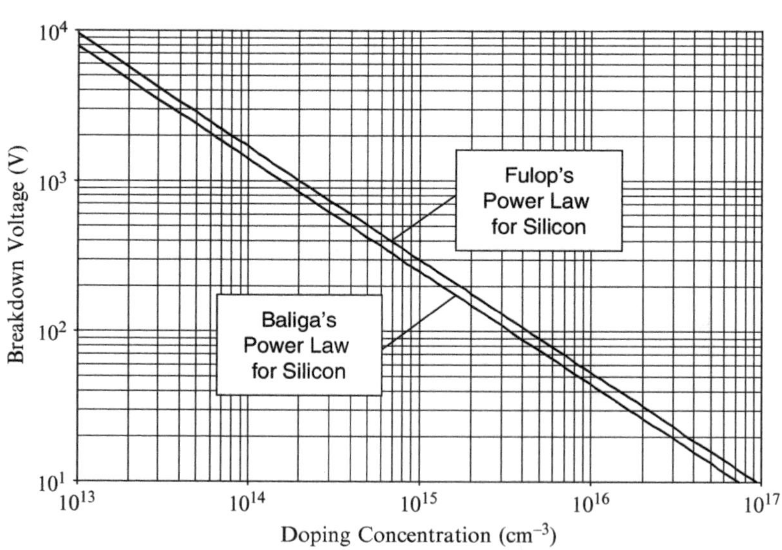


**Fig. 1.14** One-dimensional parallel-plane breakdown voltages in silicon

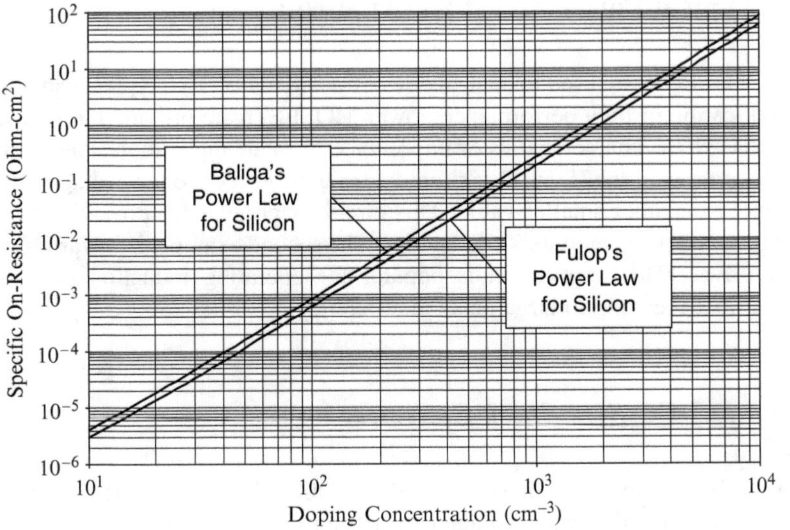


**Fig. 1.15** Ideal specific on-resistances for silicon

It is important to include the dependence of the mobility on the doping concentration when computing the ideal specific on-resistance. The ideal specific on-resistance for silicon devices is slightly larger when the Baliga's power law for the impact ionization is utilized as compared with Fulop's power law as shown in Fig. 1.15. At breakdown voltages above 40 V, the mobility in silicon can be assumed to be independent of the doping concentration. The ideal specific on-resistance obtained by using Baliga's power law for the impact ionization coefficients is then given by:

$$R_{on,sp,B}(Si) = 8.37 \times 10^{-9}\, BV^{2.5} \tag{1.24}$$

In contrast, the ideal specific on-resistance obtained by using Fulop's power law for the impact ionization coefficients is given by:

$$R_{on,sp,F}(Si) = 5.93 \times 10^{-9}\, BV^{2.5} \tag{1.25}$$

The ideal specific on-resistances for silicon devices computed using Baliga's power law are 40% larger than those predicted by Fulop's power law.

The revised information provided above based up on using Baliga's power law for the impact ionization coefficients is intended to bring the analytical calculations of breakdown voltages in silicon devices more in line with the results of numerical simulations. This information can be used to compute the doping concentration and thickness of the drift region to achieve a desired breakdown voltage for power devices.

## 1.7 Typical Power MOSFET Applications

Power MOSFETs are used to control power flow to loads. Two typical examples for the application of high performance power MOSFETs are provided in this section to emphasize the characteristics of importance from an application standpoint. The first example is in the 'Sync-Buck' converter used to reduce the voltage from one DC level to another smaller value. This circuit is popular for providing power to microprocessors in computers and laptops. The second example is in variable speed motor drives. This application is popular for operating induction motors with variable loads leading to large gains in efficiency.

### *1.7.1 DC-DC Sync-Buck Converter*

The sync-buck converter can be regarded as a DC-to-DC transformer because of its ability to reduce a DC input voltage to a smaller DC output voltage. As mentioned above, one popular application of the sync-buck converter is to provide power to various loads inside a computer from the back-plane DC power supply. The back-plane power supply has a typical voltage range of 17–20 V depending on the type of computer. Loads, such as disk drives require a DC voltage of 5–12 V. In contrast, integrated circuits in the computer, such as the microprocessor and graphics chips, require a lower DC voltage in the range of 1–2 V.

The commonly used sync-buck topology used for the DC-to-DC voltage conversion in computers is shown in Fig. 1.16. Due to the relatively low operating voltage in this circuit, a power MOSFET is typically used as the switch in the high-side location. When the transistor is turned on by the control circuit, current flows from the DC input source through the inductor to the load connected at the output terminals. When the transistor is switched off by the control circuit, the load current

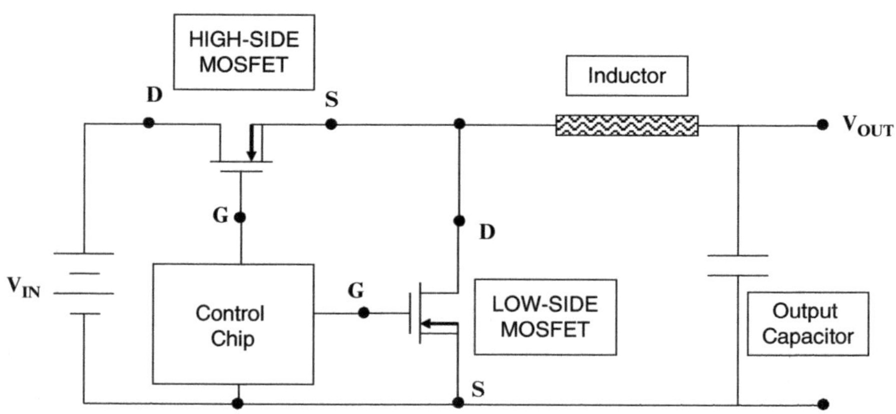


**Fig. 1.16** Sync-buck DC-DC converter circuit

circulates through the MOSFET connected on the low-side and the inductor. The regulation of the DC output voltage can be achieved by adjusting the on-time of the transistor [10]. The switching waveforms for the high-side transistor are similar to those shown in Fig. 1.2. The waveforms for the low-side transistor are a complement of those for the high-side transistor. The efficiency of the sync-buck converter is determined by the on-resistance and switching speed of the two transistors.

The low-side MOSFET is also referred to as a synchronous rectifier. The current flow in the low-side MOSFET is in the opposite direction to the normal operation of power MOSFET devices. If the low-side power MOSFET is turned on by the control circuit, the current flow occurs with a voltage drop determined by the on-resistance of the MOSFET structure. This voltage drop is usually much smaller than the built-in potential of the junction between the P-base region and the N-drift region (called the body-diode of the MOSFET). However, if the low-side MOSFET is not turned-on, the current will flow through the body-diode. This not only increases power losses due to the high on-state voltage drop of the P-N junction, it also severely slows down the switching time due to the long reverse recovery time for the body-diode.

The high-side and low-side MOSFET devices cannot be simultaneously turned-on during circuit operation to avoid short-circuiting the input power source. This requires delaying the turn-on of the low-side transistor after the gate bias to the high-side transistor has been turned-off. During this delay-time interval, the current in the low-side MOSFET flows through its body-diode. This upsets high frequency operation of the circuit. One method to overcome this problem is to connect an external Schottky rectifier across the low-side switch. This has not been found to be an effective solution because of the inductance between the low-side MOSFET and the Schottky diode in the packages and circuit boards. An elegant solution to this problem is to integrate the Schottky rectifier into the power MOSFET structure. These MOSFET structures are also discussed in this monograph in Chap. 8.

## 1.7.2 Variable-Frequency Motor Drive

A significant increase in the efficiency for running motors can be achieved by using variable-frequency motor drives, in place of constant speed drives, with dampers to regulate the output. The most commonly used topology converts the constant frequency input AC power to a DC bus voltage and then use an inverter stage to produce the variable frequency output power [11]. The circuit diagram for a three-phase motor drive system is shown in Fig. 1.17. Six IGBTs are used with six fly-back rectifiers in the inverter stage to deliver the variable frequency power to the motor windings. A pulse-width-modulation (PWM) scheme is used to generate the variable frequency AC voltage waveform that is fed to the motor windings [12].

During each cycle of the PWM period, the current in the motor winding can be considered to remain approximately constant. This allows linearization of the waveforms for the current and voltage experienced by the IGBTs and the rectifiers. Typical waveforms for the transistor and the fly-back diode are illustrated in

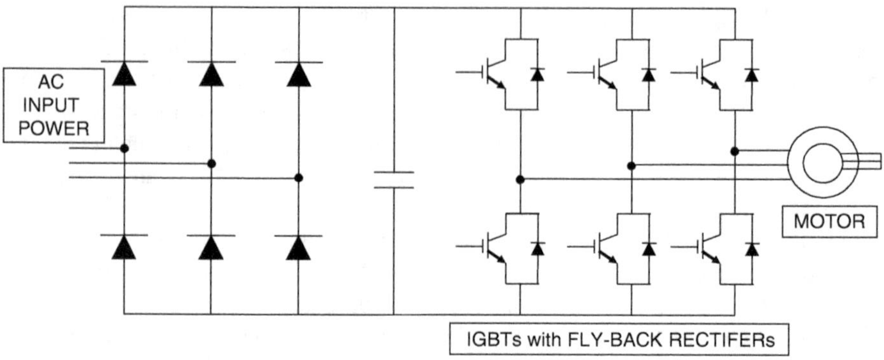


**Fig. 1.17** Variable frequency motor drive circuit

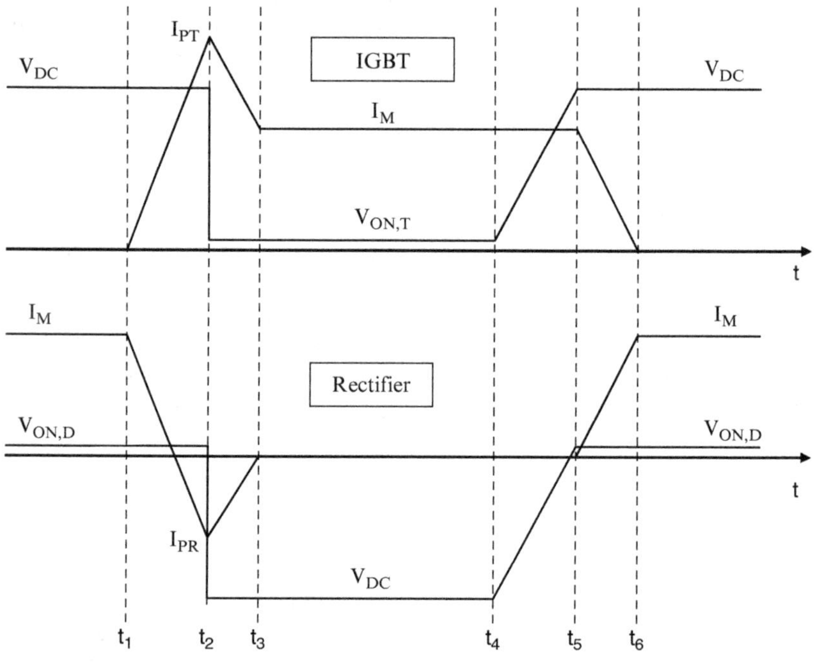


**Fig. 1.18** Linearized waveforms for the PWM motor drive circuit

Fig. 1.18. The large reverse recovery current typically observed in silicon P-i-N rectifiers during the time interval from $t_1$ to $t_3$ produces high power dissipation not only in the diodes but also in the transistors [1]. This power loss can be eliminated by replacing the silicon P-i-N rectifiers with silicon carbide Schottky rectifiers.

With the availability of high voltage silicon carbide Schottky rectifiers, the power loss in the motor control application becomes dominated by the on-state and turn-off losses in the IGBTs [13]. The development of high performance power MOSFETs can reduce these power losses making the motor drive more efficient.

The on-resistances of the conventional silicon power MOSFET structures shown in Fig. 1.5 are too large for typical high voltage motor drive applications, such as for electric vehicles with a DC-bus voltage of 400 V, due to their high (600 V) blocking voltage capability. The on-resistance of silicon power MOSFETs can be reduced by using the charge coupling concept as discussed in this monograph. A reduction in the on-resistance can also be achieved by replacing silicon with silicon carbide based power MOSFET structures. Power MOSFET structures that are suitable for implementation in silicon carbide material are therefore also discussed in this monograph.

## 1.8 Summary

The desired characteristics for power MOSFET structures have been reviewed in this chapter. The characteristics of typical devices have been compared with those for the ideal case. The ability to reduce the on-resistance by using the two-dimensional charge coupling effect has been described. Improved analytical expressions for computation of the breakdown voltage in silicon devices have been derived using a proposed Baliga's power law for impact ionization coefficients. Typical applications for the power MOSFETs have also been discussed to highlight the device characteristics that are desired for the applications. Various power MOSFET structures are discussed in detail in subsequent chapters of this book.

## References


1. B.J. Baliga, "Fundamentals of Power Semiconductor Devices", Springer Science, New York, 2008.
2. B.J. Baliga, "Silicon Carbide Power Devices", World Scientific Press, Singapore, 2005.
3. B.J. Baliga, "Silicon RF Power Devices", World Scientific Press, Singapore, 2005.
4. B.J. Baliga, "Modern Power Devices", Wiley, New York, 1987.
5. B.J. Baliga, "Semiconductors for High Voltage Vertical Channel Field Effect Transistors", Journal of Applied Physics, Vol. 53, pp. 1759–1764, 1982.
6. B.J. Baliga, et al, "Gallium Arsenide Schottky Power Rectifiers", IEEE Transactions on Electron Devices, Vol. ED-32, pp. 1130–1134, 1985.
7. M. Bhatnagar, P.M. McLarty, and B.J. Baliga, "Silicon Carbide High Voltage (400 V) Schottky Barrier Diodes", IEEE Electron Device Letters, Vol. EDL-13, pp. 501–503, 1992.
8. W. Fulop, "Calculation of Avalanche Breakdown in Silicon P-N Junctions", Solid State Electronics, Vol. 10, pp. 39–43, 1967.
9. R. Van Overstraeten and H. DeMan, "Measurements of the Ionization Rates in Diffused Silicon P-N Junctions", Solid State Electronics, Vol. 13, pp. 583–608, 1970.
10. N. Mohan, T.M. Undeland, and W.P. Robbins, "Power Electronics", pp. 164–172, Wiley, New York, 1995.
11. J.D. van Wyk, "Power Electronic Converters for Drives", pp. 81–137, in 'Power Electronics and Variable Frequency Drives', Edited by B.K. Bose, IEEE Press, New Jersey, 1997.
12. J. Holtz, "Pulse Width Modulation for Electronic Power Conversion", pp. 138–208, in 'Power Electronics and Variable Frequency Drives', Edited by B.K. Bose, IEEE Press, 1997.
13. B.J. Baliga, "Advanced Power Rectifier Concepts", Springer-Science, New York, 2009.

# Chapter 2
# D-MOSFET Structure

The first power MOSFET structure commercially introduced by the power semiconductor industry was the double-diffused or D-MOSFET structure. The channel length of this device could be reduced to sub-micron dimensions by controlling the diffusion depths of the P-base and $N^+$ source regions without resorting to expensive lithography tools [1]. The device fabrication process relied up on the available planar gate technology used to manufacture CMOS integrated circuits. These devices found applications in power electronic circuits that operated at low (<100 V) voltages. The fast switching speed and ruggedness of the D-MOSFET structure were significant advantages compared with the performance of the available bipolar power transistor.

The physics of operation of the power D-MOSFET structure has been analyzed in detail in the textbook [1]. In this chapter of the monograph, a brief description of the device operation is provided for completeness and for contrasting it to the advanced device structures that are discussed in later chapters. However, a detailed discussion of the characteristics of a device rated for blocking 30 V is provided here based up on the results of two-dimensional numerical simulations. These characteristics will be used as a bench-mark for comparison with the characteristics of the advanced power MOSFET structures described in other chapters.

## 2.1 The D-MOSFET Structure

A cross-section of the basic cell structure for the D-MOSFET structure is illustrated in Fig. 2.1. This device structure is fabricated by starting with an N-type epitaxial layer grown on a heavily doped $N^+$ substrate. The channel is formed by the difference in lateral extension of the P-base and $N^+$ source regions produced by their diffusion cycles. Both regions are self-aligned to the left-hand-side and right-hand-side of the gate region during ion-implantation to introduce the respective dopants. A refractory gate electrode, such as polysilicon, is required to allow diffusion of the dopants under the gate electrode at elevated temperatures.

B.J. Baliga, *Advanced Power MOSFET Concepts*,
DOI 10.1007/978-1-4419-5917-1_2, © Springer Science+Business Media, LLC 2010

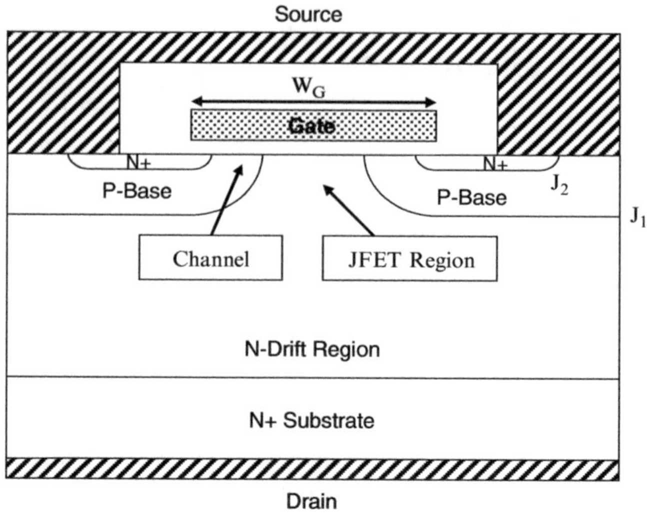


**Fig. 2.1** The D-MOSFET structure

Without the application of a gate bias, a high voltage can be supported in the D-MOSFET structure when a positive bias is applied to the drain. In this case, junction $J_1$ formed between the P-Base region and the N-drift region becomes reverse biased. The voltage is supported mainly within the thick lightly doped N-drift region. Drain current flow in the D-MOSFET structure is induced by the application of a positive bias to the gate electrode. This produces an inversion layer at the surface of the P-base region under the gate electrode. This inversion layer channel provides a path for transport of electrons from the source to the drain when a positive drain voltage is applied.

After transport from the source region through the channel, the electrons enter the N-drift region at the upper surface of the device structure. They are then transported through a relatively narrow JFET region located between the adjacent P-base regions within the D-MOSFET structure. The constriction of the current flow through the JFET region substantially increases the internal resistance of the D-MOSFET structure. A careful optimization of the gate width ($W_G$) is required, in order to minimize the internal resistance for this structure as discussed in the textbook. In addition, it is customary to enhance the doping concentration in the JFET region to reduce the resistance to current flow through this portion of the device structure.

After being transported through the JFET region, the electrons enter the N-drift region below junction $J_1$. The current spreads from the relatively narrow JFET region to the entire width of the cell cross-section. This non-uniform current distribution within the drift region enhances its resistance making the internal resistance of the D-MOSFET structure larger than the ideal specific on-resistance of the drift region. The large internal resistance for the D-MOSFET structure has provided motivation for the development of the trench-gate power MOSFET structure in the 1990s and the advanced power MOSFET structures discussed in this monograph.

## 2.2 Power D-MOSFET On-Resistance

The power D-MOSFET structure is shown in Fig. 2.2 with its internal resistance components. There are eight resistances that must be analyzed in order to obtain the total on-resistance between the source and drain electrodes when the device is turned-on. It is customary to analyze not only the resistance for a particular cell design but also the specific resistance for each of the components by multiplying the cell resistance with the cell area. The total on-resistance for the power MOSFET structure is obtained by the addition of all the resistances because they are considered to be in series in the current path between the source and the drain electrodes:

$$R_{ON} = R_{CS} + R_{N+} + R_{CH} + R_A + R_{JFET} + R_D + R_{SUB} + R_{CD} \quad (2.1)$$

Each of the resistances within the power D-MOSFET structure is analyzed below by using the procedure described in the textbook. In the textbook, it was demonstrated that the contributions from the source contact resistance ($R_{CS}$), the source resistance ($R_{N+}$), and the drain contact resistance ($R_{CD}$) are very small and will therefore be neglected in this and subsequent chapters.

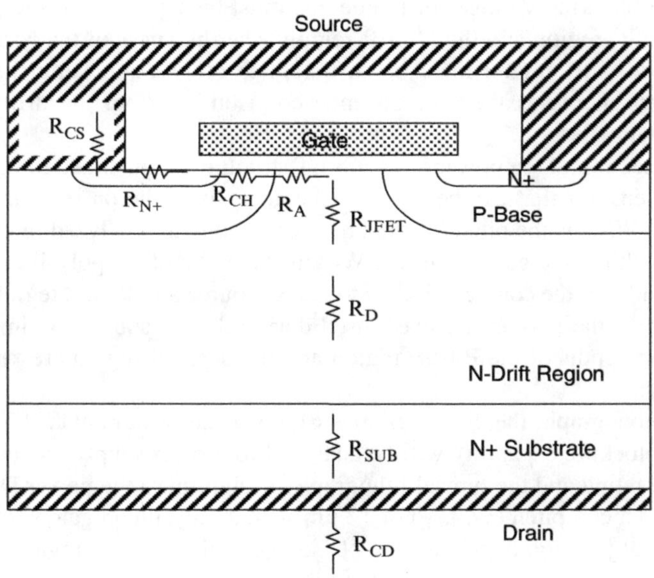


**Fig. 2.2** Power D-MOSFET structure with its internal resistances

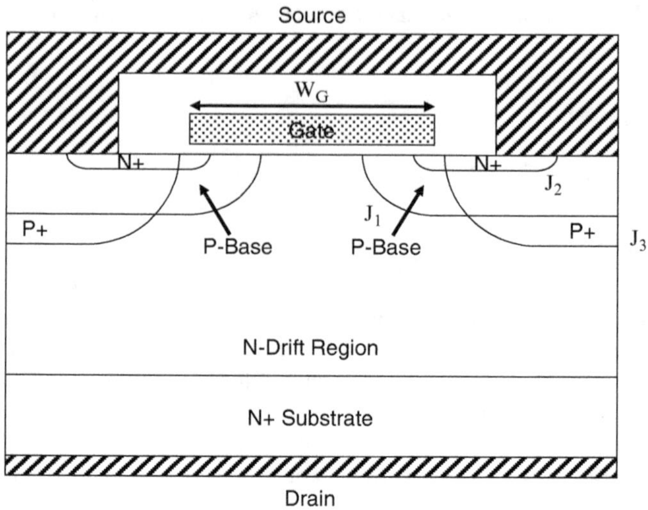


**Fig. 2.3** Power D-MOSFET structure with deep P+ region

The on-resistance of the basic structure for the power D-MOSFET device has been analyzed in the textbook. Typical power D-MOSFET products include a deep $P^+$ region as illustrated in Fig. 2.3 to improve their ruggedness and safe-operating-area as discussed in the textbook. The analysis of the power D-MOSFET structure in this monograph includes the presence of this deep $P^+$ region. The blocking voltage for the device must be supported at the junction $J_3$ between the $P^+$ region and the N-drift region. The thickness of the N-drift region below the deep $P^+$ region must be sufficient to allow supporting the blocking voltage. This increases the on-resistance contribution from the drift region as discussed below.

A cross-section of the power D-MOSFET structure is illustrated in Fig. 2.4 with various dimensions that can be used for the analysis of the on-resistance components. Here, $W_{Cell}$ is the pitch for the linear cell geometry analyzed in this section; $W_G$ is the width of the gate electrode; $W_{PW}$ is the width of the polysilicon window; $W_C$ is the width of the contact window to the $N^+$ source and P-base regions; and $W_S$ is the width of the photoresist mask used during the $N^+$ source ion-implantation. The junction depths of the P-base region and the deep $P^+$ region are $x_{JP}$ and $x_{JP+}$, respectively.

In this monograph, the characteristics of advanced power MOSFET structures with 30-V blocking capability will be analyzed for power supply applications. For this voltage rating and the typical lithography design rules, the power D-MOSFET structure has a cell pitch ($W_{CELL}$) of 12 μm with a polysilicon gate width ($W_G$) of 8 μm. Typical junction depths for the $N^+$ source region, P-base region, and the $P^+$ region are 0.5, 1.5, and 2.5 μm, respectively. The doping concentration of the N-drift region required to achieve a 30-V blocking voltage capability, with an

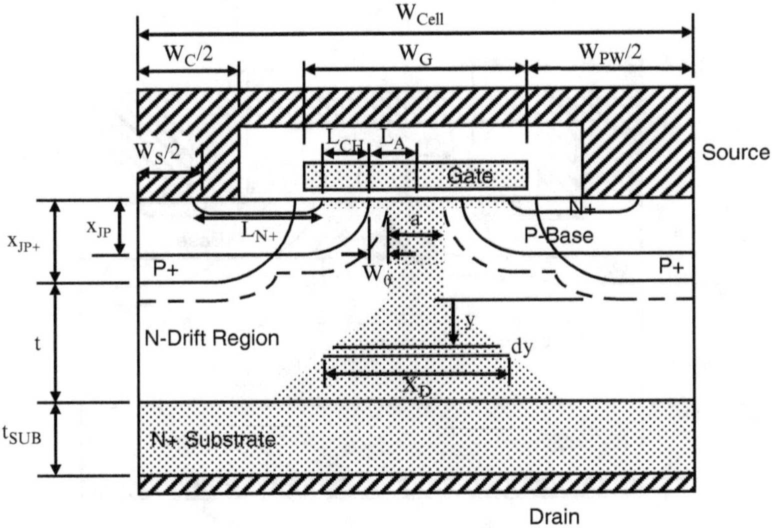


**Fig. 2.4** Power D-MOSFET structure with current flow model A used for analysis of its internal resistances

80% reduction due to the edge termination, is $1.6 \times 10^{16}/\text{cm}^3$. The thickness of the N-drift region required below the deep $P^+$ region is 2.6 μm. It is worth pointing out that the entire blocking voltage is not supported within the N-drift region because of the graded doping profile of the P-base and the $P^+$ regions. This allows increasing the doping concentration and reducing the thickness of the N-drift region to achieve a smaller specific on-resistance for the D-MOSFET structure.

The current flow pattern in the power D-MOSFET structure is indicated by the shaded area in Fig. 2.4. In the first Model A for the specific on-resistance, it will be assumed that the JFET region extends to the bottom of the deep $P^+$ region. Consequently, this model assumes that there is current constriction by the deep $P^+$ regions. In the model, it is assumed that the current spreads at a 45° angle in the drift region and then becomes uniformly distributed when it enters the $N^+$ substrate. For the dimensions provided above for the D-MOSFET structure with 30-V blocking voltage capability, the current does not spread across the entire region below the windows in the gate electrode resulting in the current distribution shown in Fig. 2.4 by the shaded area.

An alternate Model B for the current distribution in the D-MOSFET structure is illustrated in Fig. 2.5. In this model, it is assumed that the junction formed by the $P^+$ regions ($J_3$) is too far from the current path in the JFET region to contribute to current constriction. The JFET current constriction is then assumed to be associated only with the junction ($J_1$) formed by the P-base region. The JFET region is then assumed to extend to the bottom of the P-base region as shown in the figure by the shaded area. This model produces a smaller contribution to the specific

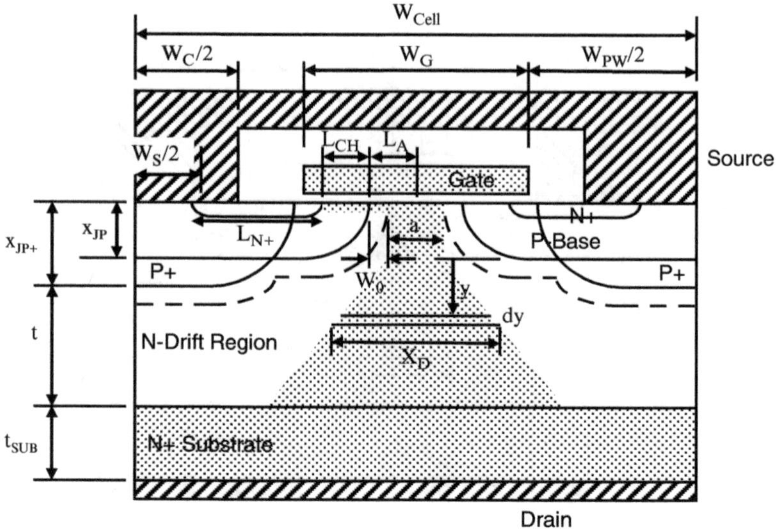


**Fig. 2.5** Power D-MOSFET structure with current flow model B used for analysis of its internal resistances

on-resistance from the JFET region than Model A but the contribution from the drift region is enhanced.

### *2.2.1 Channel Resistance*

The contribution to the specific on-resistance from the channel in the D-MOSFET structure is the same for models A and B. As derived in the textbook [1], the specific on-resistance contributed by the channel in the power D-MOSFET structure is given by:

$$R_{CH,SP} = \frac{L_{CH} W_{Cell}}{2\mu_{ni} C_{OX}(V_G - V_{TH})} \tag{2.2}$$

In the case of the 30-V power MOSFET structures used for power supply applications, it is customary to provide the on-resistance at a gate bias of 4.5 and 10 V. Assuming a gate oxide thickness is 500 Å, an inversion layer mobility of 450 cm$^2$/V s (to match the mobility used in the numerical simulations discussed later in this section), and a threshold voltage of 2 V in the above equation for the power D-MOSFET design with a cell width of 12 μm and gate electrode width of 8 μm, the specific resistance contributed by the channel at a gate bias of 4.5 V is found to be 0.784 mΩ cm$^2$. The specific on-resistance of the D-MOSFET structure is reduced to 0.245 mΩ cm$^2$ when the gate bias is increased to 10 V.

### 2.2.2 Accumulation Resistance

In the power MOSFET structure, the current flowing through the inversion channel enters the drift region at the edge of the P-base junction. The current then spreads from the edge of the P-base junction into the JFET region. The current spreading phenomenon is aided by the formation of an accumulation layer in the semiconductor below the gate oxide due to the positive gate bias applied to turn-on the device. The specific on-resistance contributed by the accumulation layer in the power D-MOSFET structure is given by [1]:

$$R_{A,SP} = K_A \frac{(W_G - 2x_{JP})W_{Cell}}{4\mu_{nA}C_{OX}(V_G - V_{TH})} \tag{2.3}$$

In writing this expression, a coefficient $K_A$ has been introduced to account for the current spreading from the accumulation layer into the JFET region. A typical value for this coefficient is 0.6 based upon the current flow observed from numerical simulations of power D-MOSFET structures. The threshold voltage in the expression is for the on-set of formation of the accumulation layer. A zero threshold voltage will be assumed here when performing the analytical computations. Note that the junction depth of the P-base region (and not the $P^+$ region) defines the length of the accumulation region.

For the 30-V power D-MOSFET design with a cell width of 12 μm and gate width of 8 μm, the specific resistance contributed by the accumulation layer at a gate bias of 4.5 V is 0.294 mΩ cm$^2$ if the P-base junction depth ($x_{JP}$) is 1.5 μm and the gate oxide thickness is 500 Å. An accumulation layer mobility of 1,000 cm$^2$/V s was used in this calculation to match the mobility used in the numerical simulations (discussed later in this section). When the gate bias is increased to 10 V, the specific resistance contributed by the accumulation layer is reduced to 0.132 mΩ cm$^2$.

### 2.2.3 JFET Resistance

The electrons entering from the channel into the drift region are distributed into the JFET region via the accumulation layer formed under the gate electrode. The spreading of current in this region was accounted for by using a constant $K_A$ of 0.6 for the accumulation layer resistance. Consequently, the current flow through the JFET region can be treated with a uniform current density. In the power D-MOSFET structure, the cross-sectional area for the JFET region increases with distance below the semiconductor surface due to the planar shape of the P-base junction. However, in order to simplify the analysis, a uniform cross-section for the current flow with a width 'a' will be assumed for the JFET region as illustrated by the shaded area in Figs. 2.4 and 2.5. The width of the current flow is related to the device structural parameters:

$$a = (W_G - 2x_{JP} - 2W_0) \tag{2.4}$$

where $W_0$ is the zero-bias depletion width for the JFET region. The depletion region boundary is indicated in the figures with the dashed lines. In the models, it is assumed that no current can flow through the depleted region because all the free carriers have been swept out by the prevailing electric field across the junction. The zero-bias depletion width ($W_0$) in the JFET region can be computed by using the doping concentrations on both sides of the junction:

$$W_0 = \sqrt{\frac{2\varepsilon_S N_A V_{bi}}{q N_{DJ}(N_A + N_{DJ})}} \tag{2.5}$$

where $N_A$ is the doping concentration in the P-base region and $N_{DJ}$ is the doping concentration in the JFET region. In the above equation, the built-in potential is also related to the doping concentrations on both sides of the junction:

$$V_{bi} = \frac{kT}{q} \ln\left(\frac{N_A N_{DJ}}{n_i^2}\right) \tag{2.6}$$

In practical devices, the P-base region is diffused into the N-drift region producing a graded doping profile. However, these expressions based upon assuming a uniform doping concentration for the P-base region are adequate for analytical computations. It is common practice to enhance the doping concentration for the JFET region above that for the drift region. It is therefore appropriate to use the enhanced doping concentration ($N_{DJ}$) of the JFET region in the above expressions.

In Model A, the JFET region is assumed to extend to the bottom the $P^+$ region. The specific on-resistance contributed by the JFET region in the power D-MOSFET structure can then be obtained by using:

$$R_{JFET,SP} = \frac{\rho_{JFET} x_{JP+} W_{Cell}}{(W_G - 2x_{JP} - 2W_0)} \tag{2.7}$$

where $\rho_{JFET}$ is the resistivity of the JFET region given by:

$$\rho_{JFET} = \frac{1}{q\mu_n N_{DJ}} \tag{2.8}$$

where $\mu_n$ is the bulk mobility appropriate to the doping level of the JFET region. Typical power D-MOSFET structures are fabricated by using a diffused JFET doping profile with its maximum concentration at the surface and a depth approximately the same as the junction depth of the P-base region. In an analytical model, it is convenient to use an average doping concentration for the JFET region. The resistivity for the JFET region is found to be 0.269 Ω cm corresponding to an average JFET doping concentration of $2.0 \times 10^{16}/\text{cm}^3$. The zero-bias depletion width in the JFET region for this JFET doping concentration is 0.228 μm. For the 30-V power D-MOSFET design with a cell width of 12 μm and gate width of 8 μm, the specific resistance contributed by the JFET region is found to be 0.178 mΩ cm$^2$ based up on using the above parameters.

In Model B, the JFET region is assumed to extend to the bottom the P-base region. The specific on-resistance contributed by the JFET region in the power D-MOSFET structure can then be obtained by using:

$$R_{JFET,SP} = \frac{\rho_{JFET} x_{JP} W_{Cell}}{(W_G - 2x_{JP} - 2W_0)} \tag{2.9}$$

For the 30-V power D-MOSFET design with a cell width of 12 μm and gate width of 8 μm, the specific resistance contributed by the JFET region is found to be 0.107 mΩ cm$^2$ based up on using the same parameters as for Model A. It can be observed that the JFET resistance is smaller in Model B than for Model A due to the shorter path for current flow in the vertical direction.

### 2.2.4 Drift Region Resistance

The resistance contributed by the drift region in the power D-MOSFET structure is enhanced well above that for the ideal drift region due to current spreading from the JFET region. The cross-sectional area for the current flow in the drift region increases from the width 'a' of the JFET region as illustrated in Figs. 2.4 and 2.5 by the shaded area. Several models for this current spreading have been proposed in the literature [2]. The current distribution model used in this monograph is based up on a spreading angle of 45°.

In the case of Model A, the current spreads from the bottom of the P$^+$ region as shown in Fig. 2.4. The specific on-resistance contributed by the drift region in the power D-MOSFET structure with this model is given by:

$$R_{D,SP} = \frac{\rho_D W_{Cell}}{2} \ln\left[\frac{a + 2t}{a}\right] \tag{2.10}$$

For the parameters given above for this structure, the dimension 'a' in the equation is found to be 2.54 μm. For the 30-V power MOSFET design with a cell width of 12 μm and gate width of 8 μm, the specific resistance contributed by the drift region is then found to be 0.149 mΩ cm$^2$ by using a resistivity of the drift region of 0.325 Ω cm based upon a doping concentration of $1.6 \times 10^{16}/\text{cm}^3$.

In the case of Model B, the current spreads from the bottom of the P-base region as shown in Fig. 2.5. The specific on-resistance contributed by the drift region in the power D-MOSFET structure with this model is given by:

$$R_{D,SP} = \frac{\rho_D W_{Cell}}{2} \ln\left[\frac{a + 2[t+(x_{JP+} - x_{JP})]}{a}\right] \tag{2.11}$$

For the parameters given above for the 30-V power MOSFET design with a cell width of 12 μm and gate width of 8 μm, the specific resistance contributed by the drift region is then found to be 0.185 mΩ cm$^2$ by using a resistivity of the drift region of 0.325 Ω cm based upon a doping concentration of $1.6 \times 10^{16}$/cm$^3$. Consequently, the drift region resistance contribution is larger for Model B than for Model A.

## 2.2.5 $N^+$ Substrate Resistance

When the current reaches the bottom of the N-drift region, it is very quickly distributed throughout the heavily doped $N^+$ substrate. The current flow through the substrate can therefore be assumed to occur with a uniform cross-sectional area. Under this assumption, the specific resistance contributed by the $N^+$ substrate is given by:

$$R_{SUB,SP} = \rho_{SUB} t_{SUB} \tag{2.12}$$

where $\rho_{SUB}$ and $t_{SUB}$ are the resistivity and thickness of the $N^+$ substrate. The contribution from the $N^+$ substrate is therefore dependent up on the available technology for reducing its thickness and resistivity while maintaining good manufacturability. Since many manufacturers have reduced this contribution to well below that from the other resistance contributions, the substrate contribution will be assumed to be negligible in this monograph for all the power MOSFET structures.

## 2.2.6 Drain and Source Contact Resistance

During the initial development of power MOSFET products, the contact resistance to the source region was performed using aluminum metallization. This method resulted in relatively high contact resistance to the $N^+$ source region. With the advent of metal-silicides for ohmic contacts, the source contact resistance is now much smaller than the other resistance components and can be neglected for all the power MOSFET structures discussed in this monograph. The process technology for making contacts to the drain regions has also evolved to the state that this resistance can be neglected for power MOSFET structures.

### 2.2.7 *Total On-Resistance*

The total specific on-resistance for the power D-MOSFET structure can be computed by adding all the above components for the on-resistance. For the case of the 30-V power D-MOSFET design with a cell pitch ($W_{Cell}$) of 12 μm and gate width of 8 μm, the total specific on-resistance is found to be 1.405 mΩ cm$^2$ at a gate bias of 4.5 V and 0.704 mΩ cm$^2$ at a gate bias of 10 V by using Model A. For this cell design, the total specific on-resistance is found to be 1.370 mΩ cm$^2$ at a gate bias of 4.5 V and 0.669 mΩ cm$^2$ at a gate bias of 10 V by using Model B. Consequently, Model B predicts a smaller total specific on-resistance than Model A. The contributions from each of the components of the on-resistance are summarized in Figs. 2.6 and 2.7 for the two models.

The ideal specific on-resistance for a drift region is given by:

$$R_{IDEAL,SP} = \frac{W_{PP}}{q\,\mu_n N_D} \tag{2.13}$$

where $W_{PP}$ is the parallel-plane depletion width at breakdown, $N_D$ is the doping concentration of the drift region to sustain the blocking voltage, and $\mu_n$ is the mobility for electrons corresponding to this doping concentration. For the case of

| Resistance | $V_G$ = 4.5 V (mΩ-cm$^2$) | $V_G$ = 10 V (mΩ-cm$^2$) |
|---|---|---|
| Channel ($R_{CH,SP}$) | 0.784 | 0.245 |
| Accumulation ($R_{A,SP}$) | 0.294 | 0.132 |
| JFET ($R_{JFET,SP}$) | 0.178 | 0.178 |
| Drift ($R_{D,SP}$) | 0.149 | 0.149 |
| Total ($R_{T,SP}$) | 1.405 | 0.704 |

**Fig. 2.6** On-resistance components within the 30-V power D-MOSFET structure using model A

| Resistance | $V_G$ = 4.5 V (mΩ-cm$^2$) | $V_G$ = 10 V (mΩ-cm$^2$) |
|---|---|---|
| Channel ($R_{CH,SP}$) | 0.784 | 0.245 |
| Accumulation ($R_{A,SP}$) | 0.294 | 0.132 |
| JFET ($R_{JFET,SP}$) | 0.107 | 0.107 |
| Drift ($R_{D,SP}$) | 0.185 | 0.185 |
| Total ($R_{T,SP}$) | 1.370 | 0.669 |

**Fig. 2.7** On-resistance components within the 30-V power D-MOSFET structure using model B

a blocking voltage of 30 V, the depletion width and doping concentration are found to be $2.2 \times 10^{16}/\text{cm}^3$ and 1.4 μm, respectively. Using the electron mobility for this doping level, the ideal specific on-resistance is found to be 0.034 mΩ cm$^2$. Since the device is constrained by the impact of an 80% reduction of breakdown voltage due to the edge termination, it is worth computing the ideal specific on-resistance for this case for comparison with the device. For the case of a blocking voltage of 37.5 V, the depletion width and doping concentration are found to be $1.6 \times 10^{16}/\text{cm}^3$ and 1.8 μm, respectively. Using the electron mobility for this doping level, the ideal specific on-resistance is found to be 0.059 mΩ cm$^2$. It can be observed that the specific on-resistance for the D-MOSFET structure is far greater than these ideal specific on-resistances.

#### 2.2.7.1 Simulation Results

The results of two-dimensional numerical simulations on the 30-V power D-MOSFET structure are described here to provide a more detailed understanding of the underlying device physics and operation. The structure used for the numerical simulations had a drift region thickness of 2.6 μm below the $P^+$ region with a doping concentration of $1.6 \times 10^{16}/\text{cm}^3$. The P-base region and $N^+$ source regions had depths of 1.7 and 0.7 μm, respectively. The doping concentration in the JFET region was enhanced by using an additional n-type doping concentration of $1 \times 10^{16}/\text{cm}^3$ with a depth of 1.5 μm. For the numerical simulations, half the cell (with a width of 6 μm) shown in Fig. 2.1 was utilized as a unit cell that is representative of the structure.

A three dimensional view of the doping distribution in the D-MOSFET structure is shown in Fig. 2.8 from the left hand edge of the structure to the center of the polysilicon gate region. The $N^+$ source and P-base regions are aligned to the gate edge, which is located at 2 μm from the left hand side. The structure also includes a $P^+$ region located at the left hand side to suppress the parasitic bipolar transistor. The enhancement of the doping in the N-drift region near the surface is due to the additional JFET doping.

The lateral doping profile taken along the surface under the gate electrode is shown in Fig. 2.9. From the profile, it can be observed that the doping concentration of the JFET region has been increased to $2.3 \times 10^{16}/\text{cm}^3$ due to the additional doping. The lateral extension of the P-base and $N^+$ source regions are 1.7 and 0.7 μm leading to a channel length of 1.0 μm. The surface concentration for the P-base region was chosen to obtain a maximum compensated P-type doping concentration in the channel of $1.0 \times 10^{17}/\text{cm}^3$.

The transfer characteristics for the D-MOSFET structure were obtained using numerical simulations with a drain bias of 0.1 V at 300 and 400°K. The resulting transfer characteristics are shown in Fig. 2.10. From this graph, a threshold voltage of 3.1 and 2.6 V can be extracted at 300 and 400°K, respectively. The threshold voltage decreases by 16% when the temperature increases. The specific on-resistance can be obtained from the transfer characteristics at any gate bias voltage.

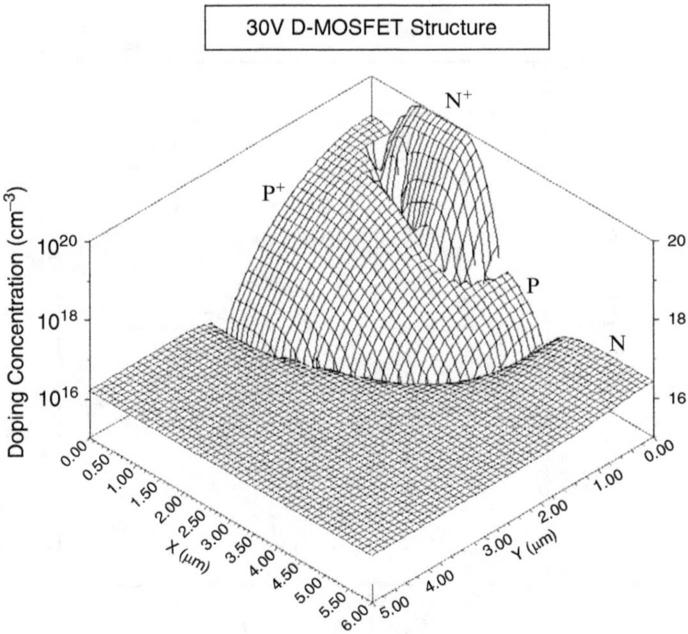


**Fig. 2.8** Doping distribution for the D-MOSFET structure

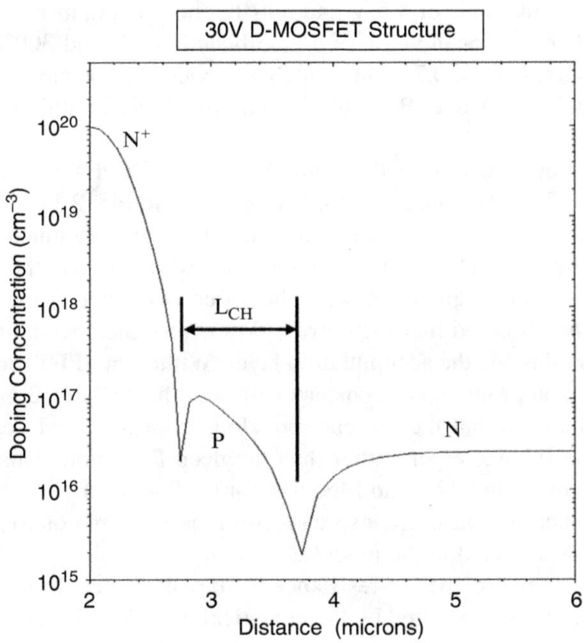


**Fig. 2.9** Channel doping profile for the D-MOSFET structure

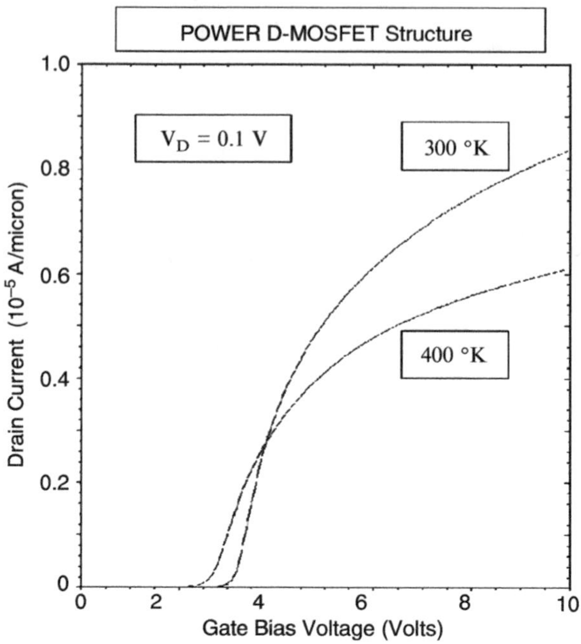


**Fig. 2.10** Transfer characteristics of the D-MOSFET structure

For the case of a gate bias of 4.5 V and 300°K, the specific in-resistance is found to be 1.382 mΩ cm$^2$. For the case of a gate bias of 10 V and 300°K, the specific in-resistance is found to be 0.717 mΩ cm$^2$. These values are in close agreement with either the Model A or Model B for the current distribution within the D-MOSFET structure.

The on-state current flow pattern within the D-MOSFET structure at a small drain bias of 0.1 V and a gate bias of 4.5 V is shown in Fig. 2.11. In the figure, the depletion layer boundary is shown by the dotted lines and the junction boundary is delineated by the dashed line. The depletion layer width ($W_0$) in the JFET region is about 0.2 μm in good agreement with the value computed using the analytical model. It can be observed that the current flows from the channel and distributes into the JFET region via the accumulation layer. Within the JFET region, the cross-sectional area is approximately constant with a width (a/2) of 2.1 μm. From the figure, it can be seen that the current spreading from the JFET region begins at the depth of the P-base region rather than the deep $P^+$ region. This indicates that the Model B is more suitable than Model A for analysis of the on-resistance of the D-MOSFET structure. The current spreads from the JFET region to the drift region at a 45° angle as assumed in the models.

In order to compare the on-resistance values extracted from the numerical simulations with those predicted by the analytical models, it is necessary to extract the mobility in the inversion and accumulation layers within the simulated D-MOSFET structure. The channel and accumulation layer mobility were extracted

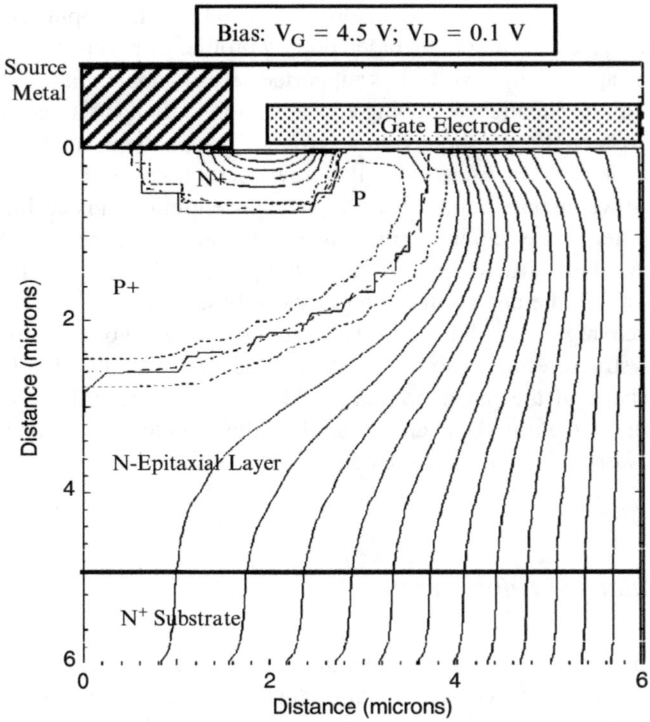


**Fig. 2.11** Current distribution in the D-MOSFET structure

by simulation of a long channel lateral MOSFET structure with the same gate oxide thickness. The inversion layer mobility was found to be 450 cm$^2$/V s while that for the accumulation layer was found to be 1,000 cm$^2$/V s at a gate bias of 10 V. These values were consequently used in the analytical models when calculating the specific on-resistance.

## 2.3 Blocking Voltage

The power D-MOSFET structure must be designed to support a high voltage in the first quadrant when the drain bias voltage is positive. During operation in the blocking mode, the gate electrode is shorted to the source electrode by the external gate bias circuit. The application of a positive drain bias voltage produces a reverse bias across junction $J_1$ between the P-base region and the N-drift region, as well as across junction $J_3$ between the deep P$^+$ region and the N-drift region. Most of the applied voltage is supported across the N-drift region. The doping concentration of donors in the N-epitaxial drift region and its thickness must be chosen to attain the desired breakdown voltage. In devices designed to support low voltages (less than 50 V), the doping concentration of the P-base region is comparable to the doping concentration of the N-drift region leading to a graded doping profile. Consequently, a significant fraction of the applied drain voltage is supported across a depletion region formed in the P-base region. The highest doping concentration in the P-base region is limited by the need to keep the threshold voltage around 2 V to achieve a low on-resistance at a gate bias of 4.5 V as discussed in the previous section.

For the allowable maximum P-base doping concentration, it is desirable to make the depth of the P-base region as small as possible to reduce the channel length in the power MOSFET structure. However, if the junction depth of the P-base region is made too small, the depletion region in the P-base region will reach through to the $N^+$ source region leading to a reduced breakdown voltage. In the power D-MOSFET structure, the gate region is not screened from the drain bias due to cylindrical shape of the planar junctions. This results in significant depletion of the P-base region making the channel length of this structure larger than that of the advanced power MOSFET structures discussed in this monograph.

### 2.3.1 *Impact of Edge Termination*

In practical devices, the maximum blocking voltage (BV) of the power MOSFET is invariably decided by the edge termination that surrounds the device cell structure. The most commonly used edge termination for power D-MOSFET devices is based up on floating field rings and field plates. The enhanced electric field at the edges limits the breakdown voltage to about 80% of the parallel-plane breakdown voltage ($BV_{PP}$):

$$BV_{PP} = \left(\frac{BV}{0.8}\right) \tag{2.14}$$

Consequently, the doping and thickness of the N-drift region must be chosen to achieve a parallel-plane breakdown voltage that is 25% larger than the blocking voltage for the device:

$$N_D = \left(\frac{4.45 \times 10^{13}}{BV_{PP}}\right)^{4/3} \tag{2.15}$$

A common design error that can occur is to make the thickness of the drift region below the P-N junction equal to the depletion width for the ideal parallel-plane junction with the above doping concentration. In actuality, the maximum depletion width is limited to that associated with the blocking voltage (BV) of the structure. The thickness of the drift region required below the P-N junction is therefore less than the depletion width for the ideal parallel-plane junction with the above doping concentration, and is given by:

$$t = W_D(BV) = \sqrt{\frac{2\varepsilon_S BV}{qN_D}} \tag{2.16}$$

The resistance of the drift region can be reduced by using this thickness rather than the maximum depletion width corresponding to the doping concentration given by (2.15).

## 2.3.2 Impact of Graded Doping Profile

For power MOSFET structures with low (less than 50 V) breakdown voltages, the doping concentration of the drift region is comparable to the doping concentration of the P-base region. This produces a graded doping profile for the junction $J_1$ between the P-base region and the N-drift region as illustrated in Fig. 2.12. In this figure, the concentrations of the donors and acceptors are shown by the solid lines while the dashed lines represent the net doping concentration after taking into account compensation near the junctions.

The electric field developed across junction $J_1$ during the blocking mode is also illustrated in Fig. 2.12. Due to the graded doping profile, the electric field extends on both sides of junction $J_1$. The electric field in the P-base region supports a portion of the applied positive drain voltage. This implies that the same breakdown voltage can be achieved with a larger doping concentration and a smaller thickness for the N-drift region. This improvement can be translated to increasing the breakdown

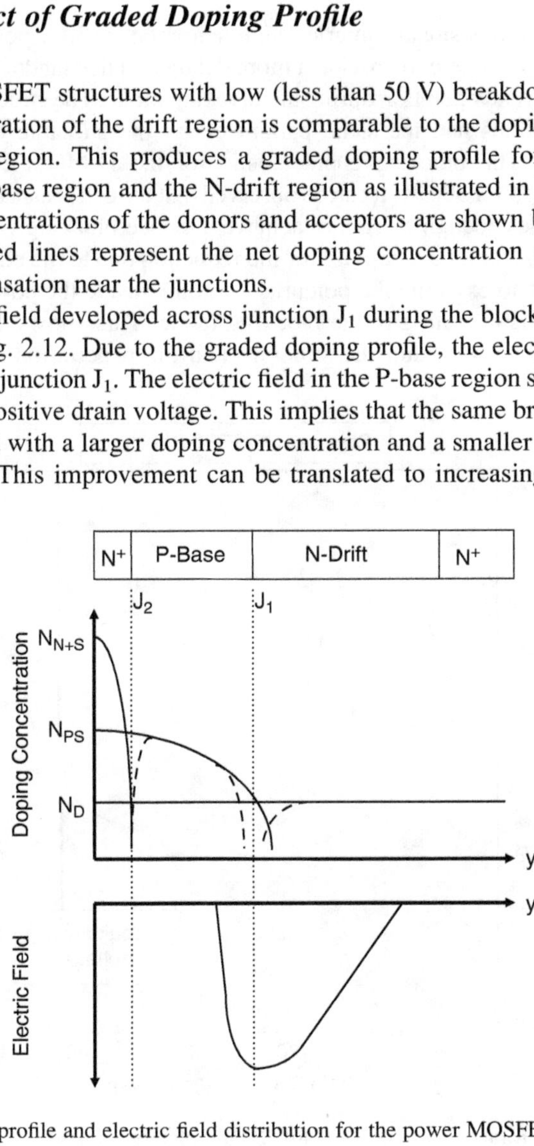


**Fig. 2.12** Doping profile and electric field distribution for the power MOSFET structure

voltage at the edge termination if the P-base region is used at the edges of the power MOSFET structure. A reduction of the resistance for the power MOSFET structure can be achieved by taking into account the voltage supported within the P-base region. An improvement in the specific on-resistance of 20% can be achieved by taking into account the graded doping profile.

#### 2.3.2.1 Simulation Results

The results of two-dimensional numerical simulations on the 30-V power D-MOSFET structure are described here to provide a more detailed understanding of the underlying device physics and operation during the blocking mode. The structure used for the numerical simulations had the same parameters as the structure described in the previous section. The blocking characteristic for the D-MOSFET cell structure is shown in Fig. 2.13 for 300°K. It can be observed that the cell is capable of supporting 42 V. This provides enough margin to achieve a device blocking voltage capability of slightly over 30 V after accounting for the reduction due to the edge termination.

It is instructive to examine the potential contours inside the power D-MOSFET structure when it is operating in the blocking mode. This allows determination of the voltage distribution within the structure and the penetration of the depletion

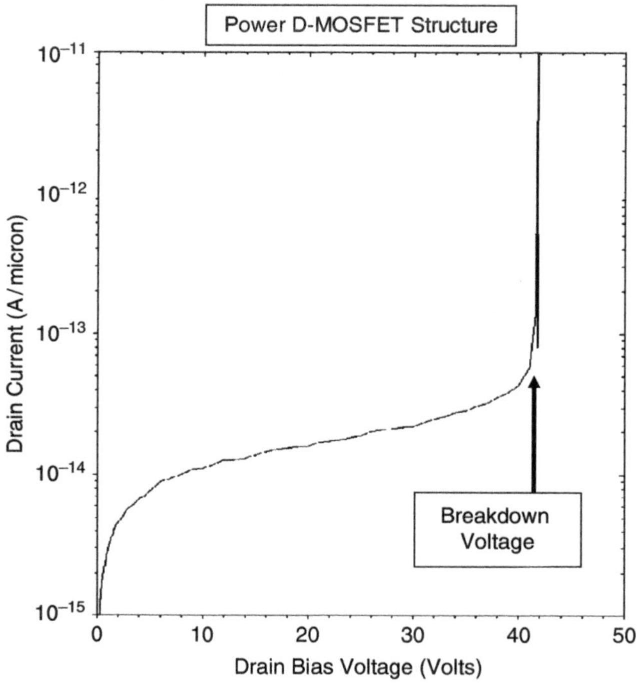


Fig. 2.13 Blocking characteristics for the D-MOSFET structure

region in the P-base region with increasing drain bias voltage. The potential contours for the D-MOSFET structure obtained using the numerical simulations with zero gate bias and various drain bias voltages are shown in Figs. 2.14–2.17. From these figures, it can be observed that the voltage distribution at the junction $J_1$ between the P-base region and the N-drift region is similar when proceeding from the surface towards the drain. This indicates that the surface region is not screened from the drain bias by the junctions. Consequently, the depletion region in the P-base region penetrates through a significant fraction of the P-base region when the drain bias is increased to 30 V. Any decrease in the doping concentration of the P-base region leads to reach-through breakdown limiting the ability to reduce the threshold voltage. The advanced power MOSFET structures discussed in subsequent chapters allow the screening of the semiconductor surface under the gate from the drain potential allowing the formation of shorter channel lengths to reduce the on-resistance.

It is insightful to also examine the electric field profile inside the power D-MOSFET structure when it is operating in the blocking mode. The electric field profile obtained through the junction between the deep $P^+$ region and the N-drift region is shown in Fig. 2.18. It can be observed that the maximum electric field occurs as expected at the junction at a depth of 2.6 μm from the surface. This figure demonstrates that a substantial portion of the voltage is supported inside the $P^+$ region due to its graded doping profile near the junction.

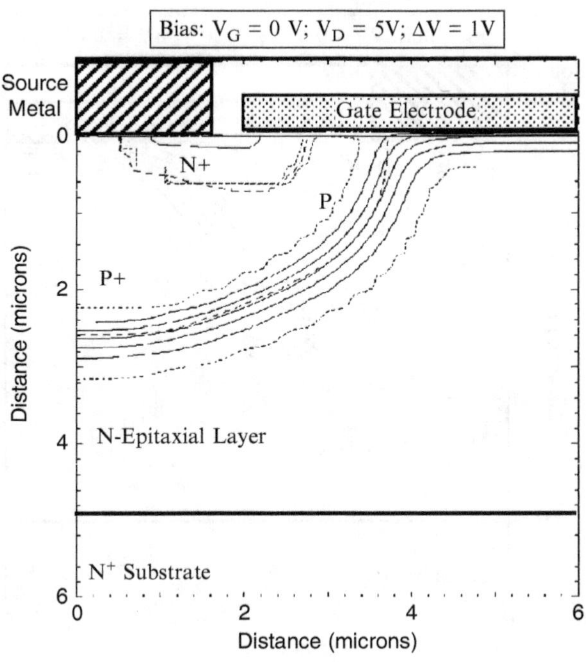


**Fig. 2.14** Potential contours in the D-MOSFET structure

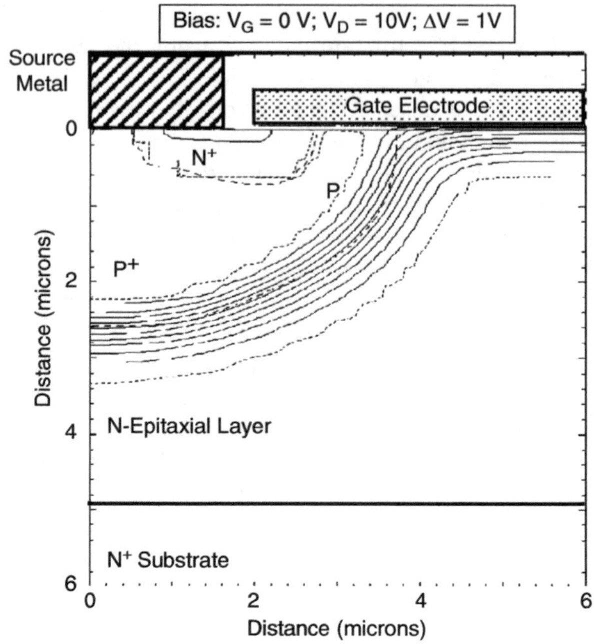


**Fig. 2.15** Potential contours in the D-MOSFET structure

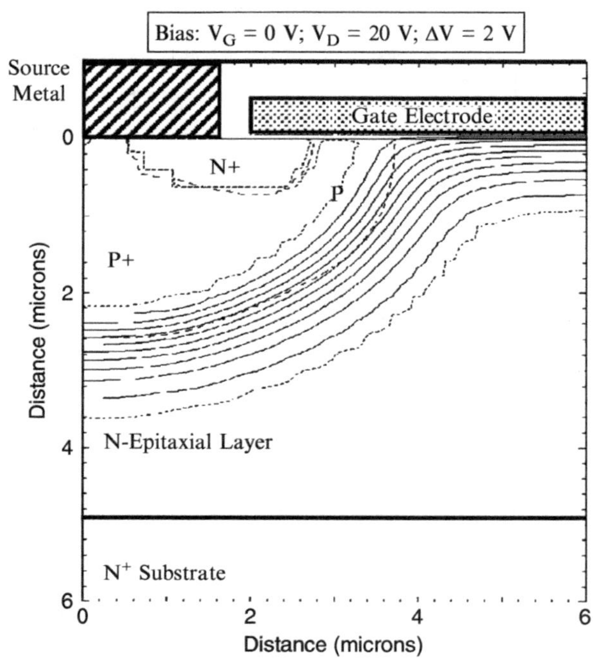


**Fig. 2.16** Potential contours in the D-MOSFET structure

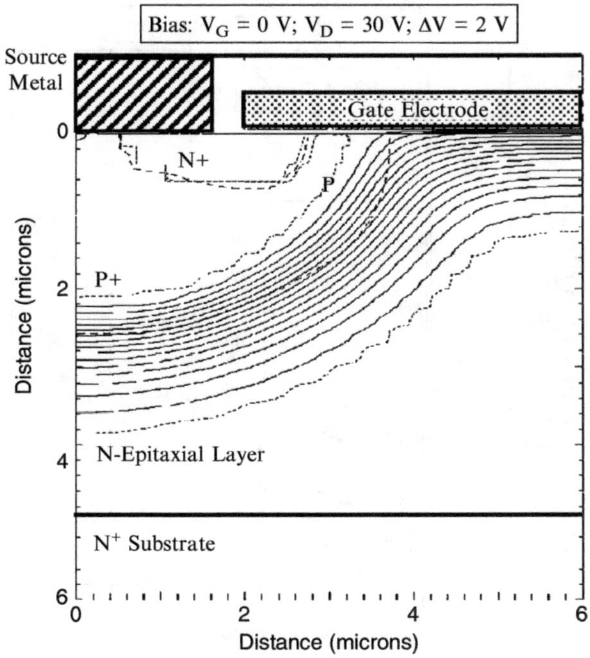


**Fig. 2.17** Potential contours in the D-MOSFET structure

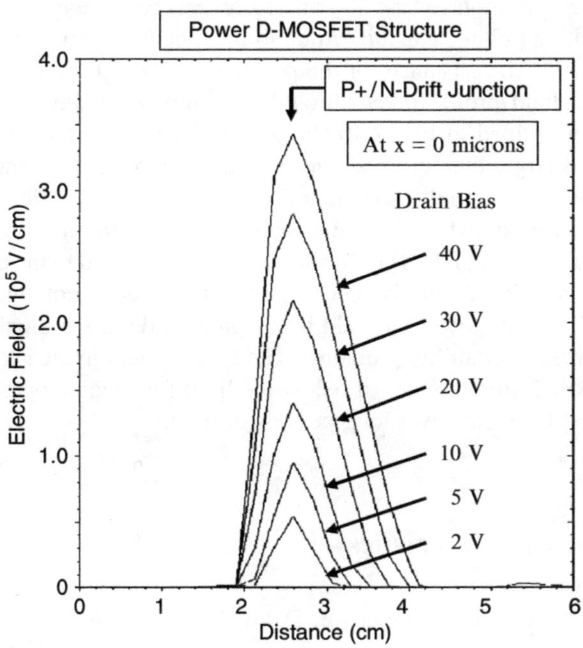


**Fig. 2.18** Electric field distribution in the D-MOSFET structure

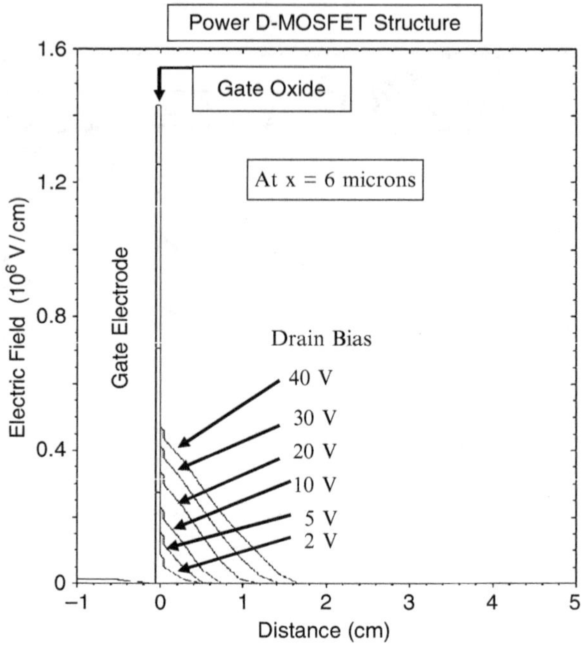


**Fig. 2.19** Electric field distribution in the D-MOSFET structure

The doping concentration for the N-drift region can be increased while achieving the target blocking voltage capability due to this phenomenon. This allows reduction of the specific on-resistance for the power D-MOSFET structure.

The electric field profile taken through the middle of the gate electrode (at $\times = 6\ \mu$m) is provided in Fig. 2.19. It can be observed that the electric field in the gate oxide is larger than in the semiconductor due to the difference in dielectric constant for the two materials. A change in the slope of the electric field profile can be observed close to the semiconductor surface due to the enhanced doping concentration arising from the JFET doping. The electric field in the oxide increases rapidly with increasing drain bias because the gate oxide is not shielded from the drain potential. The high electric field in the gate oxide in the blocking mode has been found to create reliability problems. The electric field in the gate oxide for the power D-MOSFET structure is just below the limit for reliable operation allowing stable device performance over long periods of time.

## 2.4 Output Characteristics

The output characteristics of the power D-MOSFET structure are important to the loci for the switching waveforms when it is operating in power circuits. The saturated drain current for the power MOSFET structure is given by:

$$I_{D,sat} = \frac{Z\mu_{ni}C_{OX}}{(L_{CH} - \Delta L_{CH})}(V_G - V_{TH})^2 \tag{2.17}$$

where $\Delta L_{CH}$ is reduction in the channel length due to depletion of the P-base region with increasing drain bias voltage. With sufficiently high doping concentration of the P-base region, the modulation of channel length can be made sufficiently small to ensure a high output resistance. The saturated drain current in the power D-MOSFET structure then increases as the square of the gate bias voltage.

## 2.4.1 Simulation Example

The output characteristics of the 30-V power D-MOSFET structure were obtained by using two-dimensional numerical simulations using various gate bias voltages. All the device parameters used for these numerical simulations are the same as those used in the previous sections. The output characteristics of the power D-MOSFET obtained using the simulations are shown in Fig. 2.20. The structure exhibits excellent current saturation with relatively flat output characteristics. The traces for increasing gate bias voltages are non-uniformly spaced due to the square-law behavior of the transfer characteristics.

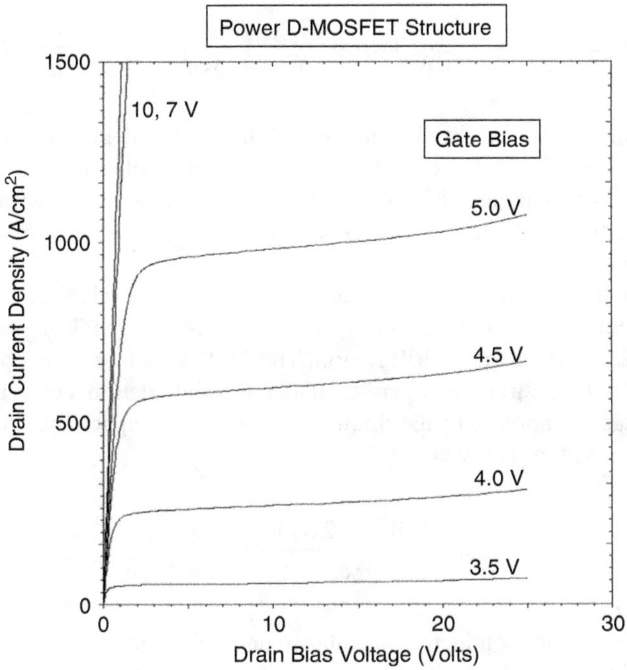


**Fig. 2.20** Output characteristics for the power D-MOSFET structure

## 2.5 Device Capacitances

One of attractive features of all power MOSFET structures is unipolar current transport. The absence of minority carrier injection allows interruption of the current flow immediately after reduction of the gate bias below the threshold voltage. Although this implies a very fast switching speed for the power MOSFET structures, in practice the switching speed is limited by the device capacitances. The input drive signal for a power MOSFET structure is applied to the gate electrode, which is a part of a Metal-Oxide-Semiconductor sandwich. Due to the small thickness of the gate oxide and large device active area, the MOS sandwich comprises a significant capacitance. Analysis of this capacitance requires taking into account the formation of a depletion layer in the semiconductor under certain bias conditions. The rate at which the power MOSFET structure can be switched between the on- and off-states is determined by the rate at which the input capacitance can be charged or discharged. In addition, the capacitance between the drain and the gate electrodes has been found to play an important role in determining the drain current and voltage transitions during the switching event.

The capacitances within the power D-MOSFET structure have been analyzed in detail in the textbook [1]. The specific input (or gate) capacitance for the power D-MOSFET structure is given by:

$$C_{IN,SP} = C_{N+} + C_{P} + C_{SM} = \frac{2x_{JP}}{W_{Cell}}\left(\frac{\varepsilon_{OX}}{t_{OX}}\right) + \frac{W_{G}}{W_{Cell}}\left(\frac{\varepsilon_{OX}}{t_{IEOX}}\right) \tag{2.18}$$

where $t_{OX}$ and $t_{IEOX}$ are the thicknesses of the gate and inter-electrode oxides, respectively. For a 30-V power D-MOSFET structure with a cell pitch ($W_{CELL}$ in Fig. 2.4) of 12 μm and gate electrode width of 8 μm, the input capacitance is found to be 22 nF/cm$^2$ for a gate oxide thickness of 500 Å and an inter-metal dielectric thickness of 5,000 Å.

The capacitance between the gate and drain electrodes (also called the reverse transfer capacitance) is determined by the width of the JFET region where the gate electrode overlaps the N-drift region. The MOS structure in this portion of the power D-MOSFET structure operates under deep depletion conditions when a positive voltage is applied to the drain. The gate-drain capacitance for the power D-MOSFET structure is given by:

$$C_{GD,SP} = \frac{(W_G - 2x_{PL})}{W_{Cell}}\left(\frac{C_{OX}C_{S,M}}{C_{OX} + C_{S,M}}\right) \tag{2.19}$$

where $C_{S,M}$ is the semiconductor capacitance under the gate oxide, which decreases with increasing drain bias voltage. The specific capacitance of the semiconductor depletion region can be obtained by computation of the depletion layer width.

The depletion layer width in the semiconductor under the gate oxide can be obtained using:

$$W_{D,MOS} = \frac{\varepsilon_S}{C_{OX}}\left\{\sqrt{1+\frac{2V_D C_{OX}^2}{q\varepsilon_S N_D}} - 1\right\} \tag{2.20}$$

The specific capacitance for the semiconductor is then obtained using:

$$C_{S,M} = \frac{\varepsilon_S}{W_{D,MOS}} \tag{2.21}$$

The gate-drain (or reverse transfer) capacitance can be computed by using (2.19) with the above equations to determine the semiconductor capacitance as a function of the drain bias voltage.

The specific gate transfer capacitance obtained by using the above analytical formulae is shown in Fig. 2.21 for the case of a power D-MOSFET structure with 12 μm cell pitch, and polysilicon window of 8 μm. This structure has a gate oxide thickness of 500 Å and a lateral junction depth of 1.7 μm for the P-base region. The gate-drain (reverse transfer) capacitance decreases with increasing drain bias voltage due to the expansion of the depletion region in the semiconductor. At a drain bias of 20 V, the specific reverse transfer capacitance predicted by the analytical model is 3.0 nF/cm$^2$ for this design.

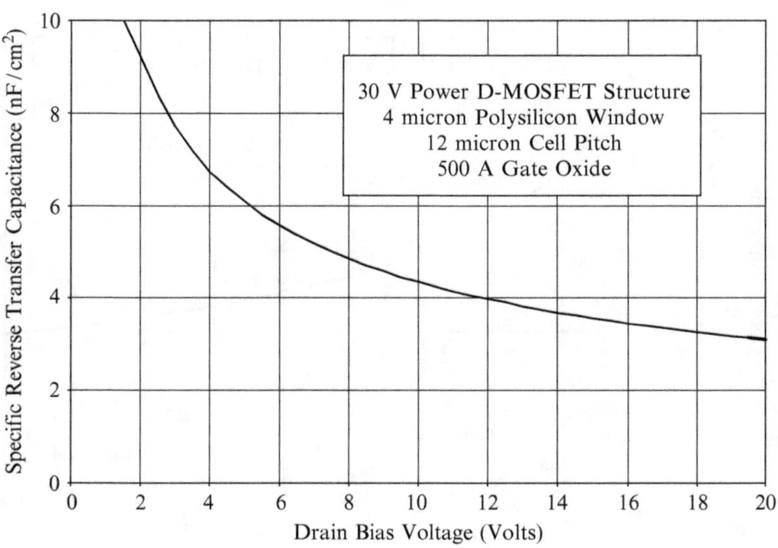


**Fig. 2.21** Gate-drain capacitance for the power D-MOSFET structure

The output capacitance for the power D-MOSFET structure is associated with the capacitance of the junction between the P-base region and the N-drift region. The specific junction capacitance is given by [1]:

$$C_{S,J} = \frac{\varepsilon_S}{W_{D,J}} \tag{2.22}$$

where the depletion region thickness at the junction ($W_{D,J}$) is related to the drain bias voltage:

$$W_{D,J} = \sqrt{\frac{2\varepsilon_S(V_D + V_{bi})}{qN_D}} \tag{2.23}$$

This depletion layer width is larger than that under the gate oxide because all the applied drain voltage must be supported across the P-N junction. The specific output capacitance for the power D-MOSFET structure can then be obtained by assuming (Model A in the textbook) that the area of the junction within the cell is ($W_{PW}$ + 2$x_{PL}$)Z:

$$C_O = \left(\frac{W_{PW} + 2x_{PL}}{W_{Cell}}\right) C_{S,J} \tag{2.24}$$

where $x_{PL}$ is the lateral extension of the P-base region under the gate electrode.

The output capacitance obtained by using the above analytical model is shown in Fig. 2.22 for the case of the 30-V power D-MOSFET structure with 12 μm cell pitch

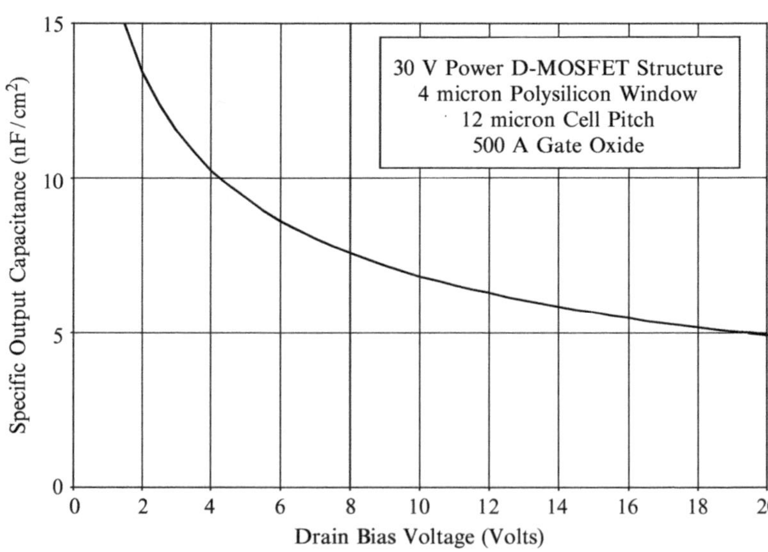


**Fig. 2.22** Output capacitance for the power D-MOSFET structure

and polysilicon window of 8 μm. This structure has a gate oxide thickness of 500 Å and a lateral junction depth of 1.7 μm for the P-base region. A built-in potential of 0.8 V was assumed for the P-base/N-drift junction, and the drift region has a doping concentration of $1.6 \times 10^{16}/cm^3$. The specific output capacitance decreases with increasing drain bias voltage due to the expansion of the depletion region under the P-base region. At a drain bias of 20 V, the specific output capacitance predicted by the analytical model for this structure is 5 nF/cm$^2$.

## 2.5.1 Simulation Example

The capacitances of the 30-V power D-MOSFET structure were extracted using two-dimensional numerical simulations on the structure with device parameters described in the previous sections. The input capacitance was extracted by performing the numerical simulations with a small AC signal superposed on the DC gate bias voltage. The input capacitances obtained for the power D-MOSFET structure are shown in Fig. 2.23 at a drain bias of 20 V. The input capacitance is comprised of two components – the first is between the gate electrode and the source electrode ($C_{GS}$) while the second is between the gate electrode and the base electrode ($C_{GB}$). The total input capacitance can be obtained by the addition of these capacitances because they are in parallel and share a common contact

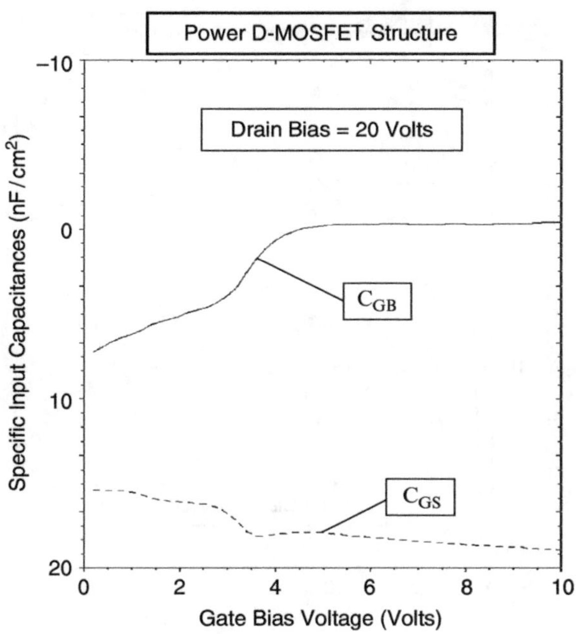


**Fig. 2.23** Input capacitances for the D-MOSFET structure

electrode in the actual power D-MOSFET structure. From the figure, a total specific input capacitance of about 20 nF/cm$^2$ is observed which is approximately independent of the gate bias voltage. At gate bias voltages below the threshold voltage, there is significant coupling between the P-base region and the gate electrode leading to a large contribution to the input capacitance from this path. When the gate bias voltage exceeds the threshold voltage, the inversion layer screens the P-base region from the gate electrode and couples it with the N$^+$ source region. Consequently, the contribution from $C_{GB}$ decreases to zero while that from $C_{GS}$ increases as the gate bias voltage is increased. The specific input capacitance extracted from the numerical simulations is in excellent agreement with that calculated with the analytical model.

The drain-gate (reverse transfer) capacitance can be extracted by performing the numerical simulations with a small AC signal superposed on the DC drain bias voltage. The values obtained for the 30-V power D-MOSFET structure are shown in Fig. 2.24. The gate-to-drain and base-to-drain capacitances are shown in the figure for comparison. Both of these capacitances decrease with increasing drain bias voltage as expected from the analytical model. For this power D-MOSFET structure, the reverse transfer (gate-drain) capacitance is comparable in magnitude to the output (base-drain) capacitance. This implies a strong feedback path between the drain and the gate electrodes which is detrimental to the switching speed and power loss for the power D-MOSFET structure. The values for the reverse transfer and output capacitances obtained by using the analytical models are in excellent agreement with the simulation values.

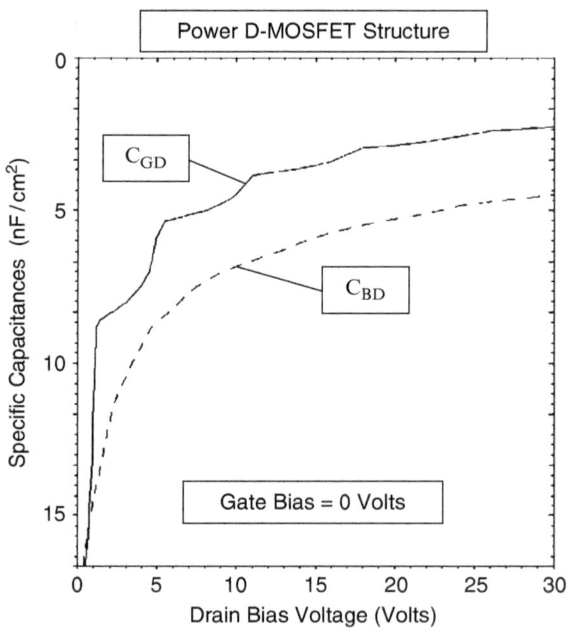


**Fig. 2.24** Reverse transfer and output capacitances for the D-MOSFET structure

## 2.6 Gate Charge

It is standard practice in the industry to provide the gate charge for power MOSFET structures as a measure of their switching performance. The gate charge can be extracted by the application of a constant current source at the gate terminal while turning-on the power MOSFET structure from the blocking state. The linearized current and voltage waveforms observed during the turn-on process are illustrated in Fig. 6.98 in the textbook [1]. The various components for the gate charge are also defined in this figure. During the turn-on process, the gate current is used to charge the capacitances $C_{GS}$ and $C_{GD}$ shown in Fig. 6.97. The most significant gate charge components for assessing the performance of the power MOSFET structures are $Q_{SW}$ (the gate switching charge), $Q_{GD}$ (the gate-drain charge), and $Q_G$ (the total gate charge). These components are given by [1]:

$$Q_{GD} = \frac{2K_G q\varepsilon_S N_D}{C_{OX}}\left[\sqrt{1+\frac{2V_{DS}C_{OX}^2}{q\varepsilon_S N_D}} - \sqrt{1+\frac{2V_{ON}C_{OX}^2}{q\varepsilon_S N_D}}\right] \tag{2.25}$$

$$Q_{SW} = [C_{GS} + C_{GD}(V_{DS})]\sqrt{\frac{J_{ON}W_{Cell}L_{CH}}{2\mu_{ni}C_{OX}}} + \frac{2K_G q\varepsilon_S N_D}{C_{OX}}\left[\sqrt{1+\frac{2V_{DS}C_{OX}^2}{q\varepsilon_S N_D}} - \sqrt{1+\frac{2V_{ON}C_{OX}^2}{q\varepsilon_S N_D}}\right] \tag{2.26}$$

$$Q_G = [C_{GS} + C_{GD}(V_{DS})]V_{GP} + \frac{2K_G q\varepsilon_S N_D}{C_{OX}}\left[\sqrt{1+\frac{2V_{DS}C_{OX}^2}{q\varepsilon_S N_D}} - \sqrt{1+\frac{2V_{ON}C_{OX}^2}{q\varepsilon_S N_D}}\right] + [C_{GS} + C_{GD}(V_{ON})](V_G - V_{GP}) \tag{2.27}$$

The gate charge values obtained for the 30-V power D-MOSFET structure by using the above equations are: $Q_{GD} = 216$ nC/cm$^2$; $Q_{SW} = 237$ nC/cm$^2$; and $Q_G = 592$ nC/cm$^2$. It can be concluded that the gate-drain charge ($Q_{GD}$) is the dominant portion (90%) of the gate switching charge ($Q_{SW}$).

Equations for the gate voltage, drain current, and drain voltage waveforms obtained by using the analytical model are provided in the textbook [1]. The waveforms obtained for the 30-V power D-MOSFET structure with 12 μm cell pitch and polysilicon gate width of 8 μm with a gate oxide thickness of 500 Å using these equations are provided in Fig. 2.25. A gate drive current density of 0.17 A/cm$^2$ was used to turn on the device from a steady-state blocking voltage of 20 V to match the results of two dimensional numerical simulations discussed below. The gate geometry factor ($K_G$) obtained for this structure using the structural dimensions is 0.38.

The gate voltage initially increases linearly with time. After reaching the threshold voltage, the drain current can be observed to increase in a non-linear manner

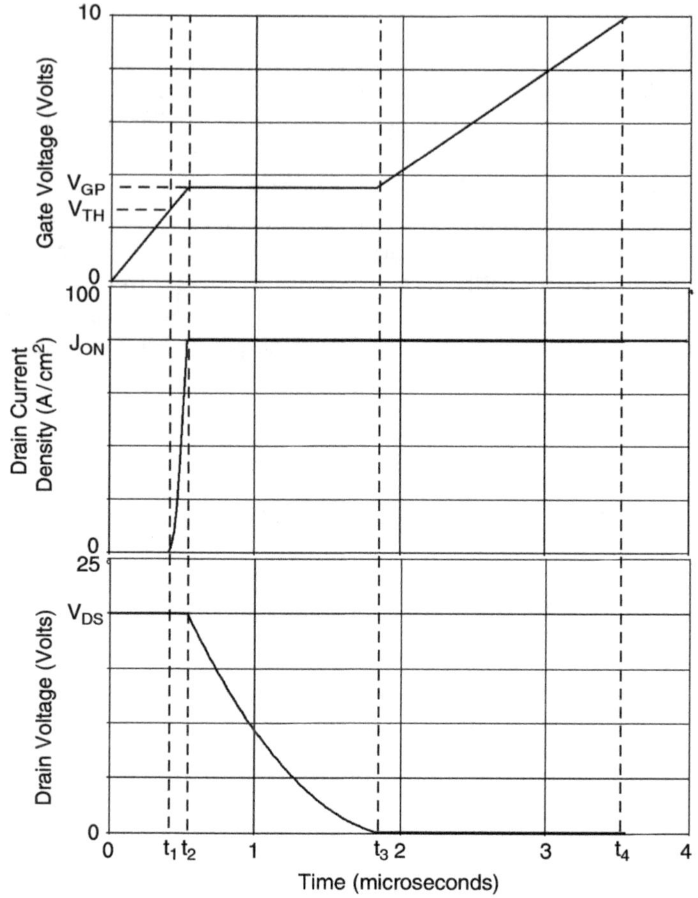


**Fig. 2.25** Analytically computed waveforms for the 30-V power D-MOSFET structure

because the transconductance is a function of the gate bias voltage. The drain current density increases until it reaches an on-state current density of 80 A/cm$^2$. This transition occurs rapidly when compared with the time taken for the drain voltage to decrease during the next time interval. The on-state current density determines the gate plateau voltage which has a value of 3.54 V. During the gate voltage plateau phase, the drain voltage decreases in a non-linear manner until it reaches the on-state voltage drop. After this time, the gate voltage again increases but at a slower rate than during the initial turn-on phase.

### *2.6.1 Simulation Example*

The gate charge for the 30-V power D-MOSFET structure was extracted by using the results of two-dimensional numerical simulations of the cell described in the previous sections. The device was turned-on from blocking state with a drain bias of 20 V by driving it using a gate current of $1 \times 10^{-8}$ A/μm (equivalent to 0.17 A/cm$^2$ for the area of $6 \times 10^{-8}$ cm$^2$). Once the drain current density reached 80 A/cm$^2$, the drain current was held constant resulting in a reduction of the drain voltage. The gate plateau voltage for this drain current density was found to be 3.5 V. Once the drain voltage reached the on-state value corresponding to the gate plateau voltage, the gate voltage increased to the steady-state value of 10 V.

The gate charge waveforms obtained by using an input gate current density of 0.17 A/cm$^2$ when turning on the power D-MOSFET structure from a blocking state with drain bias of 20 V are shown in Fig. 2.26. The on-state current density is 80 A/cm$^2$ at a DC gate bias of 10 V at the end of the turn-on transient. The gate voltage increases at a constant rate at the beginning of the turn-on process as predicted by the analytical model. When the gate voltage reaches the threshold voltage, the drain current begins to increase. The drain current increases as

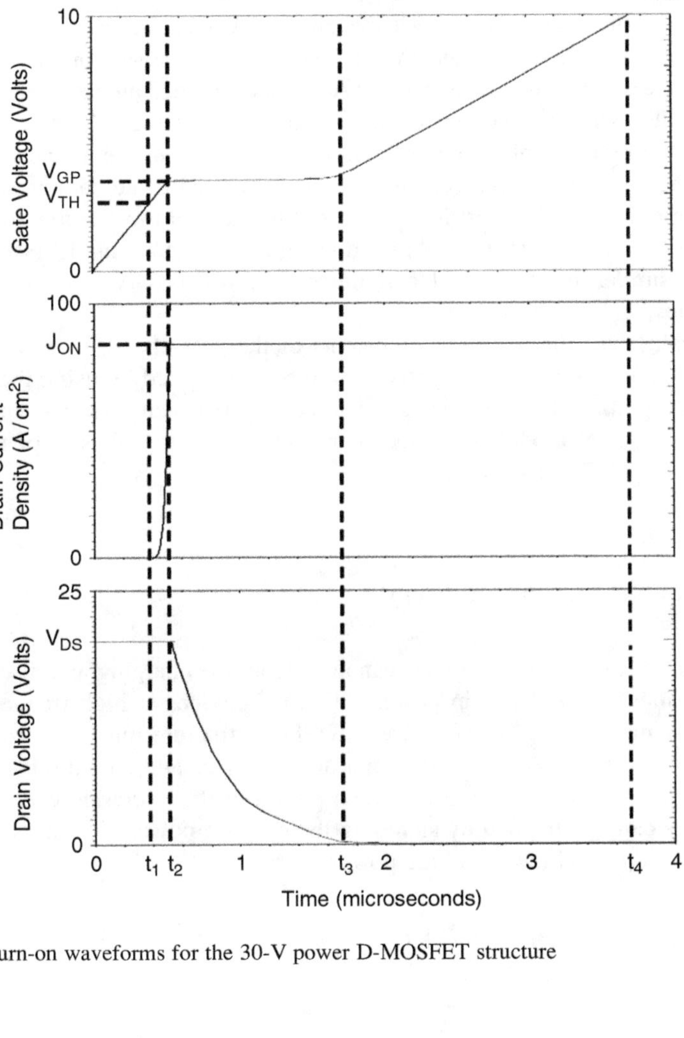


**Fig. 2.26** Turn-on waveforms for the 30-V power D-MOSFET structure

| Specific Gate Charge | Numerical Simulation (nC/cm$^2$) | Analytical Model (nC/cm$^2$) |
|---|---|---|
| $Q_{GS1}$ | 67 | 67.8 |
| $Q_{GS2}$ | 20 | 21.1 |
| $Q_{GS}$ | 87 | 88.8 |
| $Q_{GD}$ | 200 | 216 |
| $Q_{SW}$ | 220 | 237 |
| $Q_G$ | 617 | 592 |

**Fig. 2.27** Gate charge extracted from numerical simulations for the power D-MOSFET structure

predicted by the analytical model in a quadratic manner until it reaches the on-state current density of 80 A/cm$^2$.

Once the drain current reaches the on-state value, the gate voltage remains constant at the plateau voltage ($V_{GP}$). The plateau voltage for this structure is 3.5 V for the drain current density of 80 A/cm$^2$ as governed by the transconductance of the device. The drain voltage decreases during the plateau phase in a non-linear manner. After the end of the plateau phase, the gate voltage again increases until it reaches the gate supply voltage. Although the increase in gate voltage is non-linear at the beginning of this transition it becomes linear over most of the time after the plateau phase. The waveforms obtained using the analytical model (see Fig. 2.25) are very similar in shape and magnitude to those observed in the numerical simulations.

The values for the various components of the gate charge extracted from the numerical simulations are compared with those calculated by using the analytical model in Fig. 2.27. There is very good agreement between these values indicating that the analytical model is a good representation of the physics of turn-on for power D-MOSFET structure.

## 2.7 Device Figures of Merit

Significant power switching losses can arise from the charging and discharging of the large input capacitance in power MOSFET devices at high frequencies. The input capacitance ($C_{IN}$) of the power MOSFET structure must be charged to the gate supply voltage ($V_{GS}$) when turning on the device and then discharged to 0 V when turning off the device during each period of the operating cycle. The total power loss can be obtained by summing the on-state power dissipation for a duty cycle $\delta = t_{ON}/T$ and the switching power losses:

$$P_T = P_{ON} + P_{SW} = \delta R_{ON} I_{ON}^2 + C_{IN} V_{GS}^2 f \tag{2.28}$$

where $R_{ON}$ is the on-resistance of the power MOSFET structure, $I_{ON}$ is the on-state current, and f is the operating frequency. In writing this equation, the switching power losses due to the drain current and voltage transitions has been neglected. A minimum total power loss occurs for each power MOSFET structure at an optimum active area as shown in the textbook [1]. The on-state and switching power losses are equal at the optimum active area. The optimum active area at which the power dissipation is minimized is given by:

$$A_{OPT} = \sqrt{\frac{R_{ON,sp}}{C_{IN,sp}}}\left(\frac{I_{ON}}{V_{GS}}\right)\left(\sqrt{\frac{\delta}{f}}\right) \tag{2.29}$$

From the first term in this expression, a useful technology figure-of-merit can be defined:

$$FOM(A) = \frac{R_{ON,sp}}{C_{IN,sp}} \tag{2.30}$$

In the power electronics community, there is trend towards increasing the operating frequency for switch mode power supplies in order to reduce the size and weight of the magnetic components. The ability to migrate to higher operating frequencies in power conversion circuits is dependent on making enhancements to the power MOSFET technology. From the above equations, an expression for the minimum total power dissipation can be obtained [1]:

$$P_T(\min) = 2I_{ON}V_{GS}\sqrt{\delta R_{ON,sp}C_{IN,sp}f} \tag{2.31}$$

A second technology figure of merit related to the minimum power dissipation can be defined as:

$$FOM(B) = R_{ON,sp}C_{IN,sp} \tag{2.32}$$

In most applications for power MOSFET structures with high operating frequency, the switching losses associated with the drain current and voltage transitions become a dominant portion of the total power loss. The time period associated with the increase of the drain current and decrease of the drain voltage is determined by the charging of the device capacitances. It is therefore common practice in the industry to use the following figures-of-merit to compare the performance of power MOSFET products [1]:

$$FOM(C) = R_{ON,sp}Q_{GD,sp} \tag{2.33}$$

and

$$FOM(D) = R_{ON,sp}Q_{SW,sp} \tag{2.34}$$

| Figures of Merit | $V_G$ = 4.5 V | $V_G$ = 10 V |
|---|---|---|
| FOM(A) ($\Omega^2cm^4s^{-1}$) | 64,000 | 32,000 |
| FOM(B) (ps) | 30.9 | 15.5 |
| FOM(C) (mΩ*nC) | 303 | 152 |
| FOM(D) (mΩ*nC) | 333 | 167 |

**Fig. 2.28** Figures of merit for the power D-MOSFET structure

Although FOM(D) encompasses both the drain current and voltage transitions, it is customary to use FOM(C) because the gate-drain charge tends to dominate in the switching gate charge. One advantage of using these expressions is that the figure-of-merit becomes independent of the active area of the power MOSFET device.

The figures of merit computed for the power D-MOSFET structure discussed in earlier sections of this chapter are provided in Fig. 2.28. The figure of merit usually used for comparison of device technologies in the literature is FOM(C). Most often, the value for this figure of merit at a gate bias of 4.5 V is utilized for selection of devices in the voltage regulator module application.

## 2.8 Discussion

The characteristics of the power D-MOSFET structure has been reviewed in this chapter. This structure has a relatively large specific on-resistance due to its high channel and JFET resistance contributions. Moreover, its reverse transfer charge is relatively large because of the open-junction structure below the gate region than overlaps the drift region. Since the power D-MOSFET structure was developed first from a historical perspective, it will be utilized as a bench mark for assessment of the performance of the improved device structures in subsequent chapters.

For purposes of comparison with the power MOSFET structures discussed in subsequent chapters, the analysis of the power DMOSFET structure is provided here for a broad range of blocking voltages. In this analysis, the power DMOSFET structure was assumed to have the following parameters: (a) $N^+$ source junction depth of 0.5 μm; (b) P-base junction depth of 1.5 μm; (c) $P^+$ region junction depth of 2.5 μm; (d) gate oxide thickness of 500 Å; (e) polysilicon window width of 4 μm; (f) JFET region doping concentration of $2 \times 10^{16}/cm^3$; (g) threshold voltage of 2 V; (h) gate drive voltage of 10 V; (i) inversion mobility of 450 $cm^2/V$ s; (j) accumulation mobility of 1,000 $cm^2/V$ s. The contributions from the contacts and the $N^+$ substrate were neglected during the analysis. The doping concentration and thickness of the drift region were determined under the assumption that the edge termination (in conjunction with the graded junction doping profiles) limits the breakdown voltage to 90% of the parallel-plane breakdown voltage. The device parameters pertinent to each blocking voltage are provided in Fig. 2.29. It can be

| Blocking Voltage (V) | Drift Doping Concentration ($cm^{-3}$) | Drift Region Thickness (microns) |
|---|---|---|
| 30 | $1.6 \times 10^{16}$ | 3 |
| 60 | $6.0 \times 10^{15}$ | 5 |
| 100 | $3.0 \times 10^{15}$ | 8 |
| 200 | $1.2 \times 10^{15}$ | 17 |
| 300 | $6.8 \times 10^{14}$ | 26 |
| 600 | $2.7 \times 10^{14}$ | 58 |
| 1000 | $1.35 \times 10^{14}$ | 105 |

**Fig. 2.29** Device parameters for the power D-MOSFET structures

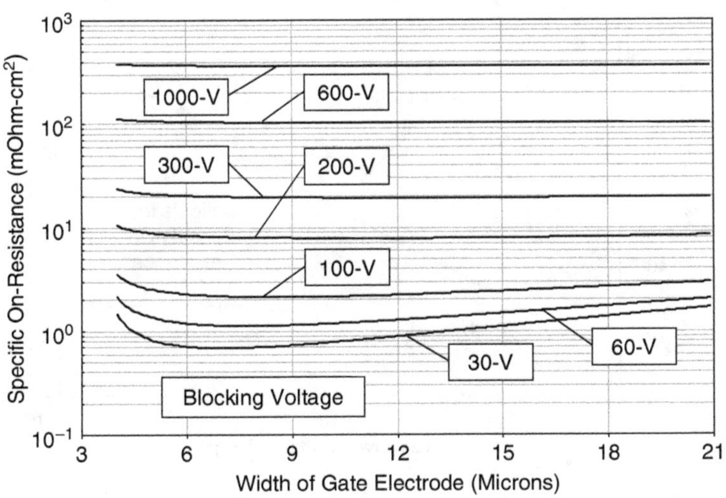


**Fig. 2.30** Optimization of the specific on-resistance for the power D-MOSFET structures

observed that the doping concentration must be greatly reduced with increasing blocking voltage capability while simultaneously increasing the thickness of the drift region.

The gate electrode width was optimized for each breakdown voltage to achieve the minimum specific on-resistance in the power D-MOSFET structure. The specific on-resistance obtained by using the analytical model for the power DMOSFET structure as a function of the width of the gate electrode is shown in Fig. 2.30 for the case of various breakdown voltages. Note that a logarithmic scale has been used for the vertical axis because of the large range of values for the specific on-resistance. The single zone model (see Fig. 6.39 in the textbook) was used for the 30-V power D-MOSFET structure while the two zone model (see Fig. 6.41 in the textbook) was

| Blocking Voltage (V) | Optimum Gate Width (microns) | Specific On-Resistance (mΩ-cm$^2$) |
|---|---|---|
| 30 | 7.0 | 0.687 |
| 60 | 7.7 | 1.13 |
| 100 | 8.7 | 2.14 |
| 200 | 10.8 | 7.85 |
| 300 | 12.7 | 19.4 |
| 600 | 17.0 | 102 |
| 1000 | 21.6 | 363 |

**Fig. 2.31** Optimized specific on-resistance for the power D-MOSFET structures

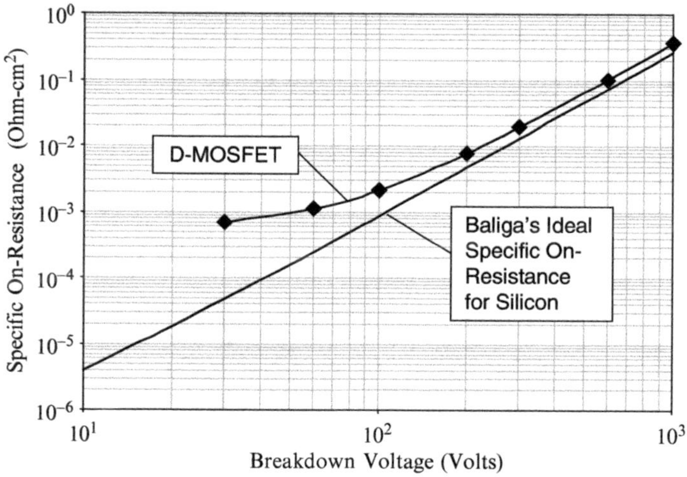


**Fig. 2.32** Specific on-resistance for the power D-MOSFET structures

used for the rest of the higher voltage power D-MOSFET structures. From Fig. 2.30, it can be observed that the minimum specific on-resistance occurs at a larger optimum gate electrode width for the power D-MOSFET structures with larger blocking voltages. The optimum gate electrode width and minimum specific on-resistances for the power D-MOSFET structures are provided in Fig. 2.31. These values can be compared with the ideal specific on-resistance obtained by using Baliga's power law for the impact ionization coefficients in Fig. 2.32. From this figure, it can be concluded that the specific on-resistance for the D-MOSFET structure is always larger than the ideal specific on-resistance. For blocking voltages below 100 V, the specific on-resistance for the power D-MOSFET structure becomes substantially larger than the ideal case.

| Blocking Voltage (V) | Drain Supply Voltage (Volts) | On-State Current Density ($A/cm^2$) | Specific Gate Transfer Charge ($nC/cm^2$) |
|---|---|---|---|
| 30 | 20 | 1720 | 199 |
| 60 | 40 | 1110 | 203 |
| 100 | 67 | 652 | 211 |
| 200 | 133 | 402 | 223 |
| 300 | 200 | 276 | 226 |
| 600 | 400 | 137 | 230 |
| 1000 | 667 | 82.5 | 227 |

**Fig. 2.33** Parameters used for gate transfer charge analysis for the power D-MOSFET structures

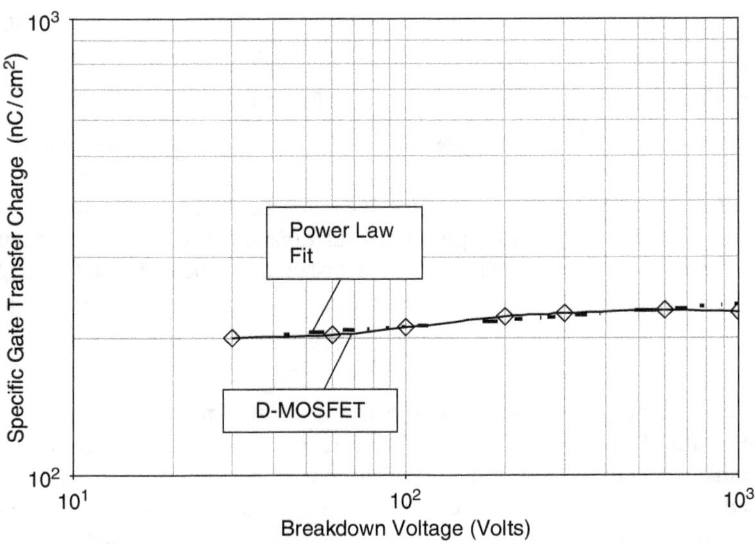


**Fig. 2.34** Specific gate transfer charge for the power D-MOSFET structures

The specific gate transfer charge for the power D-MOSFET structures was obtained by using the analytical model (see 2.25). During this analysis, the devices were assumed to be operated at an on-state current density that results in a power dissipation of 100 W/cm$^2$ as determined by the specific on-resistance for each device. The on-state current density values are provided in Fig. 2.33. The drain supply voltage used for this analysis was chosen to be two-thirds of the breakdown voltage. The drain supply voltage values are also provided in Fig. 2.33.

The specific gate transfer charge values calculated by using the analytical model for the power D-MOSFET structure are given in Fig. 2.33 as a function of the breakdown voltage. These values are also plotted in Fig. 2.34 as a function of the

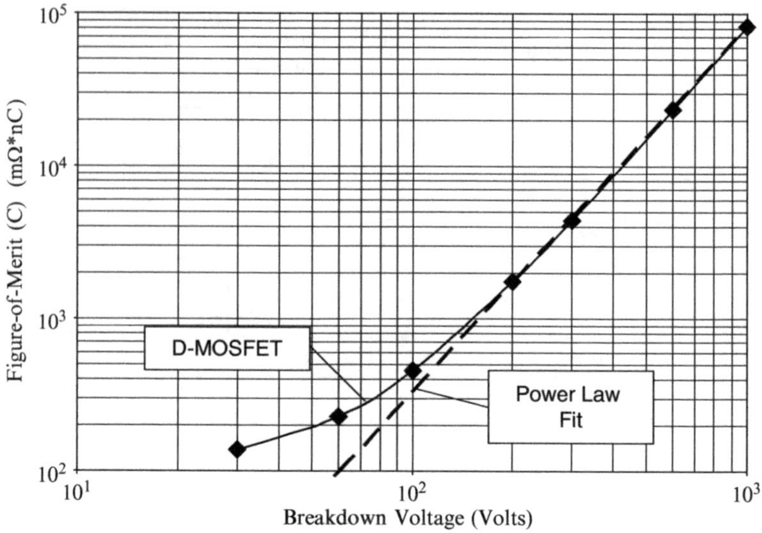


**Fig. 2.35** Figure-of-merit (C) for the power D-MOSFET structures

breakdown voltage. It can be observed that the gate transfer charge increases gradually with increasing breakdown voltage. The specific gate transfer charge increases in a linear fashion on this log-log graph for breakdown voltages ranging from 120 to 1,000 V indicating a power law relationship. The power law relationship that fits the data is shown in the figure by the dashed line. The equation for this line is:

$$Q_{GD} = 2{,}350\ BV^{-0.575} \tag{2.35}$$

where the specific gate transfer charge has units of nC/cm$^2$.

The figure-of merit (C) – product of the specific on-resistance and the specific gate transfer charge – for the optimized power D-MODSFET structures was obtained by determining the gate transfer charge for the cell with the optimum gate width for each breakdown voltage using the analytical model (see 2.25). During this analysis, the devices were assumed to be operated at an on-state current density that results in a power dissipation of 100 W/cm$^2$ as determined by the specific on-resistance for each device. The resulting values for the FOM(C) are plotted in Fig. 2.35 as a function of the breakdown voltage of the power D-MOSFET structure. It can be observed that the FOM(C) increases in a linear fashion on this log-log graph for breakdown voltages above 200 V indicating a power law relationship. The power law relationship that fits the data is shown in the figure by the dashed line. The equation for this line is:

$$\mathrm{FOM(C)} = R_{ON,sp}Q_{GD,sp} = 0.005423\ BV^{2.394} \tag{2.36}$$

where the FOM(C) has units of mΩ nC.

The values for the specific on-resistance, and FOM(C) provided here for the power D-MOSFET structure are useful as a benchmark to assess the performance of the power MOSFET structures that are discussed in subsequent chapters. Since the power D-MOSFET technology was developed in the 1970s, the manufacturing capability for these devices is mature leading to low device costs. Any advanced power MOSFET devices must provide substantial improvements in the specific on-resistance and FOM(C) in order to displace the power D-MOSFET structures in applications.

## References

1. B.J. Baliga, "Fundamentals of Power Semiconductor Devices", Springer-Science, New York, 2008.
2. S.C. Sun and J.D. Plummer, "Modeling of the On-Resistance of LDMOS, VDMOS, and VMOS Power Transistors", IEEE Transactions on Electron Devices, Vol. ED-27, pp. 356–367, 1980.

# Chapter 3
# U-MOSFET Structure

As discussed in the previous chapter, the first power MOSFET structure commercially introduced by the power semiconductor industry was the double-diffused or D-MOSFET structure. The specific on-resistance of the D-MOSFET devices designed for low blocking voltages was found to be constrained by the significant channel resistance due to the low channel density and the JFET region contribution.

With the advent of trench technology developed for applications to the Dynamic Random Access Memories (DRAMs), the power device community developed the trench gate structure for power MOSFET devices. These devices are also called U-MOSFET structures because of the shape of the gate region. The specific on-resistance of the U-MOSFET structure is substantially smaller than that of the D-MOSFET structure because the channel density can be made larger by using a smaller cell pitch. Moreover, the U-MOSFET structure does not contain a JFET region allowing further reduction of the specific on-resistance. By the mid-1990s, most power device manufacturers were building their high performance power MOSFET devices with the trench gate architecture.

## 3.1 The U-MOSFET Structure

A cross-section of the basic cell structure for the U-MOSFET structure [1] is illustrated in Fig. 3.1. This device structure is fabricated by starting with an N-type epitaxial layer grown on a heavily doped $N^+$ substrate. The P-base region is then formed across the active area of the device by ion implantation of boron followed by a drive-in cycle. The $N^+$ source regions are then produced by ion implantation of phosphorus followed by its annealing process. A mask is required during the ion-implantation of the $N^+$ source regions to create the short between the $N^+$ source and P-base regions at the top of the mesa region. The trench is next formed by reactive ion etching using a mask aligned to the $N^+$ source mask to center it in relation to the shorts. The surface of the trench must be smooth and free of damage in order to obtain a good MOS interface with high channel inversion layer mobility.

B.J. Baliga, *Advanced Power MOSFET Concepts*,
DOI 10.1007/978-1-4419-5917-1_3, 

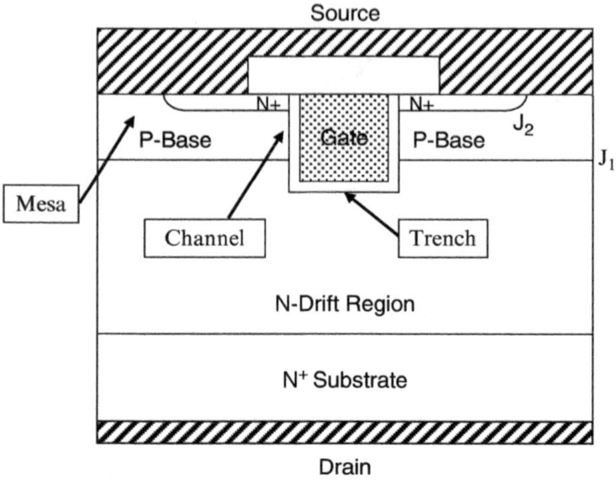


**Fig. 3.1** The U-MOSFET structure

A high voltage can be supported in the U-MOSFET structure when a positive bias is applied to the drain without the application of a gate bias (i.e. with gate shorted to the source by the external drive circuit). In this case, junction $J_1$ formed between the P-Base region and the N-drift region becomes reverse biased. The voltage is supported mainly within the thick lightly doped N-drift region. Drain current flow in the U-MOSFET structure is induced by the application of a positive bias to the gate electrode. This produces an inversion layer at the surface of the P-base region under the gate electrode along the sidewalls of the trench. This inversion layer channel provides a path for transport of electrons from the source to the drain when a positive drain voltage is applied.

After transport from the source region through the channel, the electrons enter the N-drift region at the bottom of the trench. The accumulation layer formed at the bottom of the trench surface aids in the distribution of the carriers into the drift region. The current spreads from the bottom of the trench to the entire width of the cell cross-section. This non-uniform current distribution within the drift region enhances its resistance making the internal resistance of the U-MOSFET structure larger than the ideal specific on-resistance of the drift region.

The U-MOSFET cell pitch can be made much smaller than that for the D-MOSFET structure due to the process technology and the absence of the JFET region. The smaller cell pitch increases the channel density which reduces its contribution to the specific on-resistance. However, the electric field at the junction ($J_1$) between the P-base region and the N-drift region is large in the U-MOSFET structure. The large electric field at the junction promotes the extension of the depletion region into the P-base region. Consequently, the channel length in U-MOSFET devices must be made relatively large in order to suppress reach-through breakdown which increases the channel resistance.

The large electric field at the junction produces a large electric field in the gate oxide in the trenches especially at the corners in the U-MOSFET structure. This

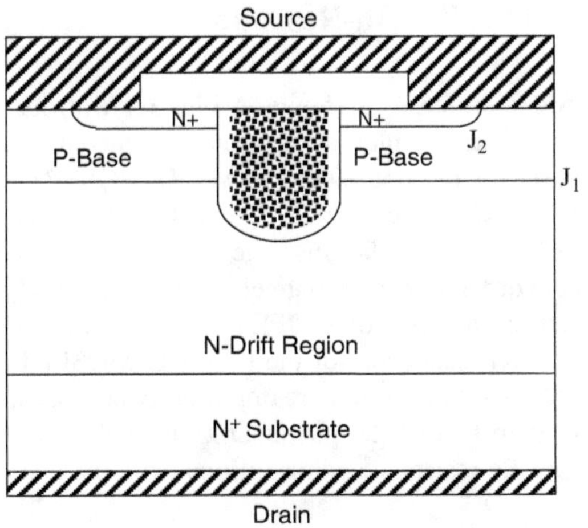


**Fig. 3.2** The U-MOSFET structure with rounded trench bottom surface

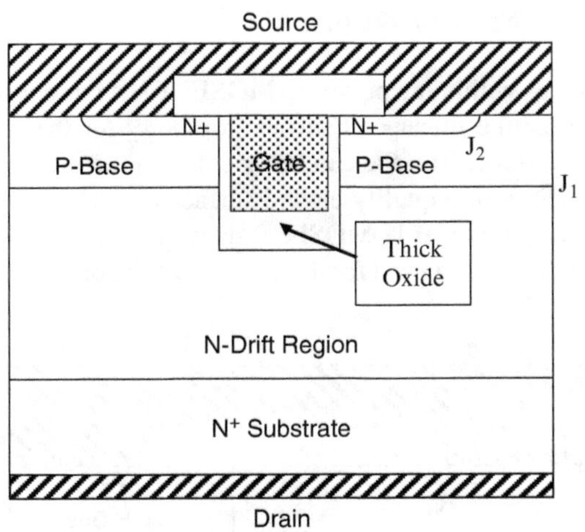


**Fig. 3.3** The U-MOSFET structure with thicker oxide at the trench bottom surface

creates reliability concerns due to generation of hot-electrons. This problem has been addressed by the industry by the development of structures with rounded surfaces at the bottom of the trench [2] as illustrated in Fig. 3.2. In spite of this improvement, the electric field in the gate oxide is relatively large in the U-MOSFET structure. In addition, the reverse transfer capacitance for the power U-MOSFET structure is much greater than for the D-MOSFET structure. This capacitance can be reduced by selectively increasing the oxide thickness at the bottom of the trench surface as illustrated in Fig. 3.3 although this adds significant complexity to the device fabrication process [3–5].

## 3.2 Power U-MOSFET On-Resistance

The power U-MOSFET structure is shown in Fig. 3.4 with its internal resistance components [1]. The same internal resistances encountered in the power D-MOSFET structure are present in the power U-MOSFET structure with the exception of the JFET region resistance. The JFET region is eliminated within the power U-MOSFET structure because the trench extends beyond the bottom of the P-base region in order to form a channel connecting the $N^+$ source region with the N-drift region. The elimination of the JFET region allows a significant reduction of the overall specific on-resistance for the power U-MOSFET structure not only because its resistance is excluded but more importantly because the cell pitch can be made much smaller than that of the power D-MOSFET structure. A smaller cell pitch reduces the specific resistance contributions from the channel, accumulation and drift regions.

The total on-resistance for the power U-MOSFET structure is obtained by the addition of all the resistances because they are considered to be in series in the current path between the source and the drain electrodes:

$$R_{ON} = R_{CS} + R_{N+} + R_{CH} + R_A + R_D + R_{SUB} + R_{CD} \tag{3.1}$$

Each of the resistances within the power U-MOSFET structure is analyzed below. It is customary to utilize the linear cell surface topology for the power U-MOSFET structure because the trench surface can be oriented in the preferred direction most favorable for producing high quality etched surfaces.

A cross-section of the power U-MOSFET structure is illustrated in Fig. 3.5 with various dimensions that can be used for the analysis of the on-resistance components.

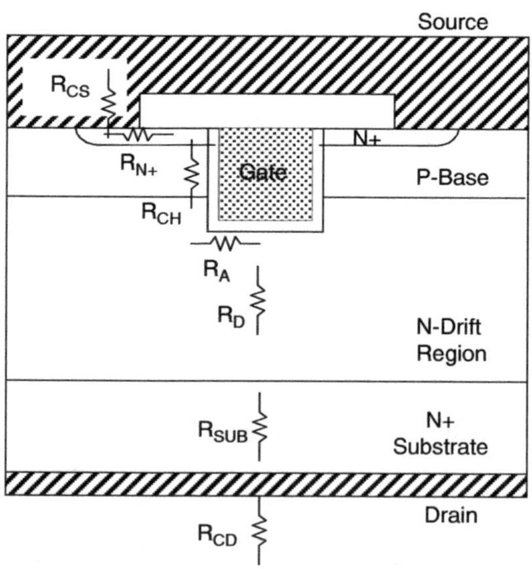


**Fig. 3.4** Power U-MOSFET structure with its internal resistances

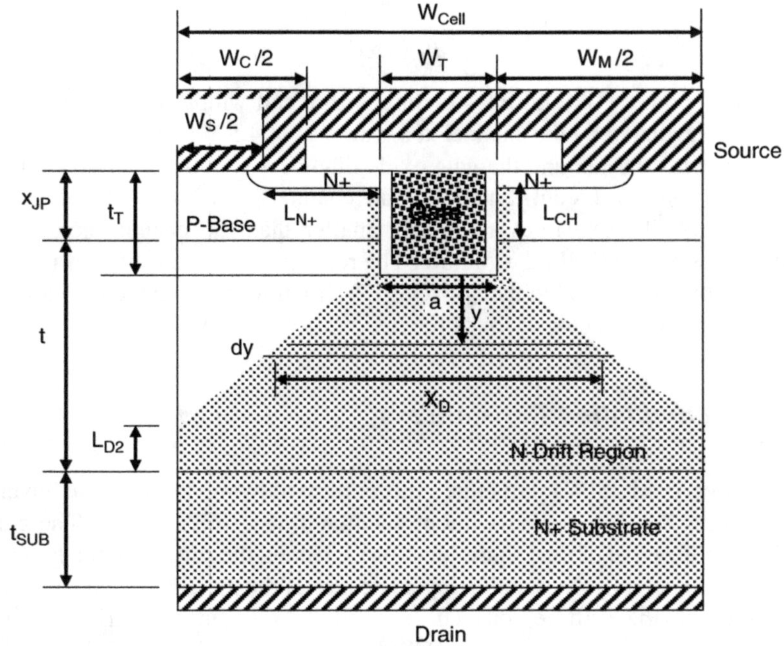


**Fig. 3.5** Power U-MOSFET structure with current flow model used for analysis of its internal resistances

The edges of the masks used to define the boundary for the $N^+$ source ion implant and the contact window are also shown in the figure by the dimension $W_S$ and $W_C$. These edges decide the area available within the structure for making contact to the $N^+$ source region. The position of the $N^+$ source definition mask also determines the length ($L_{N+}$) of the $N^+$ source region. In addition, the current flow pattern in the U-MOSFET structure is indicated by the shaded area in the figure. After entering the drift region, the current is assumed to spread from the trench bottom at a 45° angle. Since the mesa width is small, the current paths will usually overlap as illustrated in the figure. Consequently, the area for current transport varies for a portion of the drift region and then becomes uniform for the rest of the drift region.

In the textbook, it was demonstrated that the contributions from the source contact resistance ($R_{CS}$), the source resistance ($R_{SN+}$), and the drain contact resistance ($R_{CD}$) are very small for the U-MOSFET structure and will therefore be neglected in this chapter.

### *3.2.1 Channel Resistance*

The power U-MOSFET structure shown in Fig. 3.5 contains channels formed on both of the vertical sidewalls of the trench gate structure. The specific on-resistance contributed by the channel in the power U-MOSFET structure is given by [1]:

$$R_{CH,SP} = \frac{L_{CH} W_{Cell}}{2 \mu_{ni} C_{OX} (V_G - V_{TH})} \tag{3.2}$$

For the 30-V power U-MOSFET structure with a cell width of 3 μm, the specific resistance contributed by the channel at a gate bias of 4.5 V is 0.201 mΩ cm$^2$ if the channel length is 0.9 μm and the gate oxide thickness is 500 Å. For this structure, the specific resistance contributed by the channel at a gate bias of 10 V is 0.057 mΩ cm$^2$. These values are much smaller than the channel contributions within the power D-MOSFET structure. The reduction in the specific on-resistance contributed by the channel in the U-MOSFET structure is due to the increase in the channel density by a factor of four times.

### 3.2.2 *Accumulation Resistance*

In the power U-MOSFET structure, the current flowing through the inversion channel enters the drift region without encountering a JFET region. The current spreads from the edge of the P-base junction ($J_1$) along the surface of the trench due to the formation of an accumulation layer because of the positive gate bias. It is then distributed into the drift region. The specific on-resistance contributed by the accumulation layer in the power U-MOSFET structure is given by [1]:

$$R_{A,SP} = K_A \frac{L_A W_{Cell}}{2 \mu_{nA} C_{OX} (V_G - V_{TH})} \tag{3.3}$$

In this equation, a coefficient $K_A$ has been introduced to account for the current spreading from the accumulation layer into the drift region. A typical value for this coefficient is 0.6 based upon the current flow observed from numerical simulations of power U-MOSFET structures. The threshold voltage in the expression is for the on-set of formation of the accumulation layer, which will be assumed to zero in this section.

For the 30-V power U-MOSFET structure with a cell width of 3 μm, the specific resistance contributed by the accumulation layer at a gate bias of 4.5 V is 0.029 mΩ cm$^2$ if the P-base junction depth ($x_{JP}$) is 1 μm, the trench depth is 1.5 μm, and the gate oxide thickness is 500 Å. This value is an order of magnitude smaller than that for the power VD-MOSFET structure because the accumulation layer path ($L_A$) is only 1.0 μm in size and the cell pitch is much smaller for the U-MOSFET structure. The specific resistance contributed by the accumulation layer is reduced to 0.013 mΩ cm$^2$ when the gate bias is increased to 10 V.

### 3.2.3 *Drift Region Resistance*

The resistance contributed by the drift region in the power U-MOSFET structure is greater than that for the ideal drift region discussed earlier in the chapter due to current spreading from the trench surface into the drift region. The cross-sectional area for the current flow in the drift region increases from the width 'a', which is the width of the trench for the U-MOSFET structure, as illustrated in Fig. 3.5 by the shaded area. The resistance of the drift region developed in this section is based upon the current flow pattern shown in this figure. In this pattern, it is assumed that the cross-section width ($X_D$) for current flow increases at a 45° angle from the bottom of the trench. It is also assumed that the width of the mesa region is sufficiently small when compared with the thickness of the drift region to allow the current flow paths to merge before the current reaches the $N^+$ substrate. The resistance of the drift region is now determined by two portions: a first portion with a cross-sectional area that increases with the depth and a second portion with a uniform cross-sectional area for the current flow.

The specific on-resistance contributed by the drift region in the power U-MOSFET structure is given by [1]:

$$R_{D,SP} = \frac{\rho_D W_{Cell}}{2} \ln\left[\frac{W_T + W_M}{W_T}\right] + \rho_D\left(t + x_{JP} - t_T - \frac{W_M}{2}\right) \tag{3.4}$$

For the 30-V power U-MOSFET structure constructed using a drift region with doping concentration of $1.6 \times 10^{16}/\text{cm}^3$, a trench width of 1 μm, and a cell width of 3 μm, the specific resistance contributed by the drift region calculated by using the above model is 0.106 mΩ cm$^2$. This is 50% smaller than the value for the 30-V power D-MOSFET structure.

### 3.2.4 *Total On-Resistance*

The total specific on-resistance for the power U-MOSFET structure can be computed by adding all the above components for the on-resistance. For the case of the 30-V power U-MOSFET design with a cell pitch ($W_{Cell}$) of 3 μm, trench width of 1 μm, a channel length of 0.9 μm and a threshold voltage of 2.3 V, the total specific on-resistance is found to be 0.336 and 0.176 mΩ cm$^2$ at gate bias of 4.5 and 10 V, respectively. The contributions from each of the components of the on-resistance are summarized in Fig. 3.6.

| Resistance | $V_G$ = 4.5 V (mΩ–cm$^2$) | $V_G$ = 10 V (mΩ–cm$^2$) |
|---|---|---|
| Channel ($R_{CH,\,SP}$) | 0.201 | 0.057 |
| Accumulation ($R_{A,\,SP}$) | 0.029 | 0.013 |
| Drift ($R_{D,\,SP}$) | 0.106 | 0.106 |
| Total ($R_{T,\,SP}$) | 0.336 | 0.176 |

**Fig. 3.6** On-resistance components within the 30-V Power U-MOSFET structure

As discussed in Chap. 2, for the case of a blocking voltage of 30 V, the depletion width and doping concentration for the ideal drift region are found to be $2.2 \times 10^{16}/cm^3$ and 1.4 μm, respectively. Using the electron mobility for this doping level, the ideal specific on-resistance for this case is found to be 0.034 mΩ cm$^2$. Since the device is usually constrained by the impact of an 80% reduction of breakdown voltage due to the edge termination, it is worth computing the ideal specific on-resistance for this case for comparison with the device. For the case of a blocking voltage of 37.5 V, the depletion width and doping concentration are found to be $1.6 \times 10^{16}/cm^3$ and 1.8 μm, respectively. Using the electron mobility for this doping level, the ideal specific on-resistance is found to be 0.059 mΩ cm$^2$. It can be observed that the specific on-resistance for the U-MOSFET structure is only three times the ideal specific on-resistance for the blocking voltage of 37.5 V for a gate bias of 10 V. The specific on-resistance predicted by the analytical model is close to that reported for fabricated devices [2].

#### 3.2.4.1 Simulation Results

The results of two-dimensional numerical simulations on the 30-V power U-MOSFET structure are described here to provide a more detailed understanding of the underlying device physics and operation. The structure used for the numerical simulations had a drift region thickness of 3 μm below the junction ($J_1$) between the P-base region and the N-drift region with a doping concentration of $1.6 \times 10^{16}/cm^3$. The P-base region and $N^+$ source regions had depths of 1.0 and 0.1 μm, respectively. For the numerical simulations, half the cell (with a width of 1.5 μm) shown in Fig. 3.1 was utilized as a unit cell that is representative of the structure. The gate oxide was assumed to 500 Å with a uniform thickness on the trench sidewalls and bottom surfaces. A three dimensional view of the doping distribution in the U-MOSFET structure is shown in Fig. 3.7. The short between the $N^+$ source and P-base region is placed in the center of the mesa region.

The channel doping profile taken vertically along the surface of the trench is shown in Fig. 3.8. The vertical extensions of the P-base and $N^+$ source regions are 1.0 and 0.1 μm leading to a channel length of 0.9 μm. The surface concentration for the P-base region was chosen to obtain a maximum compensated P-type doping concentration in the channel of $1.0 \times 10^{17}/cm^3$. This is sufficient to prevent reach-through breakdown while obtaining an acceptable threshold voltage for the gate oxide thickness of 500 Å.

The transfer characteristics for the U-MOSFET structure were obtained using numerical simulations with a drain bias of 0.1 V at 300 and 400°K. The resulting transfer characteristics are shown in Fig. 3.9. From this graph, a threshold voltage of 2.3 and 2.0 V can be extracted at 300 and 400°K, respectively. The threshold voltage decreases by 15% when the temperature increases as in the case of the D-MOSFET structure. The specific on-resistance can be obtained from the transfer characteristics at any gate bias voltage. For the case of a gate bias of 4.5 V and 300°K, the

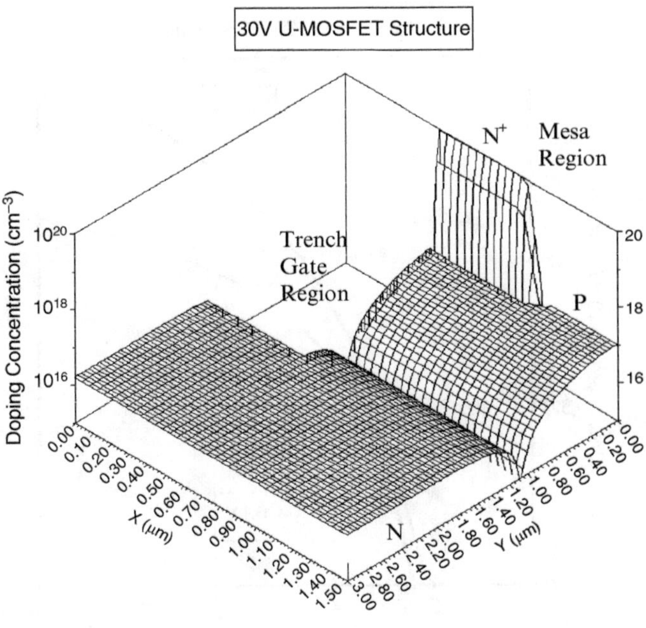


**Fig. 3.7** Doping distribution for the U-MOSFET structure

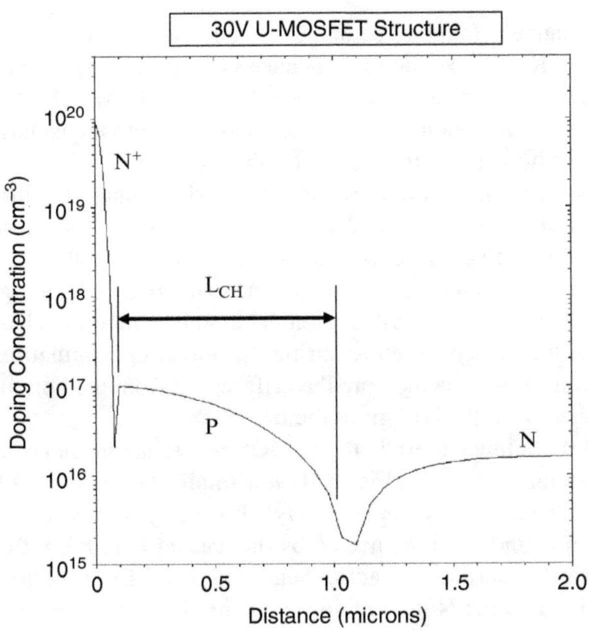


**Fig. 3.8** Channel doping profile for the U-MOSFET structure

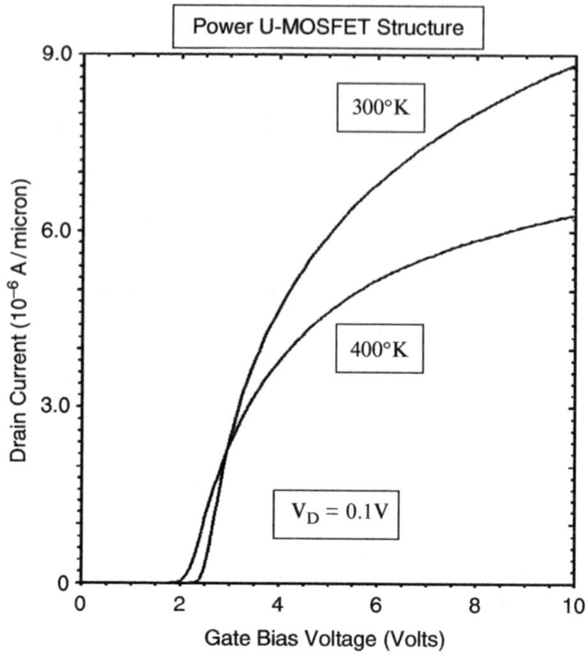


**Fig. 3.9** Transfer characteristics of the U-MOSFET structure

specific in-resistance is found to be 0.280 mΩ cm$^2$, while for the case of a gate bias of 10 V and 300°K, the specific in-resistance is found to be 0.170 mΩ cm$^2$. These values are far smaller than those observed for the D-MOSFET structure. The specific on-resistance obtained using the analytical model is larger than those obtained from the numerical simulations for the gate bias of 4.5 V. This is associated with several assumption used in the analytical model. Firstly, the threshold voltage in the analytical model is assumed to be independent of the position along the channel. In the device structure used for the numerical simulations, the threshold voltage becomes smaller along the channel due to the reduced doping concentration of the P-base region near the drift region. This will reduce the channel resistance contribution for the device structure during the numerical simulations. Second, the current spreads from the channel into the drift region from the bottom of the P-base region rather than from the bottom of the trench.

The current spreading pattern in the U-MOSFET structure can be observed using the current flow-lines shown in Fig. 3.10 at a small drain bias of 0.1 V and a gate bias of 4.5 V. In the figure, the depletion layer boundary is shown by the dotted lines and the junction boundary is delineated by the dashed line. From this figure, it can be observed that the current spreading begins to occur from the junction between the P-base region and the N-drift region reducing the drift region resistance.

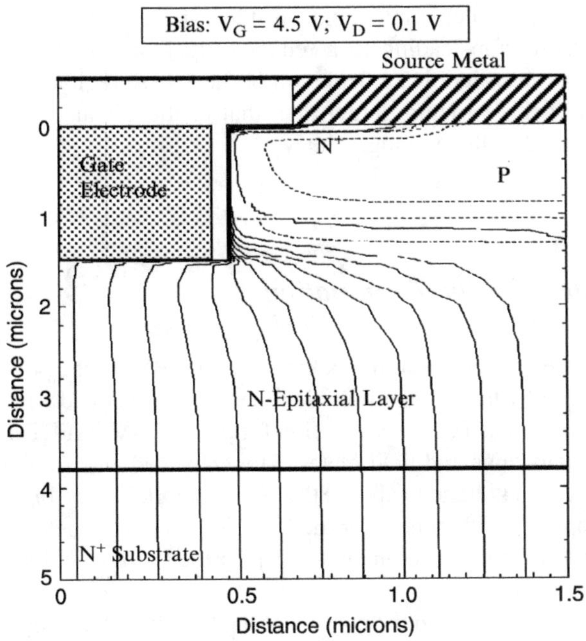


**Fig. 3.10** Current distribution in the U-MOSFET structure

## 3.3 Blocking Voltage

The power U-MOSFET structure must be designed to support a high voltage in the first quadrant when the drain bias voltage is positive. During operation in the blocking mode, the gate electrode is shorted to the source electrode by the external gate bias circuit. The application of a positive drain bias voltage produces a reverse bias across junction $J_1$ between the P-base region and the N-drift region. Most of the applied voltage is supported across the N-drift region. The doping concentration of donors in the N-epitaxial drift region and its thickness are chosen to attain the desired breakdown voltage. In devices designed to support low voltages (less than 50 V), the doping concentration of the P-base region is comparable to the doping concentration of the N-drift region leading to a graded doping profile. Consequently, a significant fraction of the applied drain voltage is supported across a depletion region formed in the P-base region. The highest doping concentration in the P-base region is limited by the need to keep the threshold voltage around 2 V to achieve a low on-resistance at a gate bias of 4.5 V as discussed in the previous section.

For the allowable maximum P-base doping concentration, it is desirable to make the depth of the P-base region as small as possible to reduce the channel length in the power MOSFET structure. However, if the junction depth of the P-base region is made too small, the depletion region in the P-base region will reach through to the $N^+$ source region leading to a reduced breakdown voltage. In the power U-MOSFET structure, significant depletion of the P-base region occurs making the channel length of this structure larger than that of the advanced power MOSFET structures discussed in this monograph.

### *3.3.1 Impact of Edge Termination*

In practical devices, the maximum blocking voltage (BV) of the power MOSFET is invariably decided by the edge termination that surrounds the device cell structure. The most commonly used edge termination for power U-MOSFET devices is based up on floating field rings and field plates. The enhanced electric field at the edges limits the breakdown voltage to about 80% of the parallel-plane breakdown voltage ($BV_{PP}$). The doping and thickness for the N-drift region must be designed using the same approach described earlier in Sect. 2.3.1 for the D-MOSFET structure in order to minimize the resistance.

It is also important to shield the corners of the trenches at the edge termination to avoid a severe reduction in the breakdown voltage [6]. The most suitable option is to envelope the corners of the trenches at the edges with a deeper $P^+$ diffusion. This requires an additional process step during device fabrication.

### *3.3.2 Impact of Graded Doping Profile*

For power MOSFET structures with low (less than 50 V) breakdown voltages, the doping concentration of the drift region is comparable to the doping concentration of the P-base region. This produces a graded doping profile for the junction $J_1$ between the P-base region and the N-drift region as illustrated in Fig. 2.12.

The electric field developed across junction $J_1$ during the blocking mode is also illustrated in Fig. 2.12. Due to the graded doping profile, the electric field extends on both sides of junction $J_1$. The electric field in the P-base region supports a portion of the applied positive drain voltage. This implies that the same breakdown voltage can be achieved with a larger doping concentration and a smaller thickness for the N-drift region. This improvement can be translated to increasing the breakdown voltage at the edge termination if the P-base region is used at the edges of the power MOSFET structure. A reduction of the resistance for the power MOSFET structure can be achieved by taking into account the voltage supported within the P-base region. An improvement in the specific on-resistance of 20% can be achieved by taking into account the graded doping profile.

#### 3.3.2.1 Simulation Results

The results of two-dimensional numerical simulations on the 30 V power U-MOSFET structure are described here to provide a more detailed understanding of the underlying device physics and operation during the blocking mode. The structure used for the numerical simulations had the same parameters as the structure described in the previous section. The blocking characteristic for the U-MOSFET cell structure is shown in Fig. 3.11 for 300°K. It can be observed that the cell is capable of supporting 42 V. This provides enough margin to achieve a device blocking voltage capability of slightly over 30 V after accounting for the reduction due to the edge termination.

It is instructive to examine the potential contours inside the power U-MOSFET structure when it is operating in the blocking mode. This allows determination of the voltage distribution within the structure and the penetration of the depletion region in the P-base region with increasing drain bias voltage. The potential contours for the U-MOSFET structure obtained using the numerical simulations with zero gate bias and various drain bias voltages are shown in Fig. 3.12–3.15. From these figures, it can be observed that the depletion region in the P-base region penetrates through a significant fraction of the P-base region when the drain bias is increased to 30 V. Any decrease in the doping concentration of the P-base region leads to reach-through breakdown limiting the ability to reduce the threshold voltage.

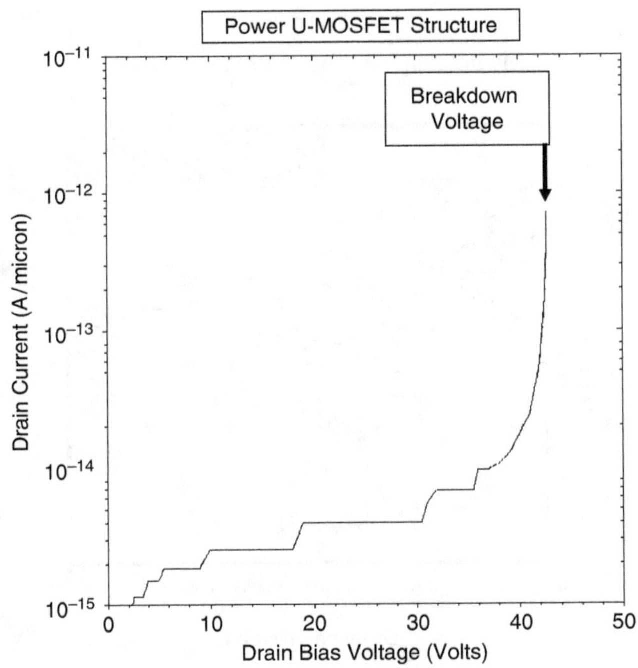


**Fig. 3.11** Blocking characteristics for the D-MOSFET structure

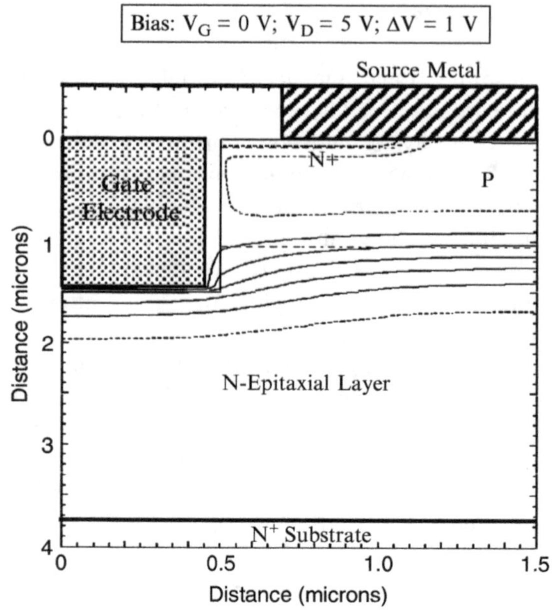


Fig. 3.12 Potential contours in the U-MOSFET structure

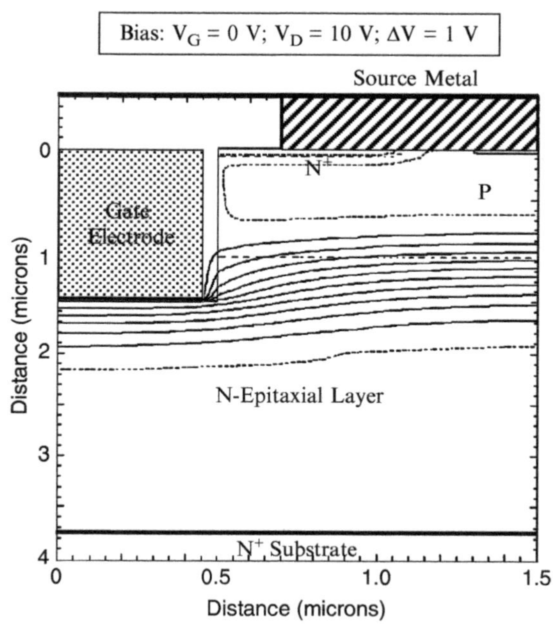


Fig. 3.13 Potential contours in the U-MOSFET structure

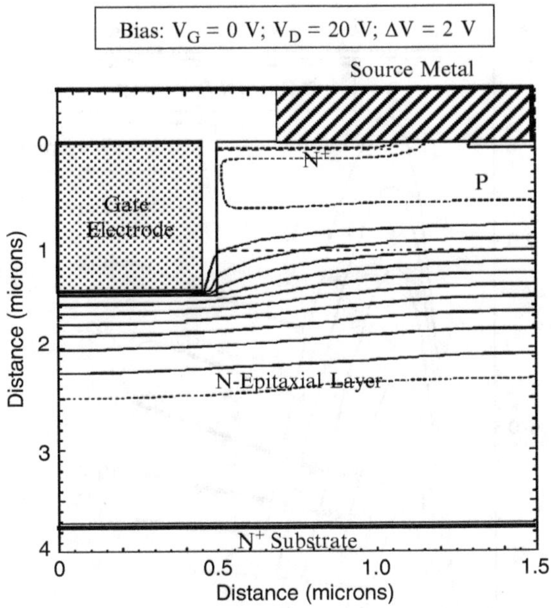


**Fig. 3.14** Potential contours in the U-MOSFET structure

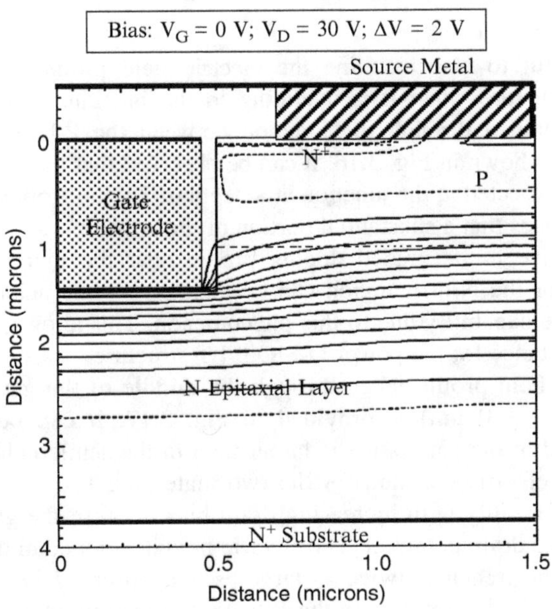


**Fig. 3.15** Potential contours in the U-MOSFET structure

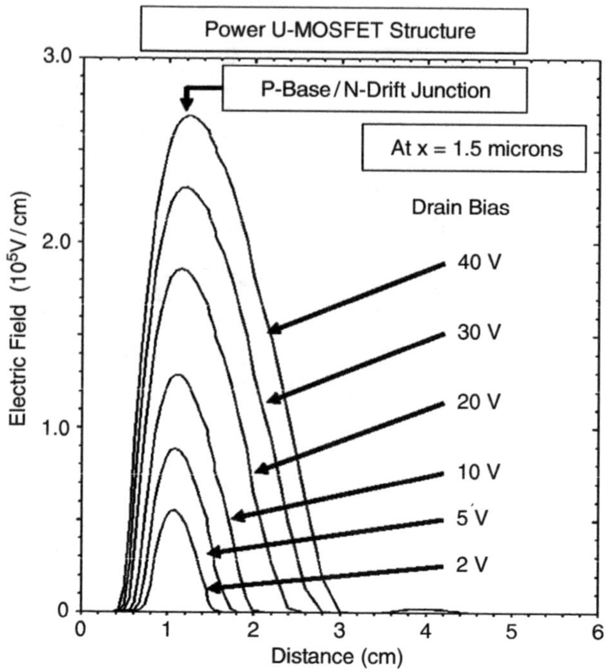


**Fig. 3.16** Electric field distribution in the U-MOSFET structure

It is insightful to also examine the electric field profile inside the power U-MOSFET structure when it is operating in the blocking mode. The electric field profile obtained through the junction between the P-base region and the N-drift region is shown in Fig. 3.16. It can be observed that the maximum electric field occurs as expected at the junction at a depth of 1.0 μm from the surface. This figure demonstrates that a substantial portion of the voltage is supported inside the P-base region due to its graded doping profile near the junction. The doping concentration for the N-drift region can be increased while achieving the target blocking voltage capability due to this phenomenon. This allows reduction of the specific on-resistance for the power D-MOSFET structure.

The electric field profile taken through the middle of the gate electrode and the trench (at $x = 0$ μm) is provided in Fig. 3.17. It can be observed that the electric field in the gate oxide is larger than in the semiconductor due to the difference in dielectric constant for the two materials. The electric field in the oxide increases rapidly with increasing drain bias because the gate oxide is not shielded from the drain potential. The electric field developed in the gate oxide at the bottom of the trench is twice as large as that observed in the D-MOSFET structure. The high electric field in the gate oxide in the blocking mode has been found to create reliability problems. The electric field in the gate oxide for this power U-MOSFET structure is above the limit for reliable operation over long periods of time.

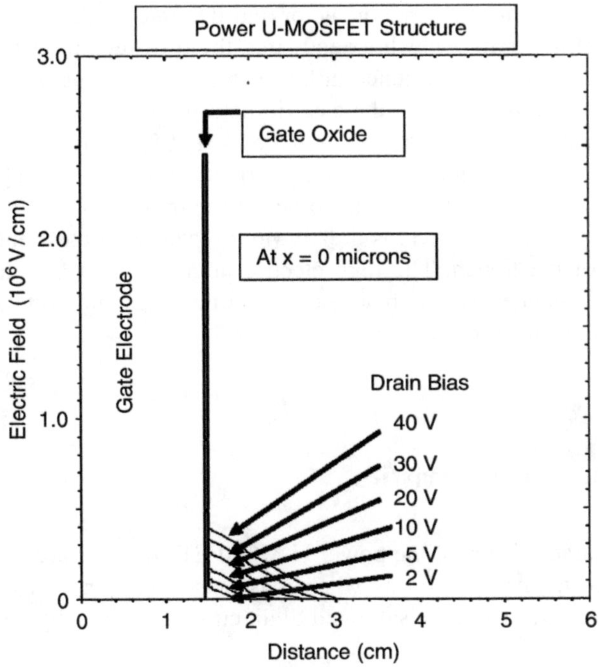


**Fig. 3.17** Electric field distribution in the U-MOSFET structure

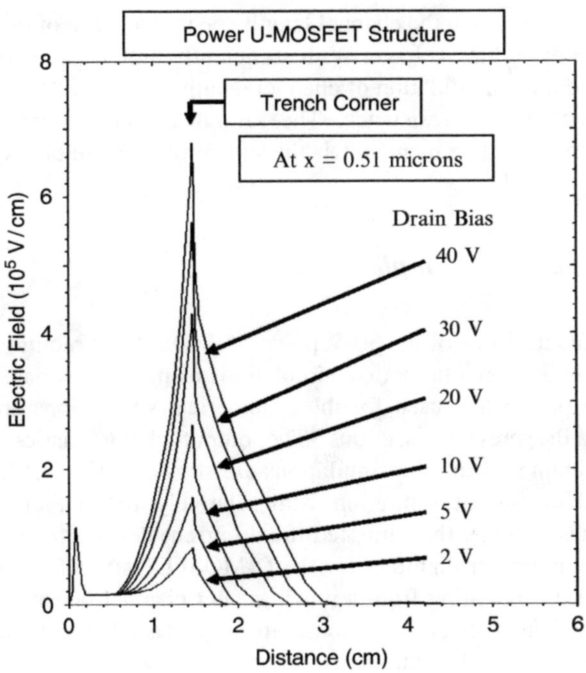


**Fig. 3.18** Electric field distribution in the U-MOSFET structure

The electric field in the semiconductor and the gate oxide is enhanced by the sharp corner of the trench which extends into the drift region near the maximum electric field generated at the junction between the P-base region and the N-drift region. This can be observed in the electric field profiles taken along the vertical direction along the surface of the trench in the U-MOSFET structure. These electric field profiles are shown in Fig. 3.18 at various drain bias voltages. It can be observed that the electric field at the corner of the trench, located at 1.5 μm from the upper surface of the device, is greatly (more than two times) enhanced by the sharp corner of the trench. The high electric field produces hot-electrons in the vicinity of the channel which creates a shift in the threshold voltage during long term operation of the device.

## 3.4 Output Characteristics

The output characteristics of the power U-MOSFET structure are important to the loci for the switching waveforms when it is operating in power circuits. In the power MOSFET structure, the saturated drain current is given by:

$$I_{D,sat} = \frac{Z\mu_{ni}C_{OX}}{(L_{CH} - \Delta L_{CH})}(V_G - V_{TH})^2 \tag{3.5}$$

where $\Delta L_{CH}$ is reduction in the channel length due to depletion of the P-base region with increasing drain bias voltage. With sufficiently high doping concentration of the P-base region, the modulation of channel length can be made sufficiently small to ensure a high output resistance. The saturated drain current in the power U-MOSFET structure then increases as the square of the gate bias voltage.

### *3.4.1 Simulation Example*

The output characteristics of the 30-V power U-MOSFET structure were obtained by using two-dimensional numerical simulations using various gate bias voltages. All the device parameters used for these numerical simulations are the same as those used in the previous sections. The output characteristics of the power U-MOSFET obtained using the simulations are shown in Fig. 3.19. The structure exhibits excellent current saturation with relatively flat output characteristics. At any gate bias voltage, the saturated drain current density for the U-MOSFET structure is much greater than that for the D-MOSFET structure due to the larger transconductance originating from a much greater channel density. The traces for increasing gate bias voltages are non-uniformly spaced due to the square-law behavior of the transfer characteristics.

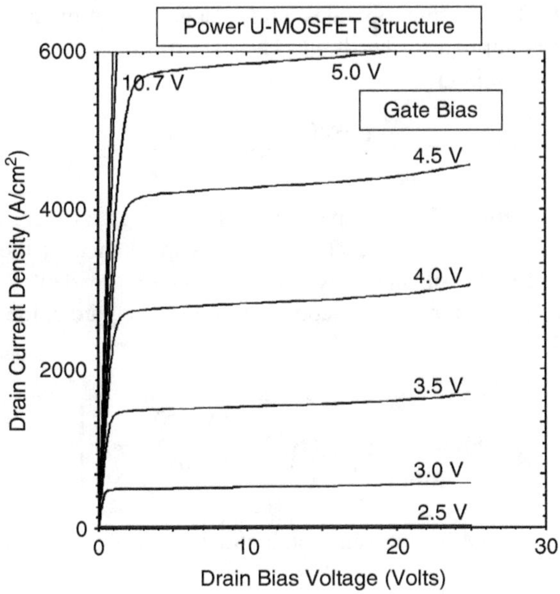


**Fig. 3.19** Output characteristics for the power U-MOSFET structure

## 3.5 Device Capacitances

As discussed in the previous chapter for the D-MOSFET structure, the rate at which the power MOSFET structure can be switched between the on- and off-states is determined by the rate at which its input capacitance can be charged or discharged. In addition, the capacitance between the drain and the gate electrodes has been found to play an important role in determining the drain current and voltage transitions during the switching event.

The capacitances within the power U-MOSFET structure have been analyzed in detail in the textbook [1]. The specific input (or gate) capacitance for the power U-MOSFET structure is given by:

$$C_{IN,SP} = C_{N+} + C_P + C_{SM} = \frac{2x_{JP}}{W_{Cell}}\left(\frac{\varepsilon_{OX}}{t_{OX}}\right) + \frac{W_T}{W_{Cell}}\left(\frac{\varepsilon_{OX}}{t_{IEOX}}\right) \tag{3.6}$$

where $t_{OX}$ and $t_{IEOX}$ are the thicknesses of the gate and inter-electrode oxides, respectively. For a 30-V power U-MOSFET structure with a cell pitch ($W_{CELL}$ in Fig. 3.5) of 3 μm and P-base junction depth width of 1.0 μm, the input capacitance is found to be 47 nF/cm$^2$ for a gate oxide thickness of 500 Å and an inter-metal dielectric thickness of 5,000 Å.

The capacitance between the gate and drain electrodes (also called the reverse transfer capacitance) is determined by the width of the trench where the gate electrode overlaps the N-drift region. The MOS structure in this portion of the power U-MOSFET structure operates under deep depletion conditions when a positive voltage is applied to the drain. The gate-drain capacitance for the power U-MOSFET structure is given by [1]:

$$C_{GD,SP} = \left[\frac{W_T + 2(t_T - x_{JP})}{W_{Cell}}\right]\left(\frac{C_{OX}C_{S,M}}{C_{OX} + C_{S,M}}\right) \tag{3.7}$$

where $C_{S,M}$ is the semiconductor capacitance under the gate oxide, which decreases with increasing drain bias voltage. The specific capacitance of the semiconductor depletion region can be obtained by computation of the depletion layer width. The depletion layer width in the semiconductor under the gate oxide can be obtained using:

$$W_{D,MOS} = \frac{\varepsilon_S}{C_{OX}}\left\{\sqrt{1 + \frac{2V_D C_{OX}^2}{q\varepsilon_S N_D}} - 1\right\} \tag{3.8}$$

The specific capacitance for the semiconductor is then obtained using:

$$C_{S,M} = \frac{\varepsilon_S}{W_{D,MOS}} \tag{3.9}$$

The gate-drain (or reverse transfer) capacitance can be computed by using (3.7) with the above equations to determine the semiconductor capacitance as a function of the drain bias voltage.

The specific gate transfer capacitance obtained by using the above analytical formulae is shown in Fig. 3.20 for the case of a power U-MOSFET structure with 3 μm cell pitch, and trench width of 1 μm. This structure has a gate oxide thickness

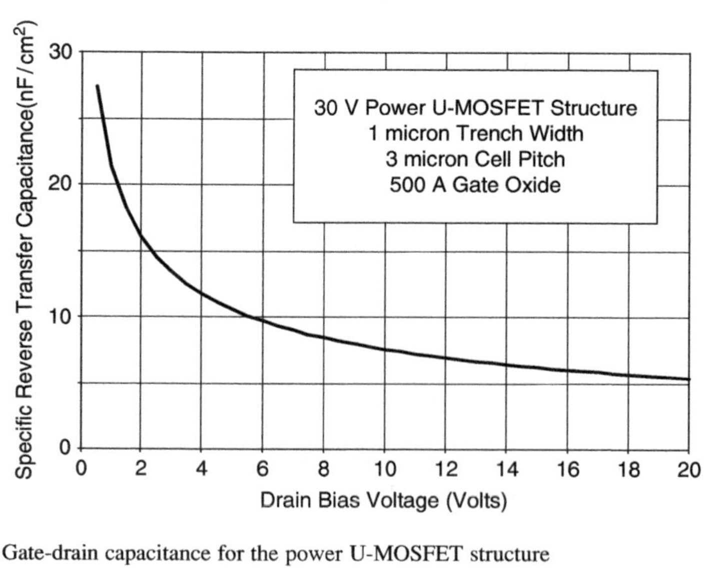


**Fig. 3.20** Gate-drain capacitance for the power U-MOSFET structure

of 500 Å, a trench depth of 1.5 μm and P-base junction depth of 1.0 μm. The gate-drain (reverse transfer) capacitance decreases with increasing drain bias voltage due to the expansion of the depletion region in the semiconductor. At a drain bias of 20 V, the specific reverse transfer capacitance predicted by the analytical model is 5.4 nF/cm$^2$ for this design.

The output capacitance for the power U-MOSFET structure is associated with the capacitance of the junction between the P-base region and the N-drift region. The specific junction capacitance is given by [1]:

$$C_{S,J} = \frac{\varepsilon_S}{W_{D,J}} \tag{3.10}$$

where the depletion region thickness at the junction ($W_{D,J}$) is related to the drain bias voltage:

$$W_{D,J} = \sqrt{\frac{2\varepsilon_S(V_D + V_{bi})}{qN_D}} \tag{3.11}$$

This depletion layer width is larger than that under the gate oxide because all the applied drain voltage must be supported across the P-N junction. The specific output capacitance for the power U-MOSFET structure can then be obtained by using [1]:

$$C_O = \left[\frac{W_M - 2K_S(t_T - x_{JP} - t_{OX})}{W_{Cell}}\right] C_{S,J} \tag{3.12}$$

where $x_{JP}$ is the junction depth of the P-base region.

The output capacitance obtained by using the above analytical model is shown in Fig. 3.21 for the case of the 30-V power U-MOSFET structure with 3 μm cell pitch and

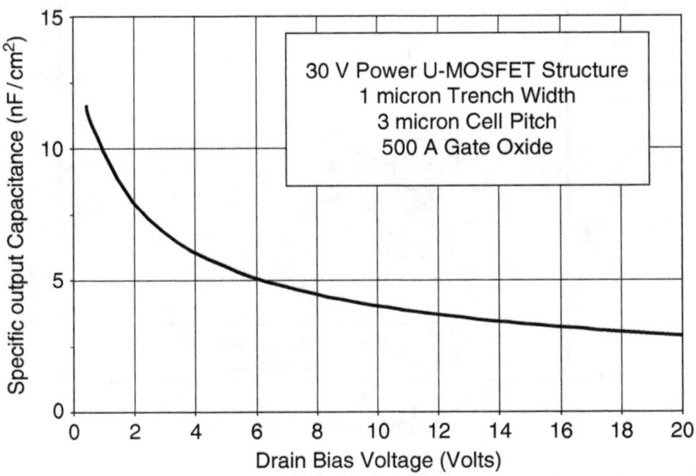


**Fig. 3.21** Output capacitance for the power U-MOSFET structure

trench width of 1 μm. This structure has a gate oxide thickness of 500 Å and a junction depth of 1.0 μm for the P-base region. A built-in potential of 0.8 V was assumed for the P-base/N-drift junction, and the drift region has a doping concentration of $1.6 \times 10^{16}$/cm$^3$. The screening factor ($K_S$) was assumed to be 1. The specific output capacitance decreases with increasing drain bias voltage due to the expansion of the depletion region under the P-base region. At a drain bias of 20 V, the specific output capacitance predicted by the analytical model for this structure is 3 nF/cm$^2$.

### 3.5.1 Simulation Example

The capacitances of the 30-V power U-MOSFET structure were extracted using two-dimensional numerical simulations on the structure with device parameters described in the previous sections. The input capacitance was extracted by performing the numerical simulations with a small AC signal superposed on the DC gate bias voltage. The input capacitances obtained for the power U-MOSFET structure are shown in Fig. 3.22 at a drain bias of 0 V. The input capacitance is comprised of two components – the first is between the gate electrode and the source electrode ($C_{GS}$) while the second is between the gate electrode and the base electrode ($C_{GB}$). The total input capacitance can be obtained by the addition of

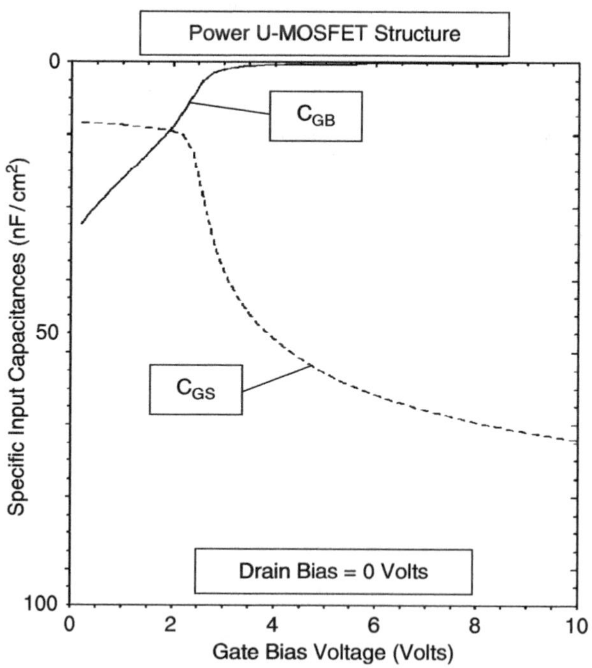


**Fig. 3.22** Input capacitances for the U-MOSFET structure

these capacitances because they are in parallel and share a common contact electrode in the actual power U-MOSFET structure. From the figure, a total specific input capacitance of about 60 nF/cm$^2$ is observed which is approximately independent of the gate bias voltage. At gate bias voltages below the threshold voltage, there is significant coupling between the P-base region and the gate electrode leading to a large contribution to the input capacitance from this path. When the gate bias voltage exceeds the threshold voltage, the inversion layer screens the P-base region from the gate electrode and couples it with the N$^+$ source region. Consequently, the contribution from $C_{GB}$ decreases to zero while that from $C_{GS}$ increases as the gate bias voltage is increased. The specific input capacitance extracted from the numerical simulations is in excellent agreement with that calculated with the analytical model.

The drain-gate (reverse transfer) capacitance can be extracted by performing the numerical simulations with a small AC signal superposed on the DC drain bias voltage. The values obtained for the 30-V power U-MOSFET structure are shown in Fig. 3.23. The gate-to-drain and base-to-drain capacitances are shown in the figure for comparison. Both of these capacitances decrease with increasing drain bias voltage as expected from the analytical model. For this power U-MOSFET structure, the reverse transfer (gate-drain) capacitance is larger in magnitude than the output (base-drain) capacitance and much larger than that for the D-MOSFET

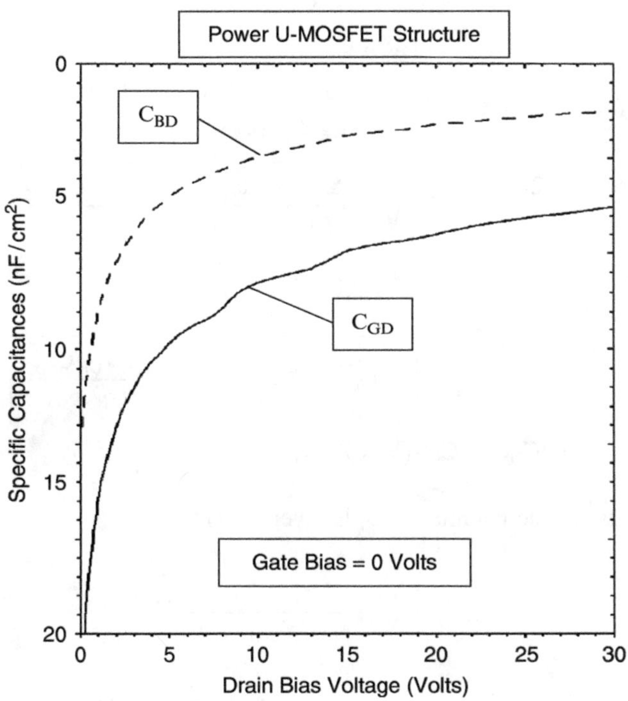


**Fig. 3.23** Reverse transfer and output capacitances for the U-MOSFET structure

structure. This implies a strong feedback path between the drain and the gate electrodes which is detrimental to the switching speed and power loss for the power U-MOSFET structure. The values for the reverse transfer and output capacitances obtained by using the analytical models are in good agreement with the simulation values.

## 3.6 Gate Charge

As mentioned in the previous chapter, it is standard practice in the industry to provide the gate charge for power MOSFET structures as a measure of their switching performance. The gate charge can be extracted by the application of a constant current source at the gate terminal while turning-on the power MOSFET structure from the blocking state. The most significant gate charge components for assessing the performance of the power MOSFET structures are $Q_{SW}$ (the gate switching charge), $Q_{GD}$ (the gate-drain charge), and $Q_G$ (the total gate charge). These components are given by [1]:

$$Q_{GD} = \frac{2K_G q\varepsilon_S N_D}{C_{OX}}\left[\sqrt{1+\frac{2V_{DS}C_{OX}^2}{q\varepsilon_S N_D}} - \sqrt{1+\frac{2V_{ON}C_{OX}^2}{q\varepsilon_S N_D}}\right] \tag{3.13}$$

$$\begin{aligned} Q_{SW} = {} & [C_{GS} + C_{GD}(V_{DS})]\sqrt{\frac{J_{ON}W_{Cell}L_{CH}}{2\,\mu_{ni}C_{OX}}} \\ & + \frac{2K_G q\varepsilon_S N_D}{C_{OX}}\left[\sqrt{1+\frac{2V_{DS}C_{OX}^2}{q\varepsilon_S N_D}} - \sqrt{1+\frac{2V_{ON}C_{OX}^2}{q\varepsilon_S N_D}}\right] \end{aligned} \tag{3.14}$$

$$\begin{aligned} Q_G = {} & [C_{GS} + C_{GD}(V_{DS})]V_{GP} \\ & + \frac{2K_G q\varepsilon_S N_D}{C_{OX}}\left[\sqrt{1+\frac{2V_{DS}C_{OX}^2}{q\varepsilon_S N_D}} - \sqrt{1+\frac{2V_{ON}C_{OX}^2}{q\varepsilon_S N_D}}\right] \\ & + [C_{GS} + C_{GD}(V_{ON})](V_G - V_{GP}) \end{aligned} \tag{3.15}$$

In these equations, the parameter $K_G$ is given by [1]:

$$K_G = \left[\frac{W_T + 2(t_T - x_P - t_{OX})}{W_{Cell}}\right] \tag{3.16}$$

where $t_{OX}$ is the gate oxide thickness. The gate geometry factor ($K_G$) obtained for this structure using the structural dimensions is 0.63.

The gate charge values obtained for the 30-V power U-MOSFET structure by using the above equations are: $Q_{GD} = 361$ nC/cm$^2$; $Q_{SW} = 376$ nC/cm$^2$; and $Q_G = 1{,}159$ nC/cm$^2$. It can be concluded that the gate-drain charge ($Q_{GD}$) is the dominant portion (90%) of the gate switching charge ($Q_{SW}$). The gate charge values for the power U-MOSFET structure are twice the values for the power D-MOSFET structure (see Chap. 2). Consequently, the improvement in the specific on-resistance achieved with the power U-MOSFET structure is offset by the much greater specific gate charge.

Equations for the gate voltage, drain current, and drain voltage waveforms obtained by using the analytical model are provided in the textbook [1]. The waveforms obtained for the 30-V power U-MOSFET structure with 3 μm cell pitch and trench width of 1 μm with a gate oxide thickness of 500 Å using these equations are provided in Fig. 3.24. A gate drive current density of 0.67 A/cm$^2$ was

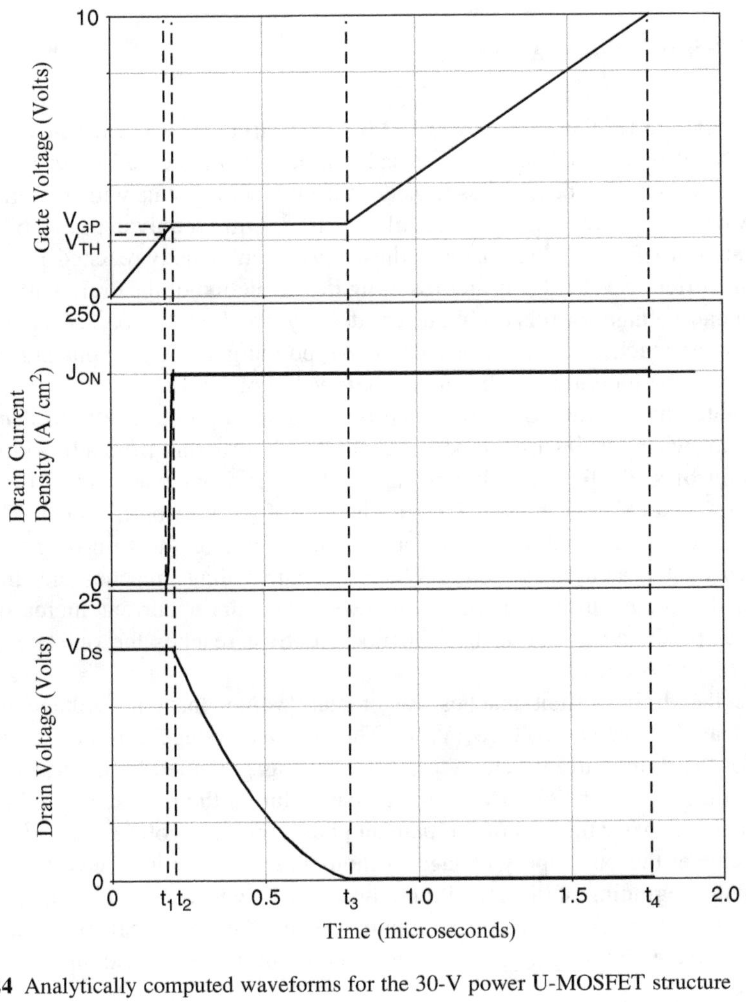


**Fig. 3.24** Analytically computed waveforms for the 30-V power U-MOSFET structure

used to turn on the device from a steady-state blocking voltage of 20 V to match the results of two dimensional numerical simulations.

The gate voltage initially increases linearly with time. After reaching the threshold voltage, the drain current can be observed to increase in a non-linear manner because the transconductance is a function of the gate bias voltage until it reaches an on-state current density of 150 A/cm$^2$. This transition occurs rapidly when compared with the time taken for the drain voltage to decrease during the next time interval. The on-state current density determines the gate plateau voltage which has a value of 2.56 V. During the gate voltage plateau phase, the drain voltage decreases in a non-linear manner until it reaches the on-state voltage drop. After this time, the gate voltage again increases but at a slower rate than during the initial turn-on phase.

### *3.6.1 Simulation Example*

The gate charge for the 30-V power U-MOSFET structure was extracted by using the results of two-dimensional numerical simulations of the cell described in the previous sections. The device was turned-on from blocking state with a drain bias of 20 V by driving it using a gate current of $1 \times 10^{-8}$ A/μm (equivalent to 0.67 A/cm$^2$ for the area of $1.5 \times 10^{-8}$ cm$^2$). Once the drain current density reached 150 A/cm$^2$, the drain current was held constant resulting in a reduction of the drain voltage. The gate plateau voltage for this drain current density was found to be 2.7 V. Once the drain voltage reached the on-state value corresponding to the gate plateau voltage, the gate voltage increased to the steady-state value of 10 V.

The gate charge waveforms obtained by using an input gate current density of 0.67 A/cm$^2$ when turning on the power U-MOSFET structure from a blocking state with drain bias of 20 V are shown in Fig. 3.25. The on-state current density is 150 A/cm$^2$ at a DC gate bias of 10 V at the end of the turn-on transient. The gate voltage increases at a constant rate at the beginning of the turn-on process as predicted by the analytical model. When the gate voltage reaches the threshold voltage, the drain current begins to increase. The drain current increases very rapidly as predicted by the analytical model until it reaches the on-state current density of 150 A/cm$^2$.

Once the drain current reaches the on-state value, the gate voltage remains constant at the plateau voltage ($V_{GP}$). The plateau voltage for this structure is 2.7 V for the drain current density of 150 A/cm$^2$ as governed by the transconductance of the device. The drain voltage decreases during the plateau phase in a non-linear manner. After the end of the plateau phase, the gate voltage again increases until it reaches the gate supply voltage. Although the increase in gate voltage is non-linear at the beginning of this transition it becomes linear over most of the time after the plateau phase. The waveforms obtained using the analytical model are very similar in shape and magnitude to those observed in the numerical simulations.

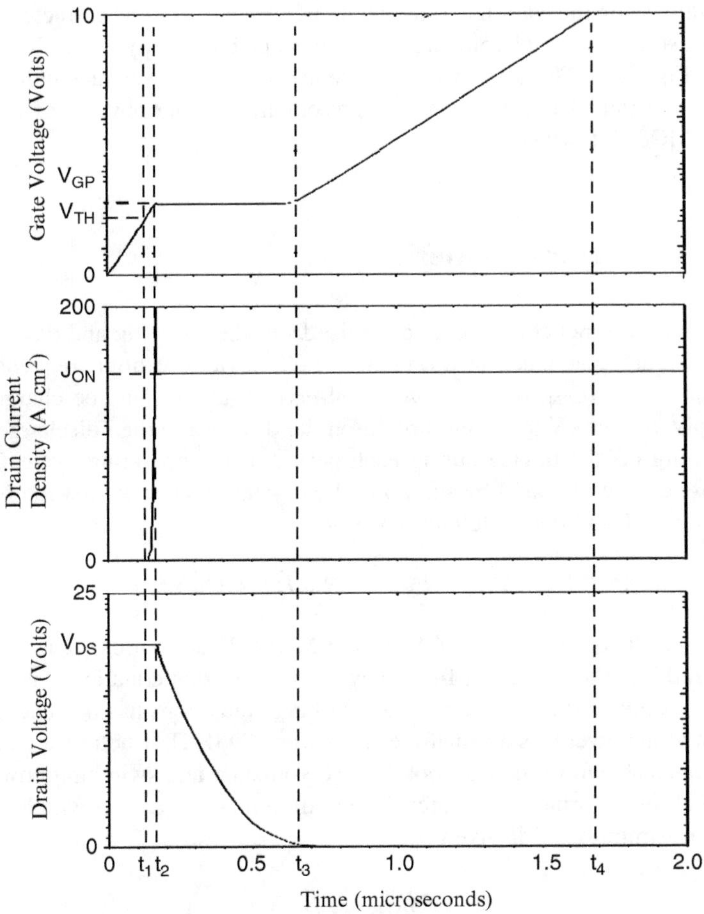


**Fig. 3.25** Turn-on waveforms for the 30-V power U-MOSFET structure

| Specific Gate Charge | Numerical Simulation ($nC/cm^2$) | Analytical Model ($nC/cm^2$) |
|---|---|---|
| $Q_{GS1}$ | 100 | 120 |
| $Q_{GS2}$ | 13 | 13.5 |
| $Q_{GS}$ | 113 | 133 |
| $Q_{GD}$ | 326 | 361 |
| $Q_{SW}$ | 339 | 376 |
| $Q_G$ | 1127 | 1159 |

**Fig. 3.26** Gate charge extracted from numerical simulations for the power U-MOSFET structure

The values for the various components of the gate charge extracted from the numerical simulations are compared with those calculated by using the analytical model in Fig. 3.26. There is good agreement between these values indicating that the analytical model is a reasonable representation of the physics of turn-on for power U-MOSFET structure.

## 3.7 Device Figures of Merit

Significant power switching losses can arise from the charging and discharging of the large input capacitance in power MOSFET devices at high frequencies. The input capacitance ($C_{IN}$) of the power MOSFET structure must be charged to the gate supply voltage ($V_{GS}$) when turning on the device and then discharged to 0 V when turning off the device during each period of the operating cycle. The total power loss can be obtained by summing the on-state power dissipation for a duty cycle $\delta = t_{ON}/T$ and the switching power losses:

$$P_T = P_{ON} + P_{SW} = \delta R_{ON} I_{ON}^2 + C_{IN} V_{GS}^2 f \tag{3.17}$$

where $R_{ON}$ is the on-resistance of the power MOSFET structure, $I_{ON}$ is the on-state current, and f is the operating frequency. In writing this equation, the switching power losses due to the drain current and voltage transitions has been neglected. A minimum total power loss occurs for each power MOSFET structure at an optimum active area as shown in the textbook [1]. The on-state and switching power losses are equal at the optimum active area. The optimum active area at which the power dissipation is minimized is given by:

$$A_{OPT} = \sqrt{\frac{R_{ON,sp}}{C_{IN,sp}}} \left(\frac{I_{ON}}{V_{GS}}\right) \left(\sqrt{\frac{\delta}{f}}\right) \tag{3.18}$$

From the first term in this expression, a useful technology figure-of-merit can be defined:

$$FOM(A) = \frac{R_{ON,sp}}{C_{IN,sp}} \tag{3.19}$$

In the power electronics community, there is trend towards increasing the operating frequency for switch mode power supplies in order to reduce the size and weight of the magnetic components. The ability to migrate to higher operating frequencies in power conversion circuits is dependent on making enhancements to the power MOSFET technology. From the above equations, an expression for the minimum total power dissipation can be obtained [1]:

$$P_T(\min) = 2 I_{ON} V_{GS} \sqrt{\delta R_{ON,sp} C_{IN,sp} f} \tag{3.20}$$

A second technology figure of merit related to the minimum power dissipation can be defined as:

$$FOM(B) = R_{ON,sp}C_{IN,sp} \tag{3.21}$$

In most applications for power MOSFET structures with high operating frequency, the switching losses associated with the drain current and voltage transitions become a dominant portion of the total power loss. The time period associated with the increase of the drain current and decrease of the drain voltage is determined by the charging of the device capacitances. It is therefore common practice in the industry to use the following figures-of-merit to compare the performance of power MOSFET products [1]:

$$FOM(C) = R_{ON,sp}Q_{GD,sp} \tag{3.22}$$

and

$$FOM(D) = R_{ON,sp}Q_{SW,sp} \tag{3.23}$$

Although FOM(D) encompasses both the drain current and voltage transitions, it is customary to use FOM(C) because the gate-drain charge tends to dominate in the switching gate charge. One advantage of using these expressions is that the figure-of-merit becomes independent of the active area of the power MOSFET device.

The figures of merit computed for the power U-MOSFET structure discussed in earlier sections of this chapter are provided in Fig. 3.27. The figure of merit usually used for comparison of device technologies in the literature is FOM(C). Most often, the value for this figure of merit at a gate bias of 4.5 V is utilized for selection of devices in the voltage regulator module application. In comparison with the power D-MOSFET structure (see Chap. 2), the power U-MOSFET structure has a FOM(C) at a gate bias of 4.5 V which is 2.4-times superior to the power D-MOSFET structure. In addition, the FOM(A) for the power U-MOSFET structure is a factor of nine times smaller than that for the power D-MOSFET structure. This implies a reduction in the active area by a factor of three times for the power U-MOSFET structure.

| **Figures of Merit** | **$V_G$ = 4.5 V** | **$V_G$ = 10 V** |
|---|---|---|
| FOM(A) ($\Omega^2$cm$^4$s$^{-1}$) | 7,085 | 3,720 |
| FOM(B) (ps) | 15.6 | 8.2 |
| FOM(C) (mΩ*nC) | 120 | 63 |
| FOM(D) (mΩ*nC) | 125 | 66 |

**Fig. 3.27** Figures of merit for the power U-MOSFET structure

## 3.8 Thick Trench Bottom Oxide Structure

The power U-MOSFET structure discussed in the previous sections has a very low specific on-resistance due to the high channel density and the elimination of the JFET resistance. However, it has a large reverse transfer capacitance and reverse transfer gate charge that has an adverse impact on the switching performance. These disadvantages can be overcome by selectively increasing the thickness of the oxide at the bottom of the trench as illustrated in the device structure shown in Fig. 3.3 [3–5]. However, the thicker oxide at the bottom of the trenches increases the accumulation layer resistance. This produces an increase in the specific on-resistance for the power U-MOSFET structure.

### *3.8.1 On-Resistance*

In the power U-MOSFET structure with the thick trench bottom oxide, only the accumulation layer resistance is impacted by the larger oxide thickness at the bottom of the trench. The specific on-resistance contributed by the accumulation layer in the power U-MOSFET structure with the thick oxide at the trench bottom can be obtained using:

$$R_{A,SP} = K_A \frac{L_A W_{Cell}}{2\mu_{nA} C_{OX,TB}(V_G - V_{TH})} \tag{3.24}$$

In this equation, the capacitance $C_{OX,TB}$ is the specific capacitance obtained using the thickness ($t_{OX,TB}$) of the trench bottom oxide. A typical trench bottom oxide thickness obtained by using the LOCOS process is 1,500 Å [3].

For the 30-V power U-MOSFET structure with a trench bottom oxide thickness of 1,500 Å and a cell width of 3 μm, the specific resistance contributed by the accumulation layer at a gate bias of 4.5 V increases to 0.087 mΩ cm$^2$ while the specific resistance contributed by the accumulation layer is increased to 0.039 mΩ cm$^2$ for the gate bias of 10 V. The total specific on-resistance for the power U-MOSFET structure with a trench bottom oxide thickness of 1,500 Å and a cell width of 3 μm then increases to 0.394 mΩ cm$^2$ for the gate bias of 4.5 V and 0.202 mΩ cm$^2$ for the gate bias of 10 V. This increase is about 15% over the power U-MOSFET structure without the thick oxide at the bottom of the trench.

### *3.8.2 Reverse Transfer Capacitance*

The capacitance between the gate and drain electrodes (the reverse transfer capacitance) is determined by the width of the trench where the gate electrode overlaps the N-drift region. The oxide capacitance for the MOS structure in this portion of the

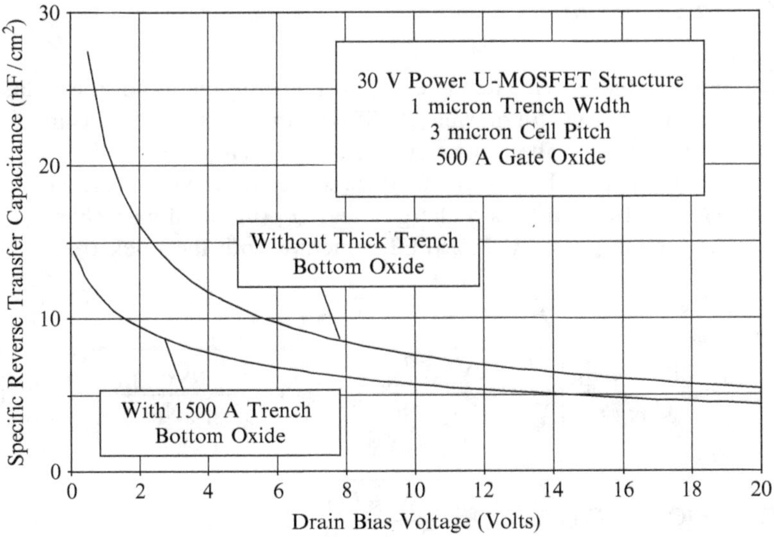


**Fig. 3.28** Gate-drain capacitance for the power U-MOSFET structures

power U-MOSFET structure with the thick oxide at the bottom of the trench is reduced. The gate-drain capacitance for the power U-MOSFET structure with the thick oxide at the bottom of the trenches can be obtained using:

$$C_{GD,SP} = \left[\frac{W_T + 2(t_T - x_{JP})}{W_{Cell}}\right]\left(\frac{C_{OX,TB}C_{S,M}}{C_{OX,TB} + C_{S,M}}\right) \tag{3.25}$$

where $C_{OX,TB}$ is the specific oxide capacitance for the thick oxide and $C_{S,M}$ is the semiconductor capacitance under the oxide, which decreases with increasing drain bias voltage. The specific capacitance of the semiconductor depletion region can be obtained by using the same formulae as in the previous U-MOSFET structure.

The specific gate transfer capacitance obtained by using the above analytical formula is shown in Fig. 3.28 for the case of a power U-MOSFET structure with 3 μm cell pitch, and trench width of 1 μm. This structure has a trench bottom oxide thickness of 1,500 Å, a trench depth of 1.5 μm and P-base junction depth of 1.0 μm. The gate-drain (reverse transfer) capacitance decreases with increasing drain bias voltage due to the expansion of the depletion region in the semiconductor. For purposes of comparison, this graph includes the reverse transfer capacitance for the power U-MOSFET structure with the gate oxide at the bottom of the trench. It can be observed that the reverse transfer capacitance is reduced by the thicker trench bottom oxide. The reduction is most effective at low drain bias voltages.

### 3.8.3 Gate Charge

As mentioned in the previous chapter, it is standard practice in the industry to provide the gate charge for power MOSFET structures as a measure of their switching performance. The most significant gate charge components for assessing the performance of the power MOSFET structures are $Q_{SW}$ (the gate switching charge), $Q_{GD}$ (the gate-drain charge), and $Q_G$ (the total gate charge). These components for the power U-MOSFET structure with the thick trench bottom oxide are:

$$Q_{GD} = \frac{2K_G q\varepsilon_S N_D}{C_{OX,TB}}\left[\sqrt{1+\frac{2V_{DS}C_{OX,TB}^2}{q\varepsilon_S N_D}} - \sqrt{1+\frac{2V_{ON}C_{OX,TB}^2}{q\varepsilon_S N_D}}\right] \quad (3.26)$$

$$\begin{aligned} Q_{SW} &= [C_{GS} + C_{GD}(V_{DS})]\sqrt{\frac{J_{ON}W_{Cell}L_{CH}}{2\mu_{ni}C_{OX}}} \\ &+ \frac{2K_G q\varepsilon_S N_D}{C_{OX,TB}}\left[\sqrt{1+\frac{2V_{DS}C_{OX,TB}^2}{q\varepsilon_S N_D}} - \sqrt{1+\frac{2V_{ON}C_{OX,TB}^2}{q\varepsilon_S N_D}}\right] \end{aligned} \quad (3.27)$$

$$\begin{aligned} Q_G &= [C_{GS} + C_{GD}(V_{DS})]V_{GP} \\ &+ \frac{2K_G q\varepsilon_S N_D}{C_{OX,TB}}\left[\sqrt{1+\frac{2V_{DS}C_{OX,TB}^2}{q\varepsilon_S N_D}} - \sqrt{1+\frac{2V_{ON}C_{OX,TB}^2}{q\varepsilon_S N_D}}\right] \\ &+ [C_{GS}+C_{GD}(V_{ON})](V_G - V_{GP}) \end{aligned} \quad (3.28)$$

In these equations, the parameter $K_G$ is altered due to the thicker oxide at the trench bottom surface to:

$$K_G = \left[\frac{W_T + 2(t_T - x_P - t_{OX,TB})}{W_{Cell}}\right] \quad (3.29)$$

where $t_{OX,TB}$ is the thickness of the trench bottom oxide.

The gate voltage waveform obtained for the power U-MOSFET structure with the thicker trench bottom oxide by using the analytical model is provided in Fig. 3.29 for the case of a gate current density of 0.67 A/cm$^2$. The time duration for the plateau in the gate voltage is reduced when compared with the structure without the thicker oxide at the trench bottom surface. The gate charge values obtained for the 30-V power U-MOSFET structure with a trench bottom oxide thickness of 1,500 Å by using the above equations are: $Q_{GD} = 258$ nC/cm$^2$; $Q_{SW} = 272$ nC/cm$^2$; and $Q_G = 849$ nC/cm$^2$. The gate-drain gate charge for the power U-MOSFET structure with the thick trench bottom oxide is reduced by 25%

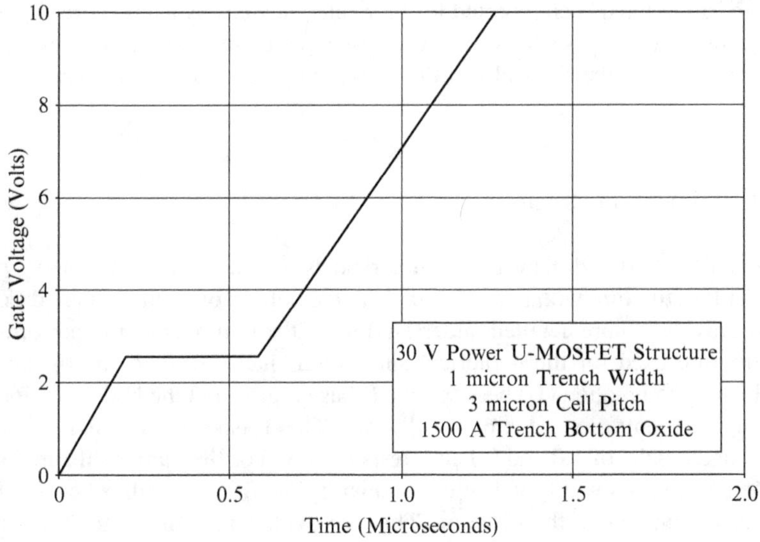


**Fig. 3.29** Analytically computed gate voltage waveform for the 30-V power U-MOSFET structure with thick bottom oxide

compared with the previous power U-MOSFET structure. A greater reduction in the gate-drain charge has been achieved by not only increasing the oxide thickness at the bottom of the trench but by increasing the depth of the P-base region relative to the trench depth [3]. This approach has to be undertaken at the peril of increasing the on-resistance because of introducing a JFET effect in the U-MOSFET structure.

## *3.8.4 Device Figures-of-Merit*

The figures of merit computed for the power U-MOSFET structure with the 1,500-Å thick oxide at the trench bottom by using the results of the analytical models are provided in Fig. 3.30. Only a modest 20% improvement in the figures-of-merit is predicted by the analytical models due to the increase in the on-resistance. However, larger improvements are observed with the simulations. Further, improved

| Figures of Merit | $V_G$ = 4.5 V | $V_G$ = 10 V |
|---|---|---|
| FOM(C) (mΩ*nC) | 101 | 52 |
| FOM(D) (mΩ*nC) | 107 | 55 |

**Fig. 3.30** Figures of merit for the power U-MOSFET structure with 1,500-Å thick oxide at the bottom of the trench

figures of merit have been reported for fabricated devices by using more aggressive processing techniques such as increased P-base region depth and increase in the oxide thickness on the sidewalls of the trench only near the bottom surface.

#### 3.8.4.1 Simulation Results

The results of two-dimensional numerical simulations on the 30-V power U-MOSFET structure with a thicker oxide at the bottom of the trench are described here to provide a more detailed understanding of the improvement in performance. The structure used for the numerical simulations had a drift region thickness of 3 μm below the junction ($J_1$) between the P-base region and the N-drift region with a doping concentration of $1.6 \times 10^{16}/cm^3$. The P-base region and $N^+$ source regions had depths of 1.0 and 0.1 μm, respectively. For the numerical simulations, half the cell (with a width of 1.5 μm) shown in Fig. 3.3 was utilized as a unit cell that is representative of the structure. The gate oxide was assumed to 500 Å on the trench sidewalls and 1,500 Å on the trench bottom surface. The doping concentration profiles for this U-MOSFET structure was kept identical to that of the structure with uniform oxide thickness in the trench.

The transfer characteristics for the U-MOSFET structure with thicker trench bottom oxide were obtained using numerical simulations with a drain bias of 0.1 V at 300 and 400°K. The resulting transfer characteristics are shown in Fig. 3.31. From this graph, a threshold voltage of 2.3 and 2.0 V can be extracted at 300 and 400°K, respectively – the same values as for the device without the thicker trench bottom oxide. The specific on-resistance can be obtained from the transfer characteristics at any gate bias voltage. For the case of a gate bias of 4.5 V and 300 °K, the specific in-resistance is found to be 0.284 mΩ cm$^2$, while for the case of a gate bias of 10 V and 300 °K, the specific in-resistance is found to be 0.172 mΩ cm$^2$. These values are very close to the values for the power U-MOSFET structure without the thick trench bottom oxide.

The specific on-resistance obtained using the analytical model indicates an increase in the specific on-resistance for the power U-MOSFET structure with the thick trench bottom oxide. This is associated with an increase in the contribution from the accumulation layer. As pointed out earlier for the power U-MOSFET structure with uniform oxide thickness, the current begins to spread into the drift region from the channel. This phenomenon ameliorates the impact of the thicker oxide at the bottom of the trench surface.

The current spreading pattern in the U-MOSFET structure with the thicker oxide on the trench bottom surface can be observed using the current flow-lines shown in Fig. 3.32 at a small drain bias of 0.1 V and a gate bias of 4.5 V. In the figure, the depletion layer boundary is shown by the dotted lines and the junction boundary is delineated by the dashed line. From this figure, it can be observed that the current spreading begins to occur from the junction between the P-base region and the N-drift region reducing the drift region resistance. Based up on these observations,

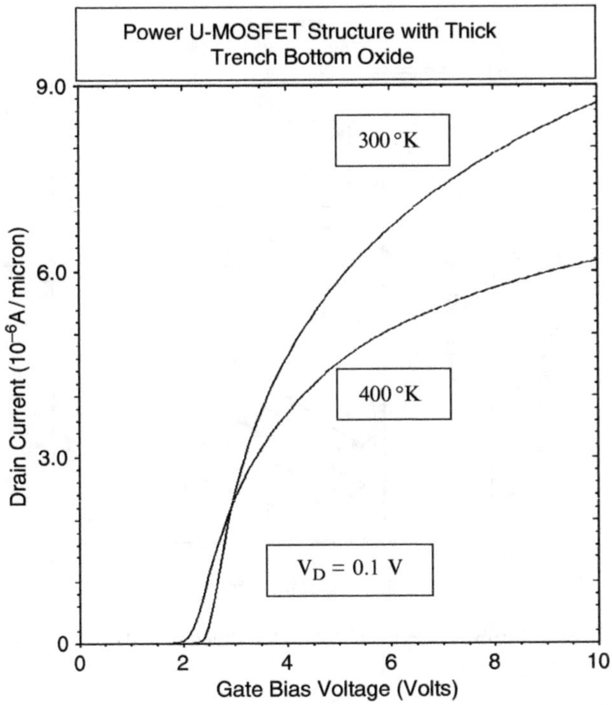


**Fig. 3.31** Transfer characteristics of the U-MOSFET structure with thick trench bottom oxide

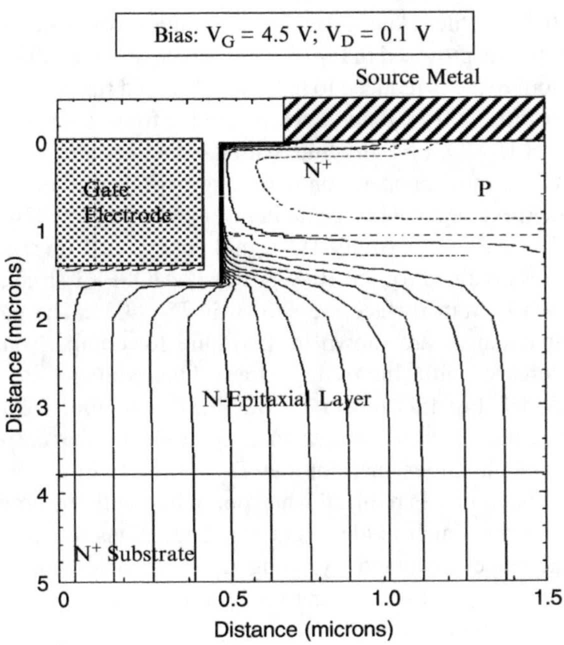


**Fig. 3.32** Current distribution in the U-MOSFET structure with the thicker trench bottom oxide

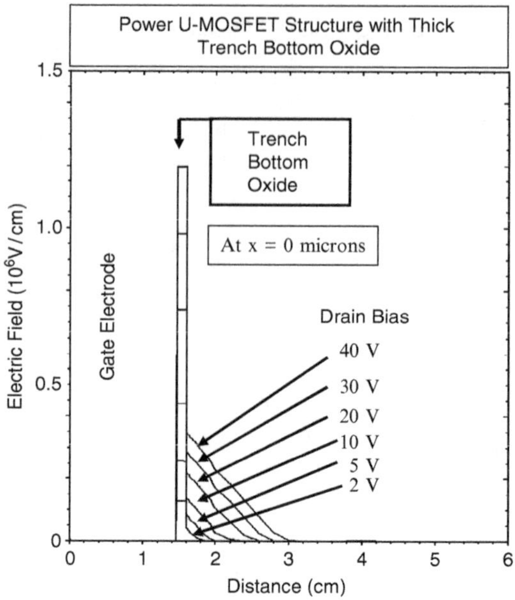


**Fig. 3.33** Electric field distribution in the U-MOSFET structure with thicker trench bottom oxide

it can be concluded that the accumulation layer resistance can be neglected in the analytical model for the power U-MOSFET structure.

The electric field profile taken through the middle of the gate electrode and the trench (at $\times = 0\,\mu\text{m}$) is provided in Fig. 3.33. It can be observed that the electric field in the trench bottom oxide is reduced to half that observed for the power U-MOSFET structure with uniform (500 Å) oxide thickness in the trench. The electric field in the oxide for the power U-MOSFET structure with the thicker oxide at the trench bottom is within the limit for reliable operation over long periods of time.

The drain-gate (reverse transfer) capacitance can be extracted by performing the numerical simulations with a small AC signal superposed on the DC drain bias voltage. The values obtained for the 30-V power U-MOSFET structure with thicker oxide at the trench bottom surface are shown in Fig. 3.34. The gate-to-drain and base-to-drain capacitances are shown in the figure for comparison. Both of these capacitances decrease with increasing drain bias voltage as expected from the analytical model. For this power U-MOSFET structure, the reverse transfer (gate-drain) capacitance is still larger in magnitude to the output (base-drain) capacitance and much larger that for the D-MOSFET structure. However, the reverse transfer capacitance is reduced when compared with the power U-MOSFET structure with uniform trench oxide thickness. The values for the reverse transfer and output capacitances obtained by using the analytical model for the power U-MOSFET structure with the thicker trench bottom oxide is in good agreement with the simulation values.

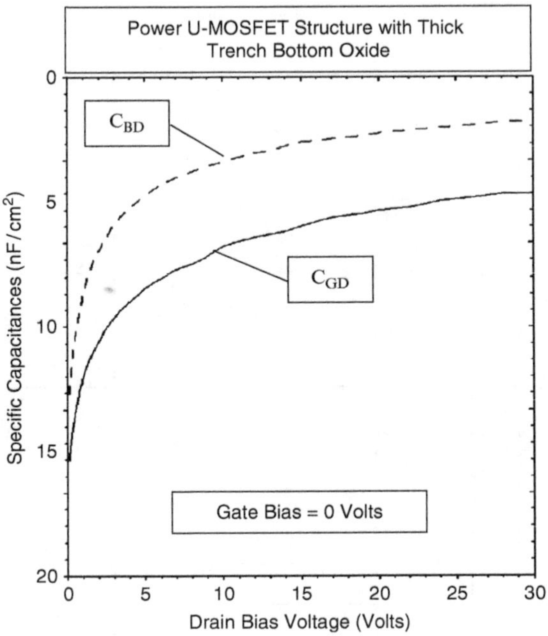


**Fig. 3.34** Reverse transfer and output capacitances for the U-MOSFET structure with the thicker trench bottom oxide

The gate charge for the 30-V power U-MOSFET structure with thick trench bottom oxide was extracted by using the results of two-dimensional numerical simulations of the cell described above. The device was turned-on from blocking state with a drain bias of 20 V by driving it using a gate current of $1 \times 10^{-8}$ A/μm (equivalent to 0.67 A/cm$^2$ for the area of $1.5 \times 10^{-8}$ cm$^2$). Once the drain current density reached 150 A/cm$^2$, the drain current was held constant resulting in a reduction of the drain voltage. The gate plateau voltage for this drain current density was found to be 2.8 V. Once the drain voltage reached the on-state value corresponding to the gate plateau voltage, the gate voltage increased to the steady-state value of 10 V.

The gate charge waveforms obtained by using an input gate current density of 0.67 A/cm$^2$ when turning on the power U-MOSFET structure from a blocking state with drain bias of 20 V are shown in Fig. 3.35. The on-state current density is 150 A/cm$^2$ at a DC gate bias of 10 V at the end of the turn-on transient. The gate voltage increases at a constant rate at the beginning of the turn-on process as predicted by the analytical model. When the gate voltage reaches the threshold voltage, the drain current begins to increase. The drain current increases very rapidly as predicted by the analytical model until it reaches the on-state current density of 150 A/cm$^2$.

Once the drain current reaches the on-state value, the gate voltage remains constant at the plateau voltage ($V_{GP}$) (Fig. 3.35). The plateau voltage for this

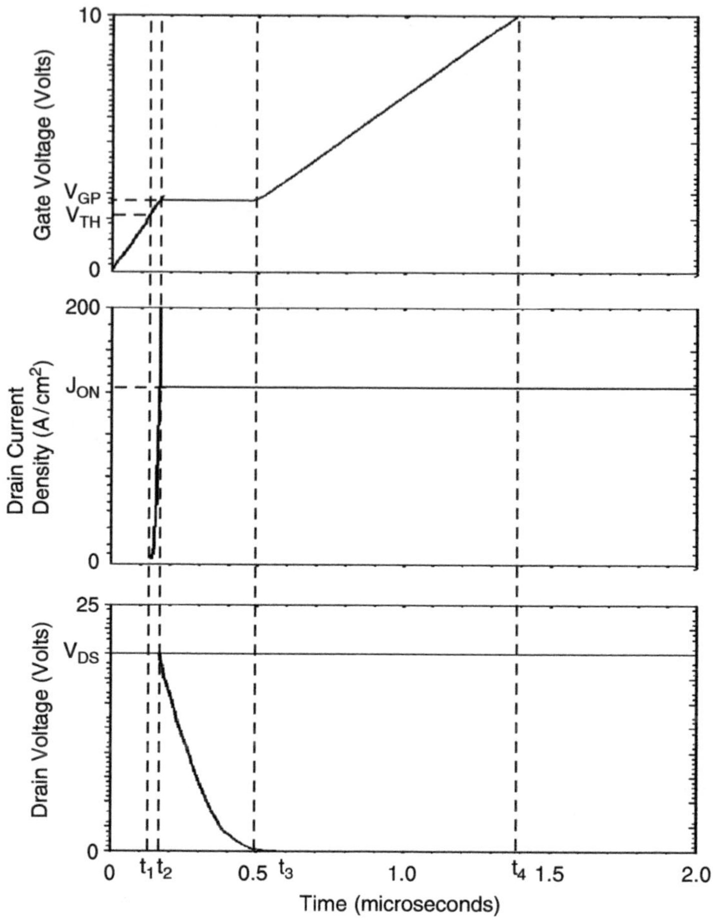


**Fig. 3.35** Turn-on waveforms for the 30-V power U-MOSFET structure with thick trench bottom oxide

structure is 2.8 V for the drain current density of 150 A/cm$^2$ as governed by the transconductance of the device. The drain voltage decreases during the plateau phase in a non-linear manner. After the end of the plateau phase, the gate voltage again increases until it reaches the gate supply voltage. Although the increase in gate voltage is non-linear at the beginning of this transition it becomes linear over most of the time after the plateau phase. The waveforms obtained using the analytical model are very similar in shape and magnitude to those observed in the numerical simulations.

The values for the various components of the gate charge extracted from the numerical simulations are compared with those calculated by using the analytical model in Fig. 3.26. The analytical model predicts a 15% larger value for the gate transfer charge than observed in the simulations. There is good agreement

| Specific Gate Charge | Numerical Simulation ($nC/cm^2$) | Analytical Model ($nC/cm^2$) |
|---|---|---|
| $Q_{GS1}$ | 100 | 118 |
| $Q_{GS2}$ | 13 | 13.5 |
| $Q_{GS}$ | 113 | 131 |
| $Q_{GD}$ | 220 | 258 |
| $Q_{SW}$ | 223 | 271 |
| $Q_G$ | 927 | 849 |

**Fig. 3.36** Gate charge extracted from numerical simulations for the power U-MOSFET structure with thick trench bottom oxide

between the values for the other gate charge components indicating that the analytical model is a good representation of the physics of turn-on for power U-MOSFET structure.

## 3.9 High Voltage Devices

The characteristics of a 600-V power U-MOSFET structure are described in this section for comparison with the performance of the power GD-MOSFET and power SJ-MOSFET structures that are discussed in later chapters. The analytical formulations presented in the previous sections for the power U-MOSFET structure are applicable for any breakdown voltage design when appropriate values for the device parameters are used during the computations. The values computed with the analytical models, provided in the last section of this chapter for devices with various blocking voltages, are in excellent agreement with the simulation results. This section will therefore focus on the results of two-dimensional numerical simulations for the 600-V power U-MOSFET structure.

### *3.9.1 Simulation Results*

The breakdown voltage for the power U-MOSFET structure can be increased by reducing the doping concentration and increasing the width of the drift region. The same cell structure can be utilized for the higher breakdown voltage structures as used previously for the 30-V device. A doping concentration of $2.7 \times 10^{14}/cm^3$ and a thickness of 60 μm is appropriate for the drift region in the case of a breakdown voltage of 600 V. The doping profile used in the simulations of the 600-V power U-MOSFET structure is shown in Fig. 3.37.

The blocking characteristic for the 600-V power U-MOSFET cell structure is shown in Fig. 3.38 at 300°K. It can be observed that the blocking voltage exceeds

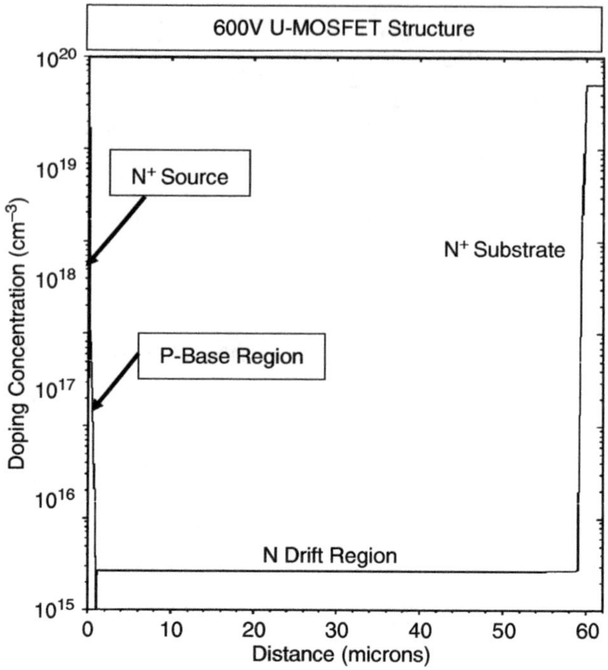


**Fig. 3.37** Vertical doping profiles in the 600-V U-MOSFET structure

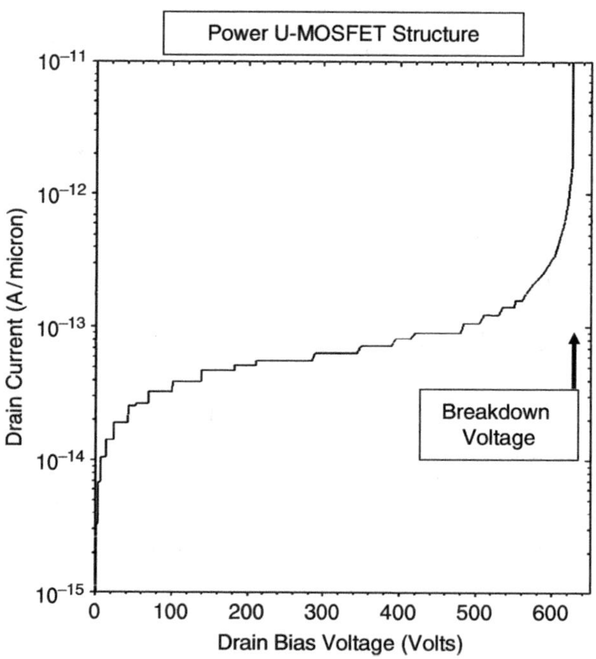


**Fig. 3.38** Blocking characteristics for the 600-V U-MOSFET structure

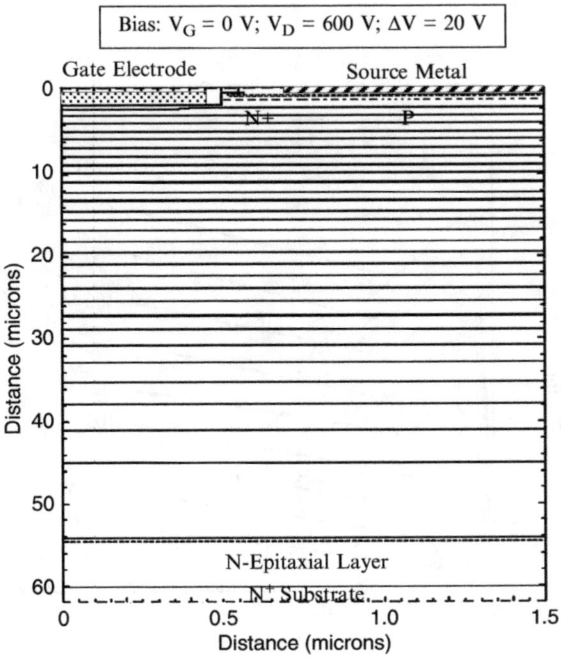


**Fig. 3.39** Potential contours in the 600-V U-MOSFET structure

600 V for the chosen doping profile. For purposes of comparison with the high voltage power MOSFET structures in subsequent chapters, it is instructive to examine the potential contours inside the power 600-V U-MOSFET structure when it is operating in the blocking mode. The potential contours obtained using the numerical simulations with zero gate bias and a drain bias voltage of 600 V for the 600-V U-MOSFET structure are shown in Fig. 3.39. It can be observed that the potential lines are flat indicating nearly parallel-plane breakdown within the device cell structure. However, the potential lines are much closer together at the top of the structure when compared with the vicinity of the $N^+$ substrate indicating a non-uniform electric field distribution.

The electric field profiles along the vertical direction through the center of the mesa region are shown in Fig. 3.40 for the 600-V power U-MOSFET structure. The electric field has a triangular shape representative of a one-dimensional junction for all applied drain bias voltages. The electric field at the vicinity of the P-N junction is $2.1 \times 10^5$ V/cm at a drain bias of 600 V (just prior to breakdown). This magnitude of the electric field is in excellent agreement with the value for the critical electric field in the case of triangular electric field distribution predicted by the analytical model (see Fig. 1.13) for a 600-V device.

The transfer characteristics for the 600-V U-MOSFET structure were obtained using numerical simulations with a drain bias of 0.1 V at 300 and 400°K. The resulting transfer characteristics are shown in Fig. 3.41. The specific on-resistance

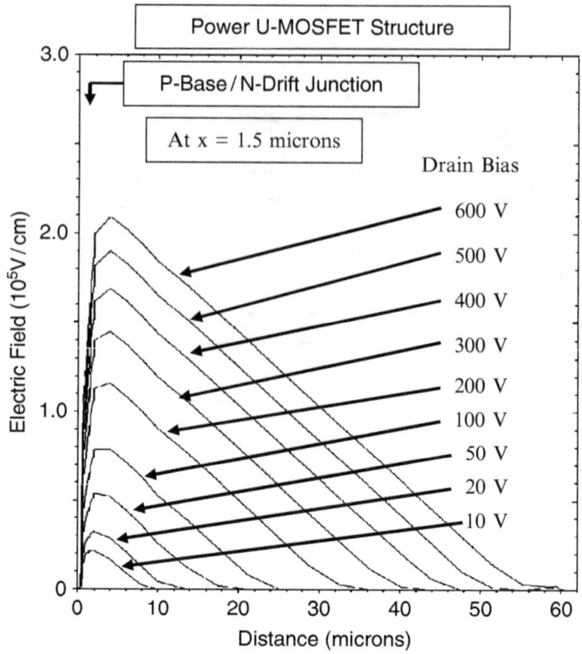


**Fig. 3.40** Electric field profiles in the 600-V U-MOSFET structure

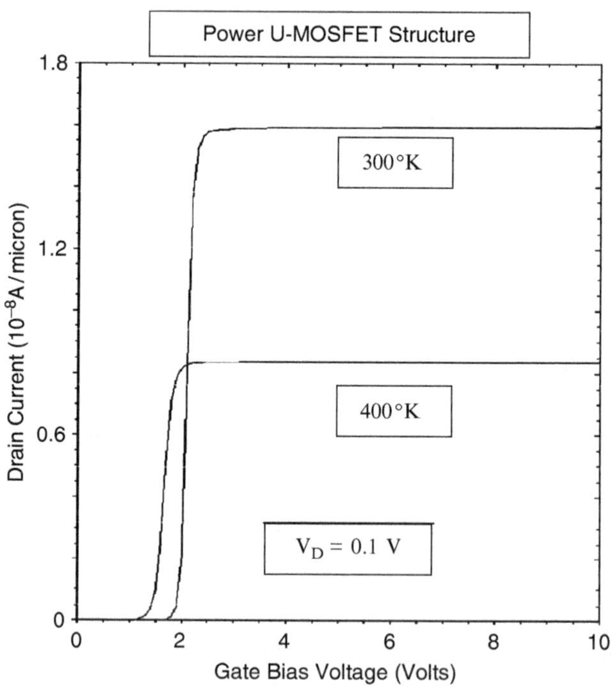


**Fig. 3.41** Transfer characteristics of the 600-V U-MOSFET structure

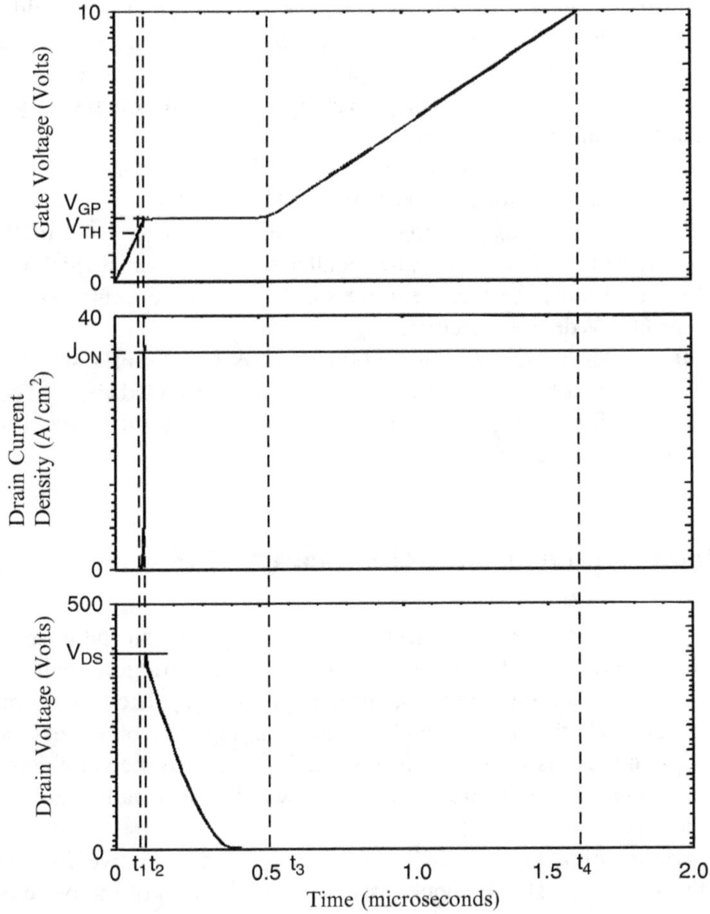


**Fig. 3.42** Turn-on waveforms for the 600-V power GD-MOSFET structure

can be obtained from the transfer characteristics at any gate bias voltage. For the case of a gate bias of 10 V and 300°K, the specific in-resistance is found to be 94.1 mΩ cm$^2$. Since the ideal specific on-resistance for a 600 V device obtained using Baliga's power law for the impact ionization coefficients is 73.9 mΩ cm$^2$, the specific on-resistance of the power U-MOSFET structure approaches close to this ideal limit for silicon. One reason for this good performance is the low channel resistance contribution making the drift region resistance dominant as can be inferred from the very flat transfer characteristics at gate bias voltages above 3 V.

The gate charges for the 600-V power U-MOSFET structure were extracted by using the results of two-dimensional numerical simulations. The device was turned-on from blocking state with a drain bias of 400 V by using a gate current of $1 \times 10^{-8}$ A/μm (equivalent to 0.67 A/cm$^2$ for the area of $1.5 \times 10^{-8}$ cm$^2$). Once the drain current density reached the on-state value (34 A/cm$^2$) corresponding to a on-state power dissipation of 100 W/cm$^2$, the drain current was held constant resulting in a reduction of the drain voltage. Once the drain voltage reached the on-state value corresponding to the gate plateau voltage, the gate voltage increased to the steady-state value of 10 V. The gate charge waveforms obtained by using the numerical simulations are shown in Fig. 3.42.

The values for the various components of the gate charge extracted from the numerical simulations for the 600-V power U-MOSFET structure are: $Q_{GS1}$ = 60 nC/cm$^2$; $Q_{GS2}$ = 7 nC/cm$^2$; $Q_{GD}$ = 267 nC/cm$^2$; $Q_{SW}$ = 67 nC/cm$^2$; and $Q_G$ = 1,070 nC/cm$^2$. These values are smaller than those obtained for the 30-V power U-MOSFET structure because of the smaller doping concentration in the drift region for the 600-V device structure.

The Figure-of-Merit FOM(C) for the 600-V power U-MOSFET structure can be obtained using the results of the numerical simulations provided above. The Figure-of-Merit FOM(C) for the 600-V power U-MOSFET structure at a gate bias of 10 V is found to be 26,630 mΩ nC.

## 3.10 Inductive Load Turn-Off Characteristics

High voltage power MOSFET devices are often used for adjustable speed motor drives which behave as inductive loads. The basic half-bridge circuit utilized in these applications is shown in Fig. 10.1 of the textbook [1]. The waveforms for the current and voltage in the power switch are shown in Fig. 10.2 of the textbook. High power dissipation occurs during each turn-off event due to the simultaneous high current and voltage during the transient. This power loss is usually characterized as an energy loss per cycle.

The operation of a power MOSFET device in an inductive load circuit is illustrated in Fig. 3.43. The textbook provides a description of the basic operation of this circuit. After carrying the load current during its duty cycle, the power MOSFET device is switched off to transfer the current back to the free wheeling diode. Prior to the turn-off transient, the device is operating in its on-state because switch $S_1$ is closed and switch $S_2$ is open. These initial conditions are defined by: $v_G = V_{GS}$; $i_D = I_L$; and $v_D = V_{ON}(V_{GS})$. In order to initiate the turn-off process, switch $S_1$ is opened and switch $S_2$ is subsequently closed by the control circuit. The gate electrode of the power MOSFET device is then connected to the source via the gate resistance to discharge its capacitances. However, no changes in the drain current or voltage can occur until the gate voltage reaches the magnitude required to operate the power MOSFET device at a saturated drain current equal to the load current. (The small increase in the drain voltage, due to the increase in on-resistance resulting from the reduction of the gate bias voltage, has been neglected here). This gate plateau voltage is given by:

$$V_{GP} = V_{TH} + \sqrt{\frac{J_{D,ON} W_{Cell} L_{CH}}{\mu_{ni} C_{OX}}} \tag{3.30}$$

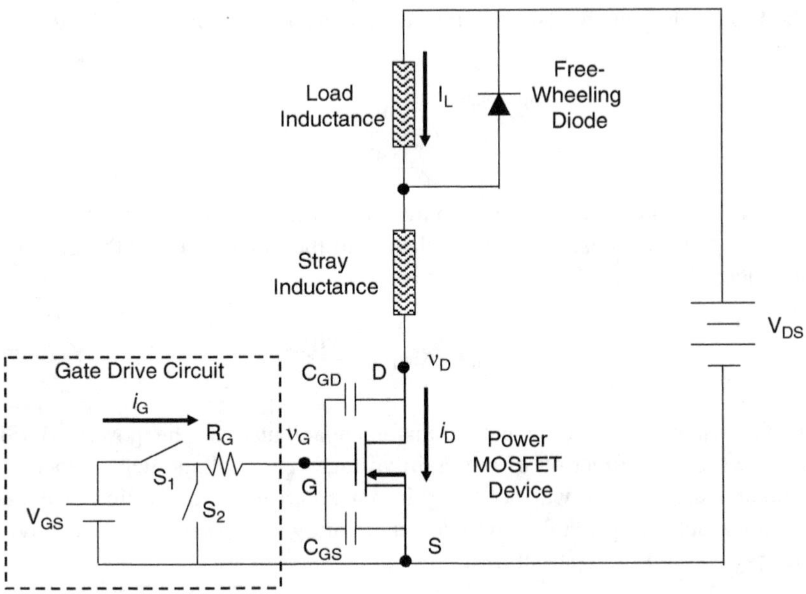


**Fig. 3.43** Power MOSFET device operating in an inductive load circuit

where $C_{OX}$ is the gate oxide capacitance. During this time interval, the gate-drain capacitance $C_{GD}(V_{ON})$ remains constant because the drain voltage is constant. Consequently, the time constant for discharging the gate of the power MOSFET device is $R_G*[C_{GS} + C_{GD}(V_{ON})]$ and the gate voltage decreases exponentially with time as given by:

$$v_G(t) = V_{GS}e^{-t/R_{G,SP}[C_{GS}+C_{GD}(V_{ON})]} \tag{3.31}$$

The time $t_4$ (using the notation from the textbook) for reaching the gate plateau voltage can be obtained by using this equation with (3.30) for the plateau voltage:

$$t_4 = R_{G,SP}[C_{GS} + C_{GD}(V_{ON})]\ln\left[\frac{V_{GS}}{V_{GP}}\right] \tag{3.32}$$

This time can be considered to a *turn-off delay time* before the drain voltage begins to increase after the turn-off is initiated by the control circuit.

The drain voltage begins to increase at time $t_4$ but the drain current remains constant at the load current $I_L$ because the current cannot be transferred to the diode until the voltage at the drain of the MOSFET device exceeds the supply voltage $V_{DS}$ by one diode drop to forward bias the diode. Since the drain current density is constant, the gate voltage also remains constant at the gate plateau voltage. Consequently:

$$J_{GP} = \frac{V_{GP}}{R_{G,SP}} \tag{3.33}$$

Since all of the gate current is used to discharge the gate-drain capacitance during the plateau phase because there is no change in the voltage across the gate-source capacitance:

$$J_{GP} = C_{GD,SP}\frac{dv_D}{dt} \tag{3.34}$$

where $C_{GD,SP}$ is the specific gate transfer capacitance of the power MOSFET structure which is a function of the drain voltage. This voltage dependence of the gate transfer capacitance was not taken into account in the derivation provided in the textbook but is important to include here to allow comparison of the behavior of various power MOSFET structures.

The specific gate transfer capacitance of the power U-MOSFET structure can be taken into account by using the gate oxide capacitance, the semiconductor capacitance under the gate oxide and the relative area of the trench region:

$$C_{GD,SP} = \left(\frac{W_T}{W_{Cell}}\right)\left(\frac{C_{OX}C_{S,M}}{C_{OX} + C_{S,M}}\right) \tag{3.35}$$

where the specific semiconductor capacitance is given by:

$$C_{S,M} = \frac{\varepsilon_S}{W_{D,M}} \tag{3.36}$$

In this equation, the depletion width under the MOS gate region can be obtained by using:

$$W_{D,M} = \frac{\varepsilon_S}{C_{OX}}\left\{\sqrt{1 + \frac{2v_D(t)C_{OX}^2}{q\varepsilon_S N_D}} - 1\right\} \tag{3.37}$$

Combining the above relationships yields the following differential equation for the voltage increase phase of the turn-off transient:

$$dt = \left(\frac{W_T}{W_{Cell}}\right)\frac{C_{OX}}{J_{GP}}\frac{dv_D}{\sqrt{1 + \frac{2v_D(t)C_{OX}^2}{q\varepsilon_S N_D}}} \tag{3.38}$$

Integration of this equation yields:

$$(t - t_4) = \left(\frac{W_T}{W_{Cell}}\right)\frac{2q\varepsilon_S N_D}{J_{GP}C_{OX}}\left[\sqrt{1 + \frac{2v_D(t)C_{OX}^2}{q\varepsilon_S N_D}} - \sqrt{1 + \frac{2V_{ON}C_{OX}^2}{q\varepsilon_S N_D}}\right] \tag{3.39}$$

The voltage rise-time, i.e. the time taken for the voltage to increase from the on-state voltage drop ($V_{ON}$) to the drain supply voltage ($V_{DS}$) can be derived from the above expression:

$$t_{V,OFF} = \left(\frac{W_T}{W_{Cell}}\right)\frac{2q\varepsilon_S N_D}{J_{GP}C_{OX}}\left[\sqrt{1 + \frac{2V_{DS}C_{OX}^2}{q\varepsilon_S N_D}} - \sqrt{1 + \frac{2V_{ON}C_{OX}^2}{q\varepsilon_S N_D}}\right] \tag{3.40}$$

The voltage waveform during the rise-time can also be derived from (3.39):

$$v_D(t) = \frac{q\varepsilon_S N_D}{J_{GP}C_{OX}^2}\left\{\left[\sqrt{1 + \frac{2V_{ON}C_{OX}^2}{q\varepsilon_S N_D}} + \left(\frac{W_{Cell}}{W_T}\right)\frac{J_{GP}C_{OX}}{2q\varepsilon_S N_D}t\right]^2 - 1\right\} \tag{3.41}$$

At the end of the plateau phase (at time $t_5$), the load current begins to transfer from the power MOSFET device to the free wheeling diode. Since the drain voltage remains constant, the gate-drain capacitance can also be assumed to remain constant during this phase. The current flowing through the gate resistance ($R_G$) discharges both the gate-drain and gate-source capacitances leading to an exponential fall in gate voltage from the plateau voltage:

$$v_G(t) = V_{GP}e^{-(t-t_5)/R_{G,SP}[C_{GS}+C_{GD}(V_{DS})]} \tag{3.42}$$

The drain current follows the gate voltage as given by:

$$J_D(t) = g_m[v_G(t) - V_{TH}] = \frac{\mu_{ni}C_{OX}}{2L_{CH}W_{Cell}}[v_G(t) - V_{TH}]^2 \tag{3.43}$$

The drain current decreases rapidly with time due to the exponential reduction of the gate voltage, as given by (3.42), during the current fall phase. The drain current becomes equal to zero when the gate voltage reaches the threshold voltage. The current fall time can therefore be obtained from (3.42):

$$t_{I,OFF} = R_{G,SP}[C_{GS} + C_{GD}(V_{DS})]\ln\left(\frac{V_{GP}}{V_{TH}}\right) \tag{3.44}$$

Specific capacitances should be used in this expression for computation of the current fall time. Beyond this point in time, the gate voltage decreases exponentially until it reaches zero. The time constant for this exponential decay is different from the initial phase due to the smaller gate-drain capacitance.

The largest power dissipation during the turn-off transient occurs during the drain voltage rise-time time interval ($t_{V,OFF}$). The turn-off energy loss per cycle can be obtained using:

$$E_{OFF} = \frac{1}{2} J_{ON} V_{DS} \left( t_{V,OFF} + t_{I,OFF} \right) \tag{3.45}$$

under the assumption that the drain current and voltage excursions are approximately linear with time.

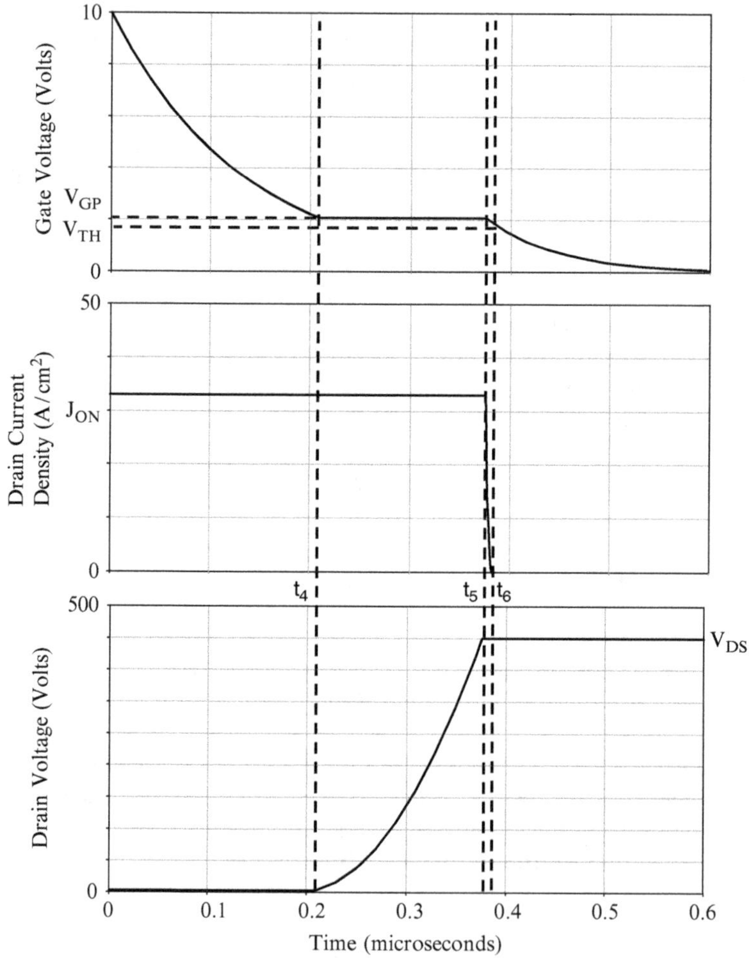


**Fig. 3.44** Analytically computed turn-off waveforms for the 600-V power U-MOSFET structure

In the case of the 600-V power U-MOSFET structure, the typical drain supply voltage is 400 V. Using the specific on-resistance for this device of 94 m$\Omega$ cm$^2$ with an on-state power dissipation of 100 W/cm$^2$, the on-state current density is found to be 33 A/cm$^2$. The on-state voltage drop at this current density is 3.1 V. The device structure has a cell pitch of 3 μm with a mesa width of 2 μm, trench depth of 1.5 μm, and a gate oxide thickness of 500 Å. The drift region has a doping concentration of $2.7 \times 10^{14}$/cm$^3$ and thickness of 60 μm. The specific gate resistance used in the turn-off circuit was assumed to have a value of 1.5 $\Omega$ cm$^2$.

Using the above parameters, the specific gate reverse transfer capacitance computed by using (3.7) at the on-state voltage drop is 41.6 and 0.16 nF/cm$^2$ at the drain supply voltage. Using these values in (3.32), the time ($t_4$) to reach the gate plateau voltage is found to be 0.209 μs. Using these parameters in (3.40), the voltage rise-time is computed as 0.167 μs. Using these parameters in (3.44), the current fall-time is computed as 0.011 μs. It can be observed that the current fall time is much smaller than the voltage rise-time. The energy loss per cycle obtained by using these values for the voltage rise-time and current fall-time in (3.45) is 1.17 mJ/cm$^2$. The waveforms that can be generated by using the above equations are shown in Fig. 3.44.

### 3.10.1 Simulation Results

The results of two-dimensional numerical simulations on the turn-off of the 600-V power U-MOSFET structure are described below. The drain supply voltage was chosen as 400 V for the turn-off analysis. During the turn-off simulations, the gate voltage was reduced to zero with a gate resistance of $1 \times 10^8$ $\Omega$ μm for the 1.5 μm half-cell structure, which is equivalent to a specific gate resistance of 1.5 $\Omega$ cm$^2$. The current density was initially held constant at an on-state current density of 33 A/cm$^2$ allowing the drain voltage to rise to the drain supply voltage. The drain supply voltage was then held constant allowing the drain current density to reduce to zero.

The turn-off waveforms obtained for the 600-V power U-MOSFET structure by using the numerical simulations are shown in Fig. 3.45. The gate voltage initially reduces to the gate plateau voltage corresponding to the on-state current density. The drain voltage then increases from the on-state voltage drop to the drain supply voltage. After this, the drain current rapidly falls to zero. The drain voltage rise-time ($t_5 - t_4$) can be observed to be much greater than the drain current fall time ($t_6 - t_5$). The drain voltage rise-time obtained from the simulations of the power U-MOSFET structure is 0.185 μs and the drain current fall-time obtained from the simulations is 0.01 μs. The shape of the waveforms predicted by the analytical model (see Fig. 3.44) for the gate voltage, drain voltage and drain current are in good agreement with the simulation results. The numerical values predicted for the analytical model for the voltage rise-time and the drain fall-time are also in good agreement with those observed in the numerical simulations

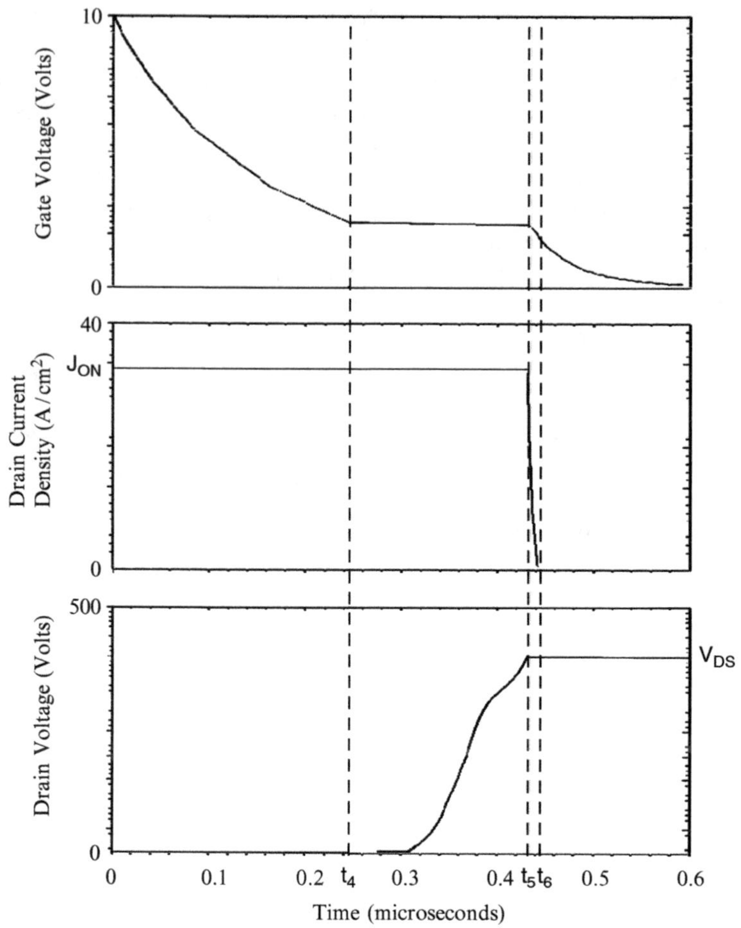


**Fig. 3.45** Turn-off waveforms for the 600-V power U-MOSFET structure

allowing an accurate computation of the energy loss per cycle for the power U-MOSFET structure.

## 3.11 Discussion

The characteristics of the power U-MOSFET structure has been reviewed in this chapter. The specific on-resistance for this device structure is significantly reduced when compared with the power D-MOSFET structure due to its high channel density and the elimination of the JFET resistance contribution. The reduced specific on-resistance for the power U-MOSFET structure allows fabrication of devices with smaller active area to achieve the same absolute value for the on-resistance.

However, the reverse transfer capacitance and gate charge for the power U-MOSFET structure are twice as large as for the power D-MOSFET structure. Despite the larger capacitance and gate charge per unit area, the figures-of-merit for the power U-MOSFET structure are superior to those for the power D-MOSFET structure. Consequently, the power U-MOSFET structure has supplanted the power D-MOSFET structure in high frequency power circuits such as the voltage regulator modules for supply power to microprocessors.

The switching performance and figures-of-merit for the power U-MOSFET structure can be enhanced by the use of a thicker oxide on the trench bottom surface. The thicker oxide serves to reduce the gate-drain capacitance especially at low drain bias voltages. This leads to a smaller gate-transfer charge which is beneficial to achieving better figures-of-merit.

For purposes of comparison with the power D-MOSFET structures in the preceding chapter, the analysis of the power U-MOSFET structure is provided here for a broad range of blocking voltages. In this analysis, the power U-MOSFET structure was assumed to have the following parameters: (a) $N^+$ source junction depth of 0.1 μm; (b) P-base junction depth of 1.0 μm; (c) gate oxide thickness of 500 Å; (d) trench width of 1.0 μm; (e) mesa width of 2.0 μm; (f) threshold voltage of 2 V; (g) gate drive voltage of 10 V; (h) inversion mobility of 450 $cm^2$/V s; (i) accumulation mobility of 1,000 $cm^2$/V s. The contributions from the contacts and the $N^+$ substrate were neglected during the analysis. The doping concentration and thickness of the drift region were determined under the assumption that the edge termination (in conjunction with the graded junction doping profiles) limits the breakdown voltage to 90% of the parallel-plane breakdown voltage. The device parameters pertinent to each blocking voltage are provided in Fig. 3.46. These values are the same as those used for the analysis of the power D-MOSFET structure in Chap. 2. It can be observed that the doping concentration must be greatly reduced with increasing blocking voltage capability while simultaneously increasing the thickness of the drift region.

| Blocking Voltage (V) | Drift Doping Concentration ($cm^{-3}$) | Drift Region Thickness (microns) |
|---|---|---|
| 30 | $1.6\times10^{16}$ | 3 |
| 60 | $6.0\times10^{15}$ | 5 |
| 100 | $3.0\times10^{15}$ | 8 |
| 200 | $1.2\times10^{15}$ | 17 |
| 300 | $6.8\times10^{14}$ | 26 |
| 600 | $2.7\times10^{14}$ | 58 |
| 1000 | $1.35\times10^{14}$ | 105 |

**Fig. 3.46** Device parameters for the power U-MOSFET structures

| Blocking Voltage (V) | Cell Pitch (microns) | Specific On-Resistance (mΩ-cm²) |
|---|---|---|
| 30 | 3.0 | 0.171 |
| 60 | 3.0 | 0.525 |
| 100 | 3.0 | 1.43 |
| 200 | 3.0 | 6.92 |
| 300 | 3.0 | 18.2 |
| 600 | 3.0 | 100 |
| 1000 | 3.0 | 360 |

**Fig. 3.47** Optimized specific on-resistance for the power U-MOSFET structures

In the power U-MOSFET structure, the smallest specific on-resistance is obtained when the trench and mesa width are minimized within the constraints of processing and lithography considerations. Consequently, a single device cell structure with a trench width of 1 μm and a mesa width of 2 μm (cell pitch of 3 μm) will be utilized here for the analysis of devices with various breakdown voltages. A two zone model for the resistance of the drift region was utilized for the analysis of the specific on-resistance. The specific on-resistances computed using the analytical model for the power U-MOSFET structures are provided in Fig. 3.47.

The specific on-resistance for the power U-MOSFET structure can be compared with the ideal specific on-resistance (obtained by using Baliga's power law for the impact ionization coefficients) in Fig. 3.48. From this figure, it can be concluded that the specific on-resistance for the U-MOSFET structure is always larger than the ideal specific on-resistance. For blocking voltages below 50 V, the specific on-resistance for the power U-MOSFET structure becomes substantially larger than the ideal case. At a blocking voltage of 30 V, the specific on-resistance is about three times larger than the ideal specific on-resistance obtained by using Baliga's power law for the impact ionization coefficients.

The specific gate transfer charge for the power U-MOSFET structures was obtained by using the analytical model (see 3.26). During this analysis, the devices were assumed to be operated at an on-state current density that results in a power dissipation of 100 W/cm$^2$ as determined by the specific on-resistance for each device. The on-state current density values are provided in Fig. 3.49. The drain supply voltage used for this analysis was chosen to be two-thirds of the breakdown voltage. The drain supply voltage values are also provided in Fig. 3.49.

The specific gate transfer charge values calculated by using the analytical model for the power U-MOSFET structure are given in Fig. 3.49 as a function of the breakdown voltage. These values are also plotted in Fig. 3.50 as a function of the breakdown voltage. It can be observed that the gate transfer charge decreases gradually with increasing breakdown voltage. The specific gate transfer charge decreases in a linear fashion on this log-log graph for breakdown voltages ranging

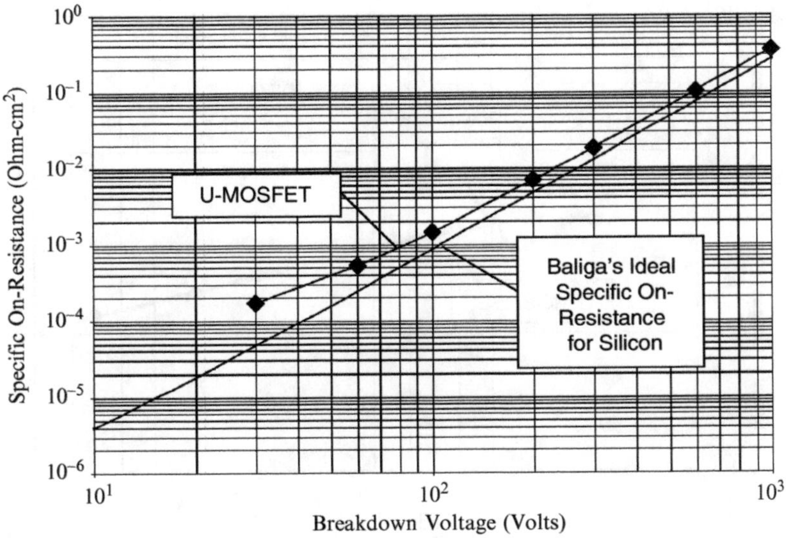


**Fig. 3.48** Specific on-resistance for the power U-MOSFET structures

| Blocking Voltage (V) | Drain Supply Voltage (Volts) | On-State Current Density ($A/cm^2$) | Specific Gate Transfer Charge ($nC/cm^2$) |
|---|---|---|---|
| 30 | 20 | 765 | 356 |
| 60 | 40 | 436 | 325 |
| 100 | 80 | 264 | 300 |
| 200 | 133 | 120 | 269 |
| 300 | 200 | 74 | 247 |
| 600 | 400 | 32 | 218 |
| 1000 | 667 | 17 | 198 |

**Fig. 3.49** Parameters used for gate transfer charge analysis for the power U-MOSFET structures

from 100 to 1,000 V indicating a power law relationship. The power law relationship that fits the data is shown in the figure by the dashed line. The equation for this line is:

$$Q_{GD} = \quad 689 BV^{-0.18} \tag{3.46}$$

where the specific gate transfer charge has units of $nC/cm^2$.

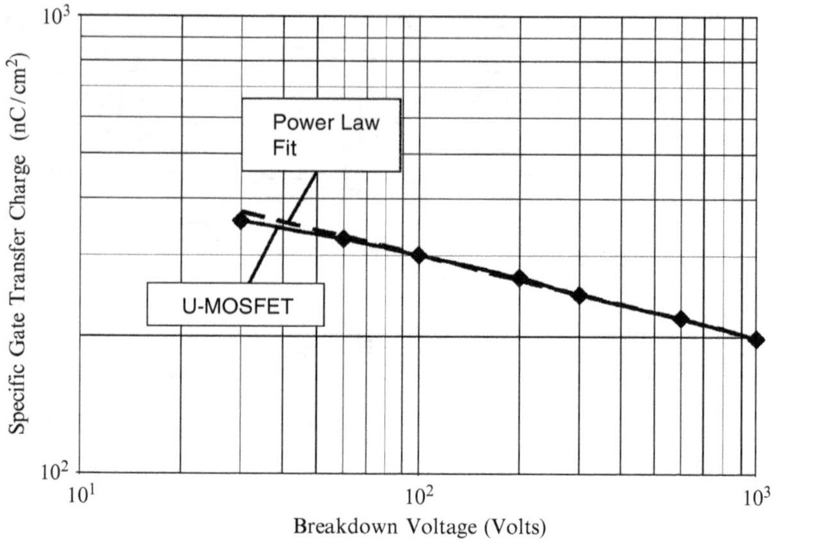


**Fig. 3.50** Specific gate transfer charge for the power U-MOSFET structures

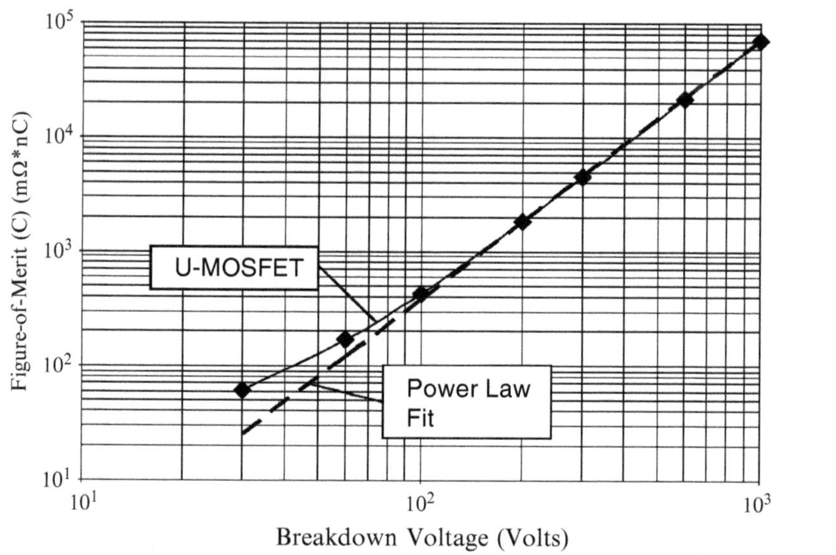


**Fig. 3.51** Figure-of-merit (C) for the power U-MOSFET structures

The figure-of merit (C) – product of the specific on-resistance and the specific gate transfer charge – for the power U-MOSFET structures was obtained by using the specific gate transfer charge and specific on-resistance values provided above. During this analysis, the devices were assumed to be operated at an on-state current density that results in a power dissipation of 100 W/cm$^2$ as determined by the specific on-resistance for each device. The resulting values for the FOM(C) are plotted in Fig. 3.51 as a function of the breakdown voltage of the power U-MOSFET structure. It can be observed that the FOM(C) increases in a linear fashion on this log-log graph for breakdown voltages above 100 V indicating a power law relationship. The power law relationship that fits the data is shown in the figure by the dashed line. The equation for this line is:

$$\mathrm{FOM(C)} = R_{ON,sp} Q_{GD,sp} = 0.0113\ BV^{2.267} \tag{3.47}$$

where the FOM(C) has units of mΩ nC. The specific on-resistance and FOM(C) discussed here for the power U-MOSFET structure will be compared with those for the power D-MOSFET and other structures in the final chapter of the book.

## References


1. B.J. Baliga, "Fundamentals of Power Semiconductor Devices", Springer-Science, New York, 2008.
2. J. Zeng, et al, "An Ultra-Dense Trench-Gated Power MOSFET Technology using a Self-Aligned Process", IEEE International Symposium on Power Semiconductor Devices, Abstract 7.3, pp. 147–150, 2001.
3. M. Darwish, et al, "A New Power W-Gated Trench MOSFET (WMOSFET) with High Switching Performance", IEEE International Symposium on Power Semiconductor Devices, Abstract 1.1, 2003, pp. 100–104.
4. T. Aoki, et al, "High Performance and Reliability Trench Gate Power MOSFET with Partially Thick Gate Oxide Film Structure", IEEE International Symposium on Power Semiconductor Devices, Abstract 6.3, pp. 85–88, 2006.
5. H. Wang, et al, "A 70V UMOS Technology with Trenched LOCOS Process to Reduce $C_{gs}$", IEEE International Symposium on Power Semiconductor Devices, Abstract P8-25, pp. 181–184, 2007.
6. N. Thapar and B. J. Baliga, "Influence of the Trench Corner Design on Edge Termination of UMOS Power Devices", Solid State Electronics, Vol. 41, pp. 1929–1936, 1997.

# Chapter 4
# SC-MOSFET Structure

As discussed in Chap. 2, the first power MOSFET structure commercially introduced by the power semiconductor industry was the D-MOSFET structure with the planar gate architecture. The fast switching speed and ruggedness of the D-MOSFET structure were significant advantages compared with the performance of the available bipolar power transistor. In order to reduce the specific on-resistance of the structure, the planar gate topology was replaced with a trench gate topology in the 1990s by creating the power U-MOSFET structure. The significant reduction of the specific on-resistance achieved using this approach has been described in Chap. 3. It has been found that the high input capacitance and large gate transfer charge for the power U-MOSFET structure offsets the benefits of the low specific on-resistance in high frequency applications such as the voltage-regulator-modules (VRMs) used to provide power to microprocessors.

Recently, significantly enhanced performance has been achieved in a planar gate power MOSFET structure by shielding the channel region from the drain. This shielded-channel power MOSFET structure (the SC-MOSFET) is the subject of this chapter. Although many aspects of operation of this device are similar to those already described for the power D-MOSFET structure, the SC-MOSFET structure contains a shielding junction with a unique JFET region doping profile to achieve a specific on-resistance close to that observed for the power U-MOSFET structure while also obtaining a reduced input capacitance and a very low gate transfer charge. This structure can be fabricated by using standard planar processes like the power D-MOSFET structure which reduces the complexity and cost in relation to the power U-MOSFET structure.

Commercially available power SC-MOSFET structures have been benchmarked against commercially available power U-MOSFET structure for the VRM applications and found to enable a significant improvement in efficiency for DC-to-DC power conversion at high frequencies. These devices are of interest for applications in laptops and servers where heat management becomes an important issue during operation.

B.J. Baliga, *Advanced Power MOSFET Concepts*,
DOI 10.1007/978-1-4419-5917-1_4, 

## 4.1 The SC-MOSFET Structure

A cross-section of the basic cell structure for the SC-MOSFET structure [1] is illustrated in Fig. 4.1. This device structure is fabricated by starting with an N-type epitaxial layer grown on a heavily doped $N^+$ substrate. As in the case of the power D-MOSFET structure, the channel is formed by the difference in lateral extension of the P-base and $N^+$ source regions produced by their diffusion cycles. Both regions are self-aligned to the left-hand-side and right-hand-side of the gate region during ion-implantation to introduce the respective dopants. A refractory gate electrode, such as polysilicon, is required to allow diffusion of the dopants under the gate electrode at elevated temperatures. The unique aspect of the power SC-MOSFET structure is the addition of a $P^+$ screening region which is located at a larger depth than the P-base region.

In the power D-MOSFET structure, it is customary to choose the energy of the ion-implants for the dopants so that their highest concentration lies at the surface of the silicon. In contrast, a retrograde doping profile, with the peak doping concentration located below the silicon surface, is employed for the $P^+$ region and the JFET region in the power SC-MOSFET structure. In addition, the $P^+$ region is driven-in further than the P-base region so that it extends further under the gate electrode than the P-base region [2]. A relatively small spacing between the $P^+$ shielding regions is utilized in the power SC-MOSFET structure when compared with the power D-MOSFET structure to reduce the cell pitch to values comparable to those for the power U-MOSFET structure. In order to prevent a high JFET resistance, the doping concentration of the JFET region is enhanced by utilizing a n-type ion-implant with a relatively high doping

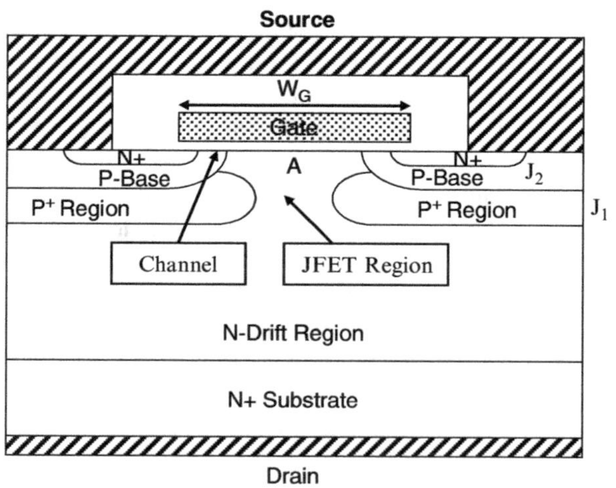


**Fig. 4.1** The SC-MOSFET structure

concentration with its peak located at the same depth as the peak doping concentration of the $P^+$ shielding regions.

The physics of operation of the power SC-MOSFET structure is similar to that for the power D-MOSFET structure. Without the application of a gate bias, a high voltage can be supported in the SC-MOSFET structure when a positive bias is applied to the drain. In this case, junction $J_1$ formed between the $P^+$ shielding region and the N-drift region becomes reverse biased. The voltage is supported mainly within the thick lightly doped N-drift region. The space between the $P^+$ shielding regions and the doping concentration of the JFET region are designed so that a potential barrier forms between the $P^+$ shielding regions at a low drain bias voltage. The potential barrier shields the region under the gate oxide (marked A in Fig. 4.1) reducing the potential supported across the P-base region. This enables reduction of the channel length without encountering reach-through breakdown in the power SC-MOSFET structure. The channel length for the power SC-MOSFET structure can be made smaller than that for the power U-MOSFET structure producing a reduction in the specific on-resistance that compensates for the resistance contribution from the JFET region.

Drain current flow in the SC-MOSFET structure is induced by the application of a positive bias to the gate electrode. This produces an inversion layer at the surface of the P-base region under the gate electrode. The threshold voltage for the power SC-MOSFET structure can be controlled by adjusting the dose for the boron ion-implantation for the P-base region. The energy for the boron ion-implantation for the P-base region is chosen to maintain its peak concentration at the surface of the silicon. The inversion layer channel provides a path for transport of electrons from the source to the drain when a positive drain voltage is applied.

After transport from the source region through the channel, the electrons enter the N-drift region at the upper surface of the device structure. They are then transported through a relatively narrow JFET region located between the adjacent $P^+$ shielding regions within the SC-MOSFET structure. The constriction of the current flow through the JFET region increases the internal resistance of the SC-MOSFET structure. A careful optimization the retrograde doping concentration profile in the JFET region is necessary to reduce the resistance to current flow through this portion of the device structure.

After being transported through the JFET region, the electrons enter the N-drift region below junction $J_1$. The current spreads from the relatively narrow JFET region to the entire width of the cell cross-section. This non-uniform current distribution within the drift region enhances its resistance making the internal resistance of the SC-MOSFET structure larger than the ideal specific on-resistance of the drift region. The small cell-pitch for the power SC-MOSFET enables reducing the drift region resistance closer to that for the ideal case than for the power D-MOSFET structure. This requires utilizing a relatively smaller window between the polysilicon gate regions than in the case of the power D-MOSFET structure, which is possible because a separate mask is not required to define the ion-implantation boundary for the $P^+$ shielding regions in the power U-MOSFET structure.

## 4.2 Power SC-MOSFET On-Resistance

The components of the on-resistance for the power SC-MOSFET structure are the same as those already described for the power D-MOSFET structure. The total on-resistance for the power SC-MOSFET structure is obtained by the addition of all the eight resistances because they are considered to be in series in the current path between the source and the drain electrodes:

$$R_{ON} = R_{CS} + R_{N+} + R_{CH} + R_{A} + R_{JFET} + R_{D} + R_{SUB} + R_{CD} \tag{4.1}$$

Each of the resistances within the power SC-MOSFET structure are analyzed below by using the procedure described in the textbook [3]. In the textbook, it was demonstrated that the contributions from the source contact resistance ($R_{CS}$), the source resistance ($R_{N+}$), and the drain contact resistance ($R_{CD}$) are very small and will therefore be neglected in this chapter. As in the case of the power D-MOSFET and U-MOSFET structures, the substrate contribution will also be excluded for the comparison of devices.

A cross-section of the power SC-MOSFET structure is illustrated in Fig. 4.2 with various dimensions that can be used for the analysis of the on-resistance components. Here, $W_{Cell}$ is the pitch for the linear cell geometry analyzed in this section; $W_G$ is the width of the gate electrode; $W_{PW}$ is the width of the polysilicon window; $W_C$ is the width of the contact window to the $N^+$ source and P-base regions; and $W_S$

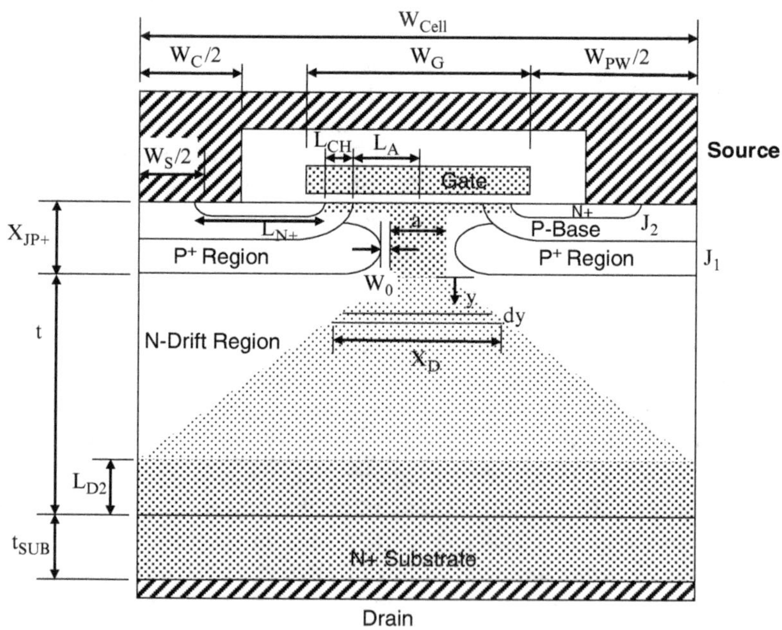


**Fig. 4.2** Power SC-MOSFET structure with current flow model used for analysis of its internal resistances

is the width of the photoresist mask used during the $N^+$ source ion-implantation. The junction depths of the P-base region and the deep $P^+$ shielding region are $x_{JP}$ and $x_{JP+}$, respectively.

In this monograph, the characteristics of power SC-MOSFET structure with 30-V blocking capability will be analyzed for power supply applications. For this voltage rating and the 1-μm lithography design rules, the power SC-MOSFET structure has a cell pitch ($W_{CELL}$) of 3 μm with a polysilicon gate width ($W_G$) of 1 μm. Typical junction depths for the $N^+$ source region and P-base region are 0.075 and 0.235 μm, respectively, leading to a channel length of only 0.16 μm. This short channel length is possible due to the shielding of the channel region in the power SC-MOSFET structure. The $P^+$ shielding region is formed by using high energy ion-implantation of boron leading to its peak concentration located below the surface at a depth of 0.3 μm. Due to the range of the ion-implantation and the subsequent diffusion during the annealing process, the vertical junction depth of the $P^+$ shielding region is typically 0.6 μm. In contrast, the lateral extension of the $P^+$ shielding region under the gate electrode is only 0.265 μm.

The doping concentration of the N-drift region required to achieve a 30-V blocking voltage capability, with an 80% reduction due to the edge termination, is $1.6 \times 10^{16}/\text{cm}^3$. The thickness of the N-drift region required below the $P^+$ shielding region is 2.6 μm. It is worth pointing out that the entire blocking voltage is not supported within the N-drift region because of the graded doping profile of the $P^+$ shielding region. This allows increasing the doping concentration and reducing the thickness of the N-drift region to achieve a smaller specific on-resistance for the D-MOSFET structure.

In the power SC-MOSFET structure, the energy for the ion-implantation of the P-type dopant (boron) for formation of the $P^+$ shielding region is chosen so that the peak doping concentration lies about 0.3 μm below the silicon surface. The polysilicon thickness must be sufficient to prevent the boron from penetration to the silicon surface during this process. The maximum concentration for the $P^+$ shielding region is limited to about $1 \times 10^{18}/\text{cm}^3$ to prevent significant contribution to the channel doping concentration. This is sufficient in magnitude to reduce the resistance of the base region for obtaining excellent ruggedness in the power SC-MOSFET structure because the $P^+$ shielding region extends below the entire length of the $N^+$ source region. The energy for the N-type dopant (phosphorus) used to form the JFET region is chosen so that its peak lies at the same depth as the $P^+$ shielding region. Its maximum doping concentration is chosen based upon a trade-off between reducing the JFET resistance contribution and shielding the P-base region from the drain potential.

The current flow pattern in the power SC-MOSFET structure is indicated by the shaded area in Fig. 4.2. In the model for the specific on-resistance, it will be assumed that the JFET region extends to the bottom of the $P^+$ shielding region. In the model, it is assumed that the current spreads at a 45° angle in the drift region. Due to the small size of the polysilicon window in the power SC-MOSFET structure, the current flow pattern overlaps for a part of the drift region similar to the case of the power U-MOSFET structure. The current is uniformly distributed in the $N^+$ substrate.

### 4.2.1 Channel Resistance

The contribution to the specific on-resistance from the channel in the SC-MOSFET structure is smaller than in the power D-MOSFET and U-MOSFET structures due to the shorter channel length. Based up on the analysis in the textbook [3] for the power D-MOSFET structure, the specific on-resistance contributed by the channel in the power SC-MOSFET structure is given by:

$$R_{CH,SP} = \frac{L_{CH} W_{Cell}}{2\mu_{ni} C_{OX}(V_G - V_{TH})} \tag{4.2}$$

In the case of the 30-V power MOSFET structures used for power supply applications, it is customary to provide the on-resistance at a gate bias of 4.5 and 10 V. Assuming a gate oxide thickness is 500 Å, an inversion layer mobility of 450 cm$^2$/V s (to match the mobility used in the numerical simulations discussed later in this section), and a threshold voltage of 2.5 V in the above equation for the power SC-MOSFET design with a cell width of 3 μm and gate electrode width of 1 μm, the specific resistance contributed by the channel at a gate bias of 4.5 V is found to be 0.039 mΩ cm$^2$. The specific on-resistance of the SC-MOSFET structure is reduced to 0.010 mΩ cm$^2$ when the gate bias is increased to 10 V. These values are much smaller than those for the power D-MOSFET and U-MOSFET structures due to the combination of a very short channel length and a small cell pitch.

### 4.2.2 Accumulation Resistance

In the power MOSFET structure, the current flowing through the inversion channel enters the drift region at the edge of the P-base junction. The current then spreads from the edge of the P-base junction into the JFET region. The current spreading phenomenon is aided by the formation of an accumulation layer in the semiconductor below the gate oxide due to the positive gate bias applied to turn-on the device. The specific on-resistance contributed by the accumulation layer in the power SC-MOSFET structure is given by the same formulation derived for the power D-MOSFET structure in the textbook [3]:

$$R_{A,SP} = K_A \frac{(W_G - 2x_{JP}) W_{Cell}}{4\mu_{nA} C_{OX}(V_G - V_{THA})} \tag{4.3}$$

In writing this expression, a coefficient $K_A$ has been introduced to account for the current spreading from the accumulation layer into the JFET region. A typical value for this coefficient is 0.6 based upon the current flow observed from numerical simulations of power D-MOSFET structures. The threshold voltage ($V_{THA}$) in the expression is for the on-set of formation of the accumulation layer. A zero threshold voltage will be assumed here when performing the analytical computations. Note that the junction depth of the P-base region (and not the $P^+$ region) defines the length of the accumulation region.

For the 30-V power SC-MOSFET design with a cell width of 3 μm and gate width of 1 μm, the specific resistance contributed by the accumulation layer at a gate bias of 4.5 V is 0.0078 mΩ cm$^2$ if the P-base junction depth ($x_{JP}$) is 0.235 μm and the gate oxide thickness is 500 Å. An accumulation layer mobility of 1,000 cm$^2$/V s was used in this calculation to match the mobility used in the numerical simulations (discussed later in this section). When the gate bias is increased to 10 V, the specific resistance contributed by the accumulation layer is reduced to 0.0035 mΩ cm$^2$. These values are also much smaller than those for the power D-MOSFET and U-MOSFET structures due to the combination of a short accumulation layer length and a small cell pitch.

### *4.2.3 JFET Resistance*

The electrons entering from the channel into the drift region are distributed into the JFET region via the accumulation layer formed under the gate electrode. The spreading of current in this region was accounted for by using a constant $K_A$ of 0.6 for the accumulation layer resistance. Consequently, the current flow through the JFET region can be treated with a uniform current density. In the power SC-MOSFET structure, the cross-sectional area for the JFET region is the smallest at the depth at which the $P^+$ shielding region has its highest concentration. Consequently, this width will be used for the analytical model in order to simplify the analysis by using a uniform cross-section for the current flow with a width 'a' for the JFET region as illustrated by the shaded area in Fig. 4.2. The width of the current flow is related to the device structural parameters:

$$a = (W_G - 2x_{JP+L} - 2W_0) \tag{4.4}$$

where $W_0$ is the zero-bias depletion width for the JFET region, and $x_{JP+L}$ is the lateral extension of the $P^+$ shielding region under the gate electrode. The depletion region boundary is indicated in the figures with the dashed lines. In the models, it is assumed that no current can flow through the depleted region because all the free carriers have been swept out by the prevailing electric field across the junction. The zero-bias depletion width ($W_0$) in the JFET region can be computed by using the doping concentrations on both sides of the junction:

$$W_0 = \sqrt{\frac{2\varepsilon_S N_A V_{bi}}{qN_{DJ}(N_A + N_{DJ})}} \tag{4.5}$$

where $N_A$ is the doping concentration in the $P^+$ shielding region and $N_{DJ}$ is the doping concentration in the JFET region. In the above equation, the built-in potential is also related to the doping concentrations on both sides of the junction:

$$V_{bi} = \frac{kT}{q} \ln\left(\frac{N_A N_{DJ}}{n_i^2}\right) \tag{4.6}$$

Since the JFET region is assumed to extend to the bottom the $P^+$ shielding region, the specific on-resistance contributed by the JFET region in the power SC-MOSFET structure can then be obtained by using:

$$R_{JFET,SP} = \frac{\rho_{JFET} x_{JP+} W_{Cell}}{(W_G - 2x_{JP+L} - 2W_0)} \tag{4.7}$$

where $\rho_{JFET}$ is the resistivity of the JFET region given by:

$$\rho_{JFET} = \frac{1}{q\mu_n N_{DJ}} \tag{4.8}$$

where $\mu_n$ is the bulk mobility appropriate to the doping level of the JFET region. In the power SC-MOSFET structure, the JFET region has a retrograde doping profile with its maximum concentration (typically $1.5 \times 10^{17}/\text{cm}^3$) at the same depth as the $P^+$ shielding region. In an analytical model, it is convenient to use an average doping concentration for the JFET region which is slightly less than its peak doping concentration. The resistivity for the JFET region is then found to be 0.066 Ω cm corresponding to an average JFET doping concentration of $1.3 \times 10^{17}/\text{cm}^3$. The zero-bias depletion width in the JFET region for this JFET doping concentration is 0.0935 μm. For the 30-V power SC-MOSFET design with a cell width of 3 μm and gate width of 1 μm, the specific resistance contributed by the JFET region is found to be 0.042 mΩ cm$^2$ based up on using the above parameters. This value is much smaller than that for the power D-MOSFET structure due to the short cell pitch.

### *4.2.4 Drift Region Resistance*

The resistance contributed by the drift region in the power SC-MOSFET structure is enhanced above that for the ideal drift region due to current spreading from the JFET region. The cross-sectional area for the current flow in the drift region increases from the width 'a' of the JFET region as illustrated in Fig. 4.2 at a spreading angle of 45°. Due to the small polysilicon window size in the power SC-MOSFET structure, the current paths overlap in the drift region.

The specific on-resistance contributed by the drift region in the power SC-MOSFET structure is given by the expression derived in the textbook [3] for high voltage D-MOSFET structures:

$$R_{D,SP} = \frac{\rho_D W_{Cell}}{2} \ln\left[\frac{W_{Cell}}{a}\right] + \rho_D\left(t - \frac{a}{2} - \frac{W_{Cell}}{2}\right) \tag{4.9}$$

For the parameters given above for this structure, the dimension 'a' in the equation is found to be 0.283 μm. For the 30-V power SC-MOSFET design with a cell width of 3 μm and gate width of 1 μm, the specific resistance contributed by the drift region is then found to be 0.1465 mΩ cm$^2$ by using a resistivity of the drift region of 0.325 Ω cm based upon a doping concentration of 1.6 × 10$^{16}$/cm$^3$.

## 4.2.5 Total On-Resistance

The total specific on-resistance for the power SC-MOSFET structure can be computed by adding all the above components for the on-resistance. For the case of the 30-V power SC-MOSFET design with a cell pitch ($W_{Cell}$) of 3 μm and gate width of 1 μm, the total specific on-resistance is found to be 0.235 mΩ cm$^2$ at a gate bias of 4.5 V and 0.202 mΩ cm$^2$ at a gate bias of 10 V by using the analytical model. The contributions from each of the components of the on-resistance are summarized in Fig. 4.3. The specific on-resistance for the power SC-MOSFET structure is found to be smaller than that for the power D-MOSFET and the U-MOSFET structures at a gate bias of 4.5 V.

The ideal specific on-resistance for a drift region is given by:

$$R_{IDEAL,SP} = \frac{W_{PP}}{q\mu_n N_D} \tag{4.10}$$

where $W_{PP}$ is the parallel-plane depletion width at breakdown, $N_D$ is the doping concentration of the drift region to sustain the blocking voltage, and $\mu_n$ is the mobility for electrons corresponding to this doping concentration. For the case of a blocking voltage of 30 V, the depletion width and doping concentration are found to be 2.2 × 10$^{16}$/cm$^3$ and 1.4 μm, respectively. Using the electron mobility for this doping level, the ideal specific on-resistance is found to be 0.034 mΩ cm$^2$. Since the device is constrained by the impact of an 80% reduction of breakdown voltage due to the edge termination, it is worth computing the ideal specific on-resistance for this case for comparison with the device. For the case of a blocking voltage of 37.5 V, the depletion width and doping concentration are found to be 1.6 × 10$^{16}$/cm$^3$ and 1.8 μm, respectively. Using the electron mobility for this doping level, the ideal

| Resistance | $V_G$ = 4.5 V (mΩ-cm$^2$) | $V_G$ = 10 V (mΩ-cm$^2$) |
|---|---|---|
| Channel ($R_{CH,SP}$) | 0.039 | 0.010 |
| Accumulation ($R_{A,SP}$) | 0.0078 | 0.0035 |
| JFET ($R_{JFET,SP}$) | 0.042 | 0.042 |
| Drift ($R_{D,SP}$) | 0.1465 | 0.1465 |
| Total ($R_{T,SP}$) | 0.235 | 0.202 |

**Fig. 4.3** On-resistance components within the 30-V power SC-MOSFET structure

specific on-resistance is found to be 0.059 m$\Omega$ cm$^2$. In spite of its planar gate topology, the specific on-resistance for the SC-MOSFET structure is much closer to these ideal specific on-resistances than the power D-MOSFET and the U-MOSFET structures for the gate bias of 4.5 V.

#### 4.2.5.1 Simulation Results

The results of two-dimensional numerical simulations on the 30-V power SC-MOSFET structure are described here to provide a more detailed understanding of the underlying device physics and operation. The structure used for the numerical simulations had a drift region thickness of 2.6 μm below the P$^+$ shielding region with a doping concentration of 1.6 × 10$^{16}$/cm$^3$. The P-base region and N$^+$ source regions had depths of 0.235 and 0.075 μm, respectively. The P$^+$ shielding region had a depth of 0.60 μm with its peak doping concentration located at 0.30 μm below the silicon surface. The doping concentration in the JFET region was enhanced by using a retrograde n-type doping profile with a peak concentration of 1.5 × 10$^{17}$/cm$^3$ with a depth of 0.3 μm. For the numerical simulations, half the cell (with a width of 1.5 μm) shown in Fig. 4.1 was utilized as a unit cell that is representative of the structure.

A three dimensional view of the doping distribution in the SC-MOSFET structure is shown in Fig. 4.4 from the left hand edge of the structure to the center of the

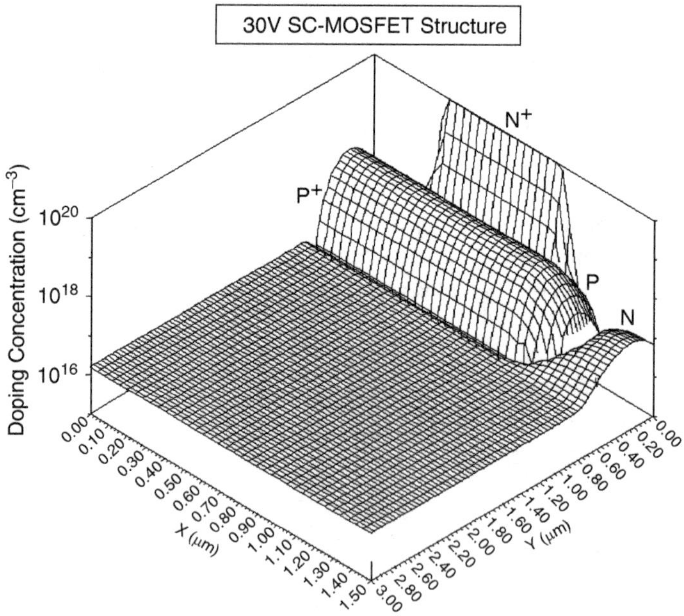


**Fig. 4.4** Doping distribution for the SC-MOSFET structure

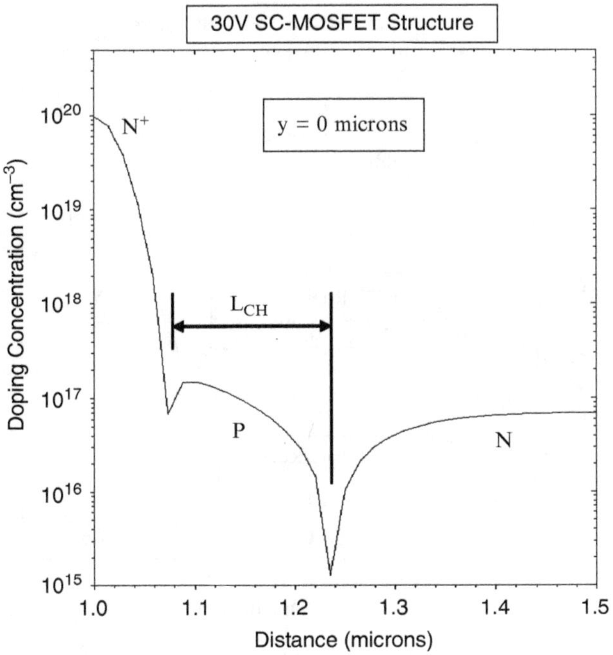


**Fig. 4.5** Channel doping profile for the SC-MOSFET structure

polysilicon gate region. The $N^+$ source, P-base, and $P^+$ shielding regions are all aligned to the gate edge, which is located at 1 μm from the left hand side. The enhancement of the doping in the N-drift region near the surface is due to the additional JFET doping which can be observed to have a retrograde profile.

The lateral doping profile taken along the surface under the gate electrode is shown in Fig. 4.5. From the profile, it can be observed that the doping concentration of the JFET region has been increased to $7 \times 10^{16}/\text{cm}^3$ at the surface due to the additional doping from the N-type ion-implant. The lateral extensions of the P-base and $N^+$ source regions under the gate electrode are 0.235 and 0.075 μm leading to a channel length of only 0.16 μm. The surface concentration for the P-base region was chosen to obtain a maximum compensated P-type doping concentration in the channel of about $1.5 \times 10^{17}/\text{cm}^3$.

The vertical doping profile taken at two positions within the power SC-MOSFET structure are provided in Fig. 4.6. From the profile taken at $\times = 0$ μm in the middle of the polysilicon window (dashed line), it can be observed that the doping concentration of the $P^+$ shielding region has a maximum value of $9 \times 10^{17}/\text{cm}^3$ at a depth of 0.3 μm. The JFET region doping profile is shown by the solid line at the middle of the gate electrode ($\times = 1.5$ μm). It can be observed that the doping concentration of the JFET region has a maximum value of $1.5 \times 10^{17}/\text{cm}^3$ at a depth of 0.3 μm.

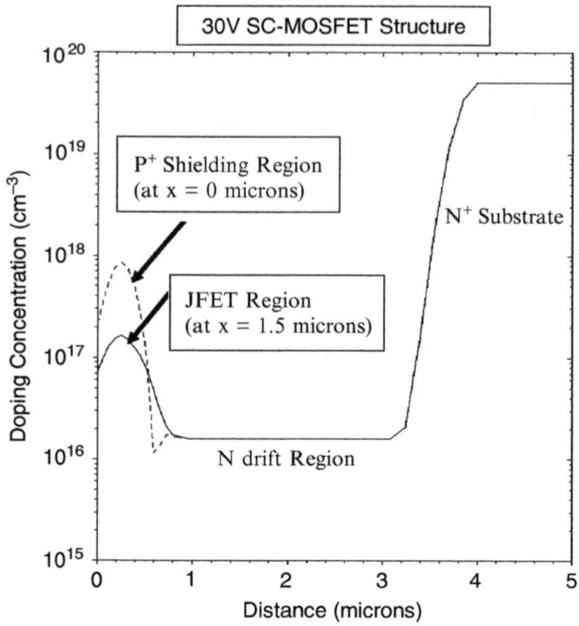


**Fig. 4.6** Vertical doping profiles in the SC-MOSFET structure

In order to analyze the space between the P$^+$ shielding regions, the lateral doping profile under the gate electrode taken at a depth of 0.3 μm within the power SC-MOSFET structure is provided in Fig. 4.7. It can be observed that the P$^+$ shielding region extends to a distance of 0.265 μm from the edge of the gate electrode. This distance is larger than the extension of 0.235 μm by the P-base region from the edge of the gate electrode. The larger extension of the P$^+$ shielding region provides the desired shielding of the surface under the gate oxide from the drain potential.

The transfer characteristics for the SC-MOSFET structure were obtained using numerical simulations with a drain bias of 0.1 V at 300 and 400°K. The resulting transfer characteristics are shown in Fig. 4.8. From this graph, a threshold voltage of 2.5 and 2.0 V can be extracted at 300 and 400°K, respectively. The threshold voltage decreases by 20% when the temperature increases similar that observed for the power D-MOSFET structure. The specific on-resistance can be obtained from the transfer characteristics at any gate bias voltage. For the case of a gate bias of 4.5 V and 300°K, the specific in-resistance is found to be 0.242 mΩ cm$^2$. For the case of a gate bias of 10 V and 300°K, the specific in-resistance is found to be 0.212 mΩ cm$^2$. These values are in very close agreement with the results obtained from the analytical model.

The on-state current flow pattern within the SC-MOSFET structure at a small drain bias of 0.1 V and a gate bias of 4.5 V is shown in Fig. 4.9. In the figure, the

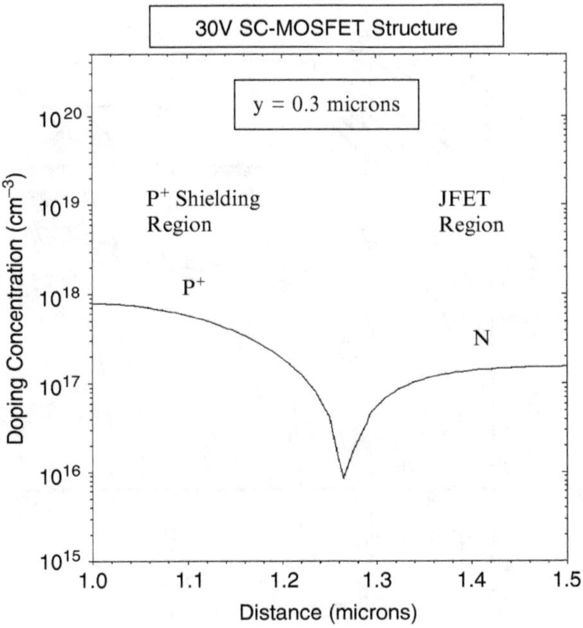


**Fig. 4.7** Lateral doping profile in the SC-MOSFET structure

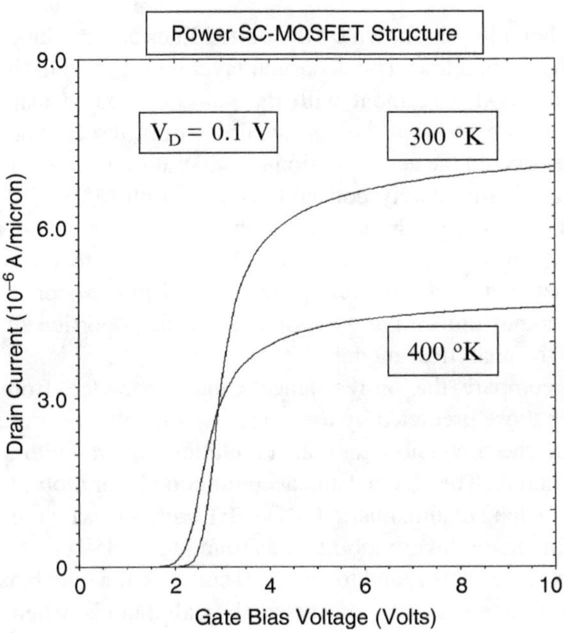


**Fig. 4.8** Transfer characteristics of the SC-MOSFET structure

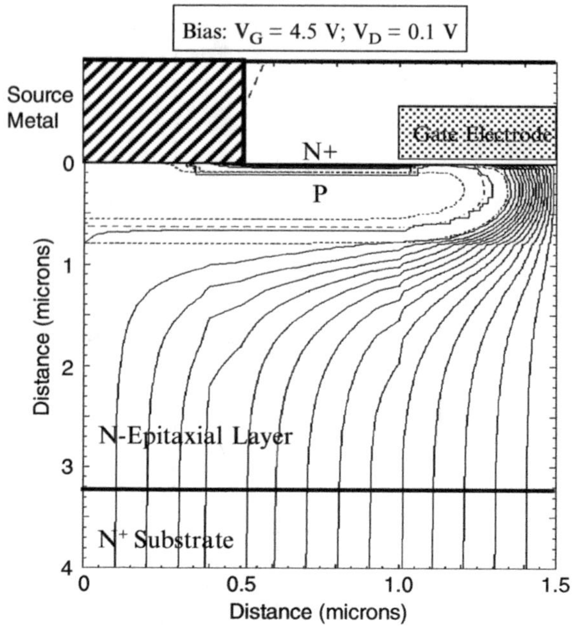


**Fig. 4.9** Current distribution in the SC-MOSFET structure

depletion layer boundary is shown by the dotted lines and the junction boundary is delineated by the dashed line. The depletion layer width ($W_0$) in the JFET region is about 0.1 μm in good agreement with the value computed using the analytical model. It can be observed that the current flows from the channel and distributes into the JFET region via the accumulation layer. Within the JFET region, the cross-sectional area is approximately constant with a width (a/2) of 0.15 μm in good agreement with the assumptions used in the analytical model. From the figure, it can be seen that the current spreads from the JFET region to the drift region at a 45° angle as assumed in the models. Due to small polysilicon window size, the current flow becomes uniform at a depth of 2 μm for a portion of the drift region consistent with the analytical model.

In order to compare the on-resistance values extracted from the numerical simulations with those predicted by the analytical models, it is necessary to extract the mobility in the inversion and accumulation layers within the simulated D-MOSFET structure. The channel and accumulation layer mobility were extracted by simulation of a long channel lateral MOSFET structure with the same gate oxide thickness. The inversion layer mobility was found to be 450 $cm^2/V$ s while that for the accumulation layer was found to be 1,000 $cm^2/V$ s at a gate bias of 10 V. These values were consequently used in the analytical models when calculating the specific on-resistance.

## 4.3 Blocking Voltage

The power SC-MOSFET structure must be designed to support a high voltage in the first quadrant when the drain bias voltage is positive. During operation in the blocking mode, the gate electrode is shorted to the source electrode by the external gate bias circuit. The application of a positive drain bias voltage produces a reverse bias across junction $J_1$ between the $P^+$ shielding region and the N-drift region. Most of the applied voltage is supported across the N-drift region. The doping concentration of donors in the N-epitaxial drift region and its thickness must be chosen to attain the desired breakdown voltage. In the power D-MOSFET structure, the gate region is not screened from the drain bias due to cylindrical shape of the planar junctions. This results in significant depletion of the P-base region making the channel length of the structure large. In contrast, the $P^+$ shielding region in the power SC-MOSFET structure creates a potential barrier in the JFET region and screens the region under the gate electrode from the drain potential. The reduced potential at the upper surface allows reduction of the channel length without reach-through problems and also a reduces the electric field in the gate oxide enhancing the reliability.

### *4.3.1 Impact of Edge Termination*

In practical devices, the maximum blocking voltage (BV) of the power MOSFET is invariably decided by the edge termination that surrounds the device cell structure. The most convenient edge termination for power SC-MOSFET devices is based up on floating field rings and field plates. With this design, the enhanced electric field at the edges limits the breakdown voltage to about 80% of the parallel-plane breakdown voltage ($BV_{PP}$):

$$BV_{PP} = \left(\frac{BV}{0.8}\right) \tag{4.11}$$

Consequently, the doping and thickness of the N-drift region must be chosen to achieve a parallel-plane breakdown voltage that is 25% larger than the blocking voltage for the device:

$$N_D = \left(\frac{4.45 \times 10^{13}}{BV_{PP}}\right)^{4/3} \tag{4.12}$$

The thickness of the drift region required below the $P^+$ shielding region is less than the depletion width for the ideal parallel-plane junction with the above doping concentration, and is given by:

$$t = W_D(BV) = \sqrt{\frac{2\varepsilon_S BV}{qN_D}} \tag{4.13}$$

The resistance of the drift region can be reduced by using this thickness rather than the maximum depletion width corresponding to the doping concentration given by (4.12).

#### 4.3.1.1 Simulation Results

The results of two-dimensional numerical simulations on the 30-V power SC-MOSFET structure are described here to provide a more detailed understanding of the underlying device physics and operation during the blocking mode. The structure used for the numerical simulations had the same parameters as the structure described in the previous section. The blocking characteristic for the SC-MOSFET cell structure is shown in Fig. 4.10 at 300°K. It can be observed that the cell is capable of supporting 41 V. This provides enough margin to achieve a device blocking voltage capability of slightly over 30 V after accounting for the reduction due to the edge termination.

It is instructive to examine the potential contours inside the power SC-MOSFET structure when it is operating in the blocking mode. This allows determination of the voltage distribution within the structure and the penetration of the depletion region in the P-base region with increasing drain bias voltage. The potential contours for the SC-MOSFET structure obtained using the numerical simulations with zero gate bias and various drain bias voltages are shown in

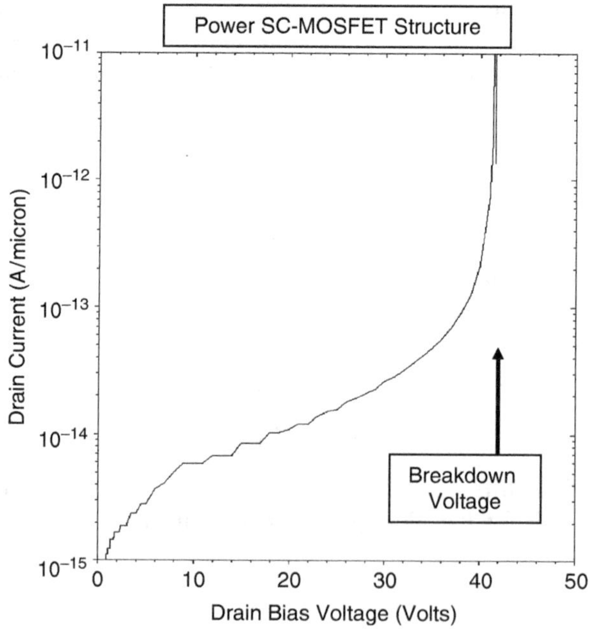


**Fig. 4.10** Blocking characteristic for the SC-MOSFET structure

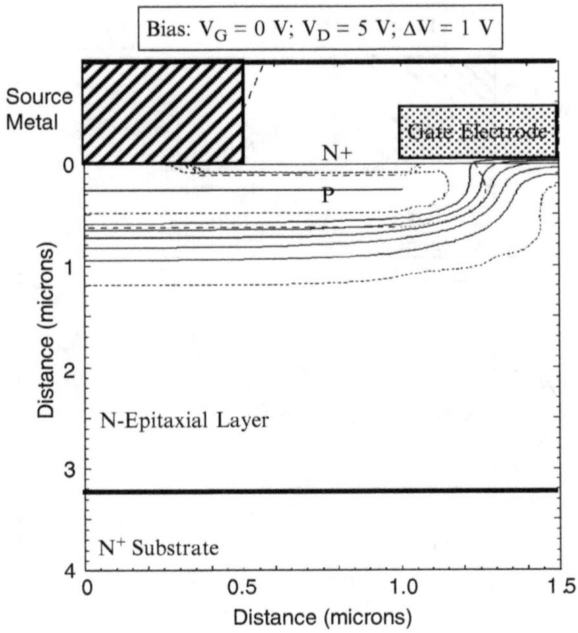


**Fig. 4.11** Potential contours in the SC-MOSFET structure

Figs. 4.11–4.14. From these figures, it can be observed that the voltage distribution at the junction $J_1$ between the $P^+$ shielding region and the N-drift region changes when proceeding from the surface towards the drain. This indicates that the surface region is screened from the drain bias by the $P^+$ shielding region. Consequently, there is very little increase in the depletion of the P-base region when the drain bias is increased to 30 V. The screening phenomenon in the power SC-MOSFET structure allows utilization of a very short channel length which reduces the specific on-resistance.

It is insightful to also examine the electric field profile inside the power SC-MOSFET structure when it is operating in the blocking mode. The electric field profile obtained through the junction between the $P^+$ shielding region and the N- drift region is shown in Fig. 4.15. It can be observed that the maximum electric field occurs as expected at the junction at a depth of 0.6 μm from the surface.

The electric field profile taken through the middle of the gate electrode (at X = 1.5 μm) is provided in Fig. 4.16. It can be observed that the electric field exhibits a maximum below the surface at a depth of about 1 μm. This reduces the electric field at the surface under the gate oxide when compared with the power D-MOSFET and U-MOSFET structures. Further, the electric field in the gate oxide is much smaller than that observed in the power D-MOSFET and U-MOSFET structures. The low electric field in the gate oxide for the power SC-MOSFET structure indicates very reliable operation allowing stable device performance over long periods of time.

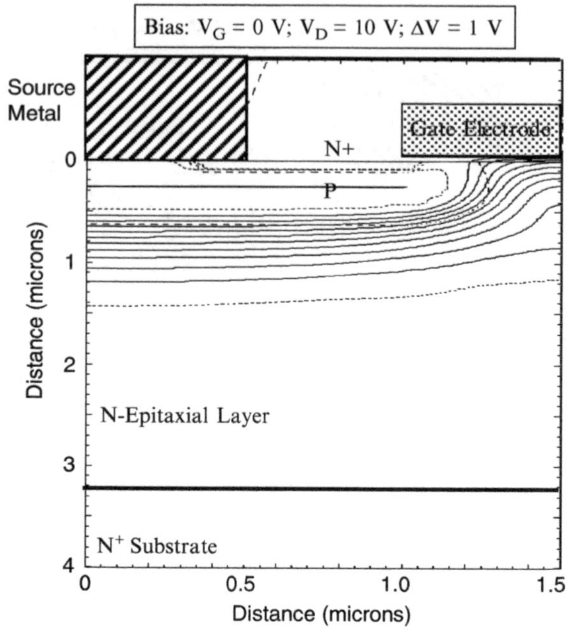


**Fig. 4.12** Potential contours in the SC-MOSFET structure

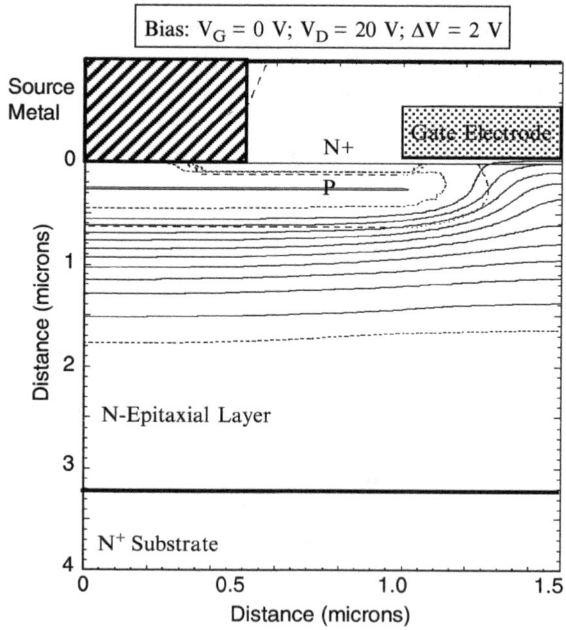


**Fig. 4.13** Potential contours in the SC-MOSFET structure

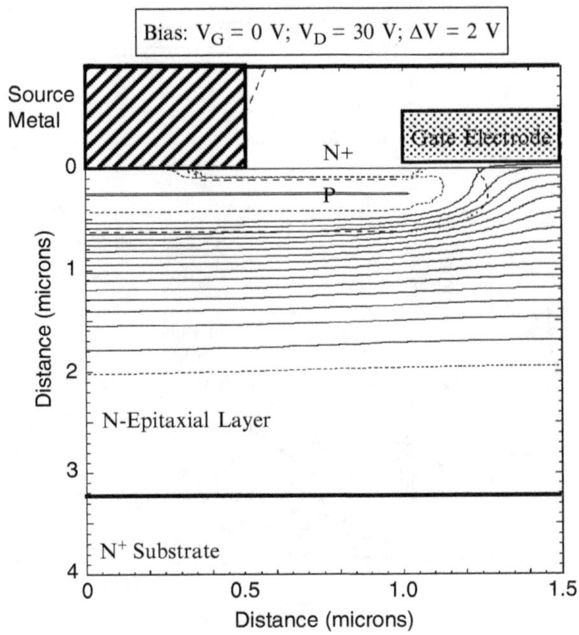


**Fig. 4.14** Potential contours in the SC-MOSFET structure

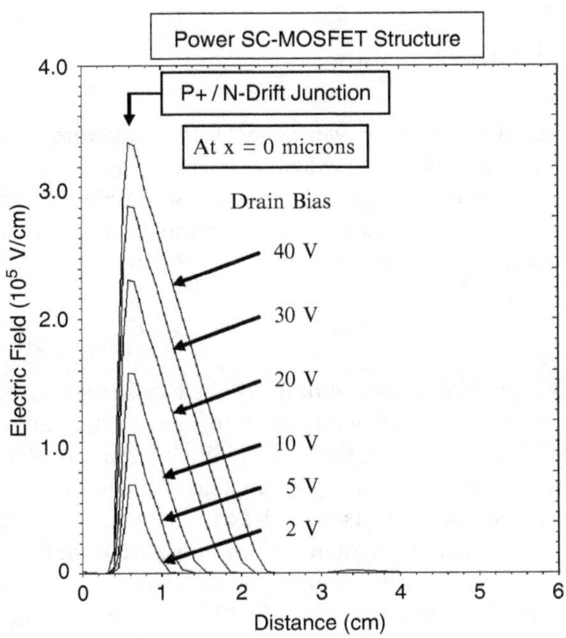


**Fig. 4.15** Electric field distribution in the SC-MOSFET structure

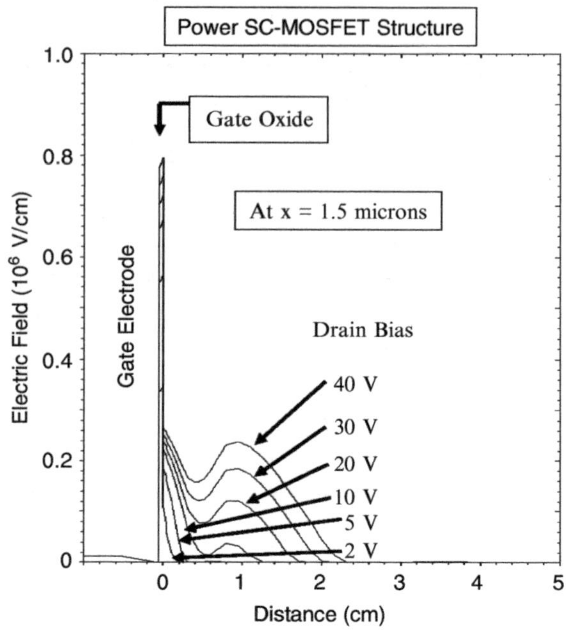


**Fig. 4.16** Electric field distribution in the SC-MOSFET structure

## 4.4 Output Characteristics

The output characteristics of the power SC-MOSFET structure are important to the loci for the switching waveforms when it is operating in power circuits. Due to shielding of the P-base region, the power SC-MOSFET structure exhibits super-linear transfer characteristics [3]. The saturated drain current for the power SC-MOSFET structure is then given by:

$$I_{Dsat} = C_{ox}(V_G - V_T)\, v_{sat,n} Z \tag{4.14}$$

With sufficiently high doping concentration of the P-base region, the modulation of channel length can be made sufficiently small to ensure a high output resistance. In the power SC-MOSFET structure, the $P^+$ shielding region suppresses the depletion of the P-base region with increasing drain bias voltage. This is beneficial for obtaining a reasonable output resistance despite the very short channel length in the device. The short channel length in the power SC-MOSFET structure produces a high transconductance which is beneficial for reducing switching losses. The saturated drain current in the power SC-MOSFET structure increases linearly with the gate bias voltage.

### *4.4.1 Simulation Example*

The output characteristics of the 30-V power SC-MOSFET structure were obtained by using two-dimensional numerical simulations using various gate bias voltages. All the device parameters used for these numerical simulations are the same as those used in the previous sections. The output characteristics of the power SC-MOSFET obtained using the simulations are shown in Fig. 4.17. The traces for increasing gate bias voltages are nearly uniformly spaced indicating super-linear [3] behavior of the transfer characteristics. When compared with the power D-MOSFET and U-MOSFET structures, the power SC-MOSFET structure exhibits a gradual transition from the on-state to the current saturation regime due to the voltage drop produced in the JFET region at high current densities.

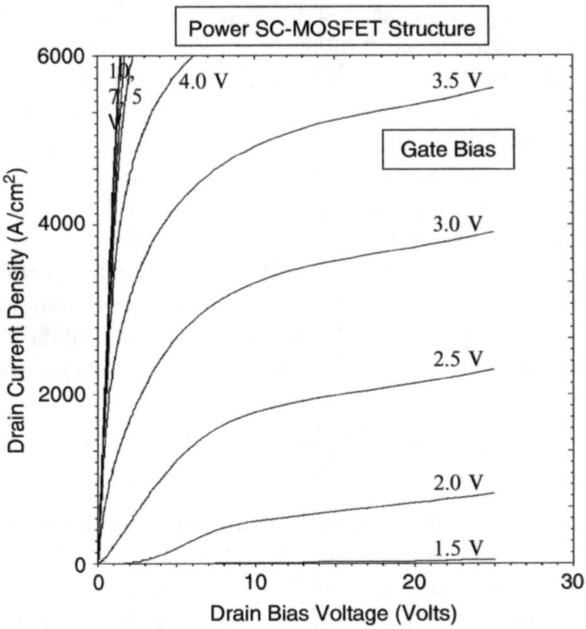


**Fig. 4.17** Output characteristics for the power SC-MOSFET structure

## 4.5 Device Capacitances

For the power MOSFET structures, the switching speed is limited by the device capacitances in practical applications as previously discussed in Chap. 2. The rate at which the power MOSFET structure can be switched between the on- and off-states is determined by the rate at which the input capacitance can be charged or discharged. In addition, the capacitance between the drain and the gate electrodes has been found to play an important role in determining the drain current and voltage transitions during the switching event.

The capacitances within the power SC-MOSFET structure can be analyzed using the same approach used in the textbook [3] for the power D-MOSFET structure. The specific input (or gate) capacitance for the power SC-MOSFET structure is given by:

$$C_{IN,SP} = C_{N+} + C_P + C_{SM} = \frac{2x_{JP}}{W_{Cell}}\left(\frac{\varepsilon_{OX}}{t_{OX}}\right) + \frac{W_G}{W_{Cell}}\left(\frac{\varepsilon_{OX}}{t_{IEOX}}\right) \tag{4.15}$$

where $t_{OX}$ and $t_{IEOX}$ are the thicknesses of the gate and inter-electrode oxides, respectively. The junction depth for the P-base region is much smaller in the power SC-MOSFET structure when compared with the power D-MOSFET structure resulting in a smaller input capacitance. For a 30-V power SC-MOSFET structure with a cell pitch ($W_{CELL}$ in Fig. 4.2) of 3 μm and gate electrode width of 1 μm, the input capacitance is found to be 13 nF/cm$^2$ for a gate oxide thickness of 500 Å and an inter-metal dielectric thickness of 5,000 Å. This value is substantially smaller than 22 nF/cm$^2$ for the power D-MOSFET structure and 47 nF/cm$^2$ for the power U-MOSFET structure. This reduces the power requirements and losses in the gate drive circuit needed for the power SC-MOSFET structure.

The capacitance between the gate and drain electrodes (also called the reverse transfer capacitance) is determined by the width of the JFET region where the gate electrode overlaps the N-drift region. The MOS structure in this portion of the power SC-MOSFET structure operates under deep depletion conditions when a positive voltage is applied to the drain. As in the case of the power D-MOSFET structure, the gate-drain capacitance for the power SC-MOSFET structure is given by:

$$C_{GD,SP} = \frac{(W_G - 2x_{PL})}{W_{Cell}}\left(\frac{C_{OX}C_{S,M}}{C_{OX} + C_{S,M}}\right) \tag{4.16}$$

where $C_{S,M}$ is the semiconductor capacitance under the gate oxide, which decreases with increasing drain bias voltage. The specific capacitance of the semiconductor depletion region can be obtained by computation of the depletion layer width. The depletion layer width in the semiconductor under the gate oxide can be obtained using:

$$W_{D,MOS} = \frac{\varepsilon_S}{C_{OX}}\left\{\sqrt{1 + \frac{2V_D C_{OX}^2}{q\varepsilon_S N_{DJ}}} - 1\right\} \tag{4.17}$$

where $N_{DJ}$ is the effective doping concentration of the semiconductor below the gate oxide for the retrograde doping profile. The specific capacitance for the semiconductor is then obtained using:

$$C_{S,M} = \frac{\varepsilon_S}{W_{D,MOS}} \tag{4.18}$$

The gate-drain (or reverse transfer) capacitance can be computed by using (4.18) with the above equations to determine the semiconductor capacitance as a function of the drain bias voltage.

However, in the case of the power SC-MOSFET structure, the above equation is only valid until the depletion region from the $P^+$ screening regions pinches-off the JFET region. Once this occurs, the gate-drain capacitance decreases drastically because the gate is screened from the drain. The drain voltage at which the JFET region is pinched-off is given by:

$$V_{P,JFET} = \frac{qN_{DJ,Peak}}{8\varepsilon_S}(W_G - 2x_{JP+L})^2 \tag{4.19}$$

where $N_{DJ,Peak}$ is the peak doping concentration for the retrograde doping profile.

For a typical power SC-MOSFET design with peak JFET region doping concentration ($N_{DJ,Peak}$) of $1.5 \times 10^{17}/cm^3$, gate width ($W_G$) of 1 μm, and $P^+$ screening region lateral junction depth of 0.265 μm, the JFET region pinch-off voltage is 6.4 V. The effective doping concentration for the JFET region in this case is $1.2 \times 10^{17}/cm^3$. After the JFET region is pinched-off, the gate-drain capacitance is determined by the edge of the depletion region located below the $P^+$ screening region. This distance below the gate oxide is given by:

$$W_S = x_{JP+} + \sqrt{\frac{2\varepsilon_S V_D}{qN_D}} \tag{4.20}$$

The specific capacitance for the semiconductor below the gate oxide can then be obtained using (4.18) with the width $W_S$. The specific gate transfer capacitance obtained by using the above analytical formulae is shown in Fig. 4.18 for the case of a power SC-MOSFET structure with 3 μm cell pitch, and polysilicon window of 2 μm. The gate-drain (reverse transfer) capacitance decreases gradually with increasing drain bias voltage due to the expansion of the depletion region in the semiconductor under the gate oxide until the drain bias reaches 6 V. Once the JFET region pinches-off at a drain bias of 6.3 V, the specific reverse transfer capacitance drops drastically which is beneficial for reducing the gate transfer charge. Despite this sharp reduction in the reverse transfer capacitance predicted by this model, the actual reverse transfer capacitance in devices is found to decrease to a much greater degree after the JFET region is pinched-off [4]. This effect can be accounted for by reducing the effective area for the gate region using a screening factor ($K_S$) as shown in Fig. 4.18.

The output capacitance for the power SC-MOSFET structure is associated with the capacitance of the junction between the P-base region and the N-drift region. Due to pinch-off of the JFET region with increasing drain bias voltage, it is necessary to examine the change in the depletion region boundary with applied voltage. The depletion layer boundary inside the power SC-MOSFET structure prior to the pinch-off of the JFET region is shown in Fig. 4.19 by the dashed lines. It can be observed that the depletion region has a vertical boundary inside the JFET

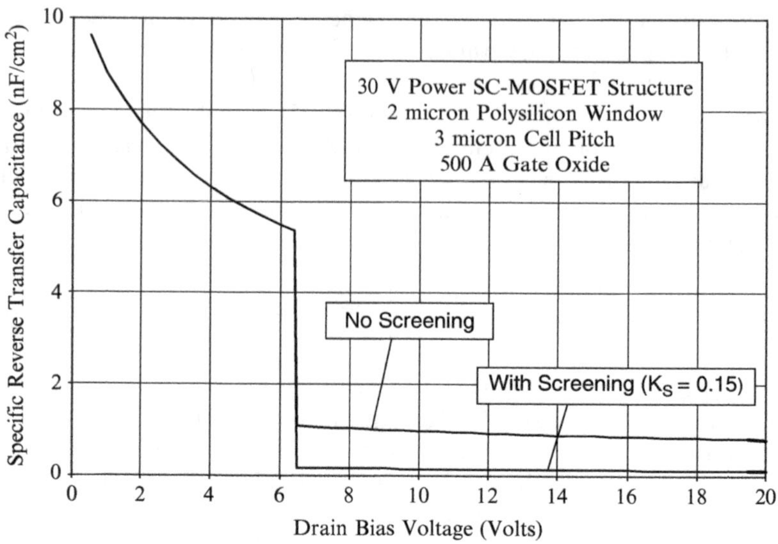


**Fig. 4.18** Gate-drain capacitance for the power SC-MOSFET structure

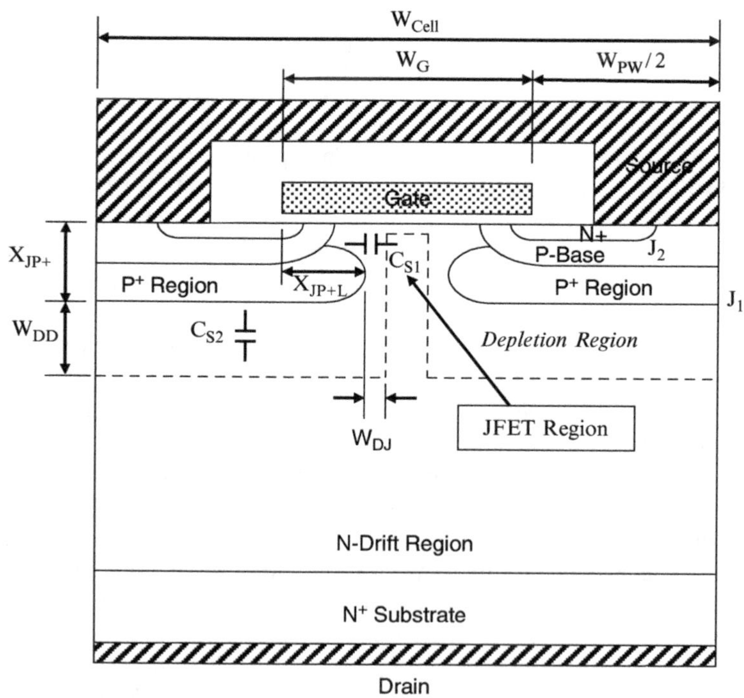


**Fig. 4.19** Depletion region boundaries for determination of the output capacitance for the power SC-MOSFET structure

region and a horizontal boundary below the $P^+$ screening region. The capacitances associated with each of these regions are indicated in the figure as $C_{S1}$ and $C_{S2}$. The specific junction capacitance associated with the JFET region is given by:

$$C_{S1,SP} = \frac{\varepsilon_S}{W_{DJ}} \left( \frac{2x_{JP+}}{W_{Cell}} \right) \tag{4.21}$$

where the depletion region thickness ($W_{DJ}$) in the JFET region is related to the drain bias voltage:

$$W_{DJ} = \sqrt{\frac{2\varepsilon_S (V_D + V_{bi})}{qN_{DJ}}} \tag{4.22}$$

where $N_{DJ}$ is the average doping concentration of the JFET region. The specific junction capacitance associated with the bottom of the $P^+$ screening region is given by [3]:

$$C_{S2,SP} = \frac{\varepsilon_S}{W_{DD}} \left( \frac{W_{PW} + 2x_{JP+L}}{W_{Cell}} \right) \tag{4.23}$$

where the depletion region thickness ($W_{DD}$) in the drift region is related to the drain bias voltage:

$$W_{DD} = \sqrt{\frac{2\varepsilon_S (V_D + V_{bi})}{qN_D}} \tag{4.24}$$

where $N_D$ is the doping concentration of the drift region.

The specific output capacitance for the power SC-MOSFET structure can then be obtained by combining the above values:

$$C_{O,SP} = C_{S1,SP} + C_{S2,SP} \tag{4.25}$$

The output capacitance obtained by using the above analytical model is shown in Fig. 4.20 for the case of the 30-V power SC-MOSFET structure with 3-μm cell pitch and polysilicon window of 2 μm. This structure has a gate oxide thickness of 500 Å and a lateral junction depth of 0.265 μm for the $P^+$ screening region. A built-in potential of 0.8 V was assumed for the P-base/N-drift junction, and the drift region has a doping concentration of $1.6 \times 10^{16}/\text{cm}^3$. An effective doping concentration of $1.2 \times 10^{17}/\text{cm}^3$ was assumed for the JFET region. The figure includes the contributions from the two components of the output capacitance. It can be observed that the specific output capacitance decreases with increasing drain bias voltage due to the expansion of the depletion region in the JFET region and under the $P^+$ screening region. At a drain bias of 6.4 V, the JFET region is pinched-off

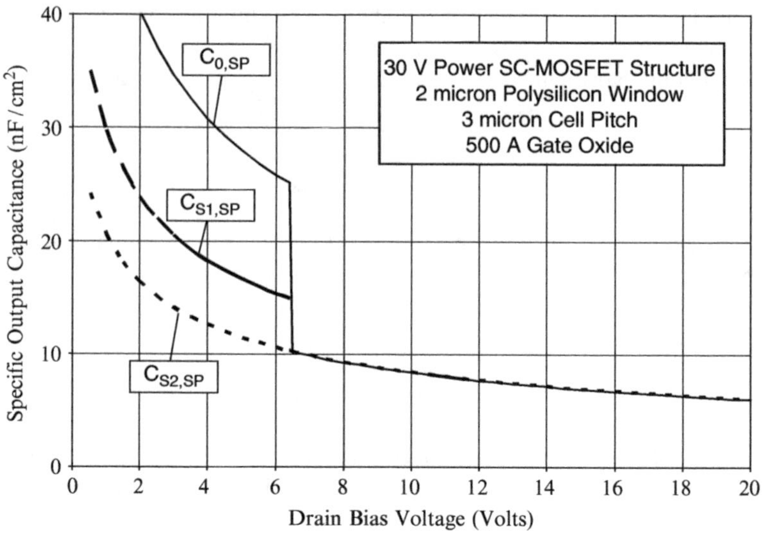


**Fig. 4.20** Output capacitance for the power SC-MOSFET structure

resulting in the merging of the depletion regions in the JFET region. The output capacitance then reduces abruptly to that associated with the $P^+$ screening region. At a drain bias of 20 V, the specific output capacitance predicted by the analytical model for this structure is 6 nF/cm$^2$.

### *4.5.1 Simulation Example*

The capacitances of the 30-V power SC-MOSFET structure were extracted using two-dimensional numerical simulations on the structure with device parameters described in the previous sections. The input capacitance was extracted by performing the numerical simulations with a small AC signal superposed on the DC gate bias voltage. The input capacitances obtained for the power SC-MOSFET structure are shown in Fig. 4.21 at a drain bias of 20 V. The input capacitance is comprised of two components – the first is between the gate electrode and the source electrode ($C_{GS}$) while the second is between the gate electrode and the base electrode ($C_{GB}$). The total input capacitance can be obtained by the addition of these capacitances because they are in parallel and share a common contact electrode in the actual power SC-MOSFET structure. From the figure, a total specific input capacitance of about 13 nF/cm$^2$ is observed at a gate bias of 4.5 V – close to that predicted by the analytical model.

The drain-gate (reverse transfer) capacitance can be extracted by performing the numerical simulations with a small AC signal superposed on the DC drain bias

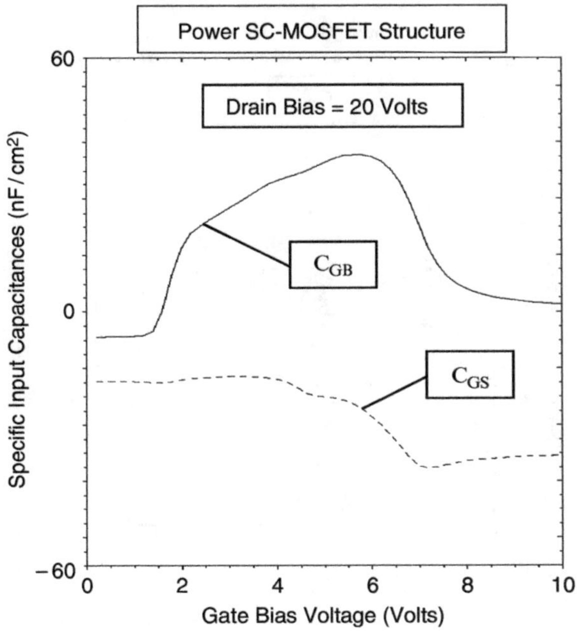


**Fig. 4.21** Input capacitances for the SC-MOSFET structure

voltage. The values obtained for the 30-V power SC-MOSFET structure are shown in Fig. 4.22. The gate-to-drain and base-to-drain capacitances are shown in the figure for comparison. Both of these capacitances decrease with increasing drain bias voltage until about 7 V and then fall abruptly as expected from the analytical model based up on the pinch-off of the JFET region. It can be observed from this figure that the reverse transfer (gate-drain) capacitance becomes extremely small in magnitude when compared to the output (base-drain) capacitance for the power SC-MOSFET structure. This implies a greatly reduced feedback path between the drain and the gate electrodes which is beneficial to the switching speed and power loss for the power SC-MOSFET structure. The values for the reverse transfer and output capacitances obtained by using the analytical models are in good agreement with the simulation values. The reverse transfer capacitance for the power SC-MOSFET structure is also shown in Fig. 4.23 to provide a more detailed view of its small values at high drain bias voltages.

The extremely small reverse transfer capacitance for the power SC-MOSFET structure has been experimentally observed in the SSCFET products [4]. The datasheets also show the abrupt reduction of the reverse transfer and output capacitances after the drain bias increases beyond the pinch-off of the JFET region. The ten-times smaller ratio of the reverse transfer to the input capacitance for the SC-MOSFET structure when compared with the D-MOSFET and U-MOSFET structures provides strong immunity to [dV/dt] induced turn-on when the device is operated in VRM applications.

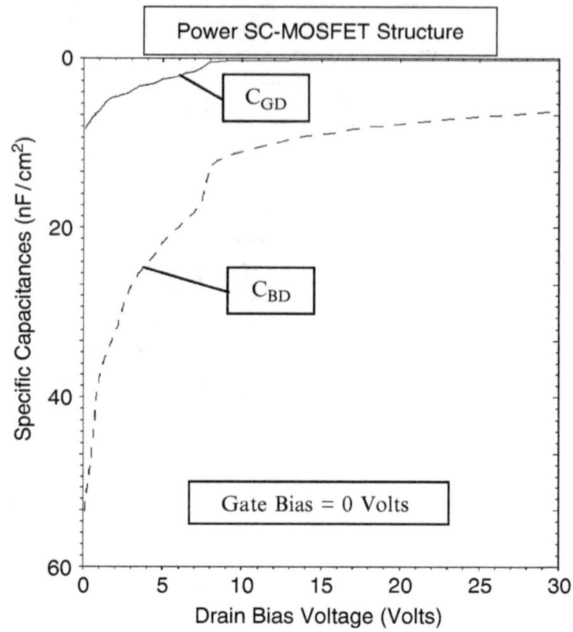


**Fig. 4.22** Reverse transfer and output capacitances for the SC-MOSFET structure

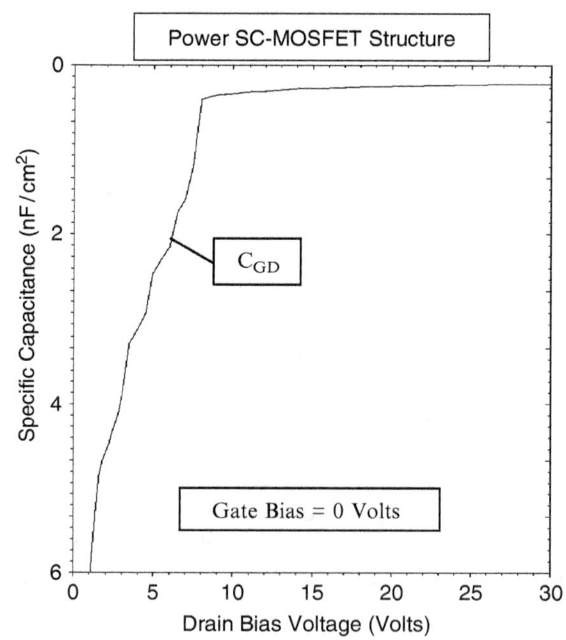


**Fig. 4.23** Reverse transfer capacitance for the SC-MOSFET structure

## 4.6 Gate Charge

It is standard practice in the industry to provide the gate charge for power MOSFET structures as a measure of their switching performance. The gate charge can be extracted by the application of a constant current source at the gate terminal while turning-on the power MOSFET structure from the blocking state. The linearized current and voltage waveforms observed during the turn-on process are illustrated in Fig. 6.98 in the textbook [3]. The various components for the gate charge are also defined in this figure.

The gate charge components for the power SC-MOSFET structure are given by similar equations. However, adjustments must be made because the gate transfer capacitance for this device structure is close to zero during the transition from a drain bias of $V_{DS}$ to the pinch-off voltage ($V_{P,JFET}$) for the JFET region. Due to the nearly zero gate transfer capacitance, an instantaneous drop in drain voltage from $V_{DS}$ to $V_{P,JFET}$ can occur at the beginning of the gate plateau phase. After this, the gate transfer capacitance is given by (4.16) including the impact of the screening factor ($K_S$). The gate transfer charge corresponding to the transition where the drain voltage changes from $V_{P,JFET}$ to $V_{ON}$ is given by:

$$Q_{GD} = \frac{2K_G q \varepsilon_S N_{DJ}}{C_{OX}} \left[ \sqrt{1 + \frac{2V_{P,JFET} C_{OX}^2}{q \varepsilon_S N_{DJ}}} - \sqrt{1 + \frac{2V_{ON} C_{OX}^2}{q \varepsilon_S N_{DJ}}} \right] \qquad (4.26)$$

The other components of the gate charge are similar to those already provided in the textbook:

$$Q_{SW} = Q_{GS2} + Q_{GD} \qquad (4.27)$$

$$Q_G = [C_{GS} + C_{GD}(V_{DS})]V_{GP} + Q_{GD} + [C_{GS} + C_{GD}(V_{ON})](V_G - V_{GP}) \qquad (4.28)$$

The gate charge values obtained for the 30-V power SC-MOSFET structure with source electrode in the trenches by using the above analytical equations are: $Q_{GD} = 90.4$ nC/cm$^2$; $Q_{SW} = 92$ nC/cm$^2$; and $Q_G = 333$ nC/cm$^2$. The gate transfer charge for the power SC-MOSFET structure is much smaller than that for the power MOSFET structures discussed in the previous chapters. It can be concluded that the power SC-MOSFET structure is well suited for high frequency applications where both the specific on-resistance and switching losses must be optimized.

Equations for the gate voltage, drain current, and drain voltage waveforms obtained by using the analytical model are provided in the textbook [3]. However, the drain voltage waveform for the power SC-MOSFET structure must be reformulated because the gradual transition occurs from the pinch-off voltage for the JFET region to the on-state voltage drop. Due to the very small reverse transfer capacitance for the power SC-MOSFET structure at larger drain bias voltages, the drain voltage first reduces abruptly from $V_{DS}$ to $V_{P,JFET}$. After this, the drain voltage is determined by the gate transfer capacitance given by (4.16):

$$v_D(t) = \frac{q\varepsilon_S N_{DJ}}{2C_{OX}^2}\left\{\left[\sqrt{1+\frac{2V_{P,JFET}C_{Ox}^2}{q\varepsilon_S N_{DJ}}} - \frac{J_G C_{OX}(t-t_2)}{2K_G q\ \varepsilon_S N_{DJ}}\right]^2 - 1\right\} \tag{4.29}$$

from $t = t_2$ to $t = t_3$.

The waveforms obtained for the 30-V power SC-MOSFET structure – using 3-μm cell pitch, gate width of 1 μm, and a gate oxide thickness of 500 Å – using these equations are provided in Fig. 4.24. A gate drive current density of 0.67 A/cm$^2$ was

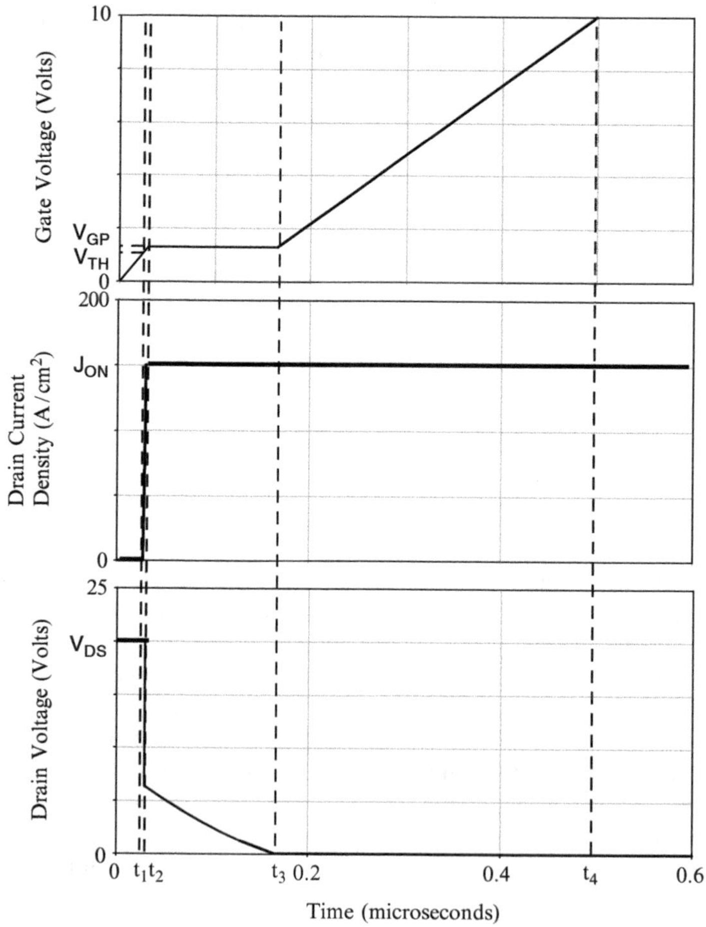


**Fig. 4.24** Analytically computed waveforms for the 30-V power SC-MOSFET structure

used to turn on the device from a steady-state blocking voltage of 20 V to match the results of two dimensional numerical simulations discussed below. The specific input capacitance was assumed to be 15 nF/cm$^2$.

The gate voltage initially increases linearly with time. After reaching the threshold voltage, the drain current can be observed to increase very quickly because of the large transconductance for this device structure. The drain current density increases until it reaches an on-state current density of 150 A/cm$^2$. The on-state current density determines the gate plateau voltage which has a value of 1.31 V for a threshold voltage of 1.2 V at this drain bias. During the gate voltage plateau phase, the drain voltage drops abruptly from the drain supply voltage of 20 V to the JFET region depletion voltage of 6.4 V. The drain voltage then decreases in a non-linear manner until it reaches the on-state voltage drop. After this time, the gate voltage again increases but at a slower rate than during the initial turn-on phase due to the larger gate transfer capacitance.

### *4.6.1 Simulation Example*

The gate charge for the 30-V power SC-MOSFET structure was extracted by using the results of two-dimensional numerical simulations of the cell described in the previous sections. The device was turned-on from blocking state with a drain bias of 20 V by driving it using a gate current of $1 \times 10^{-8}$ A/μm (equivalent to 0.67 A/cm$^2$ for the area of $1.5 \times 10^{-8}$ cm$^2$). Once the drain current density reached 150 A/cm$^2$, the drain current was held constant resulting in a reduction of the drain voltage. The gate plateau voltage for this drain current density was found to be 1.7 V. Once the drain voltage reached the on-state value corresponding to the gate plateau voltage, the gate voltage increased to the steady-state value of 10 V.

The gate charge waveforms obtained by using an input gate current density of 0.67 A/cm$^2$ when turning on the power SC-MOSFET structure from a blocking state with drain bias of 20 V are shown in Fig. 4.25. The on-state current density is 150 A/cm$^2$ at a DC gate bias of 10 V at the end of the turn-on transient. The gate voltage increases at a constant rate at the beginning of the turn-on process as predicted by the analytical model. When the gate voltage reaches the threshold voltage, the drain current begins to increase. The drain current increases very rapidly until it reaches the on-state current density of 150 A/cm$^2$.

Once the drain current reaches the on-state value, the gate voltage remains approximately constant at the plateau voltage ($V_{GP}$). The plateau voltage for this structure is 1.7 V for the drain current density of 150 A/cm$^2$ as governed by the transconductance of the device. The drain voltage drops abruptly from the supply voltage of 20 V of about 6.5 V (the JFET pinch-off voltage) as predicted by the analytical model. The drain voltage then decreases during the plateau phase in a non-linear manner. After the end of the plateau phase, the gate voltage again increases until it reaches the gate supply voltage. The waveforms obtained using

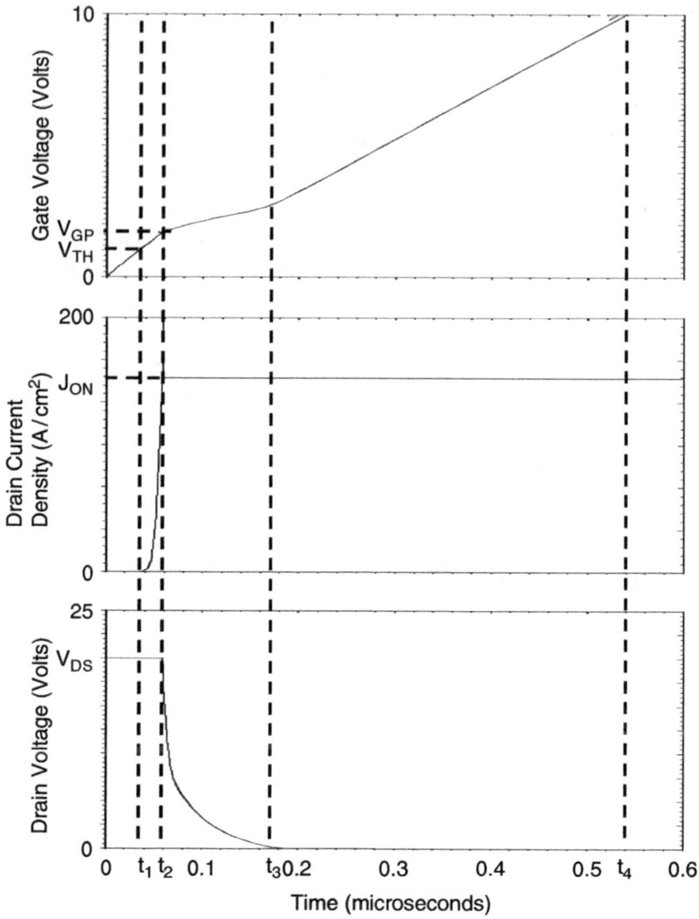


**Fig. 4.25** Turn-on waveforms for the 30-V power SC-MOSFET structure

the numerical simulations show a significant increase in the gate voltage during the plateau phase of the waveform. This indicates a reduction of the transconductance of the SC-MOSFET structure as the drain voltage is decreasing from 20 V to the on-state voltage drop. This change in transconductance with drain bias is evident in the output characteristics previously shown in Fig. 4.17.

The values for the various components of the gate charge extracted from the numerical simulations are compared with those calculated by using the analytical model in Fig. 4.26. There is excellent agreement between these values indicating that the analytical model is a good representation of the physics of turn-on for power SC-MOSFET structure. It can be concluded from these gate charge values that the SC-MOSFET structure has much superior switching performance when compared with the D-MOSFET and U-MOSFET structures.

| Specific Gate Charge | Numerical Simulation ($nC/cm^2$) | Analytical Model ($nC/cm^2$) |
|---|---|---|
| $Q_{GS1}$ | 23 | 18.1 |
| $Q_{GS2}$ | 16 | 1.6 |
| $Q_{GS}$ | 39 | 19.8 |
| $Q_{GD}$ | 73 | 90.4 |
| $Q_{SW}$ | 89 | 92.0 |
| $Q_G$ | 360 | 332.7 |

**Fig. 4.26** Gate charge extracted from numerical simulations for the power SC-MOSFET structure

## 4.7 Device Figures of Merit

Significant power switching losses can arise from the charging and discharging of the large input capacitance in power MOSFET devices at high frequencies. The input capacitance ($C_{IN}$) of the power MOSFET structure must be charged to the gate supply voltage ($V_{GS}$) when turning on the device and then discharged to 0 V when turning off the device during each period of the operating cycle. The total power loss can be obtained by summing the on-state power dissipation for a duty cycle $\delta = t_{ON}/T$ and the switching power losses:

$$P_T = P_{ON} + P_{SW} = \delta R_{ON} I_{ON}^2 + C_{IN} V_{GS}^2 f \tag{4.30}$$

where $R_{ON}$ is the on-resistance of the power MOSFET structure, $I_{ON}$ is the on-state current, and f is the operating frequency. In writing this equation, the switching power losses due to the drain current and voltage transitions has been neglected. A minimum total power loss occurs for each power MOSFET structure at an optimum active area as shown in the textbook [3]. The on-state and switching power losses are equal at the optimum active area. The optimum active area at which the power dissipation is minimized is given by:

$$A_{OPT} = \sqrt{\frac{R_{ON,sp}}{C_{IN,sp}}} \left(\frac{I_{ON}}{V_{GS}}\right) \left(\sqrt{\frac{\delta}{f}}\right) \tag{4.31}$$

From the first term in this expression, a useful technology figure-of-merit can be defined:

$$FOM(A) = \frac{R_{ON,sp}}{C_{IN,sp}} \tag{4.32}$$

In the power electronics community, there is trend towards increasing the operating frequency for switch mode power supplies in order to reduce the size and weight of the magnetic components. The ability to migrate to higher operating frequencies in power conversion circuits is dependent on making enhancements to the power MOSFET technology. From the above equations, an expression for the minimum total power dissipation can be obtained [3]:

$$P_T(\min) = 2I_{ON}V_{GS}\sqrt{\delta R_{ON,sp}C_{IN,sp}f} \tag{4.33}$$

A second technology figure of merit related to the minimum power dissipation can be defined as:

$$FOM(B) = R_{ON,sp}C_{IN,sp} \tag{4.34}$$

In most applications for power MOSFET structures with high operating frequency, the switching losses associated with the drain current and voltage transitions become a dominant portion of the total power loss. The time period associated with the increase of the drain current and decrease of the drain voltage is determined by the charging of the device capacitances. It is therefore common practice in the industry to use the following figures-of-merit to compare the performance of power MOSFET products [3]:

$$FOM(C) = R_{ON,sp}Q_{GD,sp} \tag{4.35}$$

and

$$FOM(D) = R_{ON,sp}Q_{SW,sp} \tag{4.36}$$

Although FOM(D) encompasses both the drain current and voltage transitions, it is customary to use FOM(C) because the gate-drain charge tends to dominate in the switching gate charge. One advantage of using these expressions is that the figure-of-merit becomes independent of the active area of the power MOSFET device.

The figures of merit computed for the power SC-MOSFET structure discussed in earlier sections of this chapter are provided in Fig. 4.27. The figure of merit usually

| Figures of Merit | $V_G$ = 4.5 V | $V_G$ = 10 V |
|---|---|---|
| FOM(A) ($\Omega^2$cm$^4$s$^{-1}$) | 18,080 | 15,500 |
| FOM(B) (ps) | 3.06 | 2.63 |
| FOM(C) (mΩ*nC) | 21.2 | 18.2 |
| FOM(D) (mΩ*nC) | 21.5 | 18.5 |

**Fig. 4.27** Figures of merit for the power SC-MOSFET structure

used for comparison of device technologies in the literature is FOM(C). Most often, the value for this figure of merit at a gate bias of 4.5 V is utilized for selection of devices in the voltage regulator module application. In comparison with the power D-MOSFET structure (see Chap. 2), the power SC-MOSFET structure has a FOM (C) that is 14-times smaller. In comparison with the conventional power U-MOSFET structure (see Chap. 3), the power SC-MOSFET structure has a FOM(C) that is 5.7-times smaller. In comparison with the power U-MOSFET structure with thicker oxide at the trench bottom surface (see Chap. 3), the power SC-MOSFET structure has a FOM(C) that is 4.8-times smaller. Consequently, the power SC-MOSFET structure offers significant improvement in circuit performance while utilizing a simple planar fabrication process.

## 4.8 Discussion

The physics of operation and resulting electrical characteristics of the power SC-MOSFET structure have been described in this chapter. The specific on-resistance for this device structure is significantly reduced when compared with the power D-MOSFET structure due to its very short (shielded) channel length, enhanced doping concentration in the JFET region using a retrograde doping profile, and small cell pitch. The specific on-resistance for the power SC-MOSFET structure is comparable to that of the power U-MOSFET structure. In addition, the reverse transfer capacitance and gate charge for the power SC-MOSFET structure are much smaller than those for the power D-MOSFET and U-MOSFET structures. Consequently, the figure-of-merit FOM(C) for the power SC-MOSFET structure is superior to that for the power D-MOSFET and U-MOSFET structures. The commercially developed devices [4] have been found to exhibit the characteristics described in this chapter using analytical models and numerical simulations. They have been found to improve the efficiency of Voltage-Regulator-Modules by over 5% especially under higher operating frequencies.

For purposes of comparison with the power MOSFET structures discussed in subsequent chapters, the analysis of the power SC-MOSFET structure is provided here for a broad range of blocking voltages. In this analysis, the power SC-MOSFET structure was assumed to have the following parameters: (a) $N^+$ source junction depth of 0.075 μm; (b) P-base junction depth of 0.235 μm; (c) $P^+$ region junction depth of 0.6 μm; (d) gate oxide thickness of 500 Å; (e) polysilicon window width of 2 μm; (f) JFET region doping concentration of $1.3 \times 10^{17}/\text{cm}^3$; (g) threshold voltage of 2 V; (h) gate drive voltage of 10 V; (i) inversion mobility of 450 $\text{cm}^2$/V s; (j) accumulation mobility of 1,000 $\text{cm}^2$/V s. The contributions from the contacts and the $N^+$ substrate were neglected during the analysis. The doping concentration and thickness of the drift region were determined under the assumption that the edge termination (in conjunction with the graded junction doping profiles) limits the breakdown voltage to 90% of the parallel-plane breakdown voltage. The device parameters pertinent to each blocking voltage are provided in

Fig. 4.28. It can be observed that the doping concentration must be greatly reduced with increasing blocking voltage capability while simultaneously increasing the thickness of the drift region.

The gate electrode width was optimized for each breakdown voltage to achieve the minimum specific on-resistance in the power SC-MOSFET structure. The specific on-resistance obtained by using the analytical model for the power SC-MOSFET structure as a function of the width of the gate electrode is shown in Fig. 4.29 for the case of various breakdown voltages. Note that a logarithmic scale has been used for the vertical axis because of the large range of values for the specific on-resistance. The two zone model (see Fig. 6.41 in the textbook) was used

| Blocking Voltage (V) | Drift Doping Concentration ($cm^{-3}$) | Drift Region Thickness (microns) |
|---|---|---|
| 30 | $1.6 \times 10^{16}$ | 3 |
| 60 | $6.0 \times 10^{15}$ | 5 |
| 100 | $3.0 \times 10^{15}$ | 8 |
| 200 | $1.2 \times 10^{15}$ | 17 |
| 300 | $6.8 \times 10^{14}$ | 26 |
| 600 | $2.7 \times 10^{14}$ | 58 |
| 1000 | $1.35 \times 10^{14}$ | 105 |

**Fig. 4.28** Device parameters for the power SC-MOSFET structures

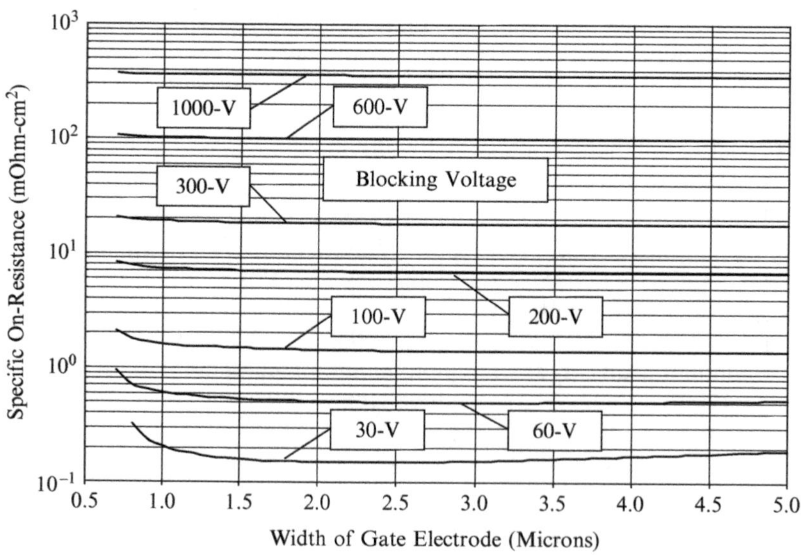


**Fig. 4.29** Optimization of the specific on-resistance for the power SC-MOSFET structures

| Blocking Voltage (V) | Optimum Gate Width (microns) | Specific On-Resistance (mΩ-cm$^2$) |
|---|---|---|
| 30 | 2.2 | 0.149 |
| 60 | 2.8 | 0.503 |
| 100 | 3.5 | 1.40 |
| 200 | 4.8 | 6.83 |
| 300 | 5.9 | 18.0 |
| 600 | 8.2 | 99.6 |
| 1000 | 10.4 | 359 |

**Fig. 4.30** Optimized specific on-resistance for the power SC-MOSFET structures

for all of the power SC-MOSFET structures because of the small width of the polysilicon window. From Fig. 4.29, it can be observed that the minimum specific on-resistance occurs at a larger optimum gate electrode width for the power SC-MOSFET structures with larger blocking voltages.

The optimum gate electrode width and minimum specific on-resistances for the power SC-MOSFET structures are provided in Fig. 4.30. The optimum gate electrode width for the power SC-MOSFET structures is much smaller than that for the power D-MOSFET structure due to the unique retrograde JFET region doping profile with much larger doping concentration. However, a small gate width (1 μm) is preferable for reducing the gate transfer charge. The specific on-resistance values for this case can be compared with the ideal specific on-resistance obtained by using Baliga's power law for the impact ionization coefficients in Fig. 4.31. From this figure, it can be concluded that the specific on-resistance for the SC-MOSFET structure is always larger than the ideal specific on-resistance. For blocking voltages below 60 V, the specific on-resistance for the power SC-MOSFET structure becomes substantially larger than the ideal case.

The specific gate transfer charge for the power SC-MOSFET structures was obtained by using the analytical model (see 4.26). During this analysis, the devices were assumed to be operated at an on-state current density that results in a power dissipation of 100 W/cm$^2$ as determined by the specific on-resistance for each device. The on-state current density values are provided in Fig. 4.32. The drain supply voltage used for this analysis was chosen to be two-thirds of the breakdown voltage. The drain supply voltage values are also provided in Fig. 4.32.

The specific gate transfer charge values calculated by using the analytical model for the power SC-MOSFET structure are given in Fig. 4.32 as a function of the breakdown voltage. These values are also plotted in Fig. 4.33 as a function of the breakdown voltage. It can be observed that the gate transfer charge decreases gradually with increasing breakdown voltage until a breakdown voltage of 300 V. The specific gate transfer charge then decreases drastically at higher breakdown voltages. This behavior occurs because the on-state voltage drop for the power SC-MOSFET begins to approach the pinch-off voltage for the JFET region as the

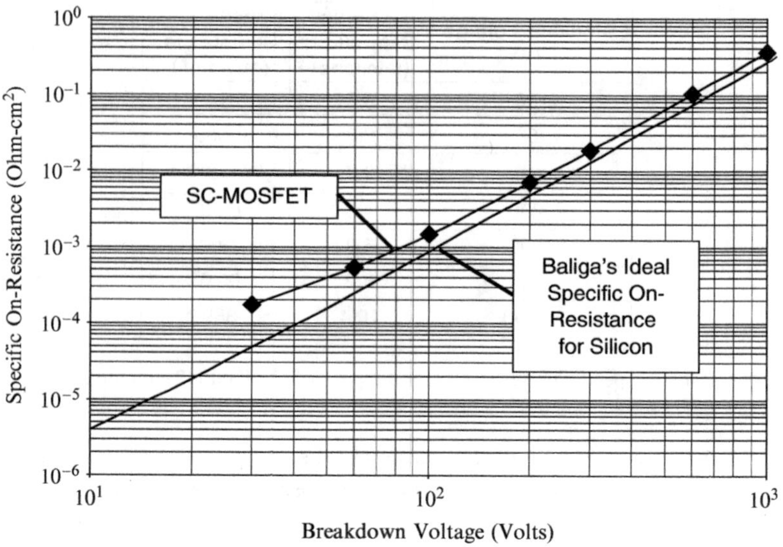


**Fig. 4.31** Specific on-resistance for the power SC-MOSFET structures

| Blocking Voltage (V) | Drain Supply Voltage (Volts) | On-State Current Density (A/cm$^2$) | Specific Gate Transfer Charge (nC/cm$^2$) |
|---|---|---|---|
| 30 | 20 | 697 | 88 |
| 60 | 40 | 407 | 86 |
| 100 | 67 | 249 | 83 |
| 200 | 133 | 116 | 74 |
| 300 | 200 | 72.5 | 65 |
| 600 | 400 | 31.3 | 38 |
| 1000 | 667 | 16.6 | 3.8 |

**Fig. 4.32** Parameters used for gate transfer charge analysis for the power SC-MOSFET structures

breakdown voltage is increased above 300 V. No power law fit is appropriate for this behavior.

The figure-of merit (C) – product of the specific on-resistance and the specific gate transfer charge – for the optimized power SC-MOSFET structures was obtained by determining the gate transfer charge for the cell with a gate width of 1 μm using the analytical model (see 4.26). During this analysis, the devices were assumed to be operated at an on-state current density that results in a power dissipation of 100 W/cm$^2$ as determined by the specific on-resistance for each

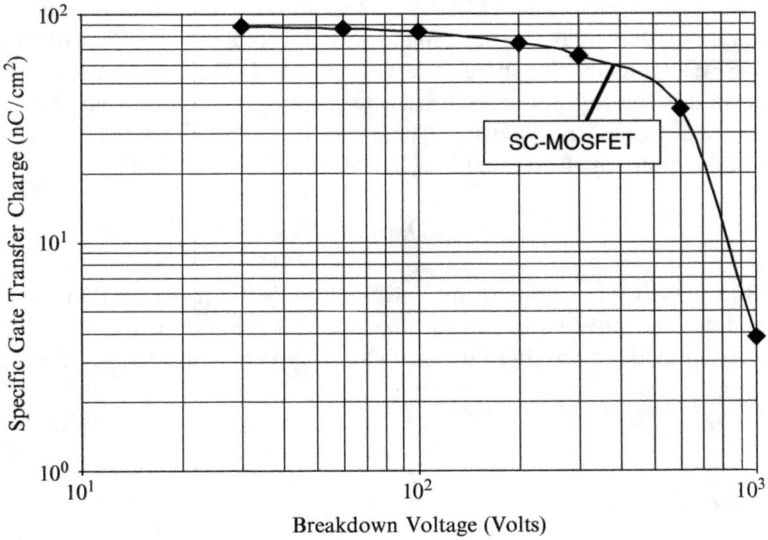


**Fig. 4.33** Specific gate transfer charge for the power SC-MOSFET structures

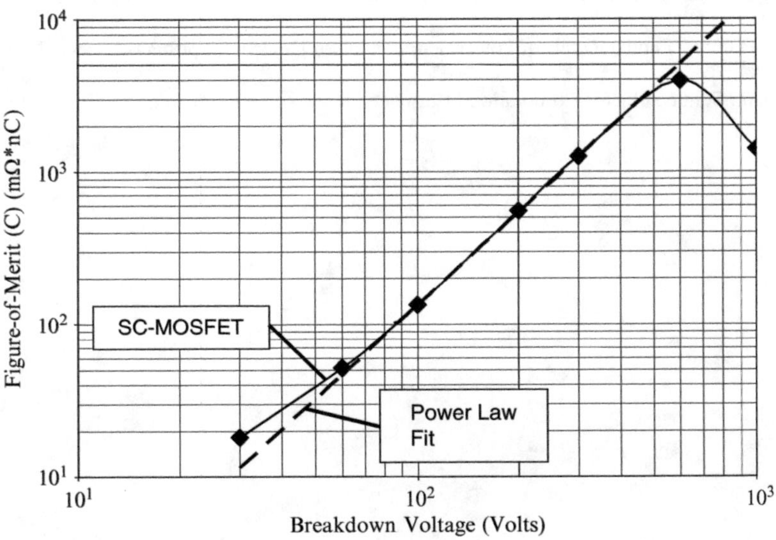


**Fig. 4.34** Figure-of-merit (C) for the power SC-MOSFET structures

device. A gate width of 1 μm was used for the power SC-MOSFET structure, rather than the optimum gate width for achieving minimum specific on-resistance, because the device was developed for applications such as VRMs where low gate charge (and consequently low FOM(C)) is required. The resulting values for the FOM(C) are plotted in Fig. 4.34 as a function of the breakdown voltage of the power SC-MOSFET structure. It can be observed that the FOM(C) increases in a linear fashion on this log-log graph for breakdown voltages ranging from 60 to 600 V indicating a power law relationship. The power law relationship that fits the data is shown in the figure by the dashed line. The equation for this line is:

$$\mathrm{FOM(C)} = R_{\mathrm{ON,sp}} Q_{\mathrm{GD,sp}} = 0.01152\ \mathrm{BV}^{2.032} \tag{4.37}$$

where the FOM(C) has units of m$\Omega$ nC. The FOM(C) has a smaller value than predicted by the power law fit for the 1,000-V device because the on-state voltage drop at the operating current density is very close to the JFET pinch-off voltage.

## References

1. B.J. Baliga and D. Girdhar, "Paradigm Shift in Planar Power MOSFET Technology", Power Electronics Technology Magazine, pp. 24–32, November 2003.
2. B.J. Baliga, "Power Semiconductor Devices having Laterally Extending Base Shielding Regions that Inhibit Base Reach-Through", U.S. Patent # 6,791,143, Issued September 14, 2004.
3. B.J. Baliga, "Fundamentals of Power Semiconductor Devices", Springer-Science, New York, 2008.
4. Silicon Semiconductor Corporation Datasheets, www.siliconsemi.com.

# Chapter 5
# CC-MOSFET Structure

The power MOSFET structures discussed in the previous chapters utilize a one-dimensional junction for supporting the drain voltage when operating in the blocking mode. As discussed and derived in Chap. 1, the smallest specific on-resistance that can be achieved in these devices is limited to the ideal specific on-resistance, which is given by:

$$R_{on-ideal} = \frac{4BV^2}{\varepsilon_S\ \mu_n E_C^3} \tag{5.1}$$

A significantly smaller specific on-resistance can be achieved by utilizing a two-dimensional charge coupling effect that alters the electric field distribution from the triangular shape in a one-dimensional case to a rectangular shape for the two-dimensional case.

Two approaches have been proposed and demonstrated for achieving the desired two-dimensional charge coupling within power MOSFET structures. The first approach [1], called the power CC-MOSFET structure, is based up on utilizing an electrode embedded within oxide coated deep trenches as a part of the drift region. This approach is discussed in this chapter because it is particularly well suited for the development of high performance power MOSFET structures with lower blocking voltage capability in the range of 30–200 V that have been the subject of the previous chapters. The second approach [2], called the power SJ-MOSFET structure, is based up on utilizing a vertical P-N junction in the drift region. This approach, generally known as a super-junction concept, will be discussed in a subsequent chapter because it is particularly well suited for the development of high performance power MOSFET structures with larger blocking voltage capability in the range of 300–1,200 V.

For the power CC-MOSFET structure, there are two options for achieving the desired charge-coupling during operation in the blocking mode. In the first case, the charge coupling electrode within the oxide coated trenches is connected to the gate electrode. Although this method achieves the desired reduction of the drift region

B.J. Baliga, *Advanced Power MOSFET Concepts*,
DOI 10.1007/978-1-4419-5917-1_5, © Springer Science+Business Media, LLC 2010

resistance, it creates a device with large input and gate transfer capacitances. Consequently, an alternative power CC-MOSFET structure [3] has been proposed with the charge coupling electrode within the oxide coated trenches connected to the source electrode. Both of these power CC-MOSFET structures are analyzed in this chapter. A further enhancement of the performance can be achieved by utilizing a graded doping profile in the drift region to homogenize the electric field. The performance of these GD-MOSFET structures is discussed in the next chapter.

Apart from the disclosure of the power CCMOSFET concept in the patents referenced above, the device structure and its superior specific on-resistance was first discussed in the archival literature in 1997 [4]. Subsequently, several papers have been published on the same concept using alternate names – Oxide-Bypassed VDMOS (OBVDMOS) structure [5], Resurf Stepped Oxide (RSO) MOSFET structure [6, 7] , Split-gate Resurf Stepped Oxide (RSO) MOSFET structure [8], and Vertical LOCOS MOSFET (VLMOS) structure [9].

## 5.1 The CC-MOSFET Structure

A cross-section of the basic cell structure for the power (charge-coupled) CC-MOSFET structure is illustrated in Fig. 5.1. This device structure can be fabricated by starting with an N-type epitaxial layer grown on a heavily doped $N^+$ substrate. The P-base region is then formed across the active area of the device by ion implantation of boron followed by a drive-in cycle. The $N^+$ source regions are then produced by ion implantation of phosphorus followed by its annealing process. A mask is required during the ion-implantation of the $N^+$ source regions to create the short between the $N^+$ source and P-base regions at the top of the mesa region. A deep trench is next formed by reactive ion etching using a mask aligned to the $N^+$ source mask to center it in relation to the shorts. The depth of the trench must be

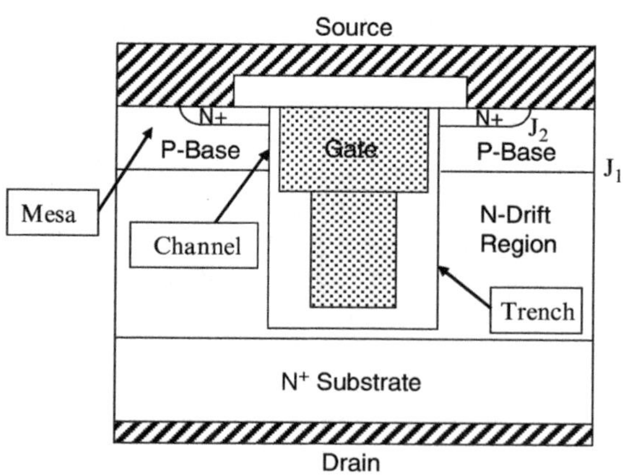


**Fig. 5.1** The CC-MOSFET structure

tailored to achieve the desired breakdown voltage. The surface of the trench must be smooth and free of damage in order to obtain a good MOS interface with high channel inversion layer mobility. A thick oxide is grown on the trench surface by thermal oxidation. The oxide thickness must be sufficient to provide the desired charge coupling as well as to support the entire drain blocking voltage.

The trench is then refilled with highly doped N-type polysilicon. The polysilicon is then etched until it is recessed below the surface to a depth slightly below the depth of the P-base region. The exposed oxide is then partially etched to leave the gate oxide on the side-walls of the trenches. This is followed by the deposition of a second highly-doped N-type polysilicon layer to refill the trench. An inter-metal dielectric is then deposited followed by etching contact windows for the $N^+$ source and P-base regions. The device fabrication is then completed by the deposition of the source metal layer. (Alternate device fabrication processes [9] based up on selective oxidation using nitride layers have also been proposed and demonstrated for this structure.)

The physics of operation of the power CC-MOSFET structure in the blocking mode is different from that for the power MOSFET structures described in the previous chapters. Without the application of a gate bias, a high voltage can be supported in the CC-MOSFET structure when a positive bias is applied to the drain. In this case, junction $J_1$ formed between the P-base region and the N-drift region becomes reverse biased. Simultaneously, the drain voltage is applied across the MOS structure formed between the electrode in the deep trenches and the N-drift region. The MOS structure operates in the deep-depletion mode due to the presence of the reverse bias across junction $J_1$ between the P-base region and the N-drift junction. Consequently, depletion regions are formed across the horizontal junction $J_1$ and the vertical trench sidewall. This two-dimensional depletion alters the electric field distribution from the triangular shape observed in conventional parallel-plane junctions to an exponential shape. This allows supporting a required blocking voltage over a shorter distance. In addition, the doping concentration in the drift region can be made much greater that predicted by the one-dimensional theory. This allows very substantial reduction of the specific on-resistance to well below the ideal specific on-resistance at any desired breakdown voltage.

Drain current flow in the CC-MOSFET structure is induced by the application of a positive bias to the gate electrode. This produces an inversion layer at the surface of the P-base region along the trench sidewalls. The threshold voltage for the power CC-MOSFET structure can be controlled by adjusting the dose for the boron ion-implantation for the P-base region. The energy for the boron ion-implantation for the P-base region is preferably chosen to maintain its peak concentration below the surface of the silicon to suppress reach-through. The inversion layer channel provides a path for transport of electrons from the source to the drain when a positive drain voltage is applied.

After transport from the source region through the channel, the electrons enter the N-drift region and are then transported through the mesa region to the $N^+$ substrate. The resistance of the drift region is very low in the power CC-MOSFET structure due to the high doping concentration in the mesa region. In addition, an accumulation

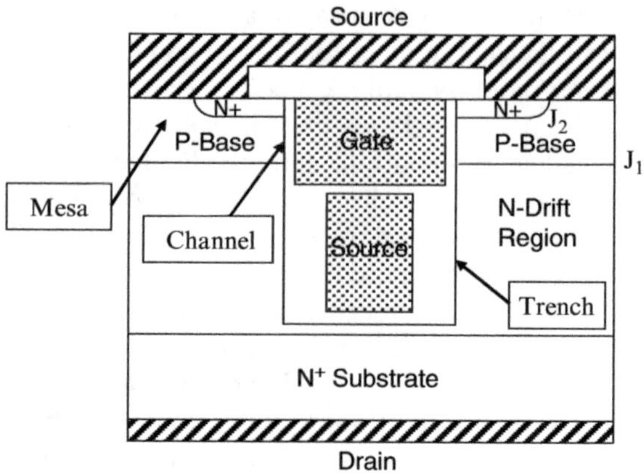


**Fig. 5.2** The CC-MOSFET structure with source electrode in trenches

layer forms on the sidewalls of the trench surface in the drift region which further reduces the resistance. The channel resistance in the power CC-MOSFET structure is also very small due to the small cell pitch or high channel density.

As discussed in earlier chapters, the input capacitance and gate charge for power MOSFET structures must be reduced to enhance their switching performance. The presence of the gate electrode within the deep trenches adjacent to the drift region increases the capacitance and gate charge degrading the switching performance of the power CC-MOSFET structure. This problem can be overcome by using a source connected electrode in the trenches adjacent to the drift region as illustrated in Fig. 5.2.

The process for the fabrication of the improved CC-MOSFET structure with source electrode in the trenches is similar to that for the basic CC-MOSFET structure. After the etching the first polysilicon refill to a depth below the P-base region, all the exposed oxide in the trenches is removed. A fresh oxide is then grown by thermal oxidation on the trench sidewalls to form the gate oxide. A thicker oxide is simultaneously formed by the thermal oxidation on the top surface of the polysilicon inside the trenches. This serves as an isolation oxide between the gate electrode and the source electrode in the trenches. A second polysilicon layer is now deposited and planarized to serve as the gate electrode. An inter-metal dielectric is then deposited followed by etching contact windows for the $N^+$ source and P-base regions. The device fabrication is then completed by the deposition of the source metal layer.

## 5.2 Charge-Coupling Physics and Blocking Voltage

A much larger doping concentration can be employed in the drift region of the power CC-MOSFET structure when compared with the D-MOSFET, U-MOSFET or SC-MOSFET structures. This enables a drastic reduction of the specific on-resistance for

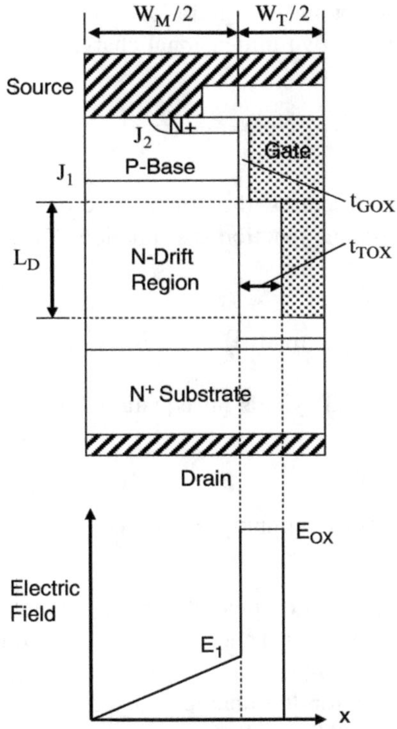


**Fig. 5.3** Electric field distribution in the CC-MOSFET structure

devices capable of supporting high voltages. The doping concentration in the drift region is determined by two-dimensional charge coupling for the power CC-MOSFET structure. In order to achieve good charge coupling, the mesa region must be completely depleted when the drain bias approaches the breakdown voltage. The electric field distribution within the power CC-MOSFET structure along the x-axis is illustrated in Fig. 5.3 under these conditions for the case of a uniformly doped drift region. The electric field varies linearly with distance in the semiconductor in accordance with the solution for Poisson's equation [10] while the electric field in the oxide is constant.

Based up on the triangular electric field distribution in the semiconductor:

$$E_1 = \frac{qN_D}{\varepsilon_S} \frac{W_M}{2} \tag{5.2}$$

where $W_M$ is the mesa width, and $N_D$ is the doping concentration in the mesa region. When the drain bias reaches the breakdown voltage, the maximum electric field in the semiconductor ($E_1$) becomes equal to the critical electric field ($E_C$) for breakdown of the semiconductor. This provides the criterion for choosing the dopant dose (product of doping concentration and mesa thickness) in the mesa region to achieve the desired two-dimensional charge coupling:

$$N_D W_M = \frac{2E_C \varepsilon_S}{q} \tag{5.3}$$

In the case of silicon structures, the critical electric field for breakdown for a one-dimensional parallel-plane junction is a function of doping level (see Chap. 1) as given by:

$$E_{C,1D,B}(\mathrm{Si}) = 3,700 N_D^{1/8} \tag{5.4}$$

Using this expression in (5.3) yields an equation for the optimum dose for the charge coupling:

$$N_D(\mathrm{Optimum}) = \left(\frac{7,400\varepsilon_S}{qW_M}\right)^{8/7} \tag{5.5}$$

The optimum doping concentration predicted by this equation is shown in Fig. 5.4 as a function of mesa width. For a mesa width of 0.5 μm, the optimum doping concentration for charge-coupling is $1.3 \times 10^{17}/\mathrm{cm}^3$.

The breakdown voltage for the charge-coupled device is determined by the length ($L_D$) of the drift region. Since the two-dimensional charge coupling produces

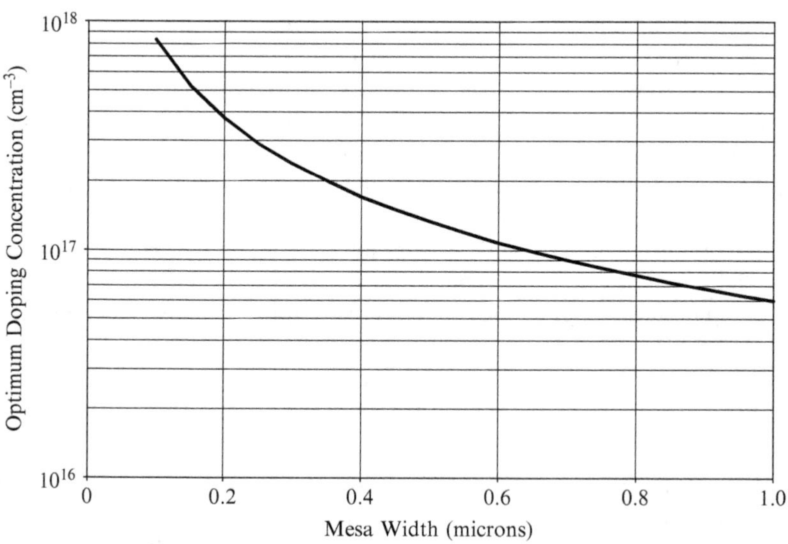


**Fig. 5.4** Optimum doping concentration for two-dimensional charge-coupling

an approximately uniform electric field in the y-direction, the breakdown voltage is given by:

$$BV = E_{C,1D,B} L_D \tag{5.6}$$

Using (5.4) for the critical electric field with (5.5) for the optimum doping concentration yields an expression for the breakdown voltage of charge coupled devices:

$$BV = 1.32 \times 10^4 L_D \left( \frac{\varepsilon_S}{q W_M} \right)^{1/7} \tag{5.7}$$

The breakdown voltage predicted by this equation is shown in Fig. 5.5 as a function of the length of the drift region with the mesa width as a parametric variable. It can be observed that the breakdown voltage increases with decreasing mesa width. This occurs because the doping concentration is larger for smaller mesa widths resulting in a slightly larger critical electric field for breakdown. It is worth pointing out that the breakdown voltage is a weak function of the mesa width. This is a favorable outcome from the processing viewpoint. For a drift length of 1.0 μm, the breakdown voltage for the charge-coupled device structure is predicted to be about 50 V for a mesa width of 0.5 μm.

The above model is based up on using the critical electric field derived in the textbook for the case of one-dimensional junctions where the electric field has a triangular profile. In the case of the devices with two-dimensional charge coupling, the electric field becomes approximately constant in the y-direction. The critical

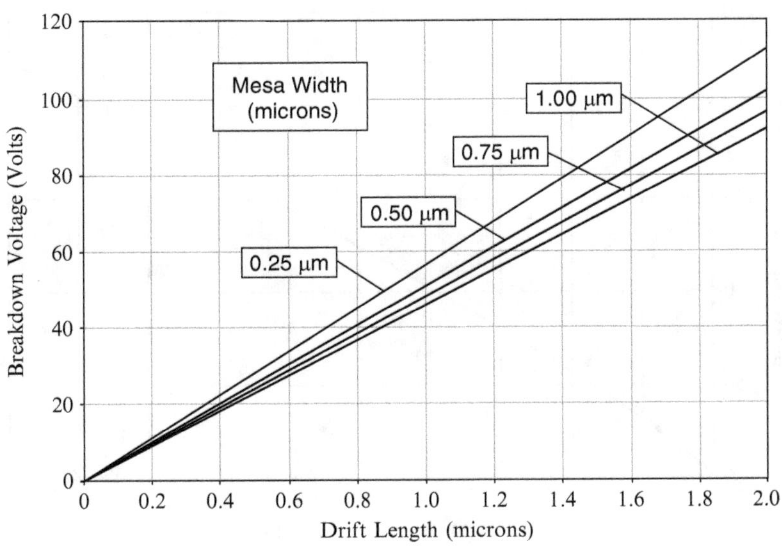


**Fig. 5.5** Breakdown voltages for two-dimensional charge-coupled devices

electric field at which breakdown occurs can be derived for this field distribution by using the criterion that the ionization integral becomes unity at breakdown:

$$\int_0^{L_D} \alpha dx = 1 \tag{5.8}$$

Using Baliga's formula for the impact ionization coefficient for silicon:

$$\alpha_B(Si) = 3.51 \times 10^{-35} E^7 \tag{5.9}$$

in (5.8), an expression for the critical electric field for the case of a uniform (or constant) electric field is obtained:

$$E_{CU} = 8.36 \times 10^4 L_D^{-1/7} \tag{5.10}$$

The breakdown voltage in this case is then given by:

$$BV_U = E_{CU} L_D = 8.36 \times 10^4 L_D^{6/7} \tag{5.11}$$

by using (5.10).

The breakdown voltage for devices with a uniform electric field can be observed in Fig. 5.6 to increase non-linearly with increasing length of the drift region. For comparison purposes, the dashed line in the figure is included to show a linear trend with a critical electric field of $3.33 \times 10^5$ V/cm independent of the length of the

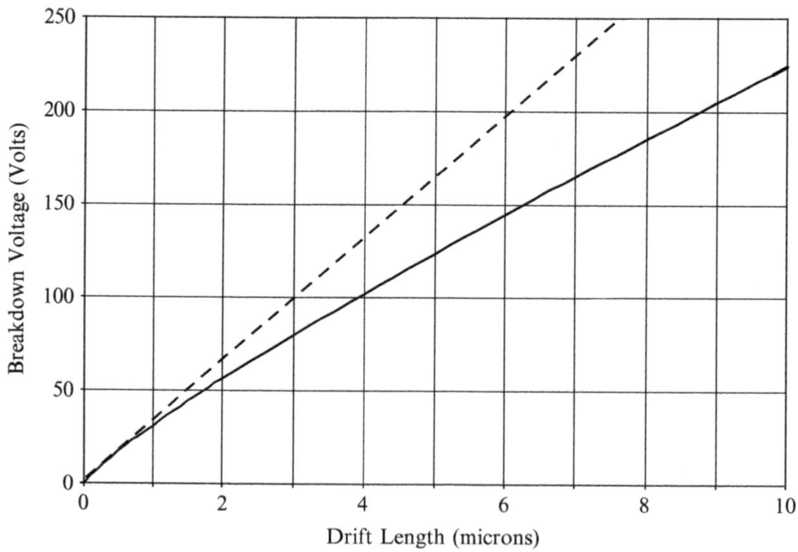


**Fig. 5.6** Breakdown voltages for the uniform electric field case

drift region. For a drift region length of 1 μm, the breakdown voltage predicted by this analytical model is 31.2 V. This value is considerably smaller than the 50 V predicted by the model for a mesa width of 0.5 μm based up on the critical electric field for the one-dimensional case.

The critical electric field for breakdown with uniform electric field can be related to the breakdown voltage by combining (5.10) and (5.11):

$$E_{CU} = 5.53 \times 10^5 BV^{-1/6} \tag{5.12}$$

In the case of the one-dimensional case with triangular electric field distribution, it has been demonstrated in Chap. 1 that the critical electric field for breakdown is given by:

$$E_{C,1D,B} = 3,700\ N_D^{1/8} \tag{5.13}$$

while the breakdown voltage is given by:

$$BV = 4.45 \times 10^{13} N_D^{-3/4} \tag{5.14}$$

Combining these relationships, the critical electric for breakdown for the one-dimensional case is given by:

$$E_{C,1D,B} = 6.97 \times 10^5 BV^{-1/6} \tag{5.15}$$

From these equations, it can be observed that the critical electric field expressions for the one- and two-dimensional cases have the same dependence on the breakdown voltage. The ratio of the critical electric field for the one-dimensional case to that for the two-dimensional case is 1.26× independent of the breakdown voltage. This difference is due to a high electric field over a longer path for impact ionization in the case of the two-dimensional charge-coupled case. The critical electric field for both cases is provided in the Fig. 5.7 as a function of the breakdown voltage.

The revised optimum dose for the dopant in the drift region for the two-dimensional charge coupled device is given by:

$$N_D W_M = \frac{2E_{CU}\varepsilon_S}{q} \tag{5.16}$$

Using (5.12):

$$N_D W_M = 1.106 \times 10^6 \frac{\varepsilon_S}{q} BV^{-1/6} \tag{5.17}$$

The optimum dose predicted by this expression is shown in Fig. 5.8 as a function of the breakdown voltage. It can be observed that the optimum dose decrease gradually with increasing breakdown voltage.

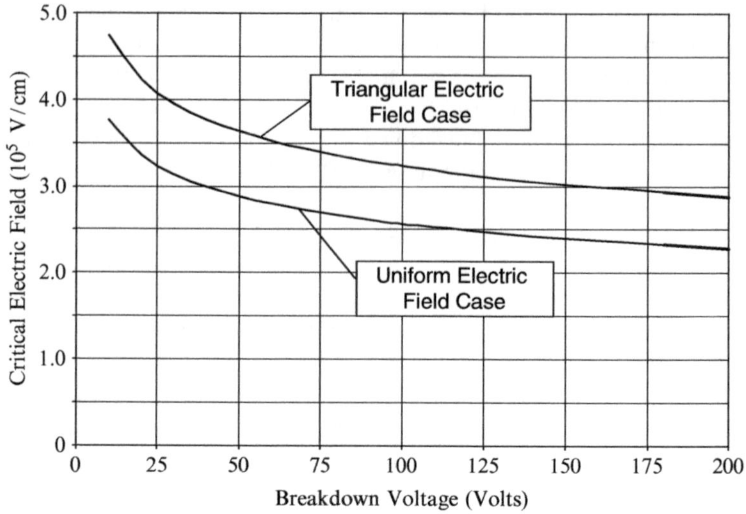


**Fig. 5.7** Critical electric field for breakdown for the triangular and uniform electric field distribution cases

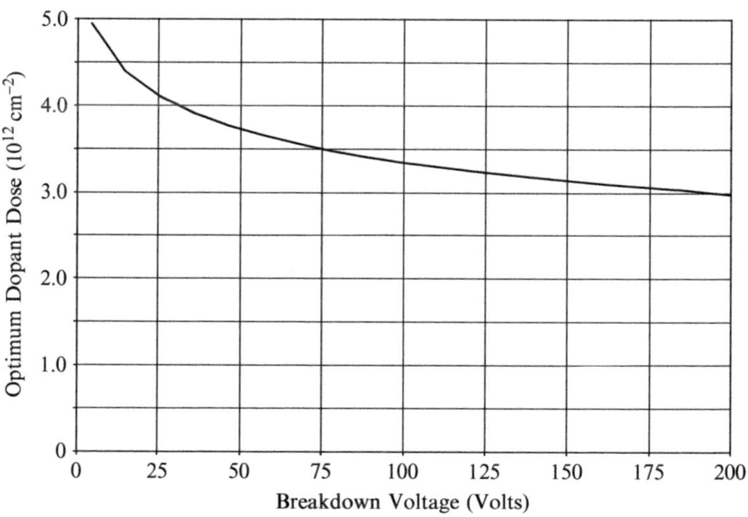


**Fig. 5.8** Optimum dose for two-dimensional charge-coupling

From (5.17), the revised optimum doping concentration for the two-dimensional charge coupled case is given by:

$$N_D = 1.106 \times 10^6 \frac{\varepsilon_S}{qW_M} BV^{-1/6} \tag{5.18}$$

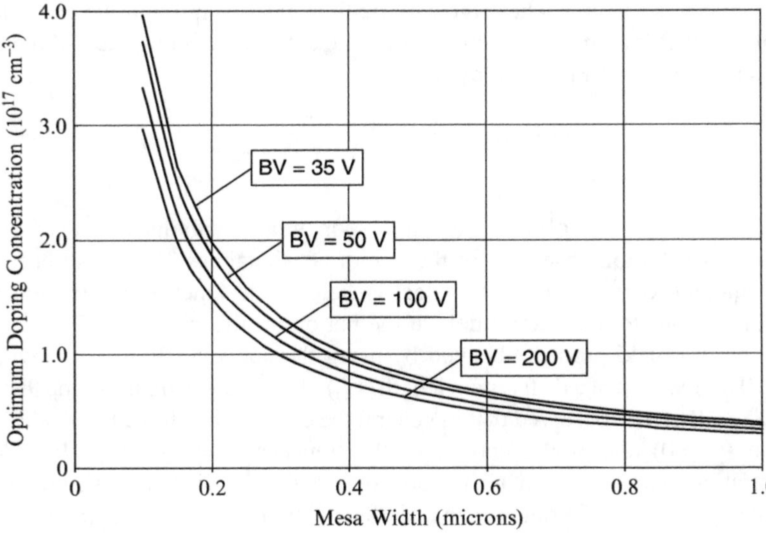


**Fig. 5.9** Optimum doping concentration for uniform electric field case

The optimum doping concentration predicted by this expression is shown in Fig. 5.9 as a function of the mesa width with the breakdown voltage as a parameter. It can be observed that the optimum dose decreases with increasing mesa width and is a weak function of the breakdown voltage. In comparison with the optimum doping concentrations obtained by using the critical electric field for the one-dimensional case (see Fig. 5.3), the optimum doping concentration for the two-dimensional charge-coupled case is a factor of about 1.5× smaller. For the case of a mesa width of 0.5 μm, the optimum doping concentration for a breakdown voltage of 35 V is found to be close to $8 \times 10^{16}/\text{cm}^3$ for the uniform electric field case.

For a deeper understanding of the charge-coupling physics, it is necessary to derive a relationship for the electric field profile in the two-dimensional structure [11]. Solving Poisson's equation with the gradual drift region approximation assuming uniform doping along the x-axis yields:

$$\frac{d^2V}{dy^2} - \frac{V}{\lambda^2} = -\frac{qN_D}{\varepsilon_S} \tag{5.19}$$

where

$$\lambda = \sqrt{\frac{W_M}{2}\left(\frac{W_M}{4} + \frac{\varepsilon_{Si}}{\varepsilon_{OX}} t_{TOX}\right)} \tag{5.20}$$

For the case of structures where the electric field extends through the entire length ($L_D$) of the drift region, the magnitude of the electric field along the y-direction in the middle of the mesa is given by:

$$E(y) = \frac{qN_D\lambda}{\varepsilon_S}e^{-y/\lambda} + \left(\frac{V_D}{\lambda} - \frac{qN_D\lambda}{\varepsilon_S}\right)e^{(y-L_D)/\lambda} \tag{5.21}$$

The electric field can be observed to vary exponentially with distance in contrast to the linear distribution observed for the one-dimensional case. The first term in the above equation corresponds to the electric field at the junction while the second term corresponds to the electric field at the bottom of the trenches.

The electric field profiles obtained by using the above analysis are provided in Fig. 5.10 for the case of a drift region length ($L_D$) of 1 μm and a trench oxide thickness of 1,000 Å. It can be observed that in general the electric field has a high value at the junction (y = 0) and at the bottom of the trenches (y = 1). For a low doping concentration ($0.8 \times 10^{17}$/cm$^3$) in the drift region, the electric field has a much larger value at the bottom of the trenches than at the junction. For a high doping concentration ($1.4 \times 10^{17}$/cm$^3$) in the drift region, the electric field has a much larger value at the junction when compared with bottom of the trenches. The electric field at the junction and the bottom of the trenches becomes nearly equal for an optimum doping concentration of $1.0 \times 10^{17}$/cm$^3$. This is consistent with the conclusions made earlier for the optimum doping concentration with the aid of Fig. 5.9 for a device with breakdown voltage of 35 V corresponding to a drift region length of 1 μm.

The electric field profile for the two-dimensional charge-coupled case is also dependent on the thickness of the oxide in the trenches. The decay length (λ) is a

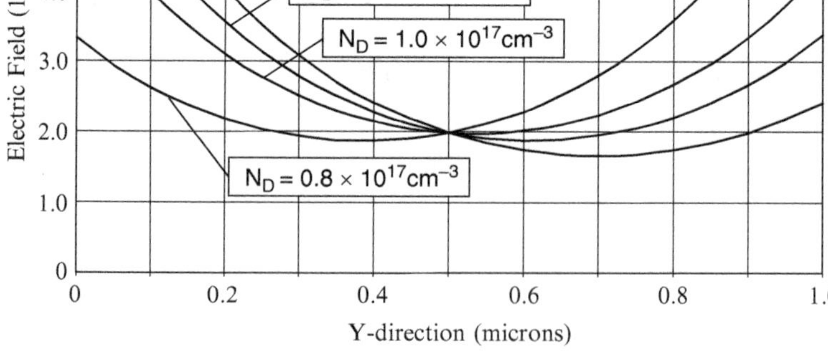


**Fig. 5.10** Electric field profile at the center of the mesa for the two-dimensional charge-coupling case with drift region length of 1 μm

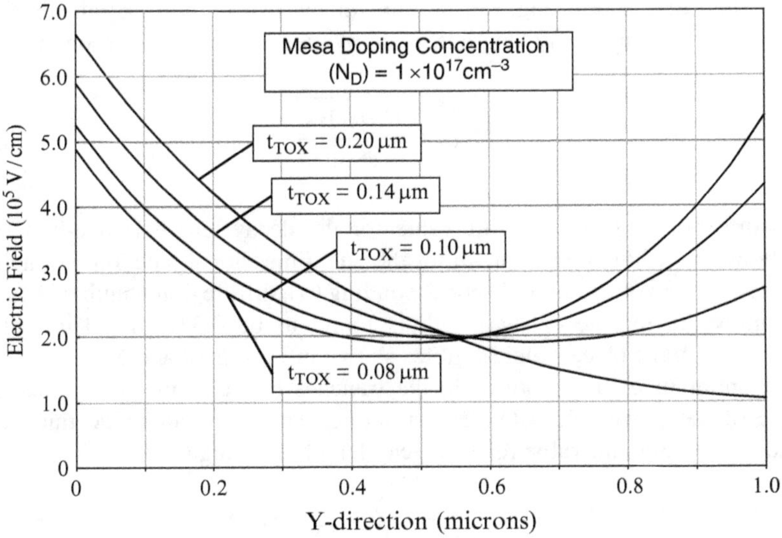


**Fig. 5.11** Electric field profile at the center of the mesa for the two-dimensional charge-coupling case with drift region length of 1 μm

function of the trench oxide thickness per (5.20). This alters the electric field distribution as illustrated in Fig. 5.11 for the case of a device with drift region length of 1 μm and doping concentration of $1.0 \times 10^{17}/\text{cm}^3$ in the mesa region. It can be observed that a high electric field occurs at the junction when the oxide thickness is large while a high electric field occurs at the bottom of the trenches when the oxide thickness is reduced. From this figure, it can be inferred that the optimum oxide thickness is 1,000 Å.

An expression for the optimum decay length (λ) can be derived under the criterion that the electric fields at the junction and at the bottom of the trench are equal at breakdown. The electric field at the junction is obtained from (5.21) with y = 0:

$$E(0) = \frac{qN_D\lambda}{\varepsilon_S} + \left(\frac{BV}{\lambda} - \frac{qN_D\lambda}{\varepsilon_S}\right)e^{-L_D/\lambda} \tag{5.22}$$

The electric field at the bottom of the trenches is obtained from (5.21) with y = $L_D$:

$$E(L_D) = \frac{qN_D\lambda}{\varepsilon_S}e^{-L_D/\lambda} + \left(\frac{BV}{\lambda} - \frac{qN_D\lambda}{\varepsilon_S}\right) \tag{5.23}$$

Equating these electric fields yields:

$$\left[e^{-L_D/\lambda} - 1\right]\left(\frac{BV}{\lambda} - \frac{2qN_D\lambda}{\varepsilon_S}\right) = 0 \tag{5.24}$$

The solution of a decay length of infinity for the first term is unrealistic. Consequently, the optimum decay length is given by the solution of the second term:

$$\lambda_{OPT} = \sqrt{\frac{\varepsilon_S BV}{2qN_D}} \tag{5.25}$$

It is worth pointing out that this value for the decay length is one-half of the depletion layer width at breakdown for the one-dimensional junction. For a device with breakdown voltage of 35 V corresponding to a drift region length of 1 μm, the optimum value for the decay length is found to be 0.337 μm. This value is consistent with the electric field profiles shown in Figs. 5.10 and 5.11.

The optimum oxide thickness in the trenches depends up on the breakdown voltage of the power CC-MOSFET structure. The optimum oxide thickness is related to the optimum value for the decay length by using (5.20):

$$t_{TOX,OPT} = \frac{\varepsilon_{OX}}{\varepsilon_S}\left(\frac{2\lambda_{OPT}^2}{W_M} - \frac{W_M}{4}\right) \tag{5.26}$$

Using (5.25) for the optimum decay length:

$$t_{TOX,OPT} = \frac{\varepsilon_{OX}}{\varepsilon_S}\left(\frac{BV}{qN_D W_M} - \frac{W_M}{4}\right) \tag{5.27}$$

Using the optimum dose ($N_D W_M$) for two-dimensional charge coupling given by (5.16) (assuming the electric field is approximately constant in the y-direction):

$$t_{TOX,OPT} = \frac{\varepsilon_{OX}}{\varepsilon_S}\left(\frac{BV}{2E_{CU}} - \frac{W_M}{4}\right) \tag{5.28}$$

If a uniform electric field along the y-direction is assumed for the breakdown calculation:

$$t_{TOX,OPT} = \frac{\varepsilon_{OX}}{\varepsilon_S}\left(\frac{L_D}{2} - \frac{W_M}{4}\right) \tag{5.29}$$

Alternately, using the critical electric field given by (5.12):

$$t_{TOX,OPT} = \frac{\varepsilon_{OX}}{\varepsilon_S}\left(9.04 \times 10^{-7} BV^{7/6} - \frac{W_M}{4}\right) \tag{5.30}$$

The optimum oxide thickness predicted by the above analysis is shown in Fig. 5.12 for various mesa widths. The trench oxide must be increased in thickness for devices with larger blocking voltages. The range of oxide thicknesses are within

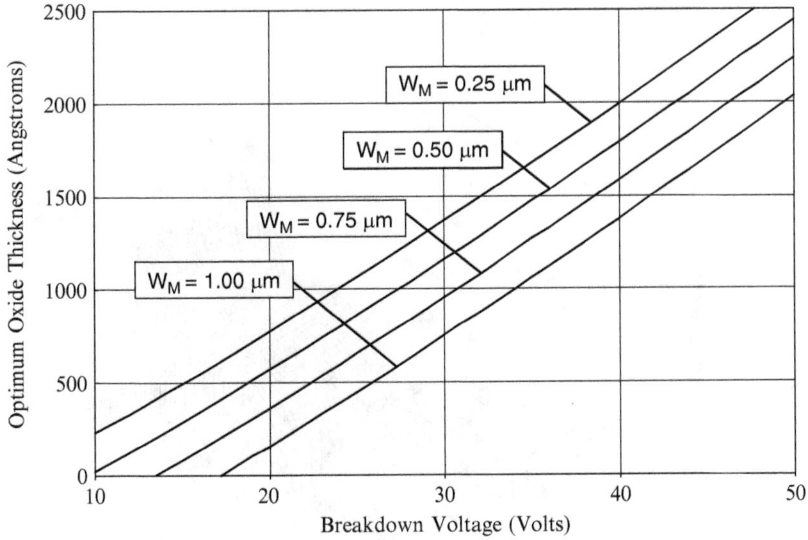


**Fig. 5.12** Optimum oxide thickness for the two-dimensional charge coupled structures

the practical limits for device processing and fabrication. In the case of a breakdown voltage of 30 V, the optimum trench oxide thickness is found to be 1,160 Å for a mesa width of 0.5 μm. This is consistent with obtaining the optimum electric field profiles shown in Fig. 5.11.

### 5.2.1 *Simulation Results*

The results of two-dimensional numerical simulations on the 30 V power CC-MOSFET structure are described here to provide a more detailed understanding of the underlying device physics and operation during the blocking mode. The structure used for the numerical simulations was designed to have the same 30-V blocking voltage as the devices in the previous chapters for comparison purposes. The baseline power CC-MOSFET structure used for the simulations had a trench width ($W_T$) of 0.5 μm and a mesa width ($W_M$) of 0.5 μm leading to a half-cell width of 0.5 μm in the simulations. The drift region consists of a total thickness of 2.5 μm with a doping concentration of $2 \times 10^{16}/\text{cm}^3$ for the upper 0.4 μm where the channel is formed and a doping concentration of $1 \times 10^{17}/\text{cm}^3$ for the rest of the drift region. The lower doping concentration in the upper portion is required to avoid compensation of the dopants in the P-base region. The P-base region is formed by using ion-implantation to create a peak doping concentration at 0.15 μm below the surface with a depth of 0.3 μm. The $N^+$ source region has a depth of 0.08 μm. The narrow width for the P-base region is possible because of the unique distribution of the electric field within the power CC-MOSFET structure.

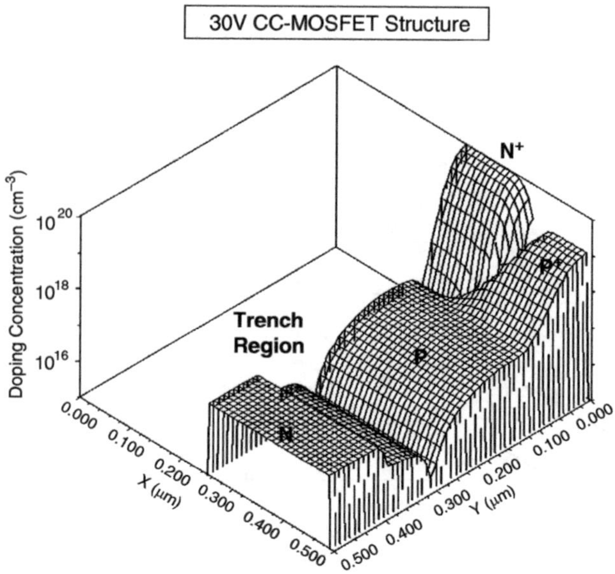


**Fig. 5.13** Doping distribution for the CC-MOSFET structure

A heavily doped $P^+$ region is also included in the structure in the center of the mesa at the upper surface to allow fabrication of ohmic contacts to the P-base region.

A three dimensional view of the doping distribution in the CC-MOSFET structure is shown in Fig. 5.13. Half of the simulated cell structure is consumed by the trench region which extends to a depth of 1.5 μm in the baseline device structure. The $N^+$ source, P-base, and $P^+$ regions can be observed on the upper right hand side in the mesa region. The lower doping concentration of the drift region can be observed to extend to a depth of 0.4 μm. After this depth, the drift region doping concentration increases to $1 \times 10^{17}/\text{cm}^3$ for the rest of the drift region.

The vertical doping profile taken along the surface of the trench is shown in Fig. 5.14. From the profile, it can be observed that the P-base region has a retrograde doping profile with a peak doping concentration of $1.5 \times 10^{17}/\text{cm}^3$ to obtain the desired threshold voltage. The vertical depths of the P-base and $N^+$ source regions are 0.30 and 0.08 μm leading to a channel length of only 0.22 μm. The drift region doping concentration increases from $2 \times 10^{16}$ to $1 \times 10^{17}/\text{cm}^3$ at a depth of 0.4 μm.

The vertical doping profile taken at two positions within the power CC-MOSFET structure are provided in Fig. 5.15. From the profile taken at $x = 0.5$ μm in the middle of the mesa region (dashed line), it can be observed that the doping concentration of the $P^+$ contact region is $1 \times 10^{19}/\text{cm}^3$ with sufficient depth to contact the P-Base region. From the profile taken at $x = 0.25$ μm along the trench sidewall, it can be observed that the drift region extends to a depth of 1.8 μm which is slightly deeper than the trench depth of 1.5 μm.

The blocking characteristic for the CC-MOSFET cell structure is shown in Fig. 5.16 at 300°K. It can be observed that the cell is capable of supporting 34 V.

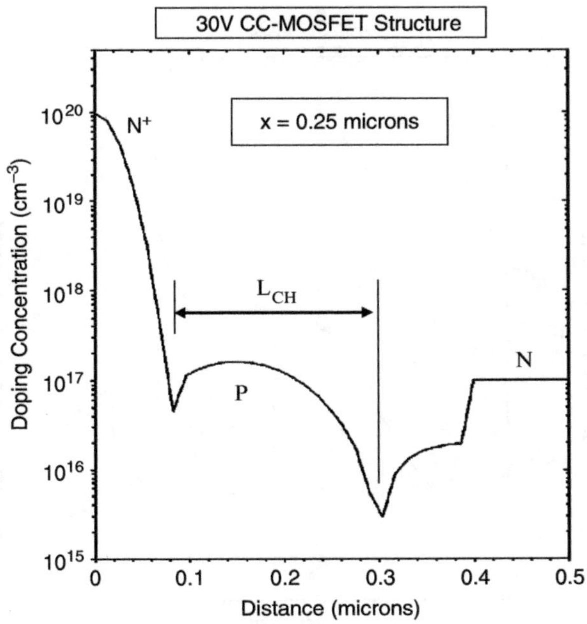


**Fig. 5.14** Channel doping profile for the CC-MOSFET structure

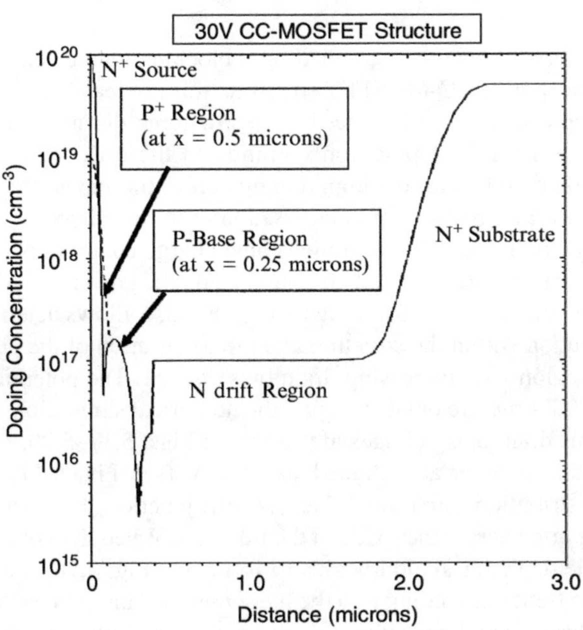


**Fig. 5.15** Vertical doping profiles in the CC-MOSFET structure

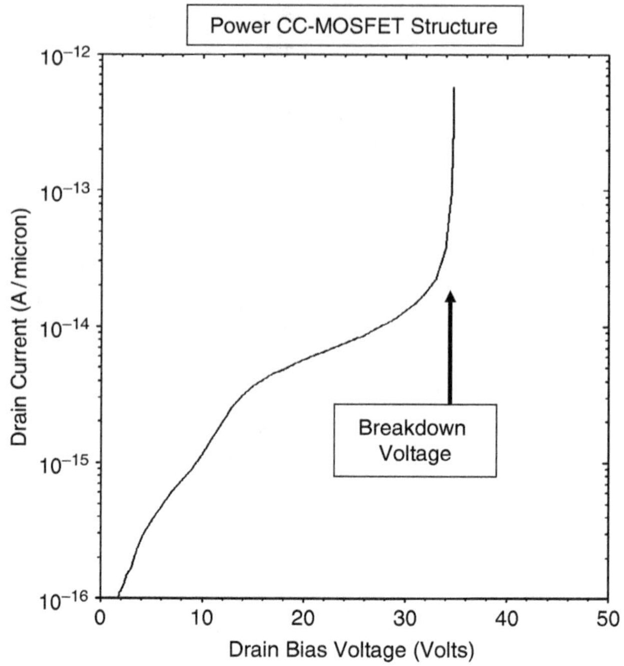


**Fig. 5.16** Blocking characteristic for the CC-MOSFET structure

This is satisfactory for a device designed for a blocking voltage capability of 30 V. In the case of the power CC-MOSFET structure, it is not feasible to utilize the edge termination approaches used for the device structures discussed in the textbook because of the very high doping concentration in the drift region. Special edge terminations compatible with the high doping concentrations in the drift region of power CC-MOSFET structures are discussed later in the chapter.

In order to demonstrate the two-dimensional charge coupling phenomenon, it is instructive to examine the potential contours inside the power CC-MOSFET structure when it is operating in the blocking mode. This also allows determination of the voltage distribution within the structure and the penetration of the depletion region in the P-base region with increasing drain bias voltage. The potential contours for the CC-MOSFET structure obtained using the numerical simulations with zero gate bias and various drain bias voltages are shown in Figs. 5.17–5.20.

The potential contours at a drain bias of 5 V (see Fig. 5.17) shown a one-dimensional distribution across the P-base/N-drift junction ($J_1$) because the voltage is initially supported across the oxide in the trenches. When the voltage is increased to 10 V, a wide depletion region begins to form (see Fig. 5.18) along the vertical sidewalls of the trench extending into the mesa region. This creates the desired two-dimensional charge coupling within the power CC-MOSFET structure. At a drain bias of 20 V, the two-dimensional charge coupling produces a fairly uniform

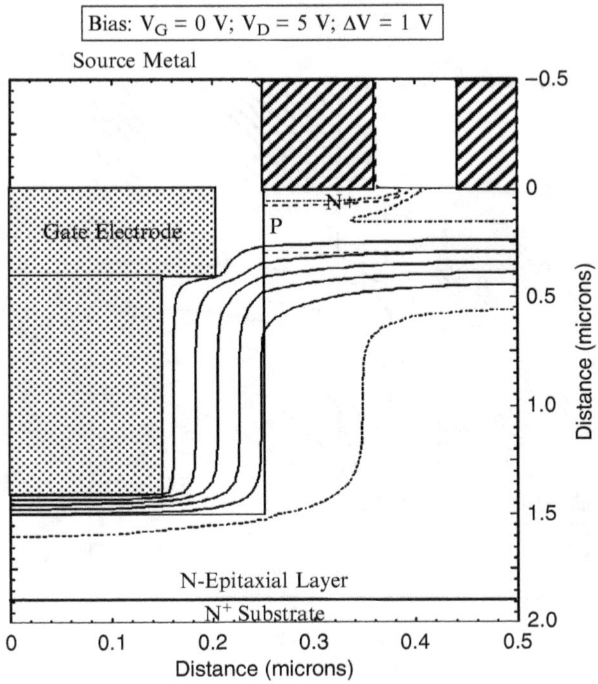


**Fig. 5.17** Potential contours in the CC-MOSFET structure

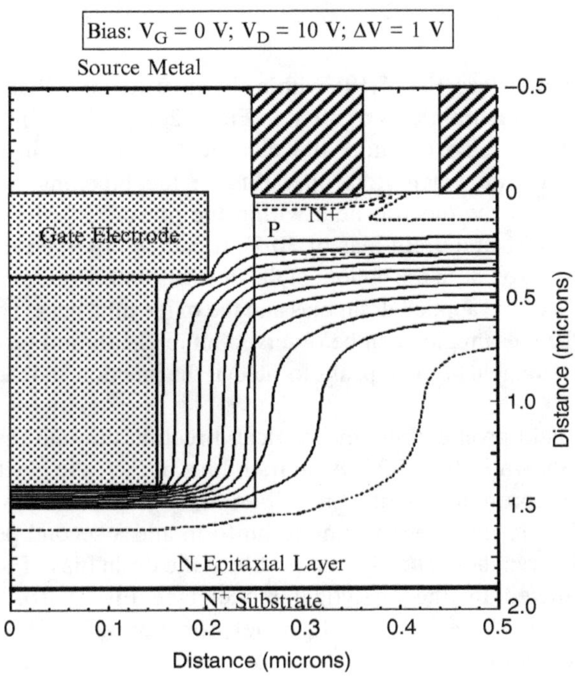


**Fig. 5.18** Potential contours in the CC-MOSFET structure

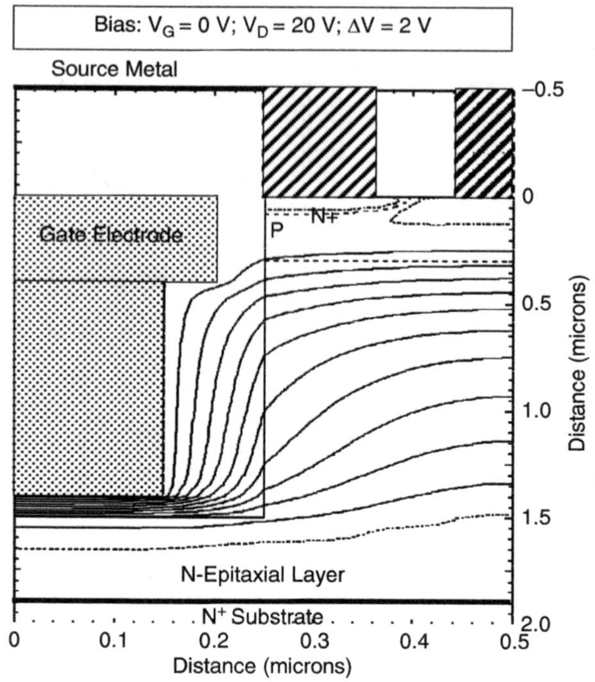


**Fig. 5.19** Potential contours in the CC-MOSFET structure

potential distribution (see Fig. 5.19) within the drift region. This feature is also visible in the case of a drain bias of 30 V in Fig. 5.20.

It is insightful to also examine the electric field profile inside the power SC-MOSFET structure when it is operating in the blocking mode. A three-dimensional view of the electric field within the power CC-MOSFET structure is shown in Fig. 5.21 at a drain bias of 30 V. It can be observed that the electric field is highest in the oxide at the bottom of the trenches with a value ($3 \times 10^6$ V/cm) given by the ratio of the applied drain bias and the oxide thickness. The electric field distribution in the mesa region can be observed more clearly in Fig. 5.22. It can be seen that the electric field has two peaks located at the junction and the bottom of the trenches.

The electric field profile along the vertical direction through the center of the mesa region is shown in Fig. 5.23. At a drain bias of 5 V, the electric field has a triangular shape representative of a one-dimensional junction. However, at larger voltages, the electric field becomes more uniform and a second peak appears at the bottom of the trenches. The electric profile at the drain bias of 30 V resembles the profile predicted by the analytical model (see Fig. 5.10). This provides validation for using the analytical model for optimization of the power CC-MOSFET structures.

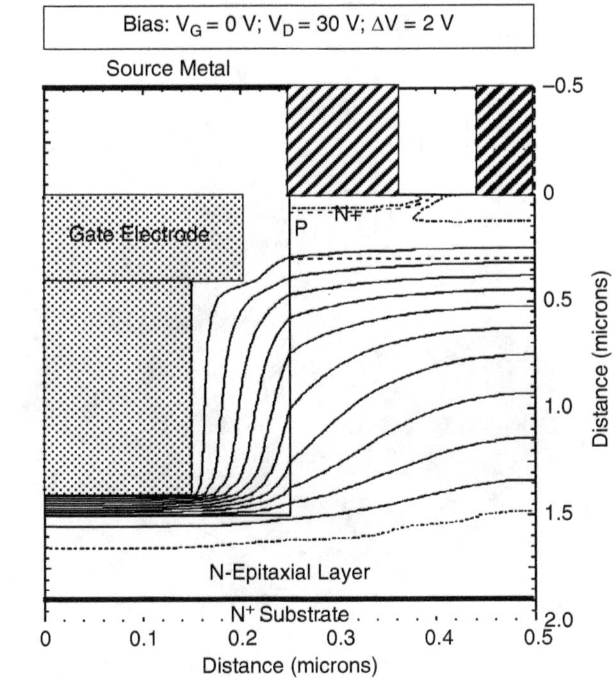


**Fig. 5.20** Potential contours in the CC-MOSFET structure

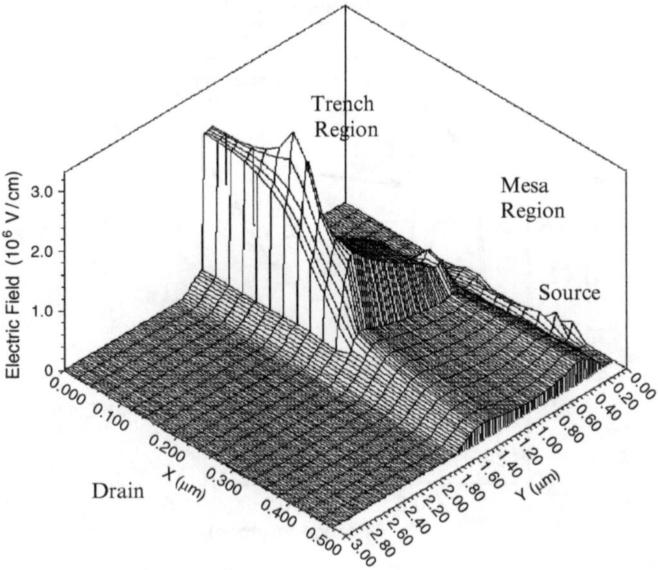


**Fig. 5.21** Three-dimensional electric field distribution in the CC-MOSFET structure

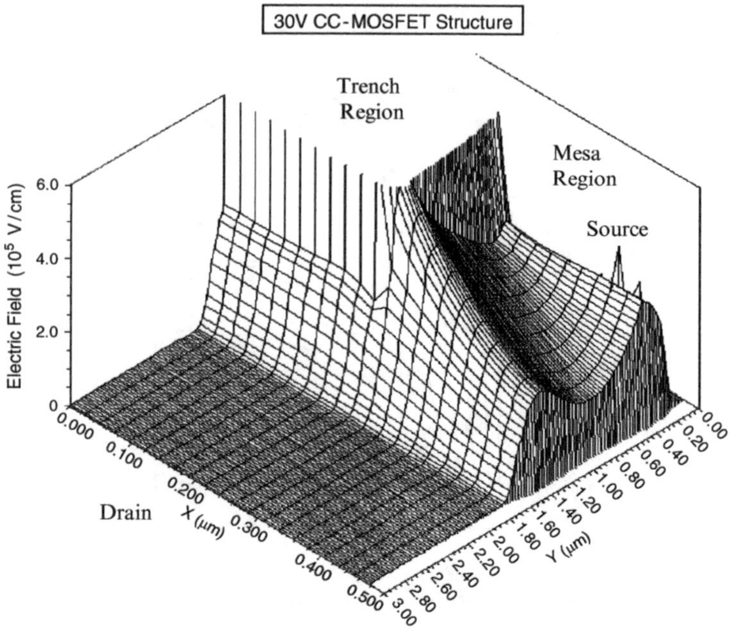


**Fig. 5.22** Three-dimensional electric field distribution in the CC-MOSFET structure

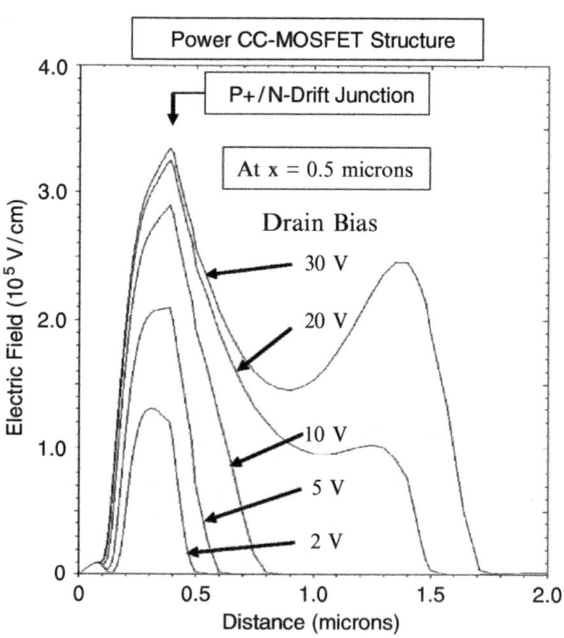


**Fig. 5.23** Electric field profiles in the CC-MOSFET structure

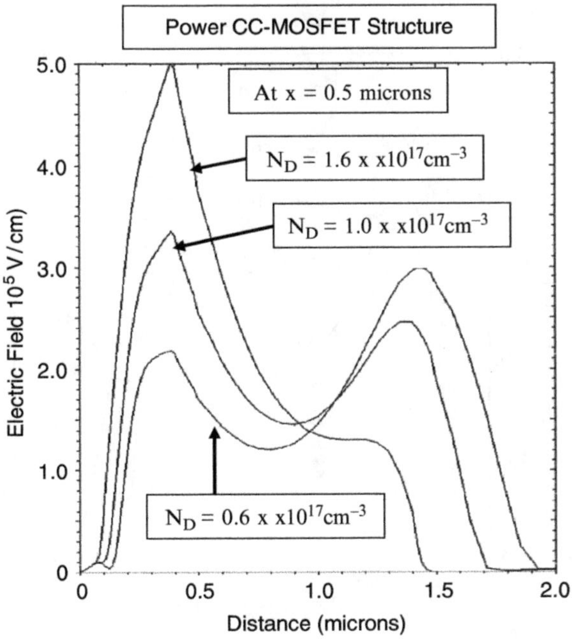


**Fig. 5.24** Impact of drift region doping concentration on the electric field profiles in the CC-MOSFET structure at a drain bias of 30 V

Numerical simulations of the power CC-MOSFET structure were performed using various doping concentrations in the drift region to study the impact on the two-dimensional charge coupling. The electric field profiles are shown in Fig. 5.24 for power CC-MOSFET structures with three doping concentrations. It can be observed that the electric field peak at the P/N junction increases when the doping concentration is increased while it increases at the bottom of the trench when the doping concentration is reduced. This change in the electric field distribution is consistent with the analytical model (see Fig. 5.10). The electric field peaks at the P/N junction and the bottom of the trench are approximately equal at a doping concentration of $1 \times 10^{17}/\text{cm}^3$ as predicted by the analytical model.

Numerical simulations of the power CC-MOSFET structure were also performed using various thicknesses for the oxide in the trench below the P/N junction to study the impact on the two-dimensional charge coupling. The electric field profiles are shown in Fig. 5.25 for power CC-MOSFET structures with four trench oxide thicknesses. It can be observed that the electric field peak at the P/N junction increases when the trench oxide thickness is increased. This change in the electric field distribution is consistent with the analytical model (see Fig. 5.11). The electric field peaks at the P/N junction and the bottom of the trench are approximately equal for a trench oxide thickness of 1,000 Å as predicted by the analytical model.

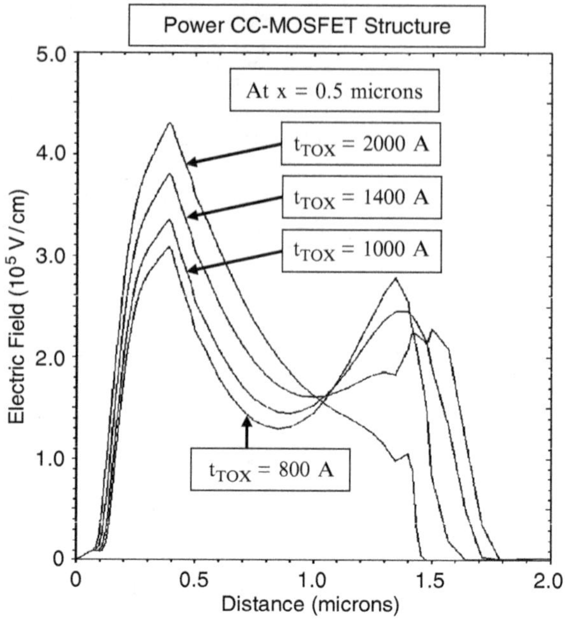


**Fig. 5.25** Impact of trench oxide thickness on the electric field profiles in the CC-MOSFET structure at a drain bias of 30 V

The changes in the electric field distribution described above have an impact on the blocking characteristics of the power CC-MOSFET structure. The blocking characteristics for the power CC-MOSFET structures with various doping concentrations in the drift region are provided in Fig. 5.26. It can be observed that the breakdown voltage, defined as the voltage at which the drain current increases abruptly, remains approximately the same (~35 V) for all the device structures. However, the leakage current prior to breakdown increases rapidly with increasing doping concentration. If the breakdown voltage is defined at a particular leakage current level as is common-place during device testing, the breakdown voltage reduces very rapidly when the doping concentration is increased beyond $1.2 \times 10^{17}/\text{cm}^3$. From these results, it can be concluded that the optimum doping concentration in the drift region is $1 \times 10^{17}/\text{cm}^3$ for a mesa width of 0.5 μm. This is consistent with the predictions of the analytical model.

The blocking characteristics for the power CC-MOSFET structures with various thicknesses for the oxide in the trenches below the P/N junction are provided in Fig. 5.27. It can be observed that the breakdown voltage, defined as the voltage at which the drain current increases abruptly, increases with increasing trench oxide thickness. However, the leakage current is much greater for the structures with the thicker trench oxide. From these results, it can be concluded that the optimum trench oxide thickness is about 1,000 Å.

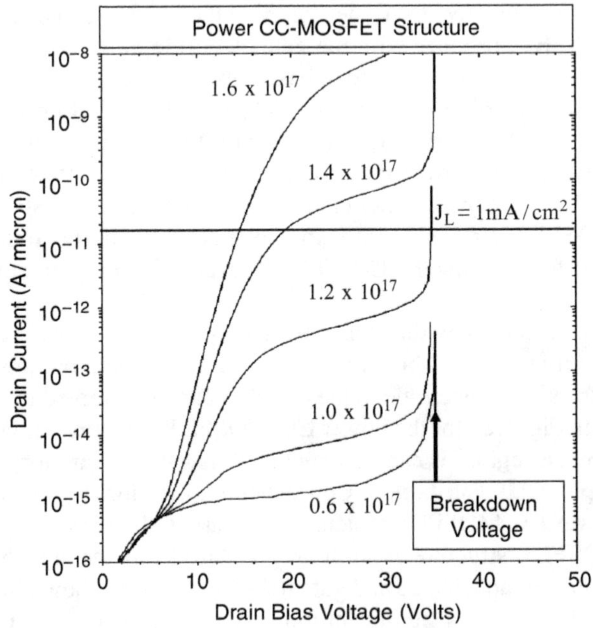


**Fig. 5.26** Blocking characteristic for the CC-MOSFET structures with various doping concentrations in the drift region

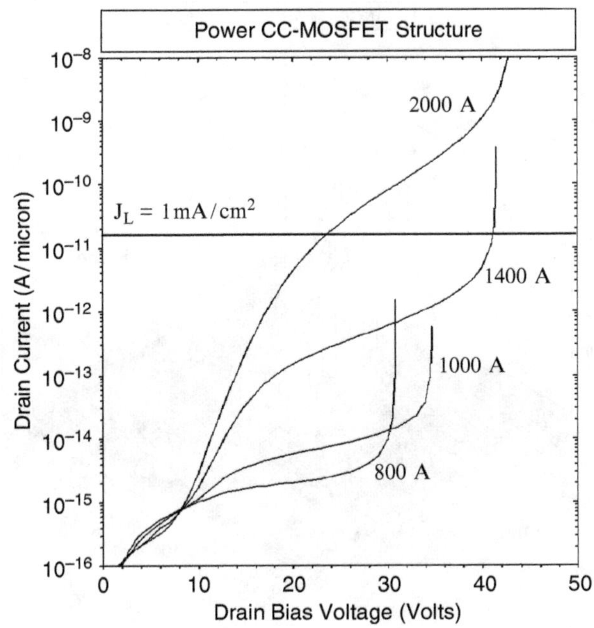


**Fig. 5.27** Blocking characteristic for the CC-MOSFET structures with various trench oxide thicknesses

The charge coupling physics for determining the blocking characteristics of the power CC-MOSFET structure with the gate electrode in the entire trench region (see Fig. 5.1) has been discussed above based up on numerical simulations. These results are also applicable for the power CC-MOSFET structure with the source electrode in the trench (see Fig. 5.2). In order to illustrate this, three-dimensional electric field distributions are provided in Figs. 5.28 and 5.29. By comparison with Figs. 5.21 and 5.22, it can be concluded that the electric field distribution for the power CC-MOSFET structure with source electrode in the trench region is identical to that for the power CC-MOSFET structure with gate electrode in the trenches.

The potential contours within the power CC-MOSFET structure with source electrode in the trenches are shown in Fig. 5.30. The separation between the gate electrode and the source electrode in the trenches can be observed in this figure. By comparison with Fig. 5.20 for the power CC-MOSFET structure with gate electrode in the entire trench region, it can be concluded that this separation has very little impact on the potential distribution. Consequently, the blocking characteristics for both of the power CC-MOSFET structures are essentially identical. However, the power CC-MOSFET structure with source electrode in the trenches has much smaller gate transfer capacitance and gate transfer charge as shown in a subsequent section of this chapter. This significant improvement in performance in the power CC-MOSFET structure with source electrode in the trench region is obtained with a slightly larger specific on-resistance.

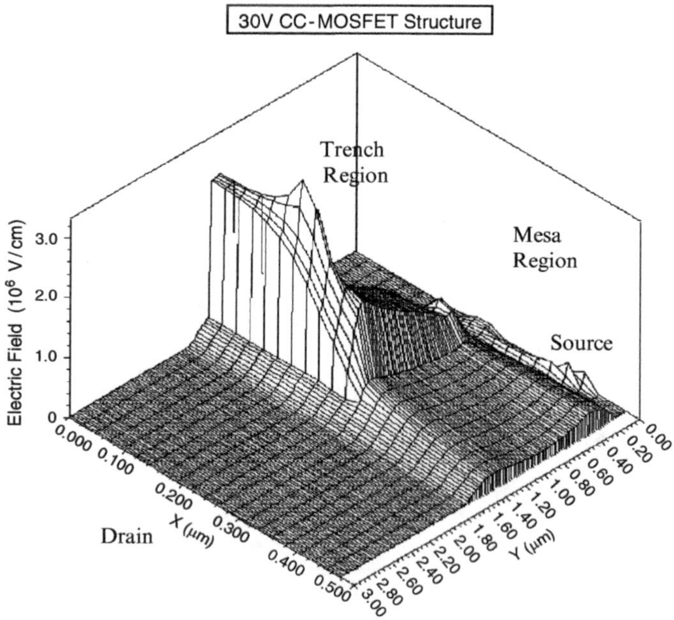


**Fig. 5.28** Three-dimensional electric field distribution in the CC-MOSFET structure with source electrode in the trenches

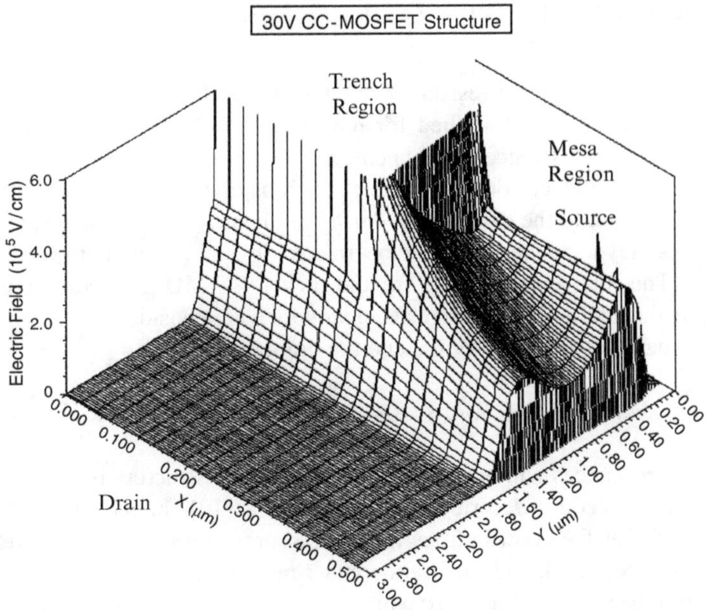


**Fig. 5.29** Three-dimensional electric field distribution in the CC-MOSFET structure with source electrode in the trenches

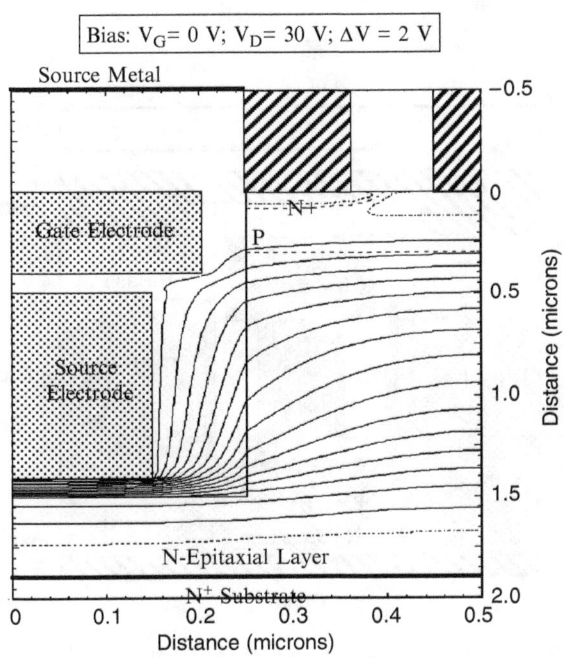


**Fig. 5.30** Potential contours in the CC-MOSFET structure with source electrode in the trenches

## 5.3 Power CC-MOSFET On-Resistance

The components of the on-resistance for the power CC-MOSFET structure are similar as those already described for the power U-MOSFET structure because there is no JFET resistance component. However, in the power CC-MOSFET structure with gate electrode in the trench regions, the accumulation layer serves as an alternate path to the drift region during drain current flow. Consequently, the accumulation layer resistance must be treated in parallel with the drift region resistance. The total on-resistance for the power CC-MOSFET structure is then obtained by the addition of seven resistances which are considered to be in series in the current path between the source and the drain electrodes:

$$R_{ON} = R_{CS} + R_{N+} + R_{CH} + R_A + R_D + R_{SUB} + R_{CD} \tag{5.31}$$

Each of the resistances within the power CC-MOSFET structure is analyzed below by using the procedure described in the textbook [10]. In the textbook, it was demonstrated that the contributions from the source contact resistance ($R_{CS}$), the source resistance ($R_{N+}$), and the drain contact resistance ($R_{CD}$) are very small and will therefore be neglected in this chapter. As in the case of the power D-MOSFET and U-MOSFET structures, the substrate contribution will also be excluded for the comparison of devices.

The power CC-MOSFET structure with gate electrode in the entire trench region is illustrated in Fig. 5.31 with the current flow path shown as the shaded area. The

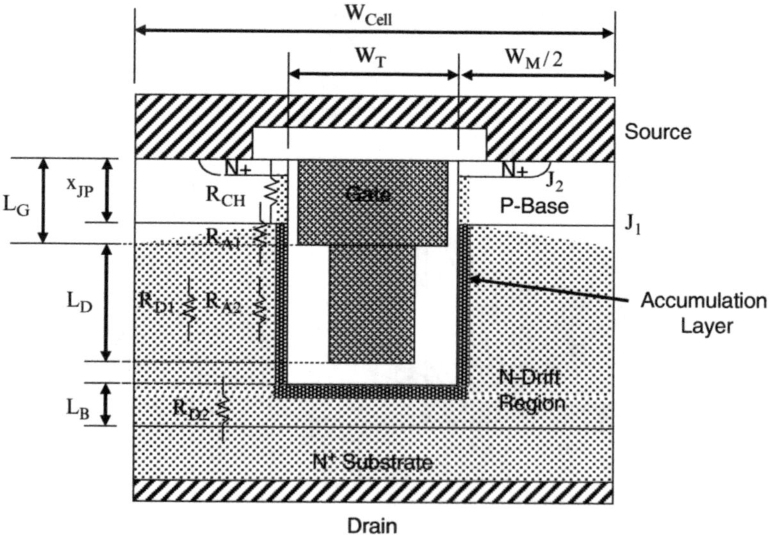


Fig. 5.31 Power CC-MOSFET structure with current flow model used for analysis of its internal resistances

accumulation layer is indicated by the darker shaded region along the boundaries of the trench. The current flow from the channel into the drift region is assumed to occur at a 45° spreading angle. Due to the small mesa width (typically 0.5 μm) for these devices, the current spreading occurs to a relatively small depth below the P-base region. This current path is aided by the formation of an accumulation layer on the trench sidewalls because the gate electrode in the portion with the thin gate oxide must extend to about 0.25 μm below the P-base region to ensure the formation of the channel region. After this depth, the current density can be assumed to be uniform in the drift region with a separate current path created by the formation of another accumulation layer. An additional resistance is contributed by the portion of the drift region, a buffer layer with thickness ($L_B$), located below the bottom of the trench because the trench must not extend into the $N^+$ substrate.

The cross-section of the power CC-MOSFET structure illustrated in Fig. 5.31 provides the various dimensions that can be used for the analysis of the on-resistance components. Here, $W_{Cell}$ is the pitch for the linear cell geometry analyzed in this section; $W_T$ is the width of the trench region; and $W_M$ is the width of the mesa region. The vertical junction depths of the P-base region and the $N^+$ source regions are $x_{JP}$ and $x_{N+}$, respectively. The various resistances that must be analyzed in the power CC-MOSFET structure are also indicated in the figure.

In this monograph, the characteristics of power CC-MOSFET structure with 30-V blocking capability will be analyzed for power supply applications. For this voltage rating and 0.5-μm lithography design rules, the power CC-MOSFET structure has a cell pitch ($W_{CELL}$) of 1.0 μm with trench and mesa widths of 0.5 μm. Typical junction depths for the $N^+$ source region and P-base region are 0.08 and 0.30 μm, respectively, leading to a channel length of only 0.22 μm. This short channel length is possible due to the shielding of the channel region in the power CC-MOSFET structure due to the charge coupling phenomenon which spreads the electric field away from the P-base/N-drift junction. Further suppression of the P-base reach-through phenomenon can be achieved by using a peak P-base doping concentration located at a depth of 0.15 μm (i.e. below the $N^+$ source/P-base junction). The peak P-base doping concentration is chosen to achieve the desired threshold voltage.

Based up on the discussion in the previous section, the optimum doping concentration of the N-drift region required to achieve a 30-V blocking voltage capability is $1.0 \times 10^{17}/cm^3$ for a mesa width of 0.5 μm. The length of the trench region ($L_D$) required below the P-base region is 1.0 μm to achieve a breakdown voltage of about 35 V.

### 5.3.1 *Channel Resistance*

The contribution to the specific on-resistance from the channel in the CC-MOSFET structure is smaller than in the power D-MOSFET and U-MOSFET structures due to the shorter channel length. Based up on the analysis in the textbook [10] for the power U-MOSFET structure, the specific on-resistance contributed by the channel in the power CC-MOSFET structure is given by:

$$R_{CH,SP} = \frac{L_{CH} W_{Cell}}{2\mu_{ni} C_{OX}(V_G - V_{TH})} \tag{5.32}$$

In the case of the 30-V power MOSFET structures used for power supply applications, it is customary to provide the on-resistance at a gate bias of 4.5 and 10 V. Assuming a gate oxide thickness is 500 Å, an inversion layer mobility of 450 $cm^2$/V s (to match the mobility used in the numerical simulations discussed later in this section), and a threshold voltage of 2.7 V (to match the numerical simulations) in the above equation for the power SC-MOSFET design with a cell width of 1 μm, the specific resistance contributed by the channel at a gate bias of 4.5 V is found to be 0.0199 mΩ $cm^2$. The specific on-resistance of the power CC-MOSFET structure is reduced to 0.0049 mΩ $cm^2$ when the gate bias is increased to 10 V. These values are an order of magnitude smaller than for the power D-MOSFET and U-MOSFET structures due to the combination of a very short channel length and a small cell pitch.

## 5.3.2 Accumulation Resistance for Current Spreading Region

In the power CC-MOSFET structure, the current flowing through the inversion channel enters the drift region at the edge of the P-base junction. The current then spreads from the edge of the P-base junction into the drift region. The current flow is strongly aided by the formation of an accumulation layer in the semiconductor at the trench sidewalls due to the positive gate bias applied to turn-on the device (as illustrated by the darker shaded regions in Fig. 5.31). The specific on-resistance contributed by current spreading resistance in the drift region can be neglected because the accumulation layer resistance ($R_{A1}$) for this portion in the power CC-MOSFET structure is much smaller. The accumulation resistance for this portion is given by the same formulation derived for the power MOSFET structures in the textbook [10]:

$$R_{A1,SP} = \frac{(L_G - x_{JP}) W_{Cell}}{2\mu_{nA} C_{OX}(V_G - V_{THA})} \tag{5.33}$$

The gate oxide thickness must be used for computation of the oxide capacitance in this case. The threshold voltage ($V_{THA}$) in the expression is for the on-set of formation of the accumulation layer. A zero threshold voltage will be assumed here when performing the analytical computations. The gate length ($L_G$) will be assumed to 0.4 μm to match the simulations.

For the 30-V power CC-MOSFET design with a cell width of 1 μm and trench width of 0.5 μm, the specific resistance contributed by the accumulation layer in the current spreading portion at a gate bias of 4.5 V is 0.0016 mΩ cm$^2$ for a gate oxide thickness of 500 Å. An accumulation layer mobility of 1,000 cm$^2$/V s was used in this calculation to match the mobility used in the numerical simulations (discussed later in this section). When the gate bias is increased to 10 V, the specific resistance contributed by the accumulation layer in the current spreading portion is reduced to 0.00073 mΩ cm$^2$.

### 5.3.3 *Drift Region Resistance*

The resistance contributed by the drift region in the power CC-MOSFET structure is reduced well below that for the ideal drift region due to the high doping concentration in the drift region. The specific on-resistance contributed by the drift region in the power CC-MOSFET structure can be computed by analysis of three resistances shown in Fig. 5.31. The accumulation resistance ($R_{A2}$) for this portion is given by:

$$R_{A2,SP} = \frac{L_D W_{Cell}}{2\mu_{nA} C_{OX,T}(V_G - V_{THA})} \tag{5.34}$$

The thicker oxide in the trenches for this portion of the structure must be used for computation of the oxide capacitance ($C_{OX,T}$) in this case. The threshold voltage ($V_{THA}$) in the expression is for the on-set of formation of the accumulation layer. A zero threshold voltage will be assumed here when performing the analytical computations. The drift region length ($L_D$) will be assumed to 1.0 μm to achieve a blocking voltage of 35 V.

For the 30-V power CC-MOSFET design with a cell width of 1 μm and trench width of 0.5 μm, the specific resistance of the accumulation layer ($R_{A2}$) at a gate bias of 4.5 V computed using the above equation is found to be 0.0326 mΩ cm$^2$ for a trench oxide thickness of 1,000 Å in this portion of the structure. An accumulation layer mobility of 1,000 cm$^2$/V s was used in this calculation to match the mobility used in the numerical simulations (discussed later in this section). When the gate bias is increased to 10 V, the specific resistance contributed by the accumulation layer in this portion of the structure is reduced to 0.0146 mΩ cm$^2$.

In this portion of the structure, the current density is uniform in the mesa region. However, the drift region contribution is enhanced because part of the device area is occupied by the trench region. When this is taken into account, the specific resistance of the drift region is given by:

$$R_{D1,SP} = \rho_D L_D \left(\frac{W_{Cell}}{W_M}\right) \tag{5.35}$$

The resistivity of the drift region for a relatively high doping concentration of $1 \times 10^{17}$/cm$^3$ is 0.0781 Ω cm. Using this value of resistivity, with the previously provided structural parameters, yields a specific drift region resistance of 0.0156 mΩ cm$^2$.

An additional resistance contribution ($R_{D2}$) in the power CC-MOSFET structure is associated with the buffer layer located below the bottom of the trenches. In this portion of the structure, the current density is uniform and current flow occurs across the entire cell width. Consequently, the specific drift region resistance for this portion is given by:

$$R_{D2,SP} = \rho_D L_B \tag{5.36}$$

For a typical buffer layer thickness of 0.5 μm, the specific resistance is found to be 0.004 mΩ cm$^2$.

## 5.3.4 Total On-Resistance

The total specific on-resistance for the power CC-MOSFET structure can be computed by combining the above components for the on-resistance. Since the resistances $R_{A2}$ and $R_{D1}$ are in parallel, the total specific on-resistance for the power CC-MOSFET structure is given by:

$$R_{T,SP} = R_{CH,SP} + R_{A1,SP} + \left(\frac{R_{A2,SP}R_{D1,SP}}{R_{A2,SP}+R_{D1,SP}}\right) + R_{D2,SP} \tag{5.37}$$

For the case of the 30-V power CC-MOSFET design with a cell pitch ($W_{Cell}$) of 1 μm and trench width of 0.5 μm, the total specific on-resistance is found to be 0.0361 mΩ cm$^2$ at a gate bias of 4.5 V and 0.0171 mΩ cm$^2$ at a gate bias of 10 V by using the analytical model. The contributions from each of the components of the on-resistance are summarized in Fig. 5.32. The specific on-resistance for the power

| Resistance | $V_G$ = 4.5 V (mΩ-cm$^2$) | $V_G$ = 10 V (mΩ-cm$^2$) |
|---|---|---|
| Channel ($R_{CH,SP}$) | 0.0199 | 0.0049 |
| Accumulation ($R_{A1,SP}$) | 0.0016 | 0.0007 |
| Accumulation ($R_{A2,SP}$) | 0.0326 | 0.0146 |
| Drift ($R_{D1,SP}$) | 0.0156 | 0.0156 |
| Drift ($R_{D2,SP}$) | 0.0039 | 0.0039 |
| Total ($R_{T,SP}$) | 0.0360 | 0.0170 |

**Fig. 5.32** Resistance components in the 30-V power CC-MOSFET structure with gate electrode in the trenches

CC-MOSFET structure is found to be an order of magnitude smaller than that for the power D-MOSFET and the U-MOSFET structures at a gate bias of 4.5 V.

From the values provided in Fig. 5.32, it can be seen that the drift region resistance for the power CC-MOSFET structure is far smaller than that for the devices discussed in the previous chapters. The very low specific resistance of the drift region is associated with the high doping concentration of the drift region enabled by the charge-coupling phenomenon. Due to this reduction in drift region resistance, the channel resistance contribution in the power CC-MOSFET structure remains a significant contributor to the total specific on-resistance. In addition, it is important to reduce the resistance contributed by the $N^+$ substrate by wafer thinning technology to realize the benefits of the charge-coupling concept.

The ideal specific on-resistance for a drift region is given by:

$$R_{IDEAL,SP} = \frac{W_{PP}}{q\mu_n N_D} \quad (5.38)$$

where $W_{PP}$ is the parallel-plane depletion width at breakdown, $N_D$ is the doping concentration of the drift region to sustain the blocking voltage, and $\mu_n$ is the mobility for electrons corresponding to this doping concentration. For the case of a blocking voltage of 30 V, the depletion width and doping concentration are found to be 1.5 μm and $1.7 \times 10^{16}/cm^3$, respectively. Using the electron mobility for this doping level, the ideal specific on-resistance is found to be 0.047 mΩ $cm^2$. Since the conventional device is constrained by the impact of an 80% reduction of breakdown voltage due to the edge termination, it is worth computing the ideal specific on-resistance for this case for comparison with the device. For the case of a blocking voltage of 37.5 V, the depletion width and doping concentration are found to be 2.0 μm and $1.3 \times 10^{16}/cm^3$, respectively. Using the electron mobility for this doping level, the ideal specific on-resistance is found to be 0.080 mΩ $cm^2$. The specific on-resistance for the 30-V CC-MOSFET structure is about five times smaller than these ideal specific on-resistances for a gate bias of 10 V. Consequently, the charge-coupling concept allows overcoming the ideal specific resistance barrier using silicon device structures.

The above analysis of the specific on-resistance was performed for the power CC-MOSFET structure with gate electrode within the entire trench region. In the case of the power CC-MOSFET structure with the source electrode inside the trenches, the specific on-resistance can be computed by using the same relationships with the exception that the accumulation layer path related to the resistance $R_{A2}$ is not present in the structure with source electrode in the trenches. The total specific on-resistance for the power CC-MOSFET structure with the source electrode inside the trenches is therefore given by:

$$R_{T,SP} = R_{CH,SP} + R_{A1,SP} + R_{D1,SP} + R_{D2,SP} \quad (5.39)$$

For the case of the 30-V power CC-MOSFET design with a cell pitch ($W_{Cell}$) of 1 μm and trench width of 0.5 μm, the total specific on-resistance is found to be

| Resistance | $V_G = 4.5$ V (mΩ-cm$^2$) | $V_G = 10$ V (mΩ-cm$^2$) |
|---|---|---|
| Channel ($R_{CH, SP}$) | 0.0199 | 0.0049 |
| Accumulation ($R_{A1, SP}$) | 0.0016 | 0.0007 |
| Drift ($R_{D1, SP}$) | 0.0156 | 0.0156 |
| Drift ($R_{D2,SP}$) | 0.0039 | 0.0039 |
| Total ($R_{T, SP}$) | 0.0410 | 0.0252 |

**Fig. 5.33** Transfer characteristics of the CC-MOSFET structure

0.041 mΩ cm$^2$ at a gate bias of 4.5 V and 0.025 mΩ cm$^2$ at a gate bias of 10 V by using the analytical model. These values are about 15% and 50% larger than for the structure with gate electrode in the trenches at a gate bias of 4.5 and 10 V, respectively. The contributions from each of the components of the on-resistance are summarized in Fig. 5.33. The specific on-resistance for the power CC-MOSFET structure with source electrode in the trenches is still an order of magnitude smaller than that for the power D-MOSFET and the U-MOSFET structures at a gate bias of 4.5 V.

#### 5.3.4.1 Simulation Results

The transfer characteristics for the CC-MOSFET structure were obtained using numerical simulations with a drain bias of 0.1 V at 300 and 400°K. The resulting transfer characteristics are shown in Fig. 5.34. From this graph, a threshold voltage of 2.7 and 2.3 V can be extracted at 300 and 400°K, respectively. The threshold voltage decreases by 15% when the temperature increases similar that observed for the power D-MOSFET structure. The specific on-resistance can be obtained from the transfer characteristics at any gate bias voltage. For the case of a gate bias of 4.5 V and 300°K, the specific in-resistance is found to be 0.0365 mΩ cm$^2$. For the case of a gate bias of 10 V and 300°K, the specific in-resistance is found to be 0.0217 mΩ cm$^2$. These values are in very close agreement with the results obtained from the analytical model.

The on-state current flow pattern within the CC-MOSFET structure at a small drain bias of 0.1 V and a gate bias of 4.5 V is shown in Fig. 5.35. In the figure, the depletion layer boundary is shown by the dotted lines and the junction boundary is delineated by the dashed line. From the figure, it can be seen that the current spreads from the channel to the drift region at a 45° angle. The current distribution appears non-uniform in the drift region due to the presence of the accumulation layer.

The transfer characteristics for the CC-MOSFET structure with source electrode in the trenches were obtained using numerical simulations with a drain bias of 0.1 V

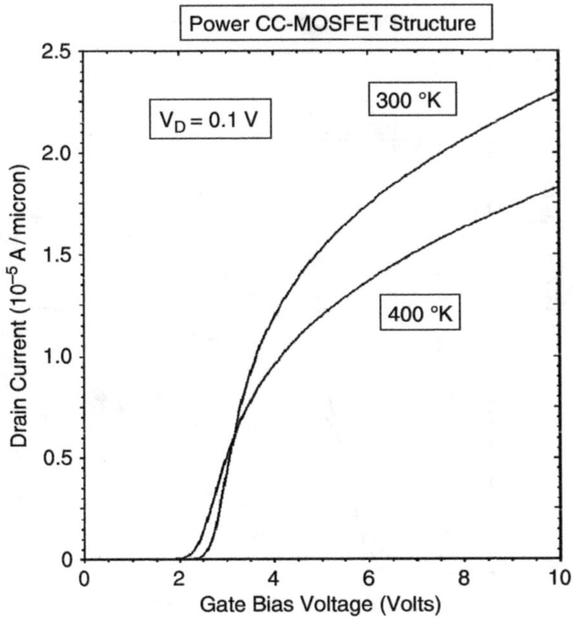


**Fig. 5.34** Current distribution in the CC-MOSFET structure

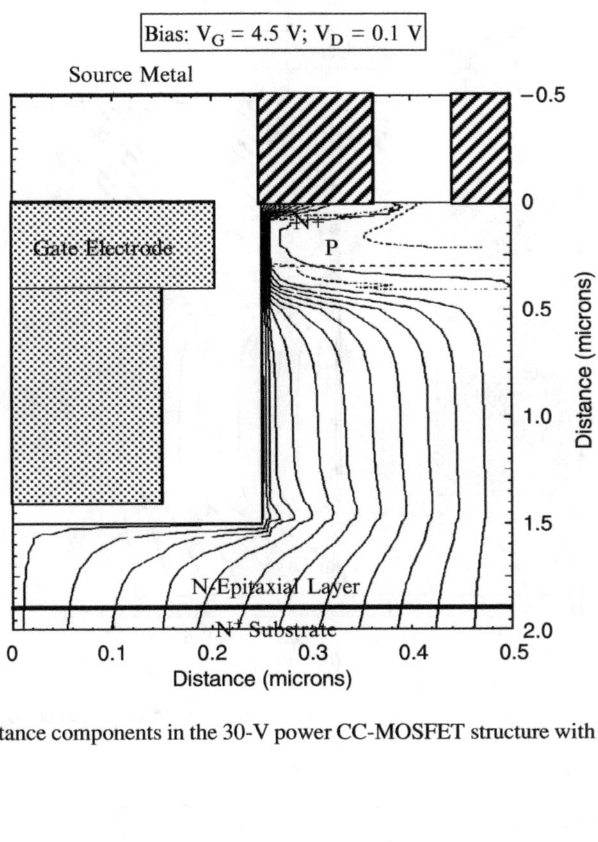


**Fig. 5.35** Resistance components in the 30-V power CC-MOSFET structure with source electrode in the trenches

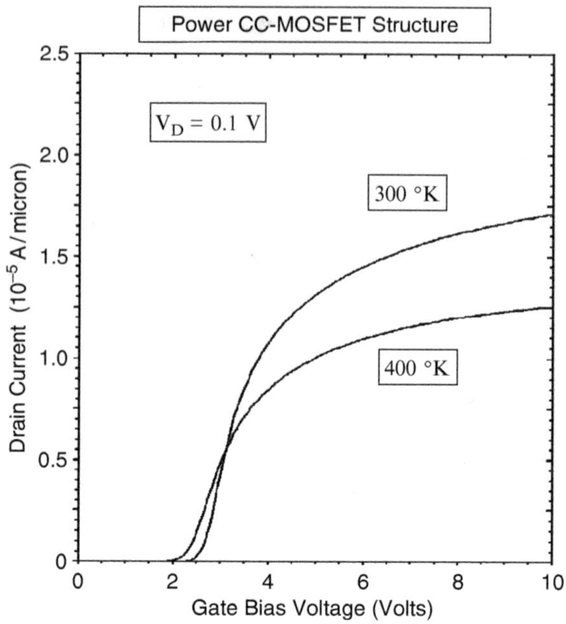


**Fig. 5.36** Transfer characteristics of the CC-MOSFET structure with source electrode in the trenches

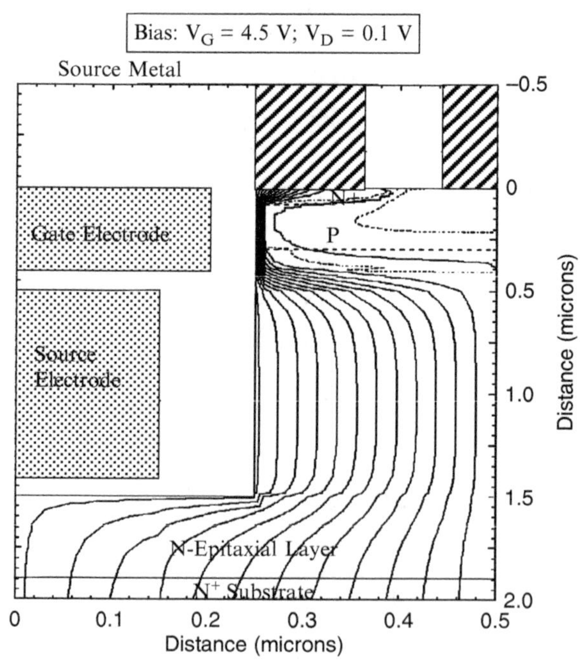


**Fig. 5.37** Current distribution in the CC-MOSFET structure with source electrode in the trenches

at 300 and 400°K. The resulting transfer characteristics are shown in Fig. 5.36. The specific on-resistance can be obtained from the transfer characteristics at any gate bias voltage. For the case of a gate bias of 4.5 V and 300°K, the specific in-resistance is found to be 0.0413 m$\Omega$ cm$^2$. For the case of a gate bias of 10 V and 300°K, the specific in-resistance is found to be 0.0292 m$\Omega$ cm$^2$. These values are in very close agreement with the results obtained from the analytical model.

The on-state current flow pattern within the CC-MOSFET structure, with source electrode in the trenches, at a small drain bias of 0.1 V and a gate bias of 4.5 V is shown in Fig. 5.37. In the figure, the depletion layer boundary is shown by the dotted lines and the junction boundary is delineated by the dashed line. From the figure, it can be seen that the current spreads from the channel to the drift region at a 45° angle. In contrast with the power CC-MOSFET structure with gate electrode in the trenches, the current distribution appears uniform in the drift region due to the absence of the accumulation layer.

## 5.4 Output Characteristics

The output characteristics of the power CC-MOSFET structure are important to the loci for the switching waveforms when it is operating in power circuits. Due to shielding of the P-base region, the power CC-MOSFET structure exhibits super-linear transfer characteristics [10]. The saturated drain current for the power CC-MOSFET structure is then given by:

$$I_{Dsat} = C_{ox}(V_G - V_T)\, v_{sat,n} Z \tag{5.40}$$

With sufficiently high doping concentration of the P-base region, the modulation of channel length can be made sufficiently small to ensure a high output resistance. In the power CC-MOSFET structure, the charge coupling phenomenon suppresses the depletion of the P-base region with increasing drain bias voltage because the voltage is spread into the drift region. This is beneficial for obtaining a reasonable output resistance despite the very short channel length in the device. The short channel length in the power CC-MOSFET structure produces a high transconductance which is beneficial for reducing switching losses. The saturated drain current in the power CC-MOSFET structure increases linearly with the gate bias voltage in this model.

### 5.4.1 Simulation Example

The output characteristics of the 30-V power CC-MOSFET structure were obtained by using two-dimensional numerical simulations using various gate bias voltages. All the device parameters used for these numerical simulations are the same as

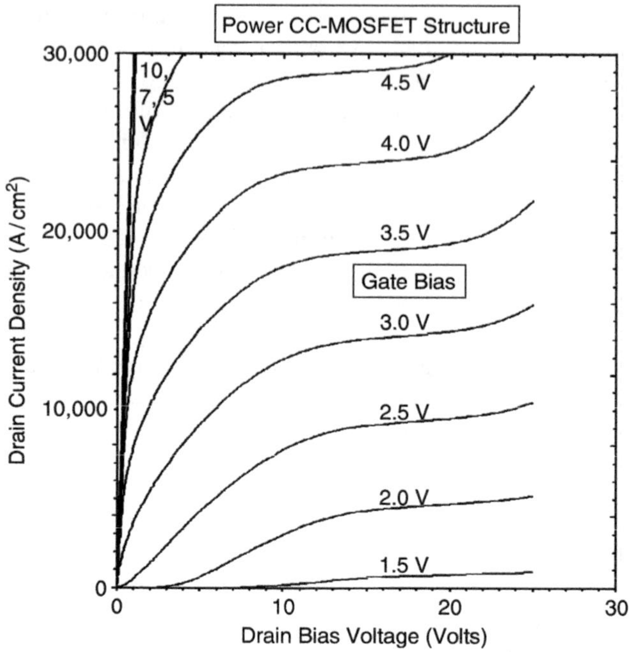


**Fig. 5.38** Output characteristics for the power CC-MOSFET structure

those used in the previous sections. The output characteristics of the power CC-MOSFET obtained using the simulations are shown in Fig. 5.38. The traces for increasing gate bias voltages are nearly uniformly spaced indicating super-linear [10] behavior of the transfer characteristics. This behavior is related to the shielding of the channel from the drain bias by the charge-coupling phenomenon, which allows the channel to operate in the linear regime even at high drain bias voltages. When compared with the power D-MOSFET and U-MOSFET structures, the power CC-MOSFET structure exhibits a gradual transition from the on-state to the current saturation regime due to the on-set of the charge-coupling effect at a drain bias above 10 V as described earlier with the aid of Figs. 5.17–5.20.

## 5.5 Device Capacitances

For the power MOSFET structures, the switching speed is limited by the device capacitances in practical applications as previously discussed in Chap. 2. The rate at which the power MOSFET structure can be switched between the on- and off-states is determined by the rate at which the input capacitance can be charged or discharged. In addition, the capacitance between the drain and the gate electrodes has been found to play an important role in determining the drain current and voltage transitions during the switching event.

The capacitances within the power CC-MOSFET structure can be analyzed using the same approach used in the textbook [10] for the power U-MOSFET structure. The specific input (or gate) capacitance for the power CC-MOSFET structure with gate electrode in the entire trench region is given by:

$$C_{IN,SP} = C_{N+} + C_P + C_{SM} = \frac{2x_{JP}}{W_{Cell}}\left(\frac{\varepsilon_{OX}}{t_{GOX}}\right) + \frac{W_T}{W_{Cell}}\left(\frac{\varepsilon_{OX}}{t_{IEOX}}\right) \tag{5.41}$$

where $t_{GOX}$ and $t_{IEOX}$ are the thicknesses of the gate and inter-electrode oxides, respectively. The smaller junction depth for the P-base region in the power CC-MOSFET structure, when compared with the power U-MOSFET structure, is beneficial for reducing the input capacitance. For a 30-V power CC-MOSFET structure with a cell pitch of 1 μm and trench width of 0.5 μm, the specific input capacitance is found to be 44.3 nF/cm$^2$ for a gate oxide thickness of 500 Å and an inter-metal dielectric thickness of 5,000 Å. This value is larger than 22 nF/cm$^2$ for the power D-MOSFET structure and approximately equal to 47 nF/cm$^2$ for the power U-MOSFET structure.

The specific input capacitance for the power CC-MOSFET structure with source electrode in the trench is the larger than that for the power CC-MOSFET structure with gate electrode in the entire trench region. In addition to the capacitance associated with the gate electrode overlap with the P-base and N$^+$ source regions on the trench sidewall, it is necessary to account for the capacitances due to overlap of the source metal electrode and the overlap of the gate electrode with the source electrode embedded within the trench. The specific input (or gate) capacitance for the power CC-MOSFET structure with source electrode in the trench is given by:

$$\begin{aligned} C_{IN,SP} &= C_{N+} + C_P + C_{SM} + C_{SG} \\ &= \frac{2x_{JP}}{W_{Cell}}\left(\frac{\varepsilon_{OX}}{t_{GOX}}\right) + \frac{W_T}{W_{Cell}}\left(\frac{\varepsilon_{OX}}{t_{IEOX}}\right) + \frac{(W_T - 2t_{TOX})}{W_{Cell}}\left(\frac{\varepsilon_{OX}}{t_{IGOX}}\right) \end{aligned} \tag{5.42}$$

where $t_{GOX}$, $t_{IEOX}$, and $t_{IGOX}$ are the thicknesses of the gate oxide and inter-electrode oxide, and the oxide between the gate and source electrodes within the trench, respectively. For a 30-V power CC-MOSFET structure with a cell pitch of 1 μm and trench width of 0.5 μm, the specific input capacitance is found to be 54.56 nF/cm$^2$ for a gate oxide thickness of 500 Å, gate to source electrode-in-trench thickness of 1,000 Å, and an inter-metal dielectric thickness of 5,000 Å. This value is larger than 22 nF/cm$^2$ for the power D-MOSFET structure and approximately equal to 47 nF/cm$^2$ for the power U-MOSFET structure.

The capacitance between the gate and drain electrodes (also called the reverse transfer capacitance) is very different for the two power CC-MOSFET structures. In the case of the power CC-MOSFET structure with gate electrode in the entire trench region, the gate transfer capacitance is large due to the overlap of the gate electrode with the drift region along the vertical trench sidewalls and at the trench bottom surface. The MOS structure in this portion of the power CC-MOSFET structure operates under deep depletion conditions when a positive voltage is applied to the drain. The depletion region boundaries for this structure are shown in Fig. 5.39 by the dashed lines together with the oxide and semiconductor capacitances needed for the analysis of the gate transfer capacitance. The illustration in Fig. 5.39a applies before the mesa region gets completely depleted by the drain bias voltage. The illustration in Fig. 5.39b applies after the mesa region gets completely depleted by the drain bias voltage.

The gate-drain capacitance for the power CC-MOSFET structure with gate electrode in the entire trench regions before the entire mesa width becomes depleted by the drain bias can be obtained by summing the capacitance along the trench sidewalls with the capacitance at the trench bottom surface:

$$C_{GD,SP} = C_{GD1} + C_{GD2} + C_{GD3} \tag{5.43}$$

where $C_{GD1}$ is the specific capacitance between the gate electrode and the drift region where the trench sidewall has the gate oxide, $C_{GD2}$ is the specific capacitance between the gate electrode and the drift region where the trench sidewall has the thicker trench oxide, and $C_{GD3}$ is the specific capacitance between the gate

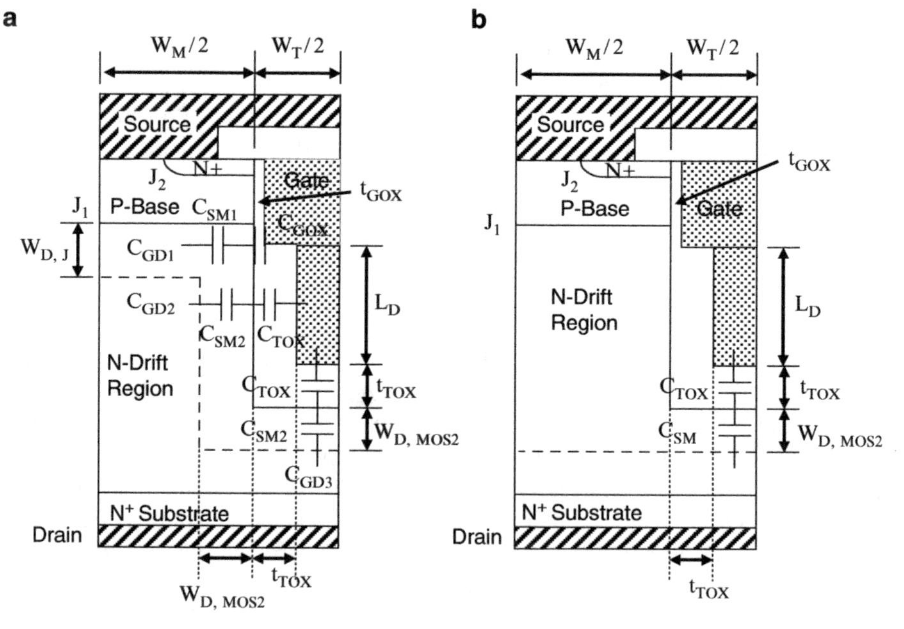


**Fig. 5.39** Depletion boundaries and capacitances within the power CC-MOSFET structure with gate electrode in the entire trench region

electrode and the drift region at the bottom of the trenches with the thicker trench oxide. Using the geometry of the device structure provided in Fig. 5.39, these capacitances are given by:

$$C_{GD1,SP} = \frac{2(L_G - x_{JP})}{W_{Cell}}\left(\frac{C_{GOX}C_{S,M1}}{C_{GOX} + C_{S,M1}}\right) \tag{5.44}$$

$$C_{GD2,SP} = \frac{2(L_D - W_{D,J})}{W_{Cell}}\left(\frac{C_{TOX}C_{S,M2}}{C_{TOX} + C_{S,M2}}\right) \tag{5.45}$$

and

$$C_{GD3,SP} = \left(\frac{W_T - 2t_{TOX}}{W_{Cell}}\right)\left(\frac{C_{TOX}C_{S,M2}}{C_{TOX} + C_{S,M2}}\right) \tag{5.46}$$

where $C_{S,M1}$ is the specific semiconductor capacitance under the gate oxide, $C_{S,M2}$ is the specific semiconductor capacitance under the trench oxide, and $C_{GOX}$ is the gate oxide specific capacitance, and $C_{TOX}$ is the thick trench oxide specific capacitance.

It is convenient to express these capacitances in terms of a gate geometry factor [10]:

$$C_{GD1,SP} = K_{G1}\left(\frac{C_{GOX}C_{S,M1}}{C_{GOX} + C_{S,M1}}\right) \tag{5.47}$$

where

$$K_{G1} = \frac{2(L_G - x_{JP})}{W_{Cell}} \tag{5.48}$$

The two terms with the thick trench capacitances can be combined to get:

$$C_{GD2,SP} + C_{GD3,SP} = K_{G2}\left(\frac{C_{TOX}C_{S,M2}}{C_{TOX} + C_{S,M2}}\right) \tag{5.49}$$

where

$$K_{G2} = \frac{2(L_D - W_{D,J})}{W_{Cell}} + \left(\frac{W_T - 2t_{TOX}}{W_{Cell}}\right) \tag{5.50}$$

The specific capacitance of the semiconductor depletion region can be obtained by computation of the depletion layer width. The depletion layer width in the semiconductor under the gate oxide can be obtained using [10]:

$$W_{D,MOS1}=\frac{\varepsilon_S}{C_{GOX}}\left\{\sqrt{1+\frac{2V_D C_{GOX}^2}{q\varepsilon_S N_D}}-1\right\} \quad (5.51)$$

The specific capacitance for the semiconductor in this region is then obtained using:

$$C_{S,M1}=\frac{\varepsilon_S}{W_{D,MOS1}} \quad (5.52)$$

The depletion layer width in the semiconductor under the trench oxide can be obtained using [10]:

$$W_{D,MOS2}=\frac{\varepsilon_S}{C_{TOX}}\left\{\sqrt{1+\frac{2V_D C_{TOX}^2}{q\varepsilon_S N_D}}-1\right\} \quad (5.53)$$

The specific capacitance for the semiconductor in this region is then obtained using:

$$C_{S,M2}=\frac{\varepsilon_S}{W_{D,MOS2}} \quad (5.54)$$

The gate-drain (or reverse transfer) capacitance can be computed by using (5.43) with the above equations to determine the semiconductor capacitance as a function of the drain bias voltage. This expression holds true only until the entire mesa region becomes depleted by the applied drain bias voltage. The drain bias voltage at which the mesa region is completely depleted can be derived from (5.53) by setting the depletion width equal to half the mesa width. This drain bias voltage is given by:

$$V_{D,M}=\frac{q\varepsilon_S N_D}{2C_{TOX}^2}\left[\left(\frac{W_M C_{TOX}}{2\ \varepsilon_S}+1\right)^2-1\right] \quad (5.55)$$

For the power CC-MOSFET structure with a mesa width ($W_M$) of 0.5 μm, trench oxide thickness of 1,000 Å, and a mesa doping concentration ($N_D$) of $1 \times 10^{17}$/cm$^3$, the mesa region is completely depleted at a drain bias voltage of 16.5 V according to this expression. The reverse transfer capacitance will reduce abruptly at this drain bias voltage.

The depletion layer boundary in the power CC-MOSFET structure after the depletion of the entire mesa region by the drain bias voltage is illustrated in

Fig. 5.39b. The gate-drain capacitance is then determined by the edge of the depletion region located across the entire cell structure:

$$C_{GD4,SP} = \left( \frac{C_{TOX} C_{S,M}}{C_{TOX} + C_{S,M}} \right) \tag{5.56}$$

The gate geometry factor ($K_{GD3}$) under these operating conditions is unity.

The specific gate transfer capacitance obtained by using the above analytical formulae is shown in Fig. 5.40 for the case of the power CC-MOSFET structure with gate electrode in the entire trench region. This structure has a cell pitch of 1 μm, trench width of 0.5 μm, gate oxide thickness of 500 Å, and a trench oxide thickness of 1,000 Å. The vertical gate length ($L_G$) of 0.4 μm extends 0.1 μm beyond the P-base junction depth of 0.3 μm. Starting from a specific capacitance of 80 nF/cm$^2$, the gate-drain (reverse transfer) capacitance decreases gradually with increasing drain bias voltage due to the expansion of the depletion region in the mesa and the trench bottom until the drain bias reaches 16.5 V (the voltage required to completely deplete the mesa region). The specific capacitance at a drain bias of 16 V is 30 nF/cm$^2$ according to the analytical model. Once the mesa region is completely depleted, the specific reverse transfer capacitance abruptly drops to 19 nF/cm$^2$ and then reduces gradually with increasing drain bias voltage.

The output capacitance for the power SC-MOSFET structure with gate electrode in the entire trench region is associated with the capacitance of the junction between the P-base region and the N-drift region. Due to lateral depletion of the mesa region with increasing drain bias voltage by the depletion region extending in the x-direction across the vertical sidewalls of the trenches, the area for the junction that produces the output capacitance decreases with increasing drain bias voltage.

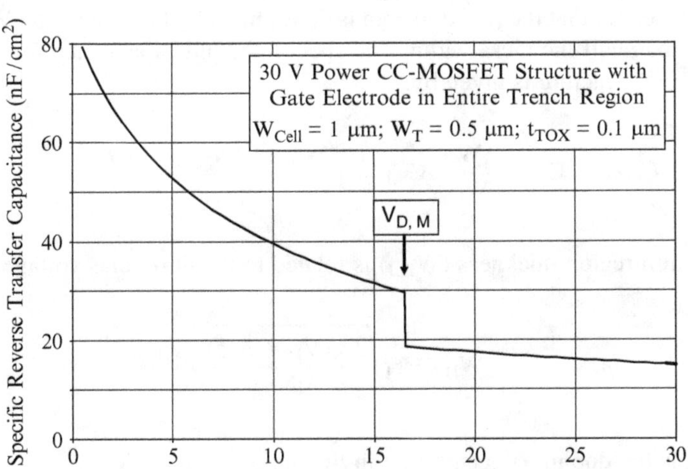


**Fig. 5.40** Gate-drain capacitance for the power CC-MOSFET structure with gate electrode in the entire trench region

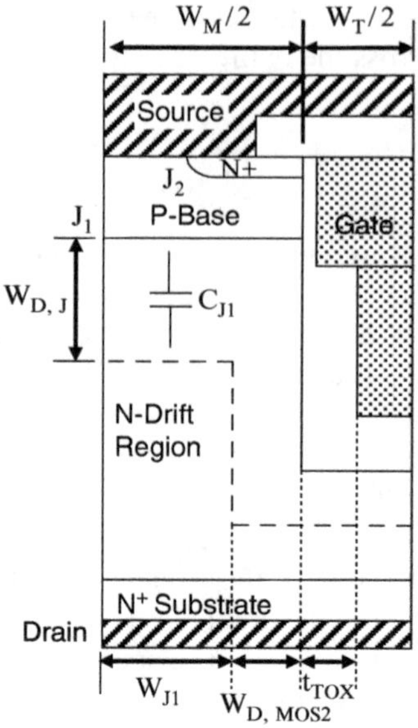


**Fig. 5.41** Depletion region boundaries for determination of the output capacitance for the power CC-MOSFET structure

At the same time, the specific junction capacitance also decreases with increasing drain bias voltage due to the larger depletion layer width at the junction.

The depletion layer boundary inside the power CC-MOSFET structure prior to the complete depletion of the mesa region is shown in Fig. 5.41 by the dashed lines. It can be observed that the junction area is determined by the dimension $W_{J1}$ which is smaller than half the mesa width. The specific output capacitance for the power CC-MOSFET structure is given by:

$$C_{O,SP} = C_{J1} = \frac{W_{J1}}{W_{Cell}} C_{S,J} = \left(\frac{W_M - 2W_{D,MOS2}}{2W_{Cell}}\right)\left(\frac{\varepsilon_S}{W_{D,J}}\right) \tag{5.57}$$

The depletion region thickness ($W_{D,J}$) is related to the drain bias voltage:

$$W_{D,J} = \sqrt{\frac{2\varepsilon_S(V_D + V_{bi})}{qN_D}} \tag{5.58}$$

where $N_D$ is the doping concentration in the mesa region.

The specific output capacitance obtained by using the above analytical model is shown in Fig. 5.42 for the case of the 30-V power CC-MOSFET structure with gate electrode in the entire trench region. This structure has a cell pitch of 1 μm, trench

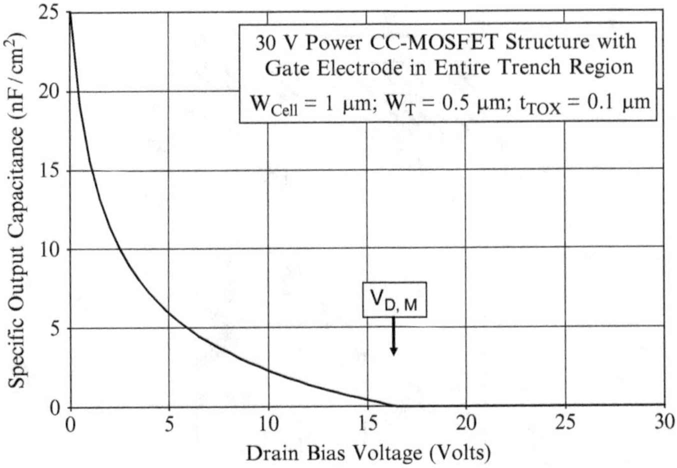


**Fig. 5.42** Output capacitance for the power CC-MOSFET structure with gate electrode in the entire trench region

width of 0.5 μm, and a trench oxide thickness of 1,000 Å. A built-in potential of 0.8 V was assumed for the P-base/N-drift junction, and the drift region had a doping concentration of $1.0 \times 10^{17}/cm^3$. It can be observed that the specific output capacitance decreases with increasing drain bias voltage until a bias of 16.5 V due to the expansion of the depletion region at the junction and a reduction of the effective area for the junction. The output capacitance becomes equal to zero at a drain bias of 16.5 V when the mesa region becomes completely depleted. At a small drain bias of 0.5 V, the specific output capacitance predicted by the analytical model for this structure is 19 $nF/cm^2$.

The input and output capacitances for the power CC-MOSFET structure with source electrode in the trenches can be modeled using the same equations derived for the power CC-MOSFET structure with gate electrode in the entire trench region. However, the gate transfer capacitance for the power CC-MOSFET structure with source electrode in the trenches is much smaller because the gate electrode does not extend into the deep trenches. Moreover, the gate electrode for power CC-MOSFET structure with source electrode in the trenches gets screened from the drain by the presence of the source electrode in the trenches.

The gate transfer capacitance for the power CC-MOSFET structure with source electrode in the trenches can be modeled using the depletion layers extending from the gate electrode into the drift region as illustrated in Fig. 5.43. Two sets of capacitances are indicated in the figure. The first set pertains to the vertical extension of the gate electrode ($L_G - x_{JP}$) into the drift region below the junction ($J_1$) between the P-base region and the drift region. The second set pertains to the bottom surface of the gate electrode ($t_{TOX} - t_{GOX}$) that does not overlap the source electrode within the trenches. Without taking the screening by the source electrode

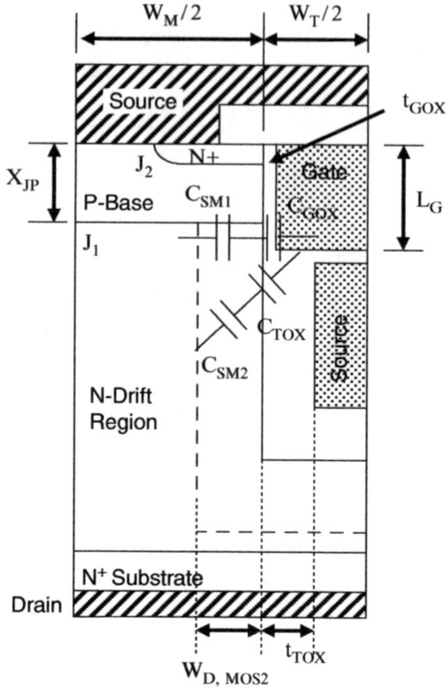


**Fig. 5.43** Depletion boundaries and capacitances within the power CC-MOSFET structure with-source electrode in the trench region

into account, the gate transfer capacitance for the power CC-MOSFET structure with source electrode in the trenches is given by:

$$C_{GD,SP} = \frac{2(L_G - x_{JP})}{W_{Cell}} \left( \frac{C_{GOX} C_{S,M1}}{C_{GOX} + C_{S,M1}} \right) + 2 \left( \frac{t_{TOX} - t_{GOX}}{W_{Cell}} \right) \left( \frac{C_{TOX} C_{S,M2}}{C_{TOX} + C_{S,M2}} \right) \quad (5.59)$$

where $C_{S,M1}$ is the semiconductor capacitance corresponding to the gate oxide thickness; $C_{GOX}$ is the gate oxide specific capacitance; $C_{S,M2}$ is the semiconductor capacitance corresponding to the trench oxide thickness; and $C_{TOX}$ is the trench oxide specific capacitance.

It is convenient to express the capacitances for the power CC-MOSFET structure with source electrode in the trenches in terms of a gate geometry factor [10]:

$$C_{GD,SP} = K_{G4} \left( \frac{C_{GOX} C_{S,M1}}{C_{GOX} + C_{S,M1}} \right) + K_{G5} \left( \frac{C_{TOX} C_{S,M2}}{C_{TOX} + C_{S,M2}} \right) \quad (5.60)$$

where

$$K_{GD4} = \frac{2(L_G - x_{JP})}{W_{Cell}} \quad (5.61)$$

and

$$K_{GD5} = \frac{2(t_{TOX} - t_{GOX})}{W_{Cell}} \tag{5.62}$$

The screening of the gate electrode by the source electrode in the trenches can be accounted for by using a screening factor that takes into account the depletion of the mesa region by the extension of a depletion layer for the MOS structure comprising the source electrode, the thick trench oxide, and the drift region. The screening factor is then given by:

$$K_{S,CC} = \frac{W_M - 2W_{D,MOS2}}{W_M} \tag{5.63}$$

The screening factor as defined above decreases with increasing drain bias voltage and becomes equal to zero at the drain bias required to completely deplete the mesa region. The gate transfer capacitance for the power CC-MOSFET structure with source electrode in the trenches with screening taken into account can be computed by multiplying the gate transfer capacitance given by (5.60) by the screening factor given by (5.63).

The specific gate transfer capacitances obtained by using the above analytical formulae are shown in Fig. 5.44 for the case of the power CC-MOSFET structure source electrode in the trenches. This structure had a cell pitch of 1 μm, a trench width of 0.5 μm, gate oxide thickness of 500 Å, and trench oxide thickness of 1,000 Å. The gate was assumed to extend to a depth of 0.4 μm which is 0.1 μm below the junction between the P-base region and the drift region. Without taking

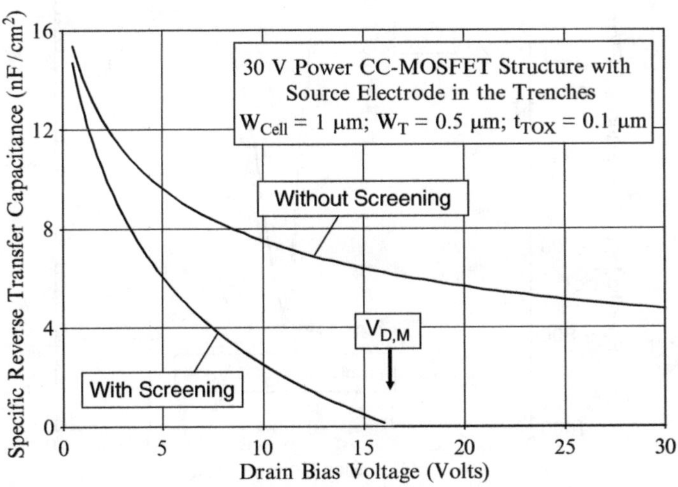


**Fig. 5.44** Gate-drain capacitance for the power CC-MOSFET structure with source electrode in the trenches

into account the screening by the source electrode, the specific gate transfer capacitance reduces from 15 to 5 nF/cm$^2$ at a drain bias of 30 V. In contrast to this, when screening by the source electrode is taken into account, the specific gate transfer capacitance has about the same value of 15 nF/cm$^2$ as without the screening at a drain bias of 0.5 V but reduces much more rapidly and becomes equal to 0 nF/cm$^2$ at a drain bias of 16.5 V when the mesa region becomes completely depleted.

### *5.5.1 Simulation Example*

The capacitances of the 30-V power CC-MOSFET structure were extracted using two-dimensional numerical simulations with a small AC signal superposed on the DC gate bias voltage. The input capacitances obtained for the power CC-MOSFET structure with gate electrode within the entire trench region are shown in Fig. 5.45 at a drain bias of 20 V. It is comprised of two components – the first is between the gate electrode and the source electrode ($C_{GS}$) while the second is between the gate electrode and the base electrode ($C_{GB}$). The total input capacitance can be obtained by the addition of these capacitances because they are in parallel and share a common contact electrode in the actual power CC-MOSFET structure. From the

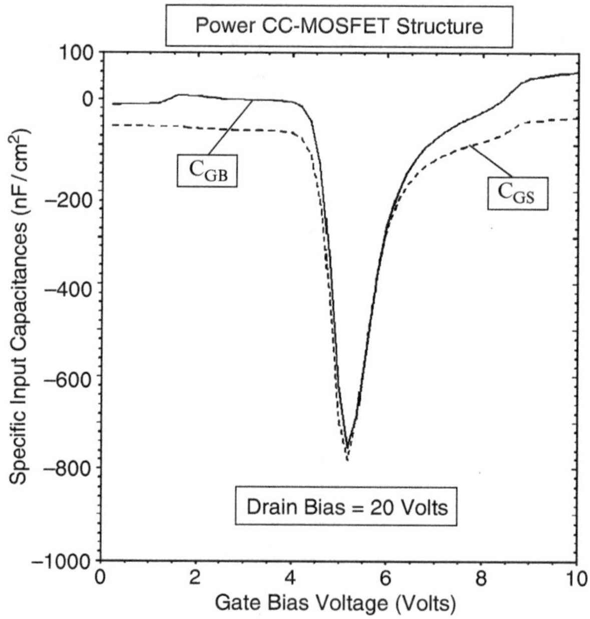


**Fig. 5.45** Input capacitances for the CC-MOSFET structure with gate electrode in the entire trench region

figure, a total specific input capacitance of about 50 nF/cm$^2$ is observed for gate bias voltages raging from zero to 4 V and from 8 to 10 V – close to that predicted by the analytical model. However, a large spike in input capacitance is observed between a gate bias of 4 and 7 V. This indicates that the effective input capacitance is larger (assumed to 75 nF/cm$^2$ when modeling the gate charge) than predicted by the simple analytical model for the power CC-MOSFET structure due to the capacitance contributed by the entire trench sidewall and bottom surfaces.

The drain-gate (reverse transfer) capacitance was extracted by performing the numerical simulations with a small AC signal superposed on the DC drain bias voltage. The values obtained for the 30-V power CC-MOSFET structure with gate electrode in the entire trench region are shown in Fig. 5.46. The gate-to-drain and base-to-drain capacitances are shown in the figure for comparison. Both of these capacitances decrease with increasing drain bias voltage. The gate transfer capacitance decreases abruptly at a drain bias of about 16 V and then reduces gradually. This behavior and the numerical values obtained from the simulations are in excellent agreement with the values predicted by the analytical model (see Fig. 5.40) for the initial point and for the transitions.

The output capacitance for the 30-V power CC-MOSFET structure with gate electrode in the entire trench region obtained from the numerical simulations is also provided shown in Fig. 5.46. It can be observed that the output capacitance is about 22 nF/cm$^2$ at a drain bias of 0.5 V and reduces to zero at a drain bias of about 16 V.

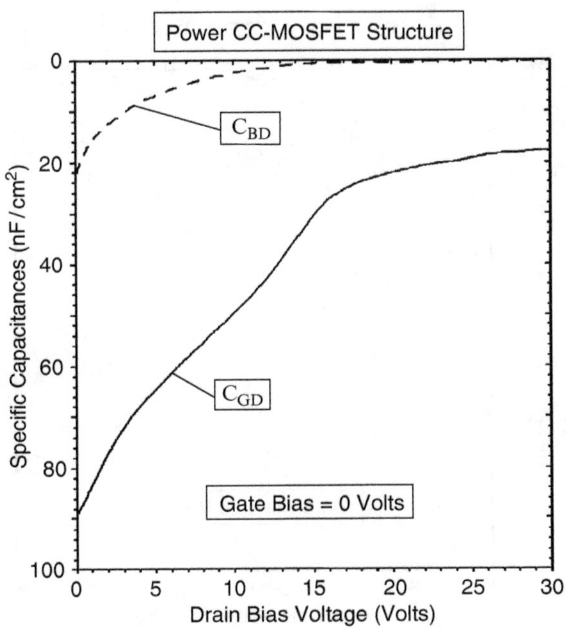


**Fig. 5.46** Reverse transfer and output capacitances for the CC-MOSFET structure with gate electrode in the entire trench region

This behavior is in excellent agreement with the values computed by using the analytical model (see Fig. 5.42). Note that the gate transfer capacitance for the power CC-MOSFET structure with gate electrode in the entire trench region is much larger than its output capacitance as well as that for other power MOSFET structures described in the previous chapters. The power CC-MOSFET structure with gate electrode in the entire trench region is therefore most suitable for applications operating at low frequencies where reducing the on-resistance is the most important criterion.

The drain-gate (reverse transfer) capacitance obtained for the 30-V power CC-MOSFET structure with source electrode in the trenches is shown in Fig. 5.47 by the solid line. The gate transfer capacitance for the structure with source electrode in the trenches can be observed to be much smaller than that for the structure with gate electrode in the entire trench region. The gate transfer capacitance for the structure with source electrode in the trenches decreases with increasing drain bias voltage and becomes close to zero at a drain bias of 16 V. This behavior and the numerical values obtained from the simulations are in excellent agreement with the values predicted by the analytical model (see Fig. 5.44) when the screening by the source electrode is taken into account. The output capacitance for the power CC-MOSFET structure with source electrode in the trenches can be observed to be very similar in behavior to the structure with gate electrode in the entire trench region.

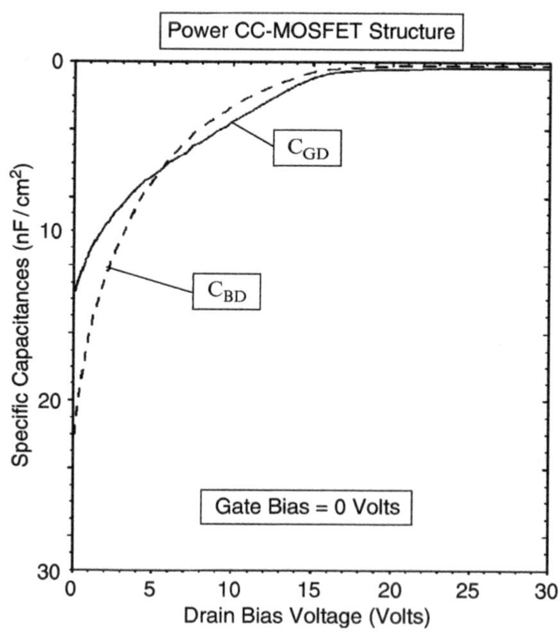


**Fig. 5.47** Reverse transfer and output capacitances for the CC-MOSFET structure with source electrode in the trenches

## 5.6 Gate Charge

The most significant gate charge components for assessing the performance of the power MOSFET structures are $Q_{SW}$ (the gate switching charge), $Q_{GD}$ (the gate-drain charge), and $Q_G$ (the total gate charge). In the case of the power CC-MOSFET structure, these components are given by the same equations derived in the textbook with adjustments made for the accounting for the trench oxide capacitance. The expressions for the gate charge $Q_{GS1}$ and $Q_{GS2}$ are the same as those provided in the textbook. However, in the case of the gate transfer charge, the gate transfer capacitance must be modeled by taking into account the portion at the gate oxide and the trench oxide separately. In addition, the voltage transition in the power CC-MOSFET structure occurs in two steps – first with a lower capacitance when the mesa region is entirely depleted at the larger drain bias voltages, and later with a larger capacitance when the entire mesa region is no longer depleted. The first transition will be defined to occur until a time $t_3$ with the second part completed at a time $t_4$.

Using the methodology provided in the textbook [10], the gate transfer charge corresponding to the first part of the transition where the drain voltage changes from $V_{DS}$ to $V_{D,M}$ is given by:

$$Q_{GD1} = \frac{2q\varepsilon_S N_D}{C_{TOX}}\left[\sqrt{1+\frac{2V_{DS}C_{TOX}^2}{q\varepsilon_S N_D}}-\sqrt{1+\frac{2V_{D,M}C_{TOX}^2}{q\varepsilon_S N_D}}\right] \quad (5.64)$$

Note that the gate geometry factor is unity during this phase of the turn-on process. The gate transfer charge corresponding to the second part of the transition where the drain voltage changes from $V_{D,M}$ to $V_{ON}$ is given by:

$$Q_{GD2} = \frac{2K_{G1}q\varepsilon_S N_D}{C_{GOX}}\left[\sqrt{1+\frac{2V_{D,M}C_{GOX}^2}{q\varepsilon_S N_D}}-\sqrt{1+\frac{2V_{ON}C_{GOX}^2}{q\varepsilon_S N_D}}\right] + \frac{2K_{G2}q\varepsilon_S N_D}{C_{TOX}}\left[\sqrt{1+\frac{2V_{D,M}C_{TOX}^2}{q\varepsilon_S N_D}}-\sqrt{1+\frac{2V_{ON}C_{TOX}^2}{q\varepsilon_S N_D}}\right] \quad (5.65)$$

The other components of the gate charge are similar to those already provided in the textbook:

$$Q_{SW} = Q_{GS2} + Q_{GD} \quad (5.66)$$

$$Q_G = [C_{GS} + C_{GD}(V_{DS})]V_{GP} + Q_{GD} + [C_{GS} + C_{GD}(V_{ON})](V_G - V_{GP}) \quad (5.67)$$

The gate charge values obtained for the 30-V power CC-MOSFET structure with gate electrode in the entire trench region by using the above analytical equations are: $Q_{GD} = 1{,}367\ \text{nC/cm}^2$; $Q_{SW} = 1{,}376\ \text{nC/cm}^2$; and $Q_G = 2{,}829\ \text{nC/cm}^2$. The gate transfer charge for the power CC-MOSFET structure with gate electrode in the entire trench region is very large when compared with the power MOSFET structures discussed in the previous chapters. Based up on this high gate transfer capacitance, it can be concluded that the power CC-MOSFET structure with gate electrode in the entire trench region is suitable for only low frequency applications where only the specific on-resistance is of importance.

Equations for the gate voltage, drain current, and drain voltage waveforms obtained by using the analytical model are provided in the textbook [10]. However, the drain voltage waveform for the power CC-MOSFET structure must be reformulated because of the more complex nature of the gate transfer capacitance. The drain voltage waveform can be derived by treating the transitions between times $(t_3 - t_2)$ and $(t_4 - t_3)$ separately. The resulting equations are:

$$v_D(t) = \frac{q\varepsilon_S N_D}{2C_{TOX}^2}\left\{\left[\sqrt{1 + \frac{2V_{DS}C_{TOX}^2}{q\varepsilon_S N_D}} - \frac{J_G C_{TOX}(t - t_2)}{2q\varepsilon_S N_D}\right]^2 - 1\right\} \quad (5.68)$$

from $t = t_2$ to $t = t_3$ and

$$v_D(t) = \left[\frac{J_G(t - t_3)}{(K_{G1} + K_{G2})\sqrt{2q\varepsilon_S N_D}} - \sqrt{V_{D,M}}\right]^2 \quad (5.69)$$

from $t = t_3$ to $t = t_4$.The waveforms obtained for the 30-V power CC-MOSFET structure with gate electrode in the entire trench region – using 1 μm cell pitch and a trench width of 0.5 μm with a gate oxide thickness of 500 Å and trench oxide thickness of 1,000 Å – using these equations are provided in Fig. 5.48. A gate drive current density of 2.0 A/cm$^2$ was used to turn on the device from a steady-state blocking voltage of 20 V to match the results of two dimensional numerical simulations discussed below.

The gate voltage initially increases linearly with time. After reaching the threshold voltage, the drain current can be observed to increase very quickly because of the large transconductance for this device structure. The drain current density increases until it reaches an on-state current density of 250 A/cm$^2$. The on-state current density determines the gate plateau voltage which has a value of 1.09 V for a threshold voltage of 1.0 V at this drain bias. During the gate voltage plateau phase, the drain voltage decreases in a non-linear manner in two phases – for time interval $(t_3 - t_2)$ and for time interval $(t_4 - t_3)$ – until it reaches the on-state voltage drop. After this time, the gate voltage again increases but at a slower rate than during the initial turn-on phase due to the larger gate transfer capacitance.

The gate charge components for the power CC-MOSFET structure with source electrode in the trenches are given by similar equations. However, adjustments must be made by accounting for the source electrode in the trenches. The gate transfer capacitance for this device structure is close to zero during the transition from a drain bias of $V_{DS}$ to a drain bias of $V_{D,M}$. This implies an instantaneous drop in drain voltage

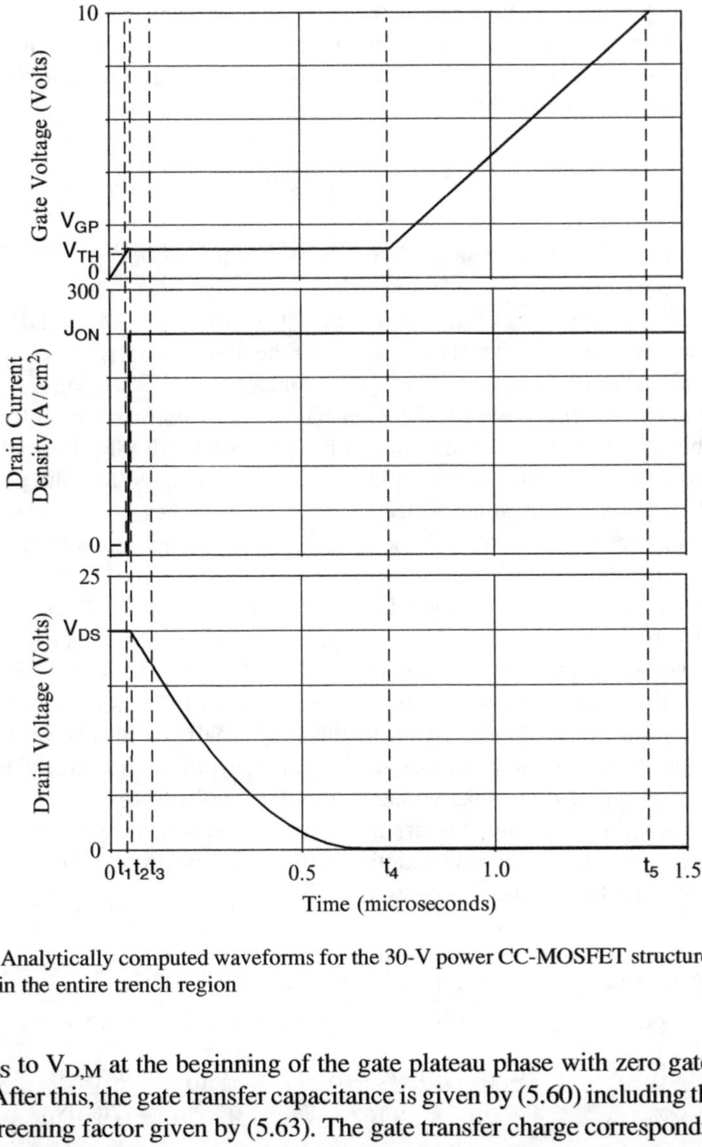


**Fig. 5.48** Analytically computed waveforms for the 30-V power CC-MOSFET structure with gate electrode in the entire trench region

from $V_{DS}$ to $V_{D,M}$ at the beginning of the gate plateau phase with zero gate transfer charge. After this, the gate transfer capacitance is given by (5.60) including the impact of the screening factor given by (5.63). The gate transfer charge corresponding to the transition where the drain voltage changes from $V_{D,M}$ to $V_{ON}$ is given by:

$$Q_{GD2} = \frac{2K_{G4}K_{SCC}q\varepsilon_S N_D}{C_{GOX}}\left[\sqrt{1+\frac{2V_{D,M}C_{GOX}^2}{q\varepsilon_S N_D}} - \sqrt{1+\frac{2V_{ON}C_{GOX}^2}{q\varepsilon_S N_D}}\right] + \frac{2K_{G5}K_{SCC}q\varepsilon_S N_D}{C_{TOX}}\left[\sqrt{1+\frac{2V_{D,M}C_{TOX}^2}{q\varepsilon_S N_D}} - \sqrt{1+\frac{2V_{ON}C_{TOX}^2}{q\varepsilon_S N_D}}\right] \quad (5.70)$$

The screening factor is a function of the drain bias voltage. However, in order to simplify the analysis, the screening factor was assumed to be unity for the gate charge analysis. The other components of the gate charge are similar to those already provided in the textbook:

$$Q_{SW} = Q_{GS2} + Q_{GD} \tag{5.71}$$

$$Q_G = [C_{GS} + C_{GD}(V_{DS})]V_{GP} + Q_{GD} + [C_{GS} + C_{GD}(V_{ON})](V_G - V_{GP}) \tag{5.72}$$

The gate charge values obtained for the 30-V power CC-MOSFET structure with source electrode in the trenches by using the above analytical equations are: $Q_{GD} = 295$ nC/cm$^2$; $Q_{SW} = 300$ nC/cm$^2$; and $Q_G = 1{,}197$ nC/cm$^2$. The gate transfer charge for the power CC-MOSFET structure with source electrode in the trenches is much smaller than that for the power CC-MOSFET structure with gate electrode in the entire trench region. It can be concluded that the power CC-MOSFET structure with source electrode in the trenches is suitable for high frequency applications where both the specific on-resistance and switching losses must be optimized.

Equations for the gate voltage, drain current, and drain voltage waveforms obtained by using the analytical model are provided in the textbook [10]. However, the drain voltage waveform for the power CC-MOSFET structure with source electrode in the trenches must be reformulated because of the more complex nature of the gate transfer capacitance. Due to the very small (or ideally zero) reverse transfer capacitance for the power CC-MOSFET structure with source electrode in the trenches at larger drain bias voltages, the drain voltage first reduces abruptly from $V_{DS}$ to $V_{D,M}$. After this, the drain voltage is determined by the gate transfer capacitance given by (5.60). The drain voltage waveform can be obtained by obtaining the solution of the following equation:

$$\begin{aligned}&\frac{K_{SCC}K_{G4}}{C_{GOX}}\sqrt{1+\frac{2C_{GOX}^2 v_D(t)}{q\varepsilon_S N_D}}+\frac{K_{SCC}K_{G5}}{C_{TOX}}\sqrt{1+\frac{2C_{TOX}^2 v_D(t)}{q\varepsilon_S N_D}}\\&=\frac{K_{SCC}K_{G4}}{C_{GOX}}\sqrt{1+\frac{2C_{GOX}^2}{q\varepsilon_S N_D}}+\frac{K_{SCC}K_{G5}}{C_{TOX}}\sqrt{1+\frac{2C_{TOX}^2}{q\varepsilon_S N_D}}-\frac{J_G(t-t_2)}{2q\varepsilon_S N_D}\end{aligned} \tag{5.73}$$

for the time interval from $t = t_2$ to $t = t_3$.

The waveforms obtained for the 30-V power CC-MOSFET structure with source electrode in the trenches – using 1 μm cell pitch and a trench width of 0.5 μm with a gate oxide thickness of 500 Å and trench oxide thickness of 1,000 Å – using these equations are provided in Fig. 5.49. A gate drive current density of 2.0 A/cm$^2$ was used to turn on the device from a steady-state blocking voltage of 20 V to match the results of two dimensional numerical simulations discussed below. The specific input capacitance was assumed to be 75 nF/cm$^2$.

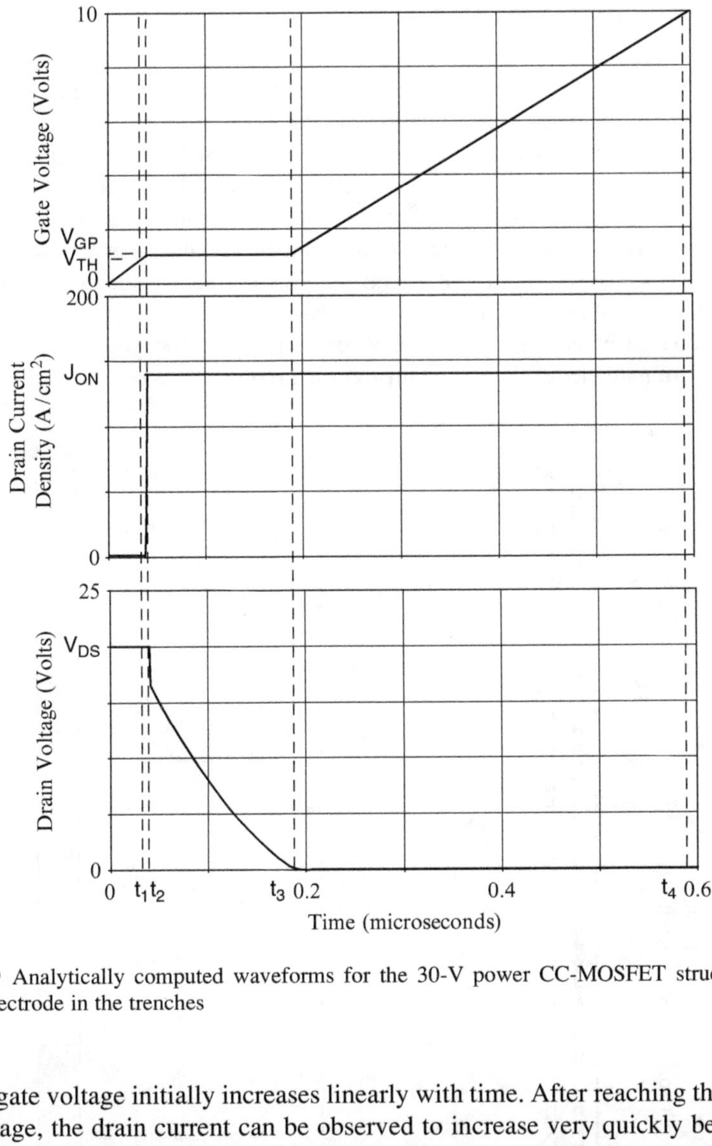


**Fig. 5.49** Analytically computed waveforms for the 30-V power CC-MOSFET structure with source electrode in the trenches

The gate voltage initially increases linearly with time. After reaching the threshold voltage, the drain current can be observed to increase very quickly because of the large transconductance for this device structure. The drain current density increases until it reaches an on-state current density of 140 A/cm$^2$. The on-state current density determines the gate plateau voltage which has a value of 1.07 V for a threshold voltage of 1.0 V at this drain bias. During the gate voltage plateau phase, the drain voltage drops abruptly from the drain supply voltage of 20 V to the mesa depletion voltage of 16.5 V. The drain voltage then decreases in a non-linear manner until it reaches the on-state voltage drop. After this time, the gate voltage again increases but at a slower rate than during the initial turn-on phase due to the larger gate transfer capacitance.

### 5.6.1 Simulation Example

The gate charges for the 30-V power CC-MOSFET structures were extracted by using the results of two-dimensional numerical simulations of the cell structures described in the previous sections. The devices were turned-on from blocking state with a drain bias of 20 V by using a gate current of $1 \times 10^{-8}$ A/μm (equivalent to 2.0 A/cm$^2$ for the area of $0.5 \times 10^{-8}$ cm$^2$). Once the drain current density reached the on-state value, the drain current was held constant resulting in a reduction of the drain voltage. Once the drain voltage reached the on-state value corresponding to the gate plateau voltage, the gate voltage increased to the steady-state value of 10 V. The gate charge waveforms obtained when turning on the power CC-MOSFET structure with gate electrode in the entire trench region are shown in Fig. 5.50. The

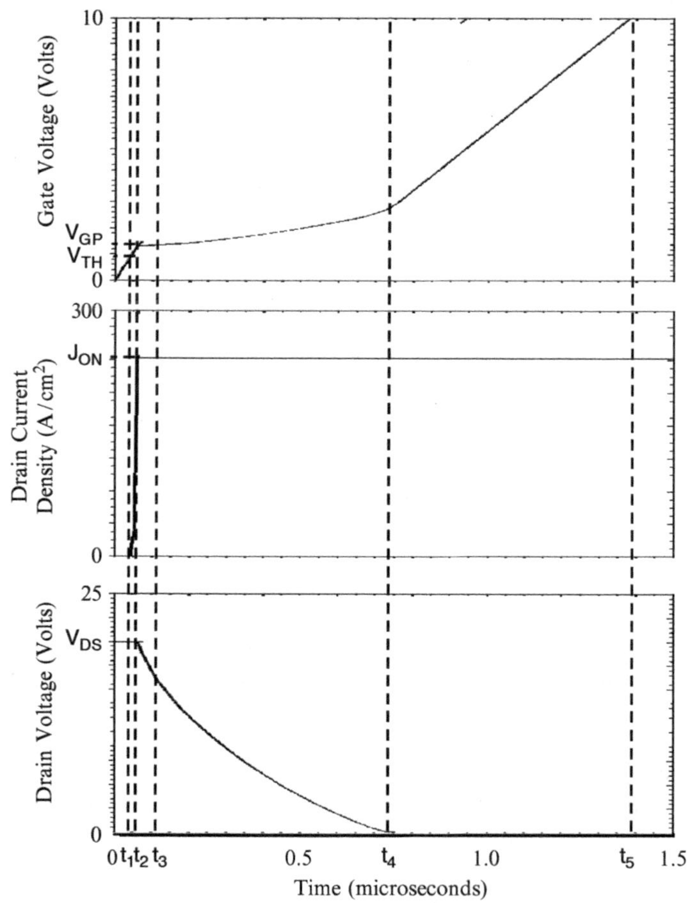


**Fig. 5.50** Turn-on waveforms for the 30-V power CC-MOSFET structure with gate electrode in the entire trench region

on-state current density in this case is 240 A/cm$^2$ at a DC gate bias of 10 V at the end of the turn-on transient. The gate voltage increases at a constant rate at the beginning of the turn-on process as predicted by the analytical model. When the gate voltage reaches the threshold voltage, the drain current begins to increase. The drain current increases very rapidly until it reaches the on-state current density of 240 A/cm$^2$.

Once the drain current reaches the on-state value, the gate voltage remains approximately constant at the plateau voltage ($V_{GP}$). The plateau voltage for this structure is 1.09 V for the drain current density of 240 A/cm$^2$ as governed by the transconductance of the device. The drain voltage decrease more rapidly at the beginning of the gate plateau phase as predicted by the analytical model until time $t_3$ and then continues to decrease during the plateau phase in a non-linear manner until time $t_4$. After the end of the plateau phase, the gate voltage again increases until it reaches the gate supply voltage. The shapes of the waveforms predicted by the analytical model (see Fig. 5.48) are in very good agreement with the results of the numerical simulations indicating that the waveforms can be predicted by properly accounting for the variation of the gate transfer capacitance for this structure.

The values for the various components of the gate charge extracted from the numerical simulations for the power CC-MOSFET structure with gate electrode in the entire trench region are compared with those calculated by using the analytical model in Fig. 5.51. There is excellent agreement between these values indicating that the analytical model is a good representation of the physics of turn-on for this device structure. Due to the large values for the gate charge, it can be concluded that the power CC-MOSFET structure with gate electrode in the entire trench region is only suitable for low frequency applications where the specific on-resistance is of the greatest importance.

| **Specific Gate Charge** | **Numerical Simulation (nC/cm$^2$)** | **Analytical Model (nC/cm$^2$)** |
|---|---|---|
| $Q_{GS1}$ | 90 | 92.5 |
| $Q_{GS2}$ | 30 | 8.8 |
| $Q_{GS}$ | 120 | 101.3 |
| $Q_{GD}$ | 1360 | 1367 |
| $Q_{SW}$ | 1390 | 1376 |
| $Q_G$ | 2780 | 2829 |

**Fig. 5.51** Gate charge extracted from numerical simulations for the power CC-MOSFET structure with gate electrode in the entire trench region

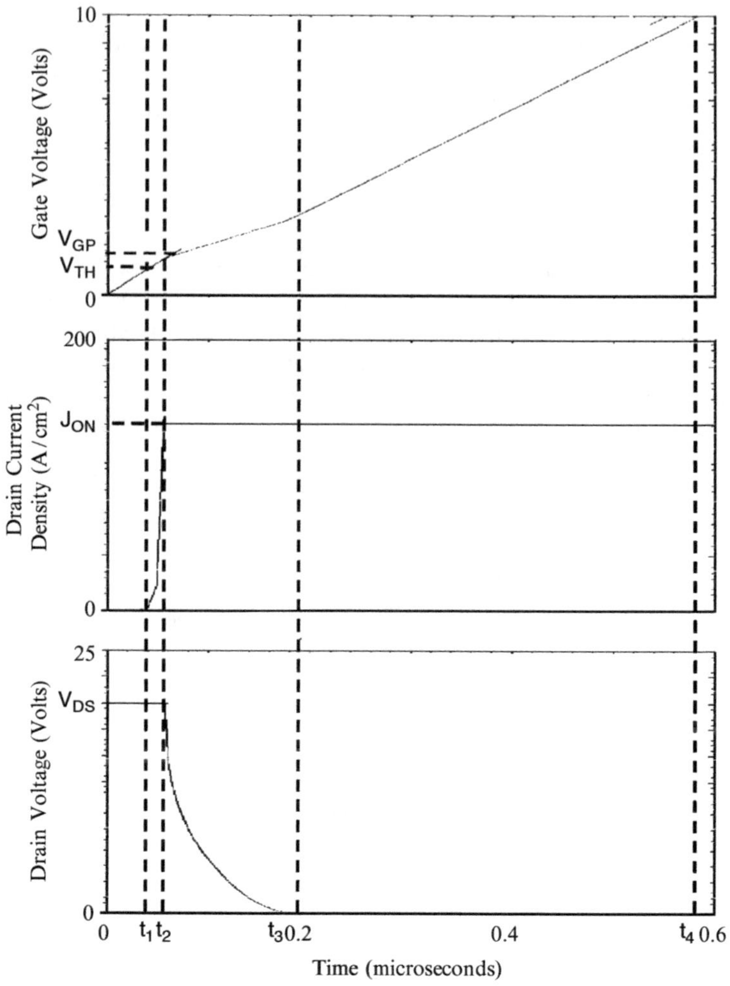


**Fig. 5.52** Turn-on waveforms for the 30-V power CC-MOSFET structure with source electrode in the trenches

The gate charge waveforms obtained by using an input gate current density of 2.0 A/cm$^2$ when turning on the power CC-MOSFET structure with source electrode in the trenches from a blocking state with drain bias of 20 V are shown in Fig. 5.52. The on-state current density in this case is 140 A/cm$^2$ at a DC gate bias of 10 V at the end of the turn-on transient. The gate voltage increases at a constant rate at the beginning of the turn-on process as predicted by the analytical model. When the gate voltage reaches the threshold voltage, the drain current begins to increase. The drain current increases very rapidly until it reaches the on-state current density of 140 A/cm$^2$.

| Specific Gate Charge | Numerical Simulation ($nC/cm^2$) | Analytical Model ($nC/cm^2$) |
|---|---|---|
| $Q_{GS1}$ | 80 | 75 |
| $Q_{GS2}$ | 30 | 5.3 |
| $Q_{GS}$ | 110 | 80.3 |
| $Q_{GD}$ | 270 | 295 |
| $Q_{SW}$ | 300 | 300 |
| $Q_G$ | 1160 | 1196 |

**Fig. 5.53** Gate charge extracted from numerical simulations for the power CC-MOSFET structure with source electrode in the trenches

Once the drain current reaches the on-state value, the gate voltage remains approximately constant at the plateau voltage ($V_{GP}$). The plateau voltage for this structure is 1.07 V for the drain current density of 140 A/cm$^2$ as governed by the transconductance of the device. At the beginning of the gate plateau phase, the drain voltage abruptly reduces from the drain supply voltage of 20 V to about 15 V. This behavior is predicted by the analytical model due to a zero gate transfer capacitance when the drain voltage is larger than the voltage required to deplete the mesa region. The drain voltage then decreases gradually as predicted by the analytical model until time $t_3$. After the end of the plateau phase, the gate voltage again increases until it reaches the gate supply voltage. The shapes of the waveforms predicted by the analytical model (see Fig. 5.49) are in very good agreement with the results of the numerical simulations indicating that the waveforms can be predicted by properly accounting for the variation of the gate transfer capacitance for this structure.

The values for the various components of the gate charge extracted from the numerical simulations for power CC-MOSFET structure with source electrode in the trenches are compared with those calculated by using the analytical model in Fig. 5.53. There is good agreement between these values indicating that the analytical model is a good representation of the physics of turn-on. Due to the smaller values for the gate charge, it can be concluded that the power CC-MOSFET structure with source electrode in the trenches is suitable for high frequency applications.

## 5.7 Device Figures of Merit

Significant power switching losses can arise from the charging and discharging of the large input capacitance in power MOSFET devices at high frequencies. The input capacitance ($C_{IN}$) of the power MOSFET structure must be charged to the gate supply voltage ($V_{GS}$) when turning on the device and then discharged to 0 V when turning off the device during each period of the operating cycle. The total power loss can be obtained by summing the on-state power dissipation for a duty cycle $\delta = t_{ON}/T$ and the switching power losses:

$$P_T = P_{ON} + P_{SW} = \delta R_{ON} I_{ON}^2 + C_{IN} V_{GS}^2 f \tag{5.74}$$

where $R_{ON}$ is the on-resistance of the power MOSFET structure, $I_{ON}$ is the on-state current, and f is the operating frequency. In writing this equation, the switching power losses due to the drain current and voltage transitions has been neglected. A minimum total power loss occurs for each power MOSFET structure at an optimum active area as shown in the textbook [10]. The on-state and switching power losses are equal at the optimum active area. The optimum active area at which the power dissipation is minimized is given by:

$$A_{OPT} = \sqrt{\frac{R_{ON,sp}}{C_{IN,sp}}} \left(\frac{I_{ON}}{V_{GS}}\right) \left(\sqrt{\frac{\delta}{f}}\right) \tag{5.75}$$

From the first term in this expression, a useful technology figure-of-merit can be defined:

$$FOM(A) = \frac{R_{ON,sp}}{C_{IN,sp}} \tag{5.76}$$

In the power electronics community, there is trend towards increasing the operating frequency for switch mode power supplies in order to reduce the size and weight of the magnetic components. The ability to migrate to higher operating frequencies in power conversion circuits is dependent on making enhancements to the power MOSFET technology. From the above equations, an expression for the minimum total power dissipation can be obtained [10]:

$$P_T(\min) = 2 I_{ON} V_{GS} \sqrt{\delta R_{ON,sp} C_{IN,sp} f} \tag{5.77}$$

A second technology figure of merit related to the minimum power dissipation can be defined as:

$$FOM(B) = R_{ON,sp} C_{IN,sp} \tag{5.78}$$

In most applications for power MOSFET structures with high operating frequency, the switching losses associated with the drain current and voltage transitions become a dominant portion of the total power loss. The time period associated with the increase of the drain current and decrease of the drain voltage is determined by the charging of the device capacitances. It is therefore common practice in the industry to use the following figures-of-merit to compare the performance of power MOSFET products [10]:

$$FOM(C) = R_{ON,sp}Q_{GD,sp} \tag{5.79}$$

and

$$FOM(D) = R_{ON,sp}Q_{SW,sp} \tag{5.80}$$

Although FOM(D) encompasses both the drain current and voltage transitions, it is customary to use FOM(C) because the gate-drain charge tends to dominate in the switching gate charge. One advantage of using these expressions is that the figure-of-merit becomes independent of the active area of the power MOSFET device.

The figures of merit computed for the power CC-MOSFET structure with gate electrode in the entire trench are provided in Fig. 5.54. The figure of merit usually used for comparison of device technologies in the literature is FOM(C). Most often, the value for this figure of merit at a gate bias of 4.5 V is utilized for selection of devices in the voltage regulator module application. In comparison with the power D-MOSFET structure (see Chap. 2), the power CC-MOSFET structure with gate electrode in the entire trench has a FOM(C) that is six times smaller. In comparison with the conventional power U-MOSFET structure (see Chap. 3), the power CC-MOSFET structure with gate electrode in the entire trench has a FOM(C) that is 2.4-times smaller. In comparison with the power U-MOSFET structure with thicker oxide at the trench bottom surface (see Chap. 3), the power CC-MOSFET structure with gate electrode in the entire trench has a FOM(C) that is 2.1-times smaller. Consequently, the power CC-MOSFET structure with gate electrode in the entire trench offers significant improvement in circuit performance when compared with these structures. However, in comparison with the power SC-MOSFET structure (see Chap. 4), the power CC-MOSFET structure with gate electrode in the entire trench has a FOM(C) that is 2.3-times larger indicating worse circuit performance at high frequencies.

The figures of merit computed for the power CC-MOSFET structure with source electrode in the trenches are provided in Fig. 5.55. The figure of merit usually used

| **Figures of Merit** | **$V_G$ = 4.5 V** | **$V_G$ = 10 V** |
|---|---|---|
| FOM(A) ($\Omega^2 cm^4 s^{-1}$) | 480 | 227 |
| FOM(B) (ps) | 2.7 | 1.3 |
| FOM(C) (mΩ*nC) | 49.2 | 23.2 |
| FOM(D) (mΩ*nC) | 49.5 | 23.4 |

**Fig. 5.54** Figures of merit for the power CC-MOSFET structure with gate electrode in the entire trench region

| Figures of Merit | $V_G = 4.5$ V | $V_G = 10$ V |
|---|---|---|
| FOM(A) ($\Omega^2 cm^4 s^{-1}$) | 547 | 333 |
| FOM(B) (ps) | 3.1 | 1.9 |
| FOM(C) (mΩ*nC) | 12.1 | 7.4 |
| FOM(D) (mΩ*nC) | 12.3 | 7.5 |

**Fig. 5.55** Figures of merit for the power CC-MOSFET structure with source electrode in the trenches

for comparison of device technologies in the literature is FOM(C). Most often, the value for this figure of merit at a gate bias of 4.5 V is utilized for selection of devices in the voltage regulator module application. In comparison with the power D-MOSFET structure (see Chap. 2), the power CC-MOSFET structure with source electrode in the trenches has a FOM(C) that is 25-times smaller. In comparison with the conventional power U-MOSFET structure (see Chap. 3), the power CC-MOSFET structure with source electrode in the trenches has a FOM(C) that is 9.9-times smaller. In comparison with the power U-MOSFET structure with thicker oxide at the trench bottom surface (see Chap. 3), the power CC-MOSFET structure with source electrode in the trenches has a FOM(C) that is 8.3-times smaller. In comparison with the power SC-MOSFET structure (see Chap. 4), the power CC-MOSFET structure with source electrode in the trenches has a FOM(C) that is 1.8-times smaller. Consequently, the power CC-MOSFET structure with source electrode in the trenches offers an extraordinary reduction of specific on-resistance, leading to smaller chip size and cost, while delivering a superior figure-of-merit for high frequency applications.

## 5.8 Edge Termination

A critical challenge for the development of the power CC-MOSFET structure, which operates with a breakdown voltage larger than the parallel-plane breakdown voltage of its drift region, is the formulation of an edge termination that can also support the blocking voltage. Fortunately, an elegant edge termination has been proposed [3] and demonstrated that can be implemented in a manner compatible with the process for the fabrication of the power CC-MOSFET structure. With this termination structure, the breakdown voltage occurs at the device cell structure rather than at the periphery resulting in performance described above for the cells.

The edge termination (labeled A) for the power CC-MOSFET structure is illustrated in Fig. 5.56 including a portion of the device with a mesa region located between two trenches as in the case of the device cell structure. The $N^+$ source region is not formed on this mesa region. In this case, the electrode in the trench at the edges of the device structure has the same features as the gate electrode, namely,

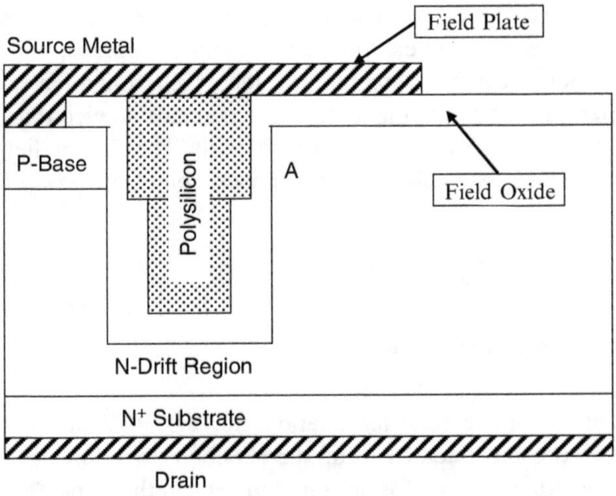


**Fig. 5.56** Edge termination A for the CC-MOSFET structure

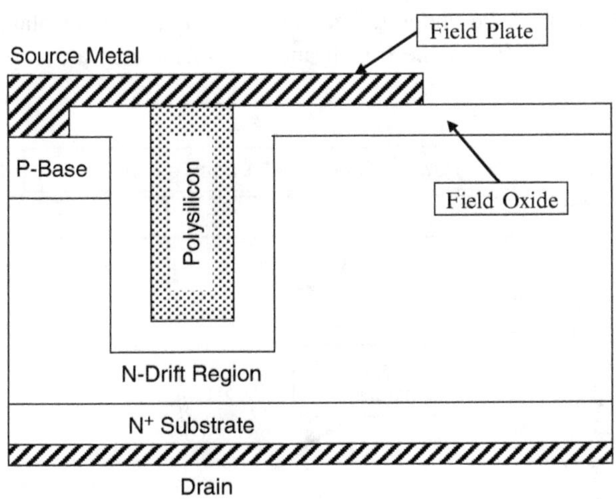


**Fig. 5.57** Edge termination B for the CC-MOSFET structure

a portion overlaps the trench sidewalls with thin gate oxide and a portion overlaps the trench sidewalls with thick trench oxide. A source connected field plate is also formed at the periphery of the device. This edge termination has the same breakdown voltage capability as the power CC-MOSFET cell structure because the applied drain bias is supported across the field oxide at the edge without creating a high electric field inside the semiconductor. However, all the applied drain bias is supported across the thin gate oxide at location A. For a device with blocking voltage rating of 30 V, an electric field of $6 \times 10^6$ V/cm is then produced across a

500-Å thick gate oxide. Although this is below the rupture strength for silicon dioxide, this high electric field can produce reliability problems.

The above problem can be overcome with the edge termination structure shown in Fig. 5.57. Here, the trench at the termination contains an electrode that overlaps the trench sidewalls with only the thick trench oxide. The electric field in the oxide is now reduced to levels suitable for reliable operation of the power CC-MOSFET structure.

### 5.8.1 Simulation Example

The viability of the above edge terminations for the 30-V power CC-MOSFET structures can be demonstrated by using the results of two-dimensional numerical simulations. The structure used for the simulations had the same doping profile for the drift region, the P-base region, and the $P^+$ contact region as shown in Fig. 5.15. The mesa width, gate oxide thickness, and trench oxide thickness, were chosen as 0.5 μm, 500 Å, and 1,000 Å, respectively, to match the device structures discussed in previous sections of the chapter. The source connected field plate extended by 1.25 μm beyond the edge of the trench and was located on a 5,000-Å thick field oxide.

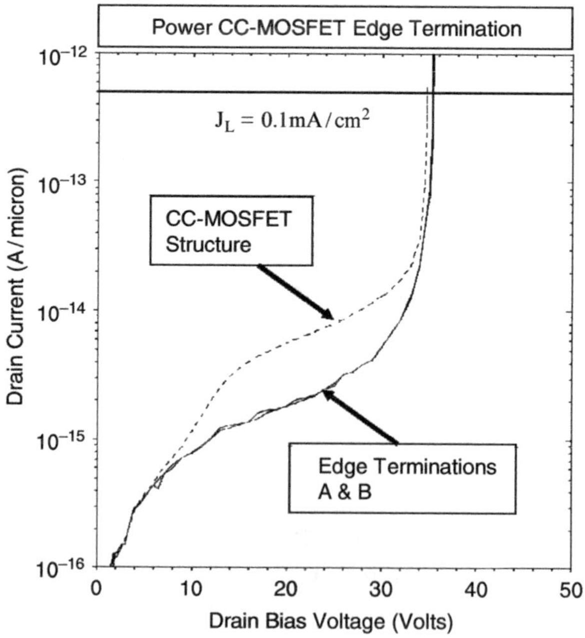


**Fig. 5.58** Blocking characteristics for the edge terminations suitable for 30-V power CC-MOSFET structures

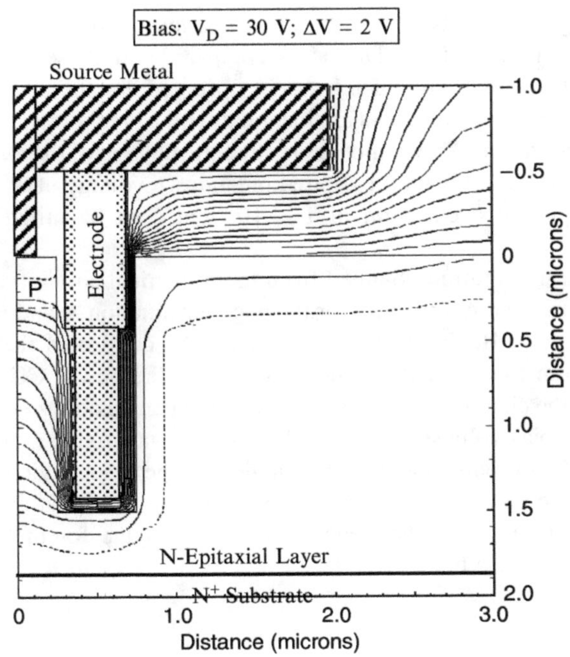


**Fig. 5.59** Potential distribution for the edge termination A at a drain bias of 30 V

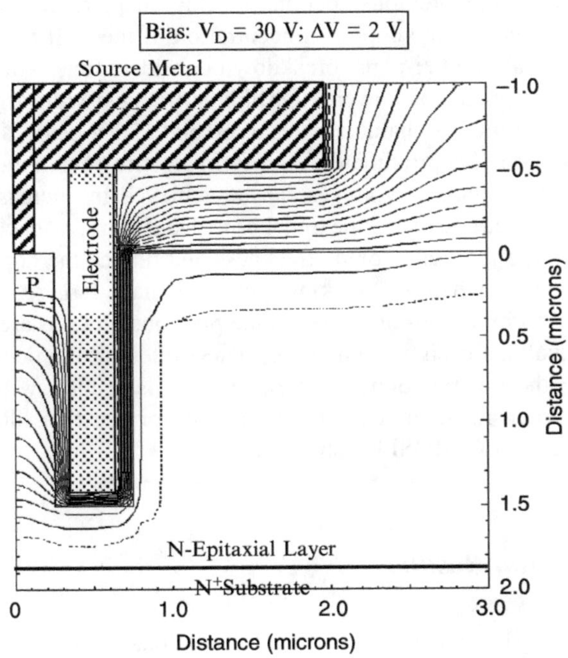


**Fig. 5.60** Potential distribution for the edge termination B at a drain bias of 30 V

The blocking characteristics for the two edge terminations obtained from the two-dimensional numerical simulations are compared with those for the power CC-MOSFET cell structure in Fig. 5.58. It can be observed that the blocking characteristics for the two edge terminations are identical with a breakdown voltage equal to that for the cell structure. This demonstrates that the physics of the two-dimensional charge coupling used to obtain the desired breakdown voltage of 35 V with a high drift region doping concentration of $1 \times 10^{17}/\text{cm}^3$ can be utilized in practical devices.

The potential distribution obtained from the numerical simulations at the edge termination is provided in Fig. 5.59 for the edge termination A and in Fig. 5.60 for the edge termination B. From the figures, it can be confirmed that the potential distribution within the mesa is identical to that for the power CC-MOSFET cell structure due to the charge-coupling phenomenon. It can be observed from Fig. 5.59 that all the drain bias voltage is supported across the gate oxide at the edge for the termination A. This produces a high electric field in the oxide which could degrade the reliability. In contrast, it can be observed from Fig. 5.60 that all the drain bias voltage is supported across the trench oxide at the edge for the termination B leading to lower electric fields.

## 5.9 High Voltage Devices

As demonstrated in the previous sections of this chapter, the charge-coupling concept allows increasing the doping concentration of the drift region well above that dictated by ideal parallel-plane breakdown considerations. This approach has utility in the development of power MOSFET structures with larger breakdown voltages. In this section, the characteristics of 60 and 120 V devices are examined. Under the assumption of approximately a uniform electric field in the vertical (or y) direction, the higher blocking voltages can be achieved by increasing the length ($L_D$) of the source electrode inside the trenches to 2 and 4 μm, respectively. It is also necessary to increase the trench oxide thickness and reduce the doping concentration of the drift region when the blocking voltage is made larger.

The analytical formulations presented in the previous sections are applicable for any breakdown voltage design when appropriate values for the device parameters are used during the computations. Consequently, this section will focus on the results of two-dimensional numerical simulations for the two blocking voltage ratings for the power CC-MOSFET structure.

### *5.9.1 Simulation Results*

The results of two-dimensional numerical simulations on the 60-V power CC-MOSFET structure are described here. The power 60 V CC-MOSFET structure used for the simulations had a mesa width ($W_M$) of 0.5 μm and a gate oxide thickness of 500 Å. The trench oxide thickness was increased to 3,000 Å due to the larger blocking voltage capability. In order to accommodate this larger trench oxide thickness, it is necessary to increase the trench width to 1 μm, leading to a half-cell width of 0.75 μm, in the simulations. The drift region consists of a total thickness of 4.5 μm with a doping concentration of $2 \times 10^{16}/cm^3$ for the upper 0.4 μm where the channel is formed. The doping concentration for the rest of the drift region was varied to study the impact on the blocking characteristics. The P-base region for this structure was formed by using ion-implantation to create a peak doping concentration at 0.15 μm below the surface with a depth of 0.3 μm. The $N^+$ source region has a depth of about 0.1 μm. The narrow width for the P-base region is possible because of the unique distribution of the electric field within the power CC-MOSFET structure. A heavily doped $P^+$ region is also included in the structure in the center of the mesa at the upper surface to allow fabrication of ohmic contacts to the P-base region.

The vertical doping profiles taken along the surface of the trench ($\times = 0.5$ μm) and the middle of the mesa region ($\times = 0.75$ μm) are shown in Fig. 5.61 for the 60 V CC-MOSFET structure. From the figure, it can be observed that the P-base region has a retrograde doping profile with a peak doping concentration of $1.5 \times 10^{17}/cm^3$ to obtain the desired threshold voltage. The vertical depths of the P-base and $N^+$ source regions are 0.30 and 0.08 μm leading to a channel length of

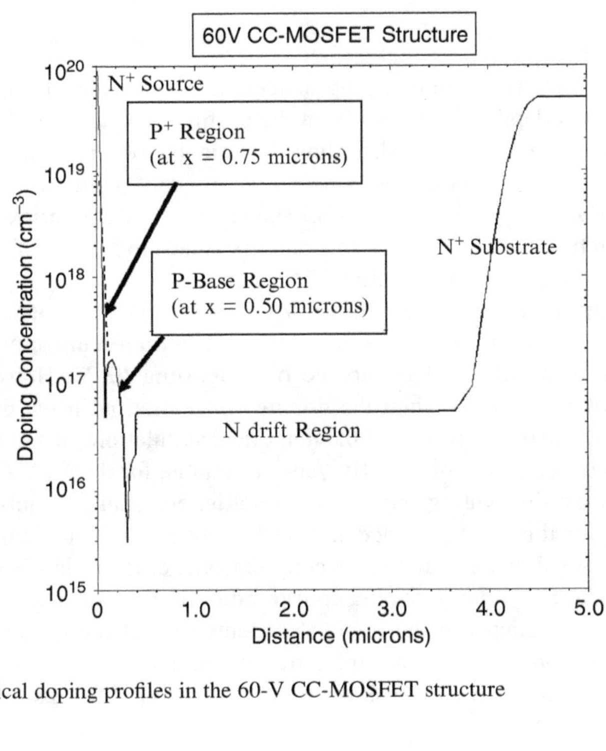


**Fig. 5.61** Vertical doping profiles in the 60-V CC-MOSFET structure

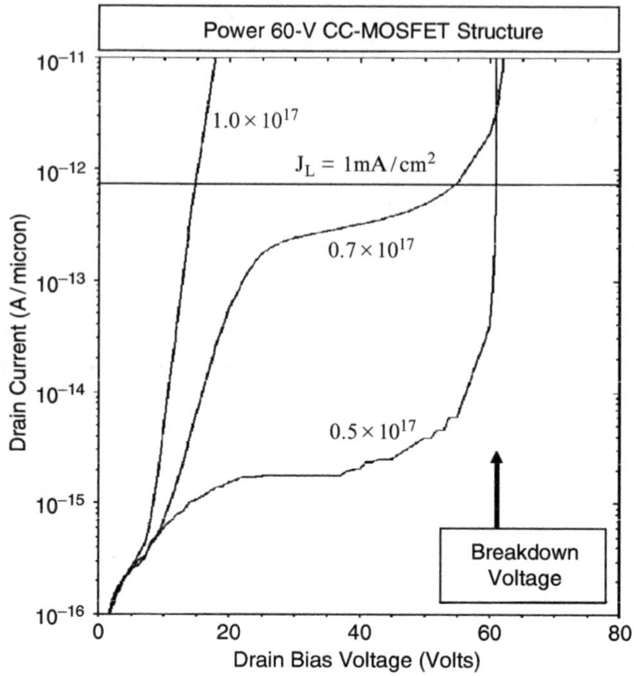


**Fig. 5.62** Blocking characteristics for the 60-V CC-MOSFET structures

only 0.22 μm. The drift region doping concentration increases from $2 \times 10^{16}$ to $5 \times 10^{16}/cm^3$ at a depth of 0.4 μm. From the profile taken at $\times = 0.75$ μm in the middle of the mesa region (dashed line), it can be observed that the doping concentration of the $P^+$ contact region is $1 \times 10^{19}/cm^3$ with sufficient depth to contact the P-Base region. From the profile taken at $\times = 0.50$ μm along the trench sidewall, it can be observed that the drift region extends to a depth of 3.8 μm which is 1 μm deeper than the trench depth of 2.8 μm.

The blocking characteristics for the 60-V CC-MOSFET cell structure are shown in Fig. 5.62 at 300°K for three cases of doping concentration in the mesa region. It can be observed that the cell is capable of supporting 62 V. However, a large leakage current is observed when the doping concentration in the drift region is increased. Based up on the results of the numerical simulations, it can be concluded that a doping concentration of $5 \times 10^{16}/cm^3$ is suitable for the 60-V CC-MOSFET structure. It is worth pointing out that the one-dimensional parallel-plane breakdown voltage for this doping concentration is only 15 V. This demonstrates the benefit of the two-dimensional charge-coupling phenomenon in obtaining a high breakdown voltage with large doping concentration in the drift region.

The impact of changes in the doping concentration of the drift region on the leakage current can be seen in Fig. 5.62. When the doping concentration is increased to $7 \times 10^{16}/cm^3$, a breakdown voltage of 60 V is again obtained but

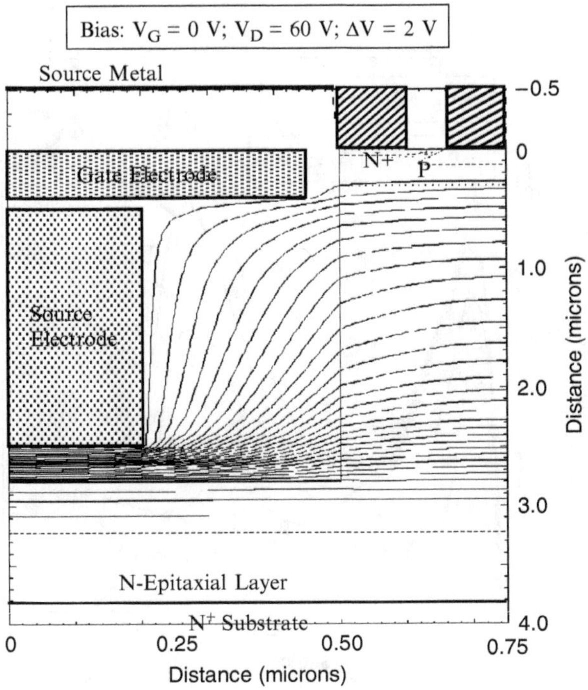


**Fig. 5.63** Potential contours in the 60-V CC-MOSFET structure

the leakage current becomes large. This effect is exacerbated when the doping concentration is increased to $1 \times 10^{17}/\text{cm}^3$ with a blocking voltage of less than 20 V for a leakage current density of 1 mA/cm$^2$.

In order to further demonstrate the two-dimensional charge coupling phenomenon, it is instructive to examine the potential contours inside the power 60-V CC-MOSFET structure when it is operating in the blocking mode. This also allows determination of the voltage distribution within the structure and the penetration of the depletion region in the P-base region. The potential contours for the 60-V CC-MOSFET structure obtained using the numerical simulations with zero gate bias and a drain bias voltage of 60 V are shown in Fig. 5.63. It can be observed that the drain bias voltage is supported in the drift region below the P-base/N-drift junction. This suppresses the depletion of the P-base region by the drain bias allowing a very short channel length for the power 60-V CC-MOSFET structure. The potential contours in Fig. 5.63 are shown for a doping concentration of $5 \times 10^{16}/\text{cm}^3$ in the drift region. It can be observed that the potential contours are closer together near the bottom of the trench when compared the vicinity of the P-base/N-drift junction. This implies that a larger doping concentration could be used in the drift region. However, this leads to an increase in the leakage current as shown in Fig. 5.62.

The electric field profiles along the vertical direction through the center of the mesa region are shown in Fig. 5.64 for the 60-V power CC-MOSFET structure with

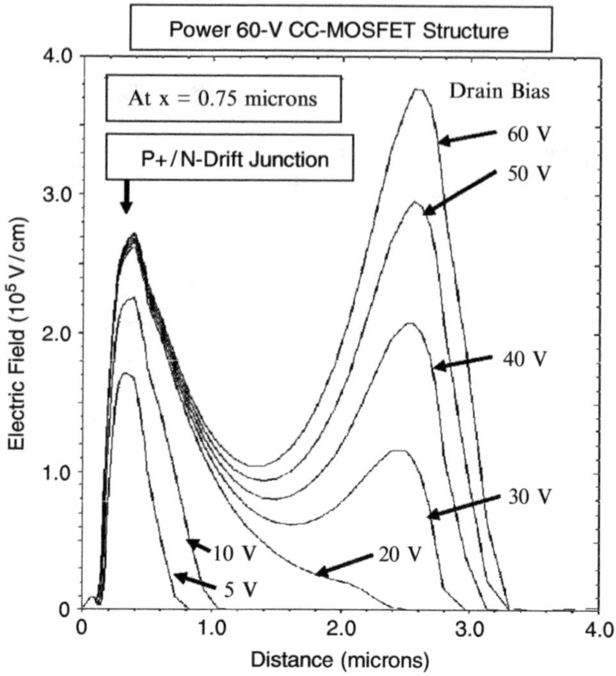


**Fig. 5.64** Electric field profiles in the 60-V CC-MOSFET structure

a drift region doping concentration of $5 \times 10^{16}/\text{cm}^3$. At a drain bias of 5 V, the electric field has a triangular shape representative of a one-dimensional junction. This shape prevails until a drain bias of 20 V, which agrees with a voltage ($V_{DM}$ = 19.99 V) for depletion of the mesa region as computed using (5.55). At drain bias voltages above 20 V, a second peak in the electric field develops near the bottom of the trenches due to the two-dimensional charge coupling phenomenon. The electric profile at the drain bias of 60 V resembles the profile predicted by the analytical model. It is worth pointing out that the electric field along the y-direction is not uniform and exhibits relatively small values at the center of the drift region along the y-direction.

The transfer characteristic for the 60 V CC-MOSFET structure with source electrode in the trenches was obtained using numerical simulations with a drain bias of 0.1 V at 300°K. The resulting transfer characteristics are shown in Fig. 5.65. The specific on-resistance can be obtained from the transfer characteristics at any gate bias voltage. For the case of a gate bias of 4.5 V and 300°K, the specific in-resistance is found to be 0.143 mΩ cm$^2$. For the case of a gate bias of 10 V and 300°K, the specific in-resistance is found to be 0.124 mΩ cm$^2$. Since these values are less than the ideal specific on-resistance of 0.247 mΩ cm$^2$ for the 60 V one-dimensional case, the power CC-MOSFET structure offers the opportunity to significantly enhance the performance of silicon devices.

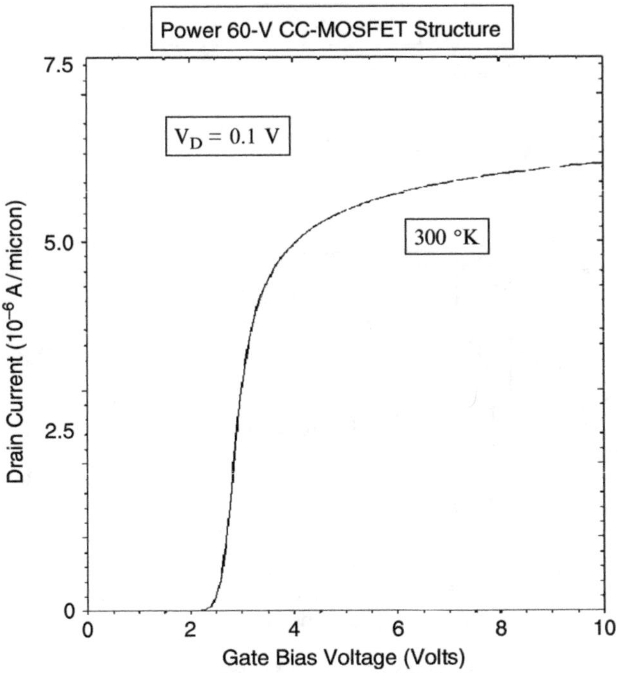


**Fig. 5.65** Transfer characteristics of the 60-V CC-MOSFET structure

The results of two-dimensional numerical simulations on the 120-V power CC-MOSFET structure are next described here. It is reasonable to assume that the larger blocking voltage capability can be achieved by increasing the depth of the trenches and the thickness of the oxide in the trenches. The power 120 V CC-MOSFET structure used for the simulations had a mesa width ($W_M$) of 0.5 μm and a gate oxide thickness of 500 Å. The trench oxide thickness was increased to 6,000 Å due to the larger blocking voltage capability. In order to accommodate this larger trench oxide thickness, it is necessary to increase the trench width to 1.6 μm, leading to a half-cell width of 1.05 μm, in the simulations.

The drift region consists of a total thickness of 6.5 μm with a doping concentration of $2 \times 10^{16}/cm^3$ for the upper 0.4 μm where the channel is formed. The doping concentration for the rest of the drift region was varied to study the impact on the blocking characteristics. The P-base region for this structure was formed by using ion-implantation to create a peak doping concentration at 0.15 μm below the surface with a depth of 0.3 μm. The $N^+$ source region has a depth of about 0.1 μm. The narrow width for the P-base region is possible because of the unique distribution of the electric field within the power CC-MOSFET structure. A heavily doped $P^+$ region is also included in the structure in the center of the mesa at the upper surface to allow fabrication of ohmic contacts to the P-base region. The vertical doping profile for the power 120 V CC-MOSFET structure is provided in

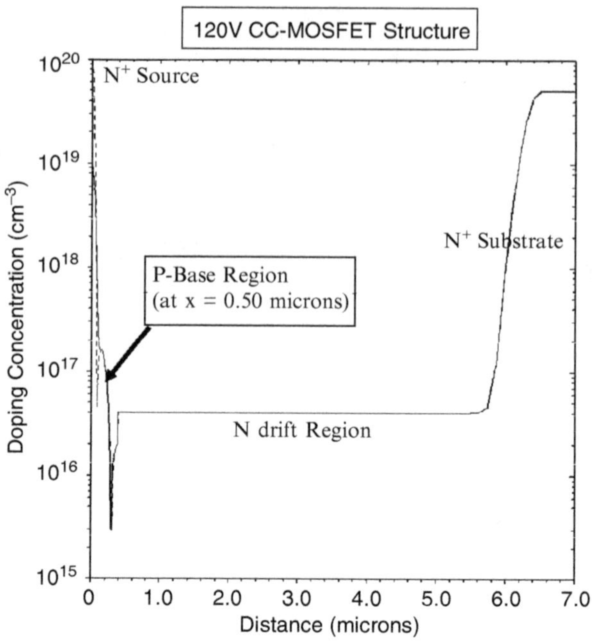


**Fig. 5.66** Vertical doping profile in the 120-V CC-MOSFET structure

Fig. 5.66. The doping profile for the channel region is the same as that used in the previous power CC-MOSFET structures. It can be observed that the drift region for the 120 V device extends to a depth of 5.6 μm which is 0.5 μm deeper than the trench depth of 5.1 μm.

The blocking characteristics for the 120-V CC-MOSFET cell structure are shown in Fig. 5.67 at 300°K for three cases of doping concentration in the mesa region. It can be observed that the cell is not capable of supporting 120 V with a trench depth scaled to 5 μm. The largest blocking voltage obtained is 86 V for a drift region doping concentration of $4 \times 10^{16}/\text{cm}^3$. A large leakage current is observed when the doping concentration in the drift region is increased.

The reason for not achieving the desired 120-V blocking voltage capability can be understood by examination of the electric field profiles along the vertical direction through the center of the mesa region as shown in Fig. 5.68 for the 120-V power CC-MOSFET structure with a drift region doping concentration of $4 \times 10^{16}/\text{cm}^3$. At a drain bias of 5 V, the electric field has a triangular shape representative of a one-dimensional junction. This shape prevails until a drain bias of 30 V, which agrees with a voltage ($V_{DM} = 30.08$ V) for depletion of the mesa region as computed using (5.55). At drain bias voltages above 30 V, a second peak in the electric field develops near the bottom of the trenches due to the two-dimensional charge coupling phenomenon. The electric field profile at the drain bias of 80 V resembles the profile predicted by the analytical model. The electric field along the y-direction is not

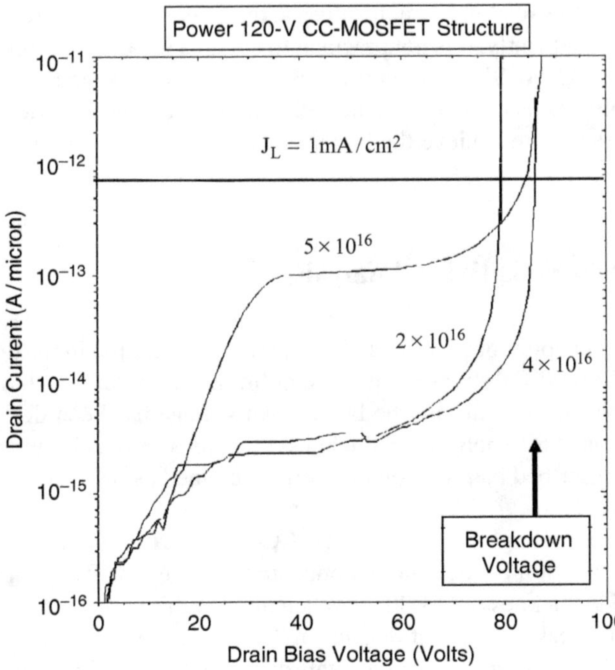


**Fig. 5.67** Blocking characteristics for the 120-V CC-MOSFET structures

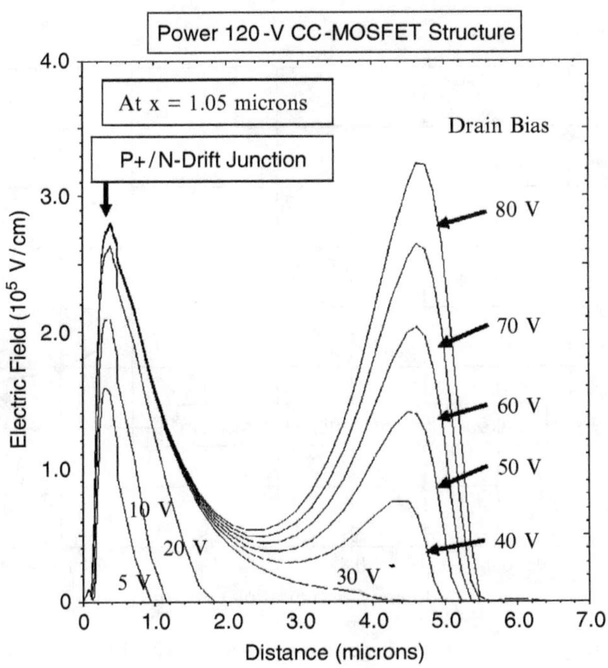


**Fig. 5.68** Electric field profiles in the 120-V CC-MOSFET structure

uniform and exhibits relatively small values at the center of the drift region along the y-direction. Consequently, it is not practical to obtain a breakdown voltage of 120 V with the power CC-MOSFET structure. It is necessary to change the doping profile in the drift region as discussed in the next chapter to make the electric field uniform along the y-direction to achieve the larger blocking voltage capability.

## 5.10 Process Sensitivity Analysis

As described in the previous sections, the doping concentration in the drift region of the power CC-MOSFET structure must be optimized in order to achieve a desired breakdown voltage. In addition, the breakdown voltage has been demonstrated to be dependent on the trench oxide thickness. In this section, the impact of these parameters is described based up on the results obtained using the two-dimensional numerical simulations.

The blocking voltage for the power CC-MOSFET structures is plotted in Fig. 5.69 as a function of the doping concentration in the drift region for the case of a trench oxide thickness of 1,000 Å. When making this plot, the blocking voltage was defined at a leakage current density of 1 mA/cm$^2$ because a high breakdown voltage (∼35 V) is observed even at high doping concentrations but the leakage current becomes very large as illustrated in Fig. 5.26. With this definition, the blocking voltage is observed to reduce rapidly when the doping concentration

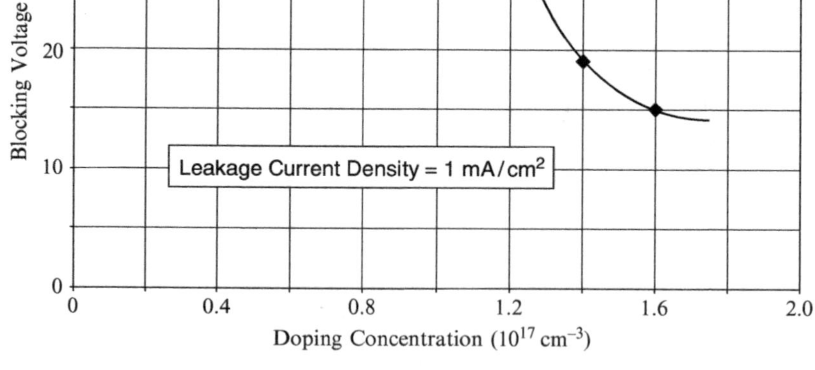


**Fig. 5.69** Sensitivity of breakdown voltage of the power CC-MOSFET structures to the doping concentration in the drift region

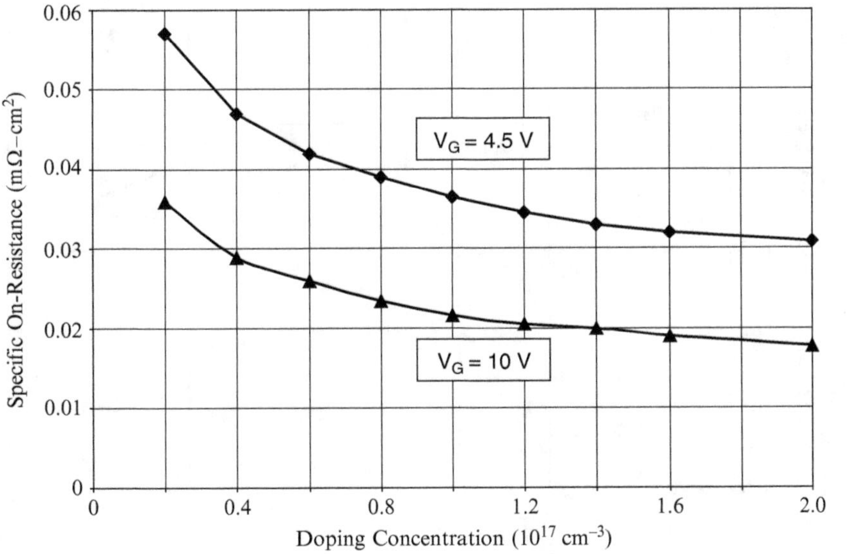


**Fig. 5.70** Sensitivity of specific on-resistance of the power CC-MOSFET structures to the doping concentration in the drift region

exceeds $1.2 \times 10^{17}/cm^3$. From the point of view of process margin, it would be adequate to use a doping concentration of $1.0 \times 10^{17}/cm^3$ in the drift region.

The specific on-resistance for the power CC-MOSFET structures is plotted in Fig. 5.70 as a function of the doping concentration in the drift region for two gate bias voltages for the case of a trench oxide thickness of 1,000 Å. As expected, the specific on-resistance reduces monotonically with increasing doping concentration of the drift region. Based up on a maximum doping concentration of $1.0 \times 10^{17}/cm^3$ in the drift region to achieve the desired blocking voltage capability with low leakage current, the smallest specific on-resistance for the power CC-MOSFET structure is found to be 0.0365 mΩ cm² at a gate bias of 4.5 V and 0.0217 mΩ cm² at a gate bias of 10 V.

An interesting trade-off curve between the specific on-resistance and the blocking voltage capability can be created by using the doping concentration in the drift region as a parametric variable. From the plot shown in Fig. 5.71, it can be observed that the blocking voltage degrades significantly while producing only a small improvement in the specific on-resistance once the doping concentration exceed $1.0 \times 10^{17}/cm^3$ in the drift region.

The blocking voltage for the power CC-MOSFET structures is plotted in Fig. 5.72 as a function of the trench oxide thickness for the case of a drift region doping concentration of $1 \times 10^{17}/cm^3$. When making this plot, the blocking voltage was defined at a leakage current density of 1 mA/cm². The blocking voltage is observed to increase up to a trench oxide thickness of 1,400 Å and then decrease with further increase in oxide thickness. A blocking voltage above 35 V can be obtained using a trench oxide thickness of between 1,000 and 1,500 Å. This is

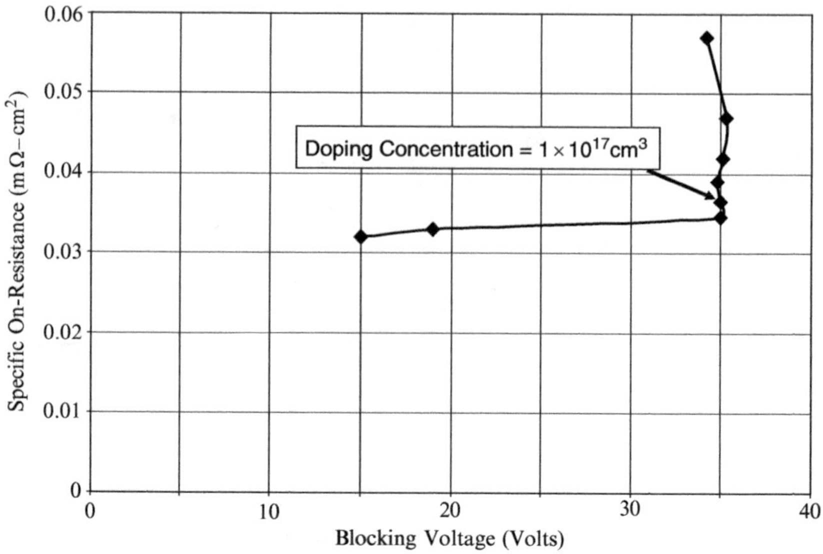


**Fig. 5.71** Trade-off curve between specific on-resistance and breakdown voltage for the power CC-MOSFET structures with doping concentration as a parameter

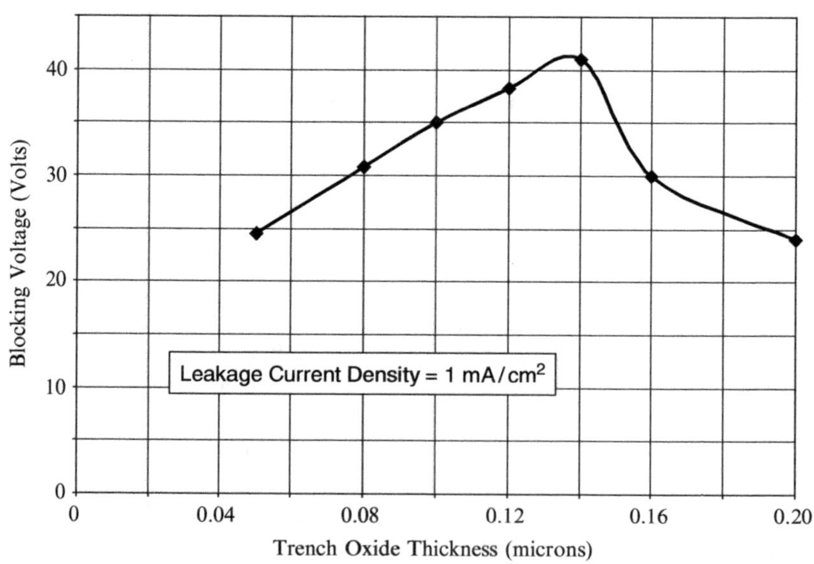


**Fig. 5.72** Sensitivity of breakdown age of the power CC-MOSFET structures to the trench oxide thickness

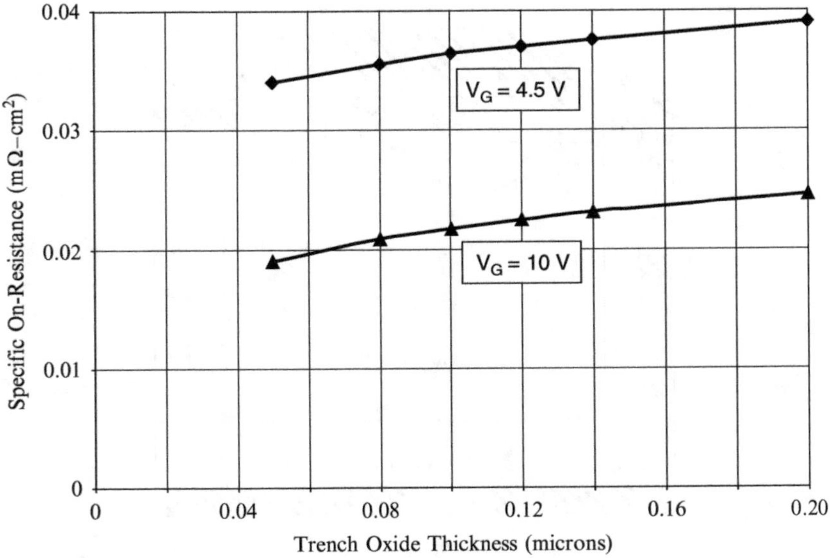


**Fig. 5.73** Sensitivity of specific on-resistance of the power CC-MOSFET structures to the trench oxide thickness

consistent with the optimum trench oxide thickness predicted by the analytical model (see Fig. 5.12). Since the trench oxide thickness can be precisely controlled by thermal oxidation, the broad maximum in blocking voltage capability is indicative of good process tolerance for the power CC-MOSFET structure.

The specific on-resistance for the power CC-MOSFET structures is plotted in Fig. 5.73 as a function of the trench oxide thickness for two gate bias voltages for the case of a drift region doping concentration of $1 \times 10^{17}/\text{cm}^3$. The specific on-resistance increases monotonically with increasing trench oxide thickness due to an increase in the resistance of the accumulation layer formed on the trench sidewalls. In the case of a trench oxide thickness of 1,000 Å, the specific on-resistance is found to be 0.0365 mΩ cm$^2$ at a gate bias of 4.5 V and 0.0217 mΩ cm$^2$ at a gate bias of 10 V.

## 5.11 Discussion

The physics of operation and resulting electrical characteristics of the power CC-MOSFET structure have been described in this chapter. The two-dimensional charge coupling phenomenon in these structures allows supporting blocking voltages well above the one-dimensional parallel-plane breakdown voltage. The specific on-resistance for this device structure is significantly reduced when compared with the power D-MOSFET and U-MOSFET structures due to its very short

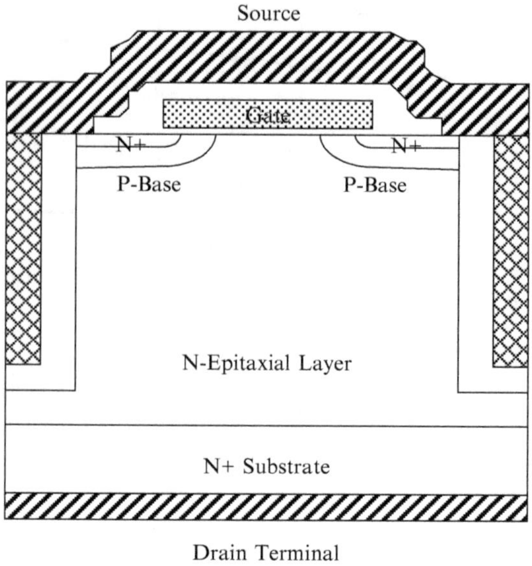


**Fig. 5.74** The CC-MOSFET structure with planar channel

channel length, small cell pitch, and high doping concentration in the drift region. With the power CC-MOSFET structure, it is possible to achieve a specific on-resistance that is smaller than the ideal specific on-resistance for silicon. In addition, small the reverse transfer capacitance and gate charge can be achieved in the power CC-MOSFET structure by utilizing a source electrode in the trenches to achieve the two-dimensional charge coupling phenomenon. Consequently, the figures-of-merit FOM(C) for the 30-V power CC-MOSFET structure with source electrode in the trenches is superior to that for all the power MOSFET structures discussed in the previous chapters.

The first commercially available power CC-MOSFET structures were developed with blocking voltage capability of 65 V for RF cellular base-station applications [12]. These devices utilize a planar gate architecture, as shown in Fig. 5.74, which is superior in terms of controlling the quality of the channel with a retrograde doping profile in the JFET region to reduce the gate transfer capacitance and charge [13]. In this structure, the source electrode in the trenches can be brought to the surface simplifying its contact to the source metal electrode. However, the cell pitch for this structure is larger than that for the power CC-MOSFET structure previously shown in Fig. 5.2 making its specific on-resistance larger.

Due to the significant reduction in the specific on-resistance achievable with the power CC-MOSFET structure, there has recently been considerable research activity around the world to develop processes for its fabrication. Devices having source electrode in the trenches with blocking voltage ratings of 35 V have been demonstrated with specific on-resistance of 0.038 m$\Omega$ cm$^2$ and a FOM(C) of 3.4 m$\Omega$ nC has been reported [8]. In addition, devices having gate electrode in the entire trench region with blocking voltage ratings of 80 V have been demonstrated with specific on-resistance of 0.46 mΩ cm$^2$ and a FOM(C) of 10 mΩ nC has been reported [7].

For purposes of comparison with the power MOSFET structures discussed in subsequent chapters, the analysis of the power CC-MOSFET structure is provided here for blocking voltages ranging from 30 to 100 V. As demonstrated in the chapter, it is not feasible to extend the breakdown voltage for the power CC-MOSFET structure beyond 100 V due to the poor electric field distribution along the drift region. In this analysis, the power CC-MOSFET structure was assumed to have the following parameters: (a) $N^+$ source junction depth of 0.08 μm; (b) P-base junction depth of 0.30 μm; (c) gate oxide thickness of 500 Å; (d) mesa width of 0.5 μm; (e) threshold voltage of 2 V; (f) gate drive voltage of 10 V; (g) inversion mobility of 450 cm$^2$/V s; (h) accumulation mobility of 1,000 cm$^2$/V s. The contributions from the contacts and the $N^+$ substrate were neglected during the analysis. The doping concentration and thickness of the drift region were determined under the assumption that the edge termination limits the breakdown voltage to the breakdown voltage of the cells, which exceeds the parallel-plane breakdown voltage. The device parameters pertinent to each blocking voltage are provided in Fig. 5.75. It can be observed that the doping concentration of the drift region is much larger than that for the previous power MOSFET structures.

The trench width in the power CC-MOSFET structure was chosen for each breakdown voltage to accommodate the thickness of the trench oxide with enough room for the polysilicon refill to form the source connected electrode in the trench. The trench oxide thickness is provided in Fig. 5.75. The cell pitch for the power CC-MOSFET structures is provided in Fig. 5.76 for the various blocking voltages. The analytical model described in Sect. 5.3 for computing the specific on-resistance was used for all of the power CC-MOSFET structures. These values are also provided in Fig. 5.76.

The specific on-resistance for the power CC-MOSFET structure can be compared with the ideal specific on-resistance obtained by using Baliga's power law for the impact ionization coefficients in Fig. 5.77. From this figure, it can be concluded that the specific on-resistance for the CC-MOSFET structure is less than the ideal specific on-resistance. The reduction of the specific on-resistance for the power CC-MOSFET structure below the ideal specific on-resistance becomes larger with increasing breakdown voltage. Unfortunately, this enhanced performance cannot

| **Blocking Voltage (V)** | **Drift Doping Concentration (cm$^{-3}$)** | **Trench Oxide Thickness (microns)** | **Drift Region Thickness (microns)** |
|---|---|---|---|
| 30 | $1.0 \times 10^{17}$ | 0.10 | 1 |
| 60 | $6.0 \times 10^{16}$ | 0.30 | 2 |
| 85 | $4.0 \times 10^{16}$ | 0.60 | 4 |

**Fig. 5.75** Device parameters for the power CC-MOSFET structures

| Blocking Voltage (V) | Cell Pitch (microns) | Specific On-Resistance (mΩ-cm²) |
|---|---|---|
| 30 | 1.0 | 0.0247 |
| 60 | 1.5 | 0.0798 |
| 85 | 2.1 | 0.272 |

**Fig. 5.76** Specific on-resistance for the power CC-MOSFET structures

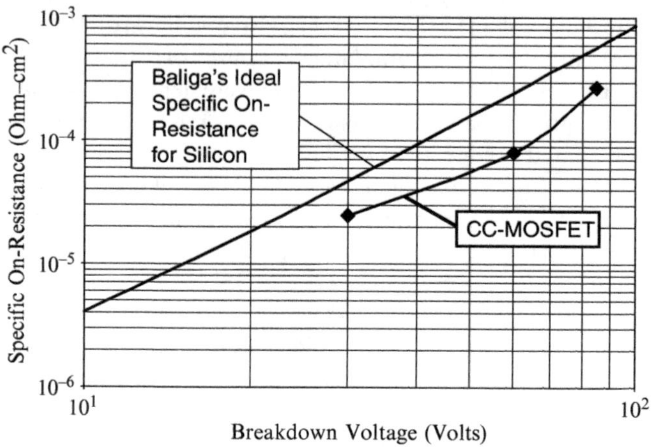


**Fig. 5.77** Specific on-resistance for the power CC-MOSFET structures

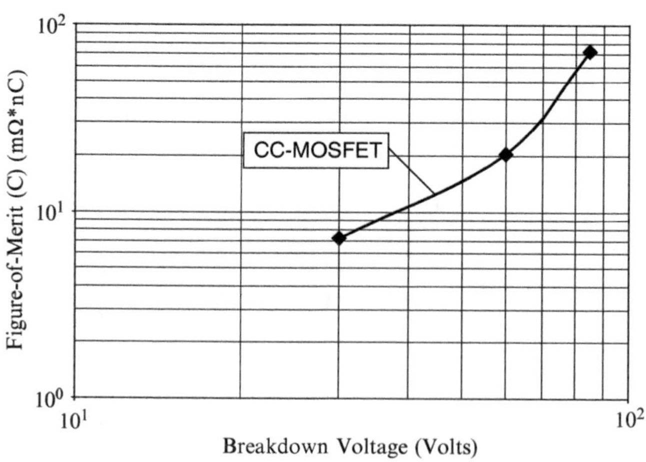


**Fig. 5.78** Figure-of-merit (C) for the power CC-MOSFET structures

be extended beyond 85 V due to the poor electric field distribution along the vertical direction. This problem can be overcome with a graded doping profile as demonstrated in the next chapter.

The figure-of merit (C) – product of the specific on-resistance and the specific gate transfer charge – for the power CC-MODSFET structures was obtained by determining the gate transfer charge using the analytical model (see 5.64 and 5.65). During this analysis, the devices were assumed to be operated at an on-state current density that results in a power dissipation of 100 W/cm$^2$ as determined by the specific on-resistance for each device. The resulting values for the FOM(C) are plotted in Fig. 5.78 as a function of the breakdown voltage of the power CC-MOSFET structure. It can be observed that the FOM(C) increases with increasing breakdown voltage. Due to the relatively small range of breakdown voltages in the plot, a power law fit is not appropriate for this device.

## References


1. B.J. Baliga, "Vertical Field Effect Transistors having improved Breakdown Voltage Capability and Low On-state Resistance", U.S. Patent # 5,637,898, Issued June 10, 1997.
2. L. Lorenz, et al, "COOLMOS – A New Milestone in High Voltage Power MOS", IEEE International Symposium on Power Semiconductor Devices and ICs, pp. 3–10, 1999.
3. B.J. Baliga, "Power Semiconductor Devices having improved High Frequency Switching and Breakdown Characteristics", U.S. Patent # 5,998,833, Issued December 7, 1999.
4. B.J. Baliga, "Trends in Power Discrete Devices", IEEE International Symposium on Power Semiconductor Devices and ICs, Abstract P-2, pp. 5–10, 1997.
5. Y.C. Liang, K.P. Gan, and S. Samudra, "Oxide-Bypassed VDMOS (OBVDMOS): An alternative to Superjunction High Voltage MOS Power Devices", IEEE Electron Device Letters, Vol. EDL-22, pp. 407–409, 2001.
6. G.E.J. Koops, et al, "Resurf Stepped Oxide (RSO) MOSFET for 85 V having a record low specific on-resistance", IEEE International Symposium on Power Semiconductor Devices and ICs, Abstract P-6, pp. 185–188, 2004.
7. M.A. Gajda, et al, "Industrialisation of Resurf Stepped Oxide Technology for Power Transistors", IEEE International Symposium on Power Semiconductor Devices and ICs, Abstract 6-10, pp. 112, 2006.
8. P. Goarin, et al, "Split-Gate Resurf Stepped Oxide (RSO) MOSFET for 25 V applications with record low gate-to-drain-charge", IEEE International Symposium on Power Semiconductor Devices and ICs, Abstract O6-2, pp. 61–64, 2007.
9. M. Kodama, et al, "Temperature Characteristics of a New 100 V Rated Power MOSFET, VLMOS (Vertical LOCOS MOS)", IEEE International Symposium on Power Semiconductor Devices and ICs, Abstract 10-4, pp. 463–466, 2004.
10. B.J. Baliga, "Fundamentals of Power Semiconductor Devices", Springer-Science, New York, 2008.
11. S. Merchant, "Analytical Model for the Electric Field Distribution in SOI RESURF and TMBS Structures", IEEE Transactions on Electron Devices, Vol. 46, pp. 1264–1267, 1999.
12. B.J. Baliga, "Silicon RF Power MOSFETs", World Scientific Publishing Company, Singapore, 2005.
13. B.J. Baliga, "MOSFET Devices having Linear Transfer Characteristics when Operating in Velocity Saturation Mode and Methods of Forming and Operating the Same", U.S. Patent # 6,545,316, Issued April 8, 2003.

# Chapter 6
# GD-MOSFET Structure

In the previous chapter, it was demonstrated that the specific on-resistance for power MOSFET structures can be greatly reduced by utilizing the two-dimensional charge-coupling concept. In these structures, a uniform doping concentration was assumed for the drift region. Although the electric field profile in this case is superior to that observed for a one-dimensional junction, the electric field was found to be non-uniform through the drift region. This non-uniformity of the electric field is relatively small for devices with low (~30 V) blocking voltage capability. However, when the desired blocking voltage is large (60–200 V), the electric field varies exponentially with distance in the uniformly doped drift region resulting in a low electric field through a large portion of the distance between the drain and source regions.

A much better electric field distribution can be achieved in the drift region by utilizing the two-dimensional charge-coupling concept with a linearly graded doping profile [1]. As in the case of the power CC-MOSFET structure, there are two options for achieving the desired charge-coupling during operation in the blocking mode. In the first case, the charge coupling electrode within the oxide coated trenches is connected to the gate electrode. Although this method achieves the desired reduction of the drift region resistance, it creates a device with large input and gate transfer capacitances. Consequently, an alternative power GD-MOSFET structure [2] has been proposed with the charge coupling electrode within the oxide coated trenches connected to the source electrode. Only the power GD-MOSFET structure with source electrode in the trenches is analyzed in this chapter due to its superior switching performance. Apart from the disclosure of the power GD-MOSFET concept in the patents referenced above, the device structure and its superior specific on-resistance was first discussed in the archival literature in 1997 [3].

B.J. Baliga, *Advanced Power MOSFET Concepts*,
DOI 10.1007/978-1-4419-5917-1_6, 

## 6.1 The GD-MOSFET Structure

A cross-section of the basic cell structure for the power graded-doped (charge-coupled) GD-MOSFET structure is illustrated in Fig. 6.1 together with the doping profile in the vertical direction. This device structure can be fabricated by starting with an N-type epitaxial layer grown with a linear doping gradient on a heavily doped $N^+$ substrate. Since the doping concentration is the largest at the $N^+$ substrate, this profile can be readily grown using computer controlled doping without auto-doping problems [4]. The P-base region is then formed across the active area of the device by ion implantation of boron followed by a drive-in cycle. The energy for the boron ion implantation is preferably chosen so that its peak doping concentration is below the depth of the $N^+$ source region. The $N^+$ source regions are then produced by ion implantation of phosphorus followed by its annealing process. A mask is required during the ion-implantation of the $N^+$ source regions to create the short between the $N^+$ source and P-base regions at the top of the mesa region. A deep trench is next formed by reactive ion etching using a mask aligned to the $N^+$ source mask to center it in relation to the shorts. The depth of the trench must be tailored to achieve the desired breakdown voltage. The surface of the trench must be smooth and free of damage in order to obtain a good MOS interface with high channel inversion layer mobility. A thick oxide is grown on the trench surface by thermal oxidation. The oxide thickness must be sufficient to provide the desired charge coupling as well as to support the entire drain blocking voltage.

The trench is then refilled with highly doped N-type polysilicon. The polysilicon is then etched until it is recessed below the surface to a depth slightly below the depth of the P-base region. All the exposed oxide in the trenches is now removed. A fresh oxide is then grown by thermal oxidation on the trench sidewalls to form the gate oxide. A thicker oxide is simultaneously formed by the thermal oxidation on the top surface of the polysilicon inside the trenches. This serves as an isolation

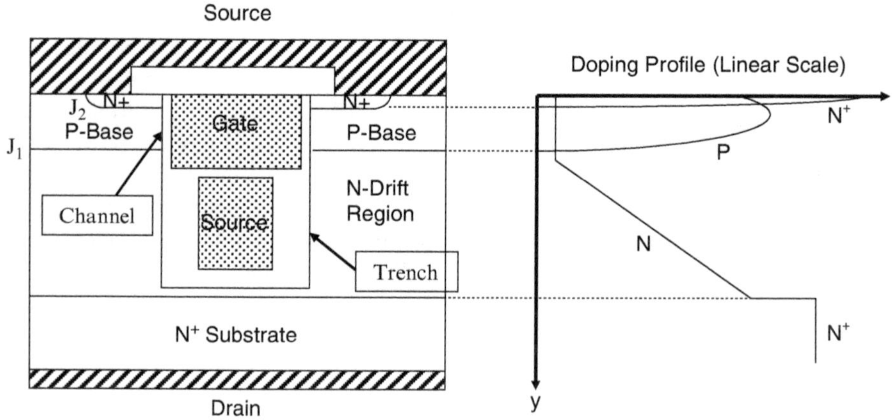


**Fig. 6.1** The GD-MOSFET structure and drift region doping profile

oxide between the gate electrode and the source electrode in the trenches. A second polysilicon layer is now deposited and planarized to serve as the gate electrode. An inter-metal dielectric is then deposited followed by etching contact windows for the $N^+$ source and P-base regions. The device fabrication is then completed by the deposition and patterning of the source metal layer.

The physics of operation of the power GD-MOSFET structure in the blocking mode is similar to that for the power CC-MOSFET structure described in the previous chapter. Without the application of a gate bias, a high voltage can be supported in the GD-MOSFET structure when a positive bias is applied to the drain. In this case, junction $J_1$ formed between the P-base region and the N-drift region becomes reverse biased. Simultaneously, the drain voltage is applied across the vertical MOS structure formed between the electrode in the deep trenches and the N-drift region. The MOS structure operates in the deep-depletion mode due to the presence of the reverse bias across junction $J_1$ between the P-base region and the N-drift junction. Consequently, depletion regions are formed across the horizontal junction $J_1$ and the vertical trench sidewall. This two-dimensional depletion alters the electric field distribution from the triangular shape observed in conventional parallel-plane junctions to a uniform distribution due to the linear doping profile. This allows supporting a required blocking voltage over a shorter distance. In addition, the doping concentration in the drift region can be made much greater that predicted by the one-dimensional theory and for the case of the uniformly doped charge-coupled device. This allows very substantial reduction of the specific on-resistance to well below the ideal specific on-resistance at any desired breakdown voltage.

Drain current flow in the GD-MOSFET structure is induced by the application of a positive bias to the gate electrode. This produces an inversion layer at the surface of the P-base region along the trench sidewalls. The threshold voltage for the power GD-MOSFET structure can be controlled by adjusting the dose for the boron ion-implantation for the P-base region. The energy for the boron ion-implantation for the P-base region is chosen to maintain its peak concentration below the surface of the silicon to suppress reach-through. The inversion layer channel provides a path for transport of electrons from the source to the drain when a positive drain voltage is applied. After transport from the source region through the channel, the electrons enter the N-drift region and are then transported through the mesa region to the $N^+$ substrate. The resistance of the drift region is very low in the power GD-MOSFET structure due to the high doping concentration in the mesa region. The channel resistance in the power GD-MOSFET structure is also very small due to the small cell pitch or high channel density.

As discussed in earlier chapters, the input capacitance and gate charge for power MOSFET structures must be reduced to enhance their switching performance. These device parameters can be reduced by using a source connected electrode in the trenches adjacent to the drift region as illustrated in Fig. 6.1. Although a power GD-MOSFET structure with gate electrode in the trenches is viable from the blocking capability stand-point, it will not be discussed in this chapter due to its inferior switching performance.

## 6.2 Charge-Coupling Physics and Blocking Voltage

The doping profile required in the drift region to achieve a uniform electric field along the y-direction is determined by two-dimensional charge coupling for the power GD-MOSFET structure. In order to achieve good charge coupling, the mesa region must be completely depleted when the drain bias approaches the breakdown voltage. The electric field distribution within the power GD-MOSFET structure along the x-axis is illustrated in Fig. 6.2 under these conditions under the assumption of a uniformly doping in the drift region in the x-direction. The electric field varies linearly with distance in the semiconductor in accordance with the solution for Poisson's equation [5] while the electric field in the oxide is constant. The electric field along the x-direction reduces to zero at the center of the mesa. However, there is a large electric field along the y-direction in the mesa region which supports the voltage between the drain and the source regions.

When the drain bias reaches the breakdown voltage, the maximum electric field in the semiconductor along the y-direction becomes equal to the critical electric field ($E_{CU}$) for breakdown of the semiconductor with uniform electric field profile. The critical electric field at which breakdown occurs can be derived for this field distribution by using the criterion that the ionization integral becomes unity at breakdown:

$$\int_0^{L_D} \alpha dx = 1 \tag{6.1}$$

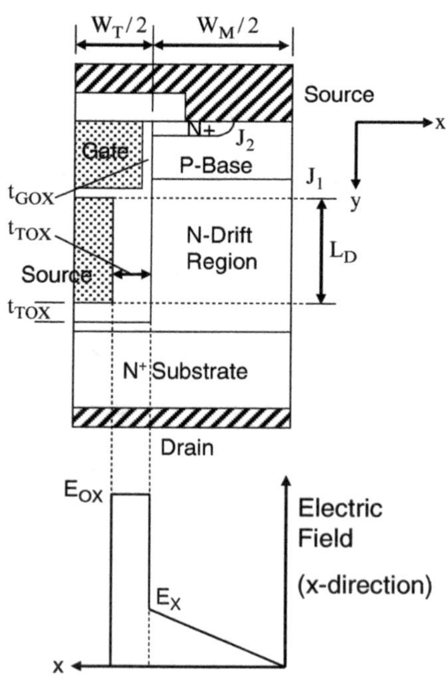


**Fig. 6.2** Electric field distribution in the GD-MOSFET structure

Using Baliga's formula for the impact ionization coefficient for silicon:

$$\alpha_B = 3.51 \times 10^{-35}\ E^7 \tag{6.2}$$

in (6.1), an expression for the critical electric field for the case of a uniform (or constant) electric field is obtained:

$$E_{CU} = 8.36 \times 10^4\ L_D^{-1/7} \tag{6.3}$$

The breakdown voltage in this case is then given by:

$$BV = E_{CU} L_D = 8.36 \times 10^4\ L_D^{6/7} \tag{6.4}$$

by using (6.3).

The breakdown voltage for charge-coupled devices with a uniform electric field can be observed in Fig. 6.3 to increase non-linearly with increasing length of the drift region. Using this analytical model, the drift region lengths required to achieve breakdown voltages of 60, 120, and 200 V are predicted to be 2.1, 4.8, and 8.8 μm, respectively. Note that the drift region length for the power GD-MOSFET structure is the length of the source region within the trenches as indicated in Fig. 6.2 and not the trench depth because the two-dimensional charge coupling occurs only between the source electrode and the drift region.

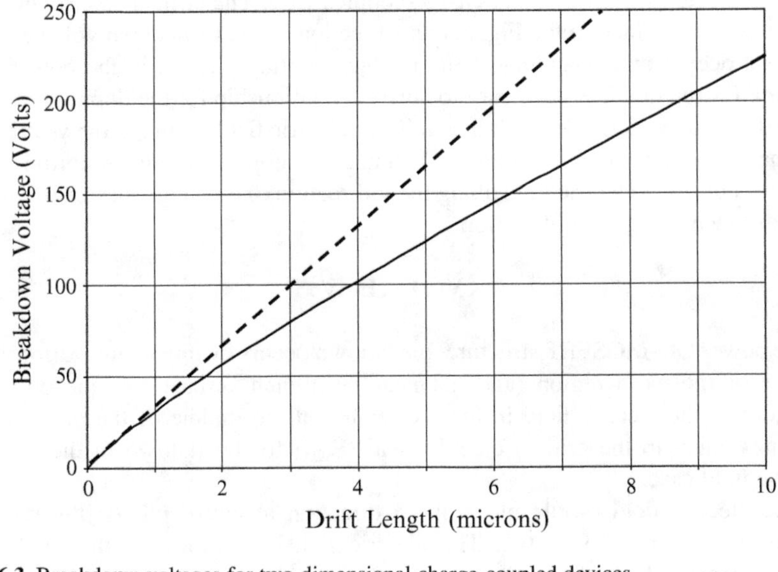


**Fig. 6.3** Breakdown voltages for two-dimensional charge-coupled devices

The critical electric field for breakdown with uniform electric field can be related to the breakdown voltage by combining (6.3) and (6.4):

$$E_{CU} = 5.53 \times 10^5\ BV^{-1/6} \tag{6.5}$$

In the case of the one-dimensional case with triangular electric field distribution, it has been demonstrated in Chap. 2 that the critical electric field for breakdown obtained by using Baliga's power law for impact ionization is given by:

$$E_C = 3,700\ N_D^{1/8} \tag{6.6}$$

while the breakdown voltage is given by:

$$BV = 4.45\ \times 10^{13}\ N_D^{-3/4} \tag{6.7}$$

Combining these relationships, the critical electric for breakdown for the one-dimensional case is given by:

$$E_C = 6.97 \times 10^5\ BV^{-1/6} \tag{6.8}$$

From these equations, it can be observed that the critical electric field expressions for the one- and two-dimensional cases have the same dependence on the breakdown voltage. The ratio of the critical electric field for the one-dimensional case to that for the two-dimensional case is 1.26× independent of the breakdown voltage. This difference is due to a high electric field over a longer path for impact ionization in the case of the two-dimensional charge-coupled case. The critical electric field for both cases is provided in the Fig. 6.4 as a function of the breakdown voltage.

For a deeper understanding of the charge-coupling physics in the power GD-MOSFET structure, it is necessary to derive a relationship for the doping profile in the drift region which will result in a uniform electric field profile in the y-direction for the two-dimensional structure [6]. Since the doping profile is optimized to achieve a uniform electric field along the y-direction, the voltage increases linearly with distance along the y-direction:

$$V(y) = E_Y\ y \tag{6.9}$$

In the power GD-MOSFET structure, breakdown occurs by impact ionization at the middle of the mesa region (and not near the trench surface). At the on-set of breakdown, the electric field in the y-direction at the middle of the mesa region becomes equal to the critical electric field ($E_{CU}$) for breakdown in the uniform electric field case.

The electric field profile along the x-direction at any depth is illustrated in the lower portion of Fig. 6.2. The electric field is uniform in the oxide and varies linearly along the x-direction in the semiconductor because the doping

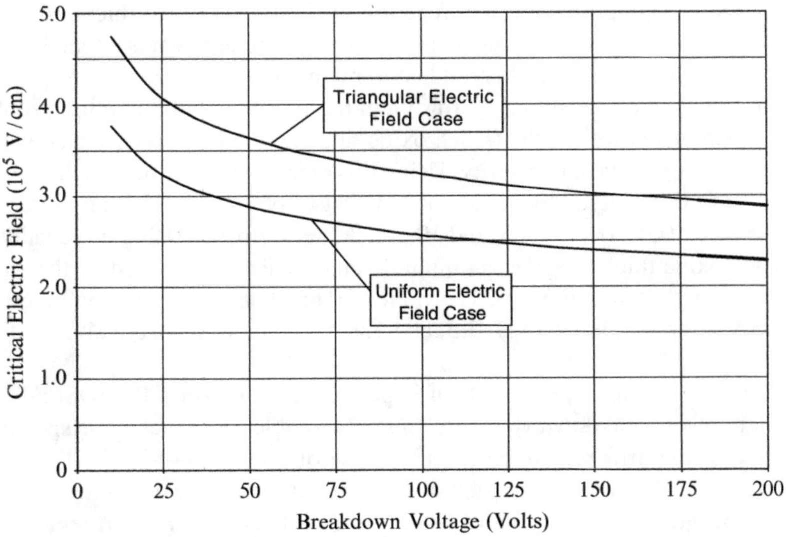


**Fig. 6.4** Critical electric field for breakdown for the triangular and uniform electric field distribution cases

concentration is constant along the x-direction. The voltage at any depth in the drift region is then given by:

$$V(y) = V_{OX} + V_S = E_X \left[ \frac{\varepsilon_{Si}}{\varepsilon_{OX}} \; t_{TOX} + \frac{W_M}{4} \right] \tag{6.10}$$

where $E_X$ is the electric field in the x-direction at the interface between the oxide and the semiconductor. Applying Gausses law at any depth in the drift region:

$$\varepsilon_{Si} \; E_X = qN_D \; (y) \; \frac{W_M}{2} \tag{6.11}$$

The optimum doping profile to achieve a uniform electric field in the y-direction is obtained by combining the above relationships:

$$N_D(y) = \frac{2\varepsilon_{Si} E_{CU}}{qW_M \left[ \frac{\varepsilon_{Si}}{\varepsilon_{OX}} t_{T,OX} + \frac{W_M}{4} \right]} y \tag{6.12}$$

It can be seen that the predicted optimum doping profile has a linear distribution in the y-direction.

The gradient (or slope) of the optimum doping profile is given by:

$$G = \frac{2\varepsilon_{Si} E_{CU}}{qW_M \left[ \frac{\varepsilon_{Si}}{\varepsilon_{OX}} t_{TOX} + \frac{W_M}{4} \right]} \tag{6.13}$$

The optimum doping gradient is a function of the trench oxide thickness and the mesa width. In addition, the critical electric field for breakdown is a function of the breakdown voltage. Further, it is prudent to maintain the electric field in the trench oxide to less than or equal to $2 \times 10^6$ V/cm in order to avoid reliability problems. Based up on this criterion, the trench oxide thickness must also be scaled with the desired blocking voltage capability. Using this electric field in the trench oxide, the trench oxide thicknesses for breakdown voltages of 30, 60, 120, and 200 V are found to be 1,500, 3,000, 6,000, and 10,000 Å, respectively. Using these values for the trench oxide thickness, the optimum doping gradients predicted by the analytical model are plotted in Fig. 6.5. The values for breakdown voltages of 30, 60, 120, and 200 V are found to be 14, 7, 3.3, and $1.9 \times 10^{20}$/cm$^4$, respectively, for a mesa width of 0.5 μm.

A larger optimum doping gradient is predicted for power GD-MOSFET structures with lower breakdown voltages. This is favorable for reducing the specific on-resistance of the drift region. Even in the case of power GD-MOSFET structures with larger breakdown voltages, the doping concentration is high in the vicinity of the $N^+$ substrate despite the smaller doping gradient because the drift region length is longer. As a typical example, the doping concentration for the power GD-MOSFET structure with breakdown voltage of 60 V increases from $1 \times 10^{16}$/cm$^3$ near the P-base/N-drift junction to nearly $2 \times 10^{17}$/cm$^3$ near the bottom of the trench. Such high doping concentrations are also required for the higher voltage power GD-MOSFET structures as shown in Fig. 6.6 where the optimum doping profiles are provided for the devices with breakdown voltages of 30, 60, 120, and 200 V. These high doping levels result in very low specific on-resistance for the power GD-MOSFET structures.

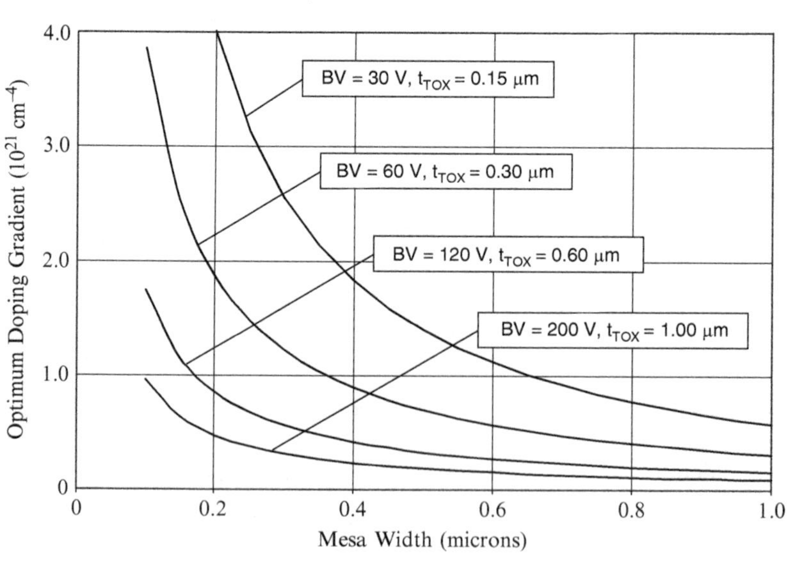


**Fig. 6.5** Optimum doping gradients for the power GD-MOSFET structures

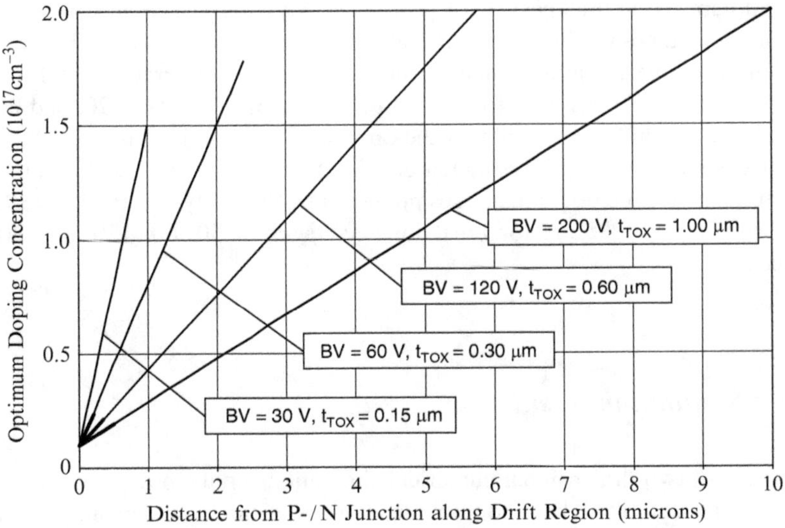


**Fig. 6.6** Optimum linear doping profiles for the power GD-MOSFET structures

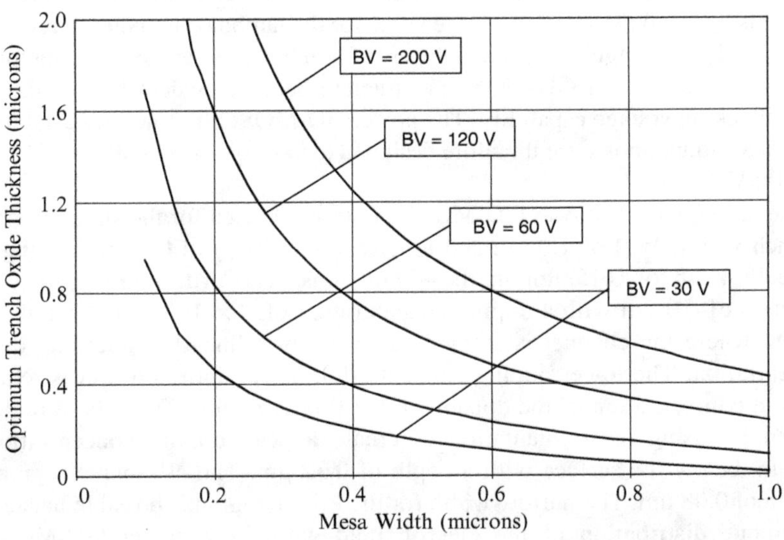


**Fig. 6.7** Optimum trench oxide thickness for the power GD-MOSFET structures

For any given breakdown voltage with a fixed doping gradient, an optimum trench oxide thickness is also predicted by the analytical model. Using (6.5) and (6.13):

$$t_{TOX}(Opt) = \frac{1.106 \times 10^6\ \varepsilon_{OX}\ BV^{-1/6}}{qW_M G} - \frac{\varepsilon_{OX}\ W_M}{4\ \varepsilon_S} \qquad (6.14)$$

The optimum oxide thicknesses for power GD-MOSFET structures with various breakdown voltages predicted by the analytical model are provided in Fig. 6.7 as a function of the mesa width. In making these plots, a doping gradient of 14, 7, 3.3 and $1.9 \times 10^{20}/\text{cm}^4$ was used for breakdown voltages of 30, 60, 120, and 200 V, respectively. It can be observed that the optimum trench oxide thickness becomes smaller when the mesa width is increased. For the case of a mesa width of 0.5 μm, the optimum trench oxide thicknesses predicted by the analytical model are 1,500, 2,994, 6,023, and 9,852 Å for breakdown voltages of 30, 60, 120, and 200 V, respectively.

## *6.2.1 Simulation Results*

The results of two-dimensional numerical simulations on the power GD-MOSFET structure are described here to provide a more detailed understanding of the underlying device physics and operation during the blocking mode. In the previous chapter, it was demonstrated that the electric field is high throughout the drift region for the 30-V power CC-MOSFET structure despite the uniform doping concentration. This is due to the relatively large value for the lambda parameter in relation to the drift region length. The advantages of the linearly graded doping profile employed in the power GD-MOSFET structure become evident for devices with larger blocking voltage capability. The power GD-MOSFET structures used for the numerical simulations were therefore designed to have blocking voltage of 60, 120, and 200 V.

The baseline 60-V power GD-MOSFET structure used for the simulations had a trench width ($W_T$) of 1.0 μm and a mesa width ($W_M$) of 0.5 μm leading to a half-cell width of 0.75 μm in the simulations. The drift region had a total thickness of 4.0 μm with a doping concentration of $1 \times 10^{16}/\text{cm}^3$ for the upper 0.4 μm where the channel is formed followed by a linearly increasing doping concentration. The lower doping concentration in the upper portion is required to avoid compensation of the dopants in the P-base region. The P-base region is formed by using ion-implantation to create a peak doping concentration at 0.15 μm below the surface with a depth of 0.32 μm. The $N^+$ source region has a depth of 0.08 μm. The narrow width for the P-base region is possible because of the unique distribution of the electric field within the power GD-MOSFET structure. A heavily doped $P^+$ region is also included in the structure in the center of the mesa at the upper surface to allow fabrication of ohmic contacts to the P-base region.

A three dimensional view of the doping distribution in the GD-MOSFET structure is shown in Fig. 6.8 for the upper portion of the structure. A large portion of the simulated cell structure is consumed by the trench region which extends to a depth of 2.8 μm in the baseline device structure. The $N^+$ source, P-base, and $P^+$ regions can be observed on the upper right hand side in the mesa region. The lower

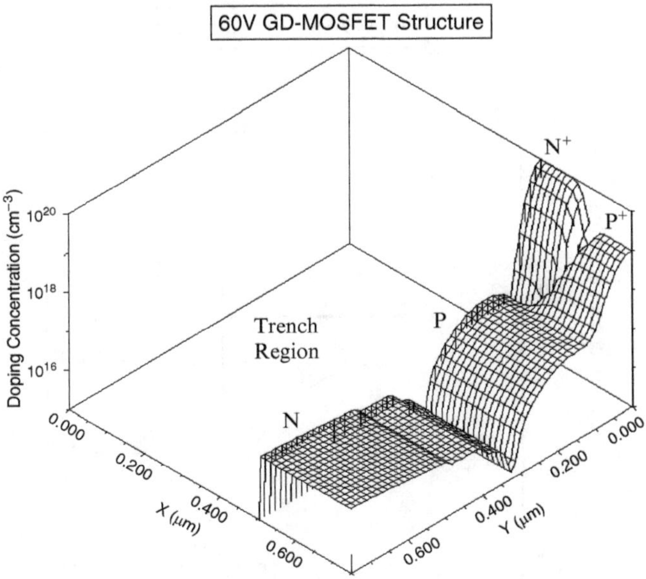


**Fig. 6.8** Doping distribution for the GD-MOSFET structure

doping concentration of the drift region can be observed to extend to a depth of 0.4 μm. After this depth, the drift region doping concentration increases due to the linear doping profile.

The channel doping profile taken along the surface of the trench to a depth of 0.5 μm is shown in Fig. 6.9. From the profile, it can be observed that the P-base region has a retrograde doping profile with a peak doping concentration of $1.5 \times 10^{17}/cm^3$ to obtain the desired threshold voltage. The vertical depths of the P-base and $N^+$ source regions are 0.32 and 0.08 μm leading to a channel length of only 0.24 μm. The drift region doping concentration is $1 \times 10^{16}/cm^3$ up to a depth of 0.4 μm. The doping concentration then increases with a gradient of $8 \times 10^{20}/cm^4$ when proceeding towards the $N^+$ substrate.

The vertical doping profiles taken at two positions within the power GD-MOSFET structure are provided in Fig. 6.10. From the profile taken at x = 0.75 μm in the middle of the mesa region (dashed line), it can be observed that the doping concentration of the $P^+$ contact region is $1 \times 10^{19}/cm^3$ with sufficient depth to contact the P-Base region. From the profile taken at x = 0.50 μm along the trench sidewall, it can be observed that the drift region extends to a depth of 4.0 μm which is beyond the trench depth of 2.8 μm. The linearly graded doping profile is clearly observed in this figure with the drift region concentration increasing from $1 \times 10^{16}/cm^3$ at the P-N junction to about $2.5 \times 10^{17}/cm^3$ at the $N^+$ substrate. (Note that the linear doping profile has non-linear shape in this figure due to the logarithmic scale used for the doping concentration).

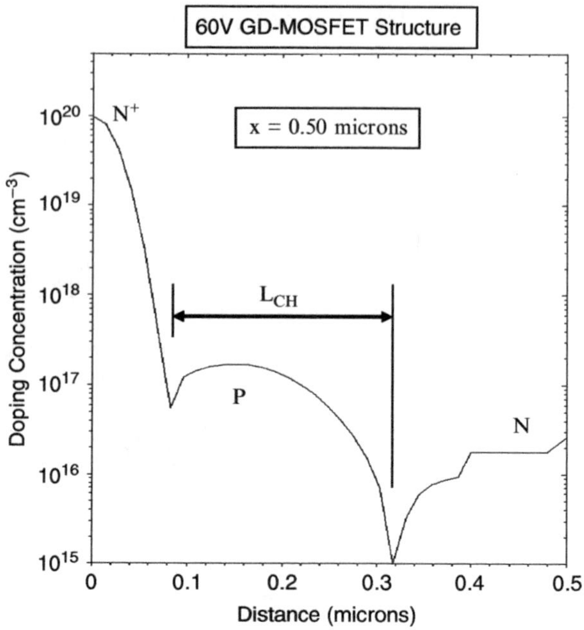


**Fig. 6.9** Channel doping profile for the GD-MOSFET structure

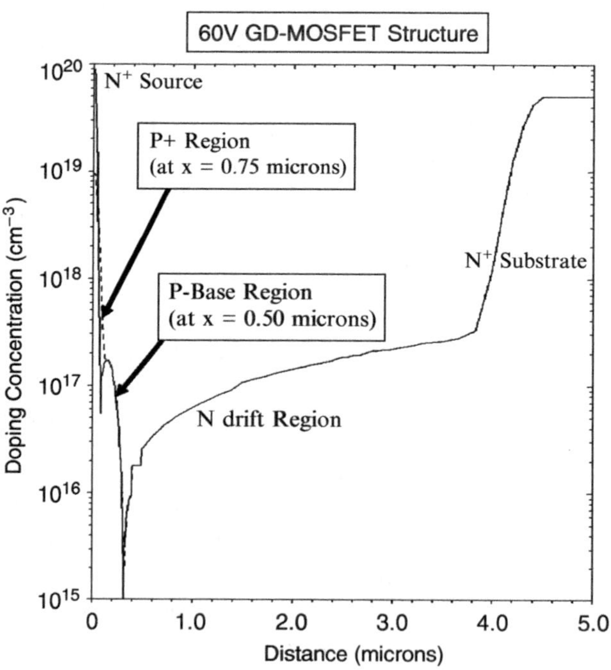


**Fig. 6.10** Vertical doping profiles in the GD-MOSFET structure

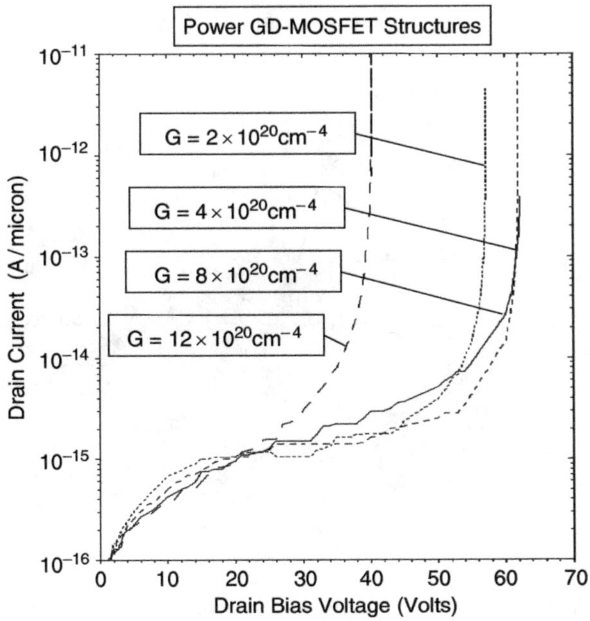


**Fig. 6.11** Blocking characteristic for the GD-MOSFET structures

The blocking characteristics for the GD-MOSFET cell structure are shown in Fig. 6.11 at 300°K for the case of various doping gradients. The trench oxide thickness and mesa width were maintained at 3,000 Å and 0.5 μm for these structures. The trench depth was 2.8 μm with the source electrode in the trench extending from a depth of 0.5 to a depth of 2.5 μm. This corresponds to a drift region length ($L_D$ as defined in Fig. 6.2) of 2.0 μm. It can be observed that the cell is capable of supporting 62 V for the case of a doping gradient of $8 \times 10^{20}/cm^4$. The breakdown voltage is greatly reduced when the doping gradient is increased to $12 \times 10^{20}/cm^4$ and slightly reduced for doping gradients below $8 \times 10^{20}/cm^4$.

It is insightful to also examine the electric field profile inside the power GD-MOSFET structure when it is operating in the blocking mode. A three-dimensional view of the electric field within the power GD-MOSFET structure with a doping gradient of $8 \times 10^{20}/cm^4$ is shown in Fig. 6.12 at a drain bias of 60 V. It can be observed that the electric field is highest in the oxide at the bottom of the trenches with a value ($2 \times 10^6$ V/cm) given by the ratio of the applied drain bias and the oxide thickness. The electric field in the oxide on the trench sidewalls increases linearly with depth because the voltage increases linearly with depth as desired. More importantly, the electric field distribution in the middle of the mesa region can be observed to be constant with depth as predicted by the analytical model for the case of a linearly graded doping profile. It can be seen that the electric field at the trench sidewall near the bottom of the trench is much larger than the electric field at the middle of the mesa region.

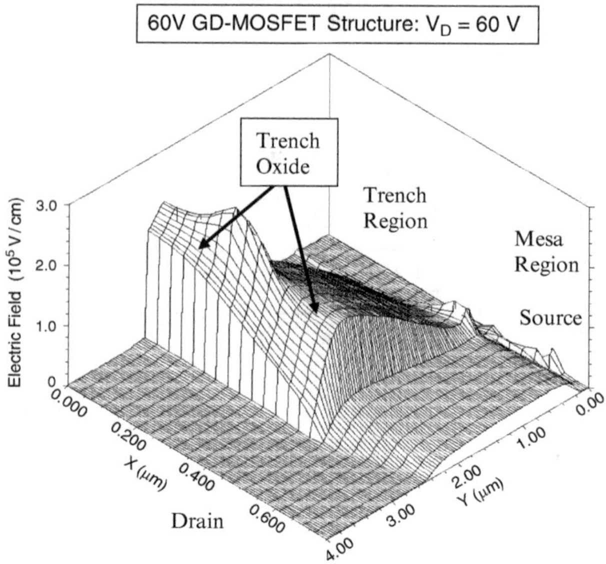


**Fig. 6.12** Three-dimensional electric field distribution in the GD-MOSFET structure

The profile for the electric field along the y-direction at the middle of the mesa (solid-line) and at the trench surface (dashed-line) in the power GD-MOSFET structure can be seen more clearly in Fig. 6.13. From this figure, it can be observed that the electric field at the middle of the mesa is fairly constant in the drift region with a value of about $3 \times 10^5$ V/cm at the drain bias of 60 V. This value for the electric field is consistent with the critical electric field ($E_{CU}$) of $2.8 \times 10^5$ V/cm for the case of uniform electric field distribution predicted for a breakdown voltage of 60 V using Baliga's power law for the impact ionization coefficient in silicon (see 6.5 and Fig. 6.4). In contrast, the electric field at the trench surface is much greater in magnitude.

The carrier generation due to impact ionization in the power GD-MOSFET structure just prior to breakdown (at a drain bias of 60 V) is shown in Fig. 6.14. It can be observed that the impact ionization is occurring at the middle of the mesa region and not at the oxide interface despite the larger absolute electric field being located at the trench sidewall. This is because the electric field at the trench sidewall is oriented orthogonal to the trench surface (i.e. along the x-direction) and impact ionization in the drift region is produced by the vertical component of the electric field (i.e. along the y-direction). This observation justifies the methodology used for developing the analytical solution for the optimum doping profile in the power GD-MOSFET structure.

In order to demonstrate the two-dimensional charge coupling phenomenon, it is instructive to examine the potential contours inside the power GD-MOSFET structure when it is operating in the blocking mode. This also allows determination of the voltage distribution within the structure and the penetration of the depletion

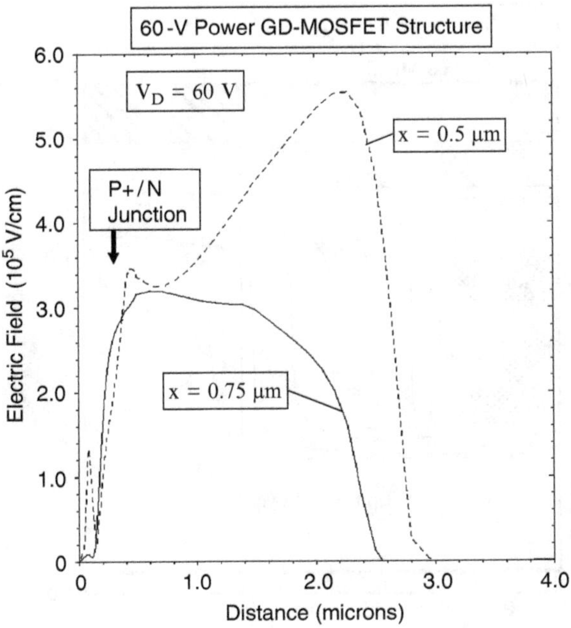


**Fig. 6.13** Electric field profiles in the GD-MOSFET structure

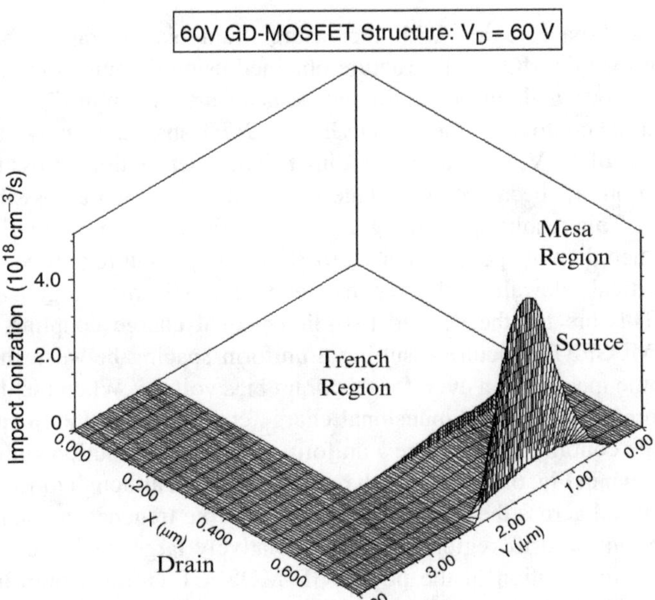


**Fig. 6.14** Distribution of impact ionization in the GD-MOSFET structure

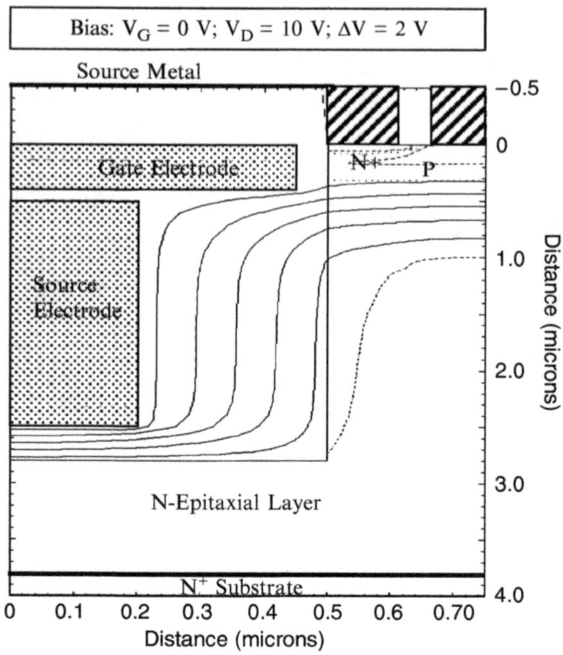


**Fig. 6.15** Potential contours in the power GD-MOSFET structure

region in the P-base region with increasing drain bias voltage. The potential contours for the GD-MOSFET structure obtained using the numerical simulations with zero gate bias and various drain bias voltages are shown in Fig. 6.15–6.17.

The potential contours in the power GD-MOSFET structure, shown in Fig. 6.15 at a drain bias of 10 V, indicate a one-dimensional distribution across the P-base/N-drift junction ($J_1$) because the voltage is initially supported across the oxide in the trenches. This results in a triangular electric field profile at small drain bias voltages. When the voltage is increased to 30 V, a depletion region begins to form along the vertical sidewalls of the trench extending into the mesa region as shown in Fig. 6.16. This creates the desired two-dimensional charge coupling within the power CC-MOSFET structure resulting a uniform spacing between the potential contours in the mesa region even for this drain bias voltage. When the drain bias is increased to 60 V, the two-dimensional charge coupling with the linearly graded doping profile continues to produce a uniform potential distribution within the drift region as shown in Fig. 6.17. It is worth pointing out that, although most of the drain bias is supported across the oxide at the bottom of the trenches, a small depletion region forms in the drift region despite its relatively large doping concentration. The potential distribution in the power GD-MOSFET is much more uniform in comparison with the power CC-MOSFET structure (see Fig. 5.63).

The electric field profile along the vertical direction through the center of the mesa region is shown in Fig. 6.18. At drain biases of 5 and 10 V, the electric field has a

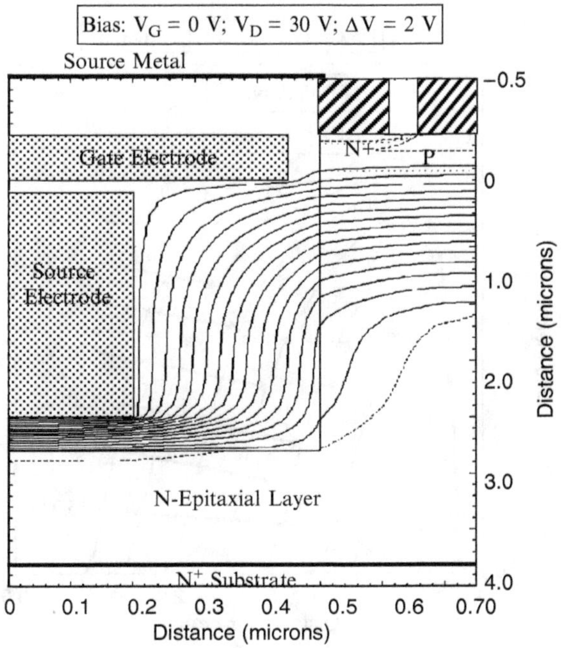


**Fig. 6.16** Potential contours in the power GD-MOSFET structure

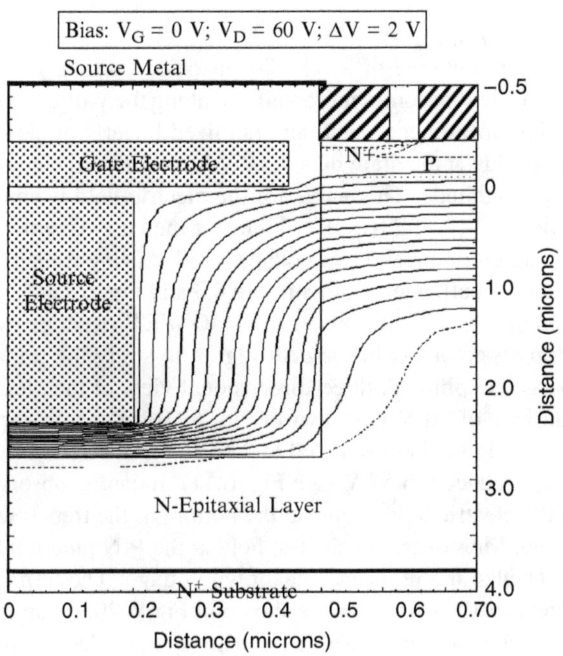


**Fig. 6.17** Potential contours in the power GD-MOSFET structure

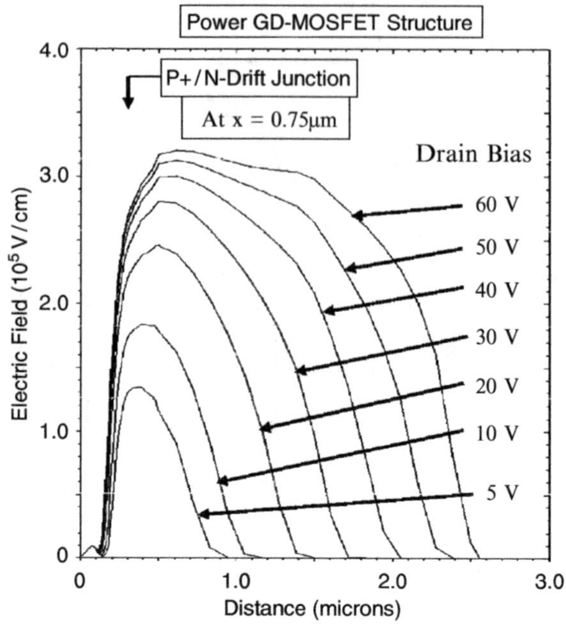


**Fig. 6.18** Electric field profiles in the power GD-MOSFET structure with a doping gradient of $8 \times 10^{20}/\text{cm}^4$

triangular shape representative of a one-dimensional junction. However, at larger voltages, the electric field becomes more uniform along the y-direction demonstrating the charge coupling phenomenon with an optimized linearly graded doping profile. The electric field profile at the drain bias of 60 V has a uniform electric field of about $3 \times 10^5$ V/cm in magnitude. This value for the electric field is consistent with the critical electric field ($E_{CU}$) of $2.8 \times 10^5$ V/cm for the case of uniform electric field distribution predicted for a breakdown voltage of 60 V using Baliga's power law for the impact ionization coefficient in silicon (see 6.3 and Fig. 6.4).

Numerical simulations of the power GD-MOSFET structure were performed using various doping gradients in the drift region to study the impact on the two-dimensional charge coupling. A three-dimensional view of the electric field distribution at a drain bias of 50 V is shown in Fig. 6.19 for the power GD-MOSFET structure with a doping gradient reduced to $2 \times 10^{20}/\text{cm}^4$. The breakdown voltage for this structure is reduced to 57 V (see Fig. 6.11). It can be observed in Fig. 6.19 that the peak in the electric field occurs at the bottom of the trench when the doping gradient is reduced. Moreover, the electric field at the P-N junction has a relatively low value. This results in a reduced breakdown voltage. The impact of increasing the doping gradient to $12 \times 10^{20}/\text{cm}^4$ is shown in Fig. 6.20. It can be observed that the electric field takes a triangular shape with a high electric field at the P-N junction resulting in reduced breakdown voltage.

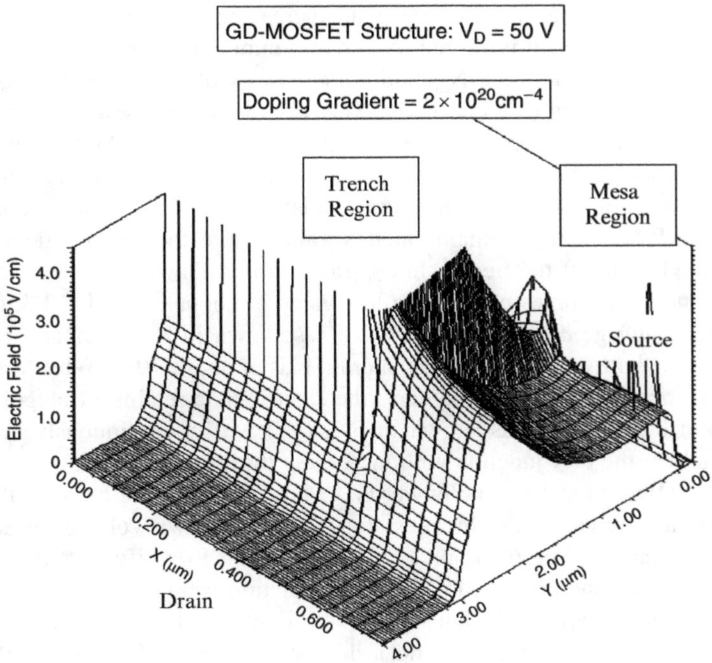


**Fig. 6.19** Three-dimensional electric field distribution in the GD-MOSFET structure

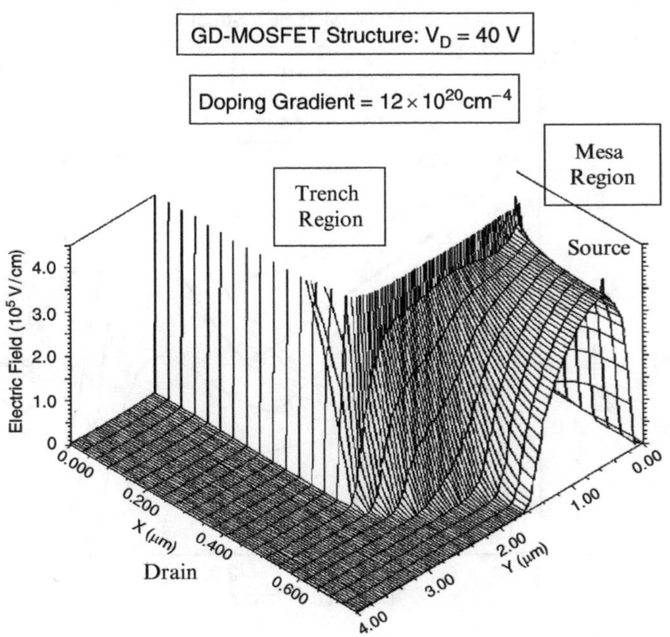


**Fig. 6.20** Three-dimensional electric field distribution in the GD-MOSFET structure

The evolution of the electric field with increasing drain bias voltage changes when the doping gradient is altered. At the optimum doping gradient of $8 \times 10^{20}/\text{cm}^4$, the electric field at the P-N junction increases with drain bias voltage until a bias of 30 V is reached (see Fig. 6.18). The electric field at the P-N junction then becomes approximately equal to the critical electric field for breakdown for the uniform electric field case ($2.8 \times 10^5$ V/cm) based up on Baliga's power law for the impact ionization coefficient in silicon. At larger drain bias voltages, the electric field at the P-N junction remains at this value while the electric field spreads towards the bottom of the trench. In contrast to this behavior, the electric field at the P-N junction becomes pinned at a low value of about $1.3 \times 10^5$ V/cm for the case of a doping gradient of $2 \times 10^{20}/\text{cm}^4$ as shown in Fig. 6.21 and begins to increase at the bottom of the trench at a drain bias of above 10 V. When the doping gradient is increased to $12 \times 10^{20}/\text{cm}^4$, the electric field retains a triangular shape at all drain bias voltages as shown in Fig. 6.22 with a continuously increasing electric field at the P-N junction with increasing drain bias voltage.

The analytical model for the power GD-MOSFET structure predicts that there is an optimum trench oxide thickness for each breakdown voltage, mesa width, and doping gradient (see 6.14 and Fig. 6.7). In order to verify this, the blocking characteristics for the power GD-MOSFET structure were obtained by numerical simulations using various thicknesses for the oxide in the trenches while maintaining a mesa width of 0.5 μm and a doping gradient of $8 \times 10^{20}/\text{cm}^4$. It can

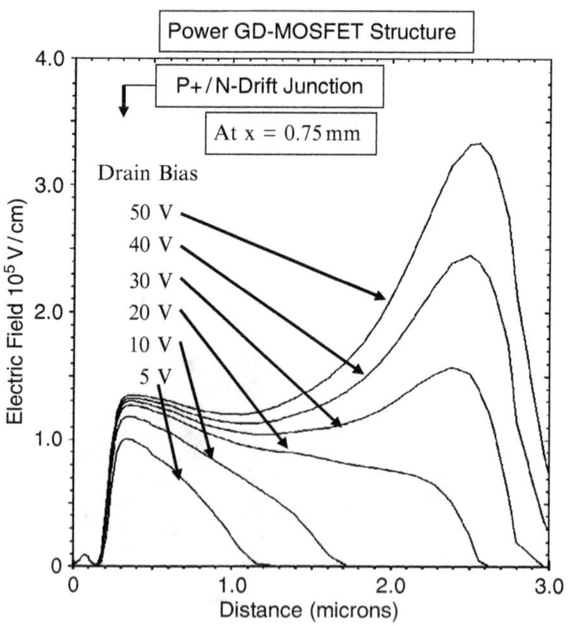


**Fig. 6.21** Electric field profiles in the power GD-MOSFET structure with a doping gradient of $2 \times 10^{20}/\text{cm}^4$

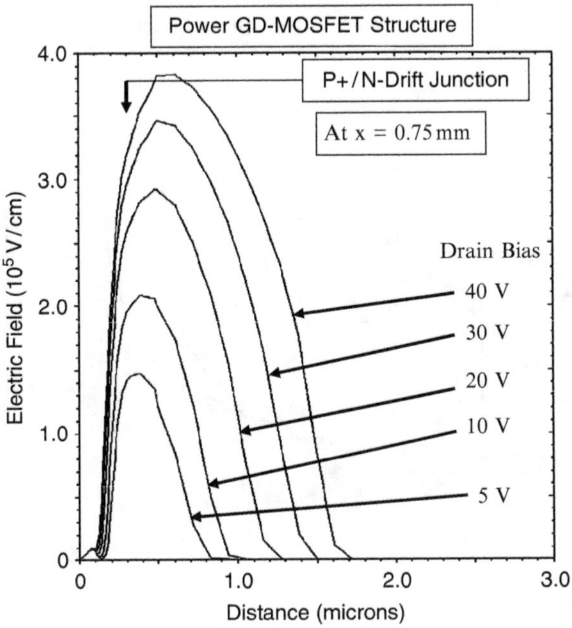


**Fig. 6.22** Electric field profiles in the power GD-MOSFET structure with a doping gradient of $12 \times 10^{20}/\text{cm}^4$

be observed in Fig. 6.23 that the breakdown voltage has the largest value when the oxide thickness lies between 2,500 and 3,000 Å. This result obtained from the numerical simulations is in excellent agreement with an optimum trench oxide thickness of 2,994 Å predicted by the analytical model.

The evolution of the electric field with increasing drain bias voltage is also a function of the trench oxide thickness. At the optimum trench oxide thickness of 3,000 Å, the electric field at the P-N junction increases with drain bias voltage until a bias of 30 V is reached (see Fig. 6.18). The electric field at the P-N junction then becomes approximately equal to the critical electric field for breakdown for the uniform electric field case ($2.8 \times 10^5$ V/cm) based up on Baliga's power law for the impact ionization coefficient in silicon. At larger drain bias voltages, the electric field at the P-N junction remains at this value while the electric field spreads towards the bottom of the trench. In contrast to this behavior, the electric field at the P-N junction becomes pinned at a lower value of about $2.3 \times 10^5$ V/cm for the case of an oxide thickness of 2,000 Å as shown in Fig. 6.24 indicating that the charge coupling effect is too strong. When the trench oxide thickness is increased to 3,500 Å, the electric field at the P-N junction continues to increase with drain bias voltage and reaches a magnitude of $3.3 \times 10^5$ V/cm as shown in Fig. 6.25 indicating that the charge coupling effect is too weak.

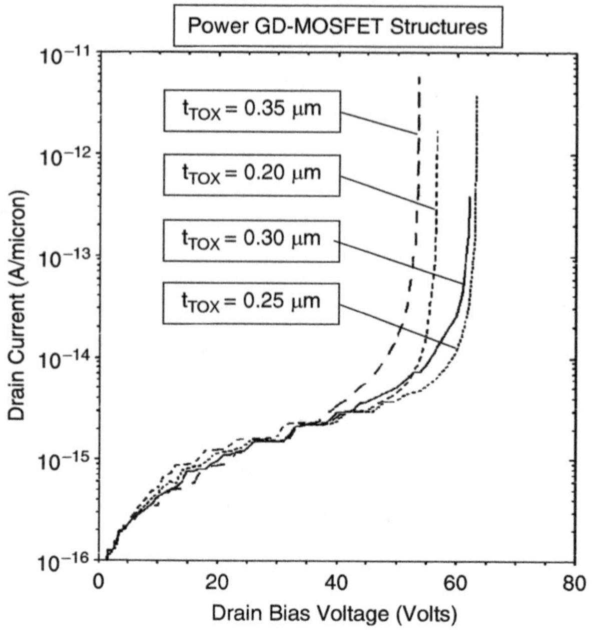


**Fig. 6.23** Blocking characteristic for the power GD-MOSFET structures with various trench oxide thicknesses

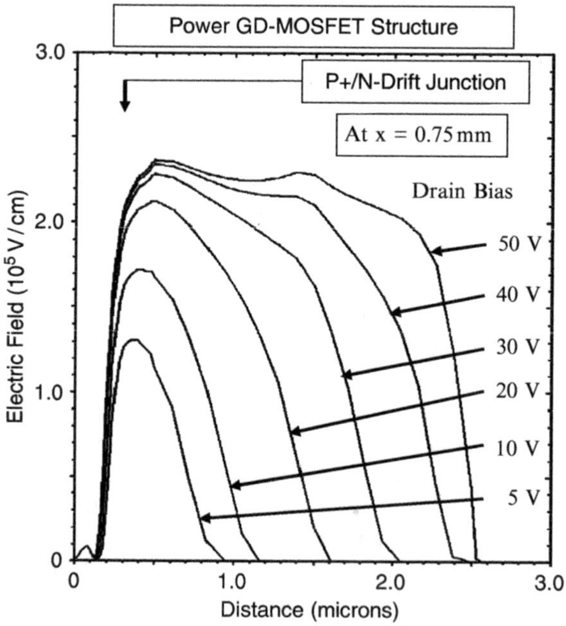


**Fig. 6.24** Electric field profiles in the power GD-MOSFET structure with a trench oxide thickness of 2,000 Å

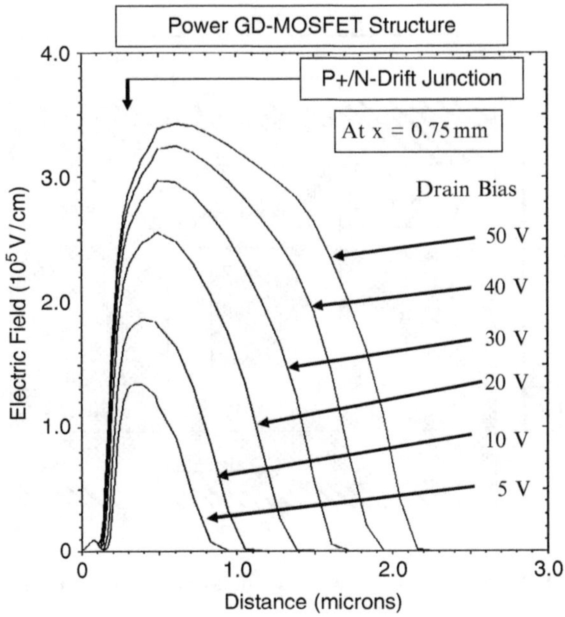


**Fig. 6.25** Electric field profiles in the power GD-MOSFET structure with a trench oxide thickness of 3,500 Å

The charge coupling physics for determining the blocking characteristics of the power GD-MOSFET structure with the source electrode in the trench region has been discussed above based up on numerical simulations. These results are also applicable for the power GD-MOSFET structure with the gate electrode in the entire trench.

## 6.3 Power GD-MOSFET On-Resistance

The components of the on-resistance for the power GD-MOSFET structure are similar as those already described for the power CC-MOSFET structure in the previous chapter. However, the drift region resistance in the power GD-MOSFET structure is even lower than that for the power CC-MOSFET structure due to the higher doping concentrations in the drift region and an improved electric field distribution. The total on-resistance for the power GD-MOSFET structure can be obtained by the addition of seven resistances which are considered to be in series in the current path between the source and the drain electrodes:

$$R_{ON} = R_{CS} + R_{N+} + R_{CH} + R_{A} + R_{D} + R_{SUB} + R_{CD} \tag{6.15}$$

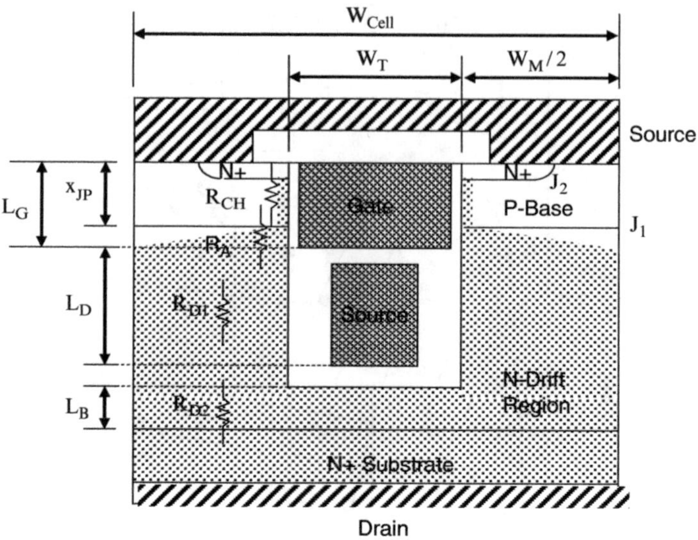


**Fig. 6.26** Power GD-MOSFET structure with current flow model used for analysis of its internal resistances

Each of the resistances within the power GD-MOSFET structure is analyzed below by using the procedure described in the textbook [5]. In the textbook, it was demonstrated that the contributions from the source contact resistance ($R_{CS}$), the source resistance ($R_{N+}$), and the drain contact resistance ($R_{CD}$) are very small and will therefore be neglected in this chapter. As in the case of the power D-MOSFET and U-MOSFET structures, the substrate contribution will also be excluded for the comparison of devices.

The power GD-MOSFET structure with source electrode in the trench region is illustrated in Fig. 6.26 with the current flow path shown as the shaded area. The current flow from the channel into the drift region is assumed to occur at a 45° spreading angle. Due to the small mesa width (typically 0.5 μm) for these devices, the current spreading occurs to a relatively small depth below the P-base region. This current path is aided by the formation of an accumulation layer on the trench sidewalls because the gate electrode in the portion with the thin gate oxide must extend to about 0.25 μm below the P-base region to ensure the formation of the channel region. After this depth, the current density can be assumed to be uniform in the drift region. An additional resistance is contributed by the portion of the drift region, a buffer layer with thickness ($L_B$), located below the bottom of the trench because the trench must not extend into the $N^+$ substrate.

The cross-section of the power GD-MOSFET structure illustrated in Fig. 6.26 provides the various dimensions that can be used for the analysis of the on-resistance components. Here, $W_{Cell}$ is the pitch for the linear cell geometry analyzed in this section; $W_T$ is the width of the trench region; and $W_M$ is the width of the mesa region. The vertical junction depths of the P-base region and the $N^+$ source regions are $x_{JP}$ and $x_{N+}$, respectively. The various resistances that must be analyzed in the power CC-MOSFET structure are also indicated in the figure.

In this monograph, the characteristics of power GD-MOSFET structure with 60-V blocking capability will be analyzed. For this voltage rating and 0.5-μm lithography design rules, the power GD-MOSFET structure has a cell pitch ($W_{CELL}$) of 1.5 μm with trench and mesa widths of 1.0 and 0.5 μm, respectively. Typical junction depths for the $N^+$ source region and P-base region are 0.08 μm and 0.32 μm, respectively, leading to a channel length of only 0.24 μm. This short channel length is possible due to the shielding of the channel region in the power GD-MOSFET structure due to the charge coupling phenomenon which spreads the electric field away from the P-base/N-drift junction. Further suppression of the P-base reach-through phenomenon can be achieved by using a peak P-base doping concentration located at a depth of 0.15 μm (i.e. below the $N^+$ source/P-base junction). The peak P-base doping concentration is chosen to achieve a desired threshold voltage.

Based up on the discussion in the previous section, the optimum doping gradient of the N-drift region required to achieve a 60-V blocking voltage capability is $8 \times 10^{20}/\text{cm}^4$ for a mesa width of 0.5 μm and a trench oxide thickness of 3,000 Å. The length of the trench region ($L_D$) required below the P-base region is 2.0 μm to achieve a breakdown voltage of 60 V.

### 6.3.1 Channel Resistance

The contribution to the specific on-resistance from the channel in the GD-MOSFET structure is the same as that for the power CC-MOSFET structure – which is smaller than in the power D-MOSFET and U-MOSFET structures due to the shorter channel length in the charge-coupled devices. Based up on the analysis in the textbook [5] for the power U-MOSFET structure, the specific on-resistance contributed by the channel in the power GD-MOSFET structure is given by:

$$R_{CH,SP} = \frac{L_{CH} W_{Cell}}{2\mu_{ni}\, C_{OX}\, (V_G - V_{TH})} \tag{6.16}$$

In the case of the power MOSFET structures, it is customary to provide the on-resistance at a gate bias of 4.5 and 10 V. Assuming a gate oxide thickness is 500 Å, an inversion layer mobility of 450 $\text{cm}^2/\text{V s}$ (to match the mobility used in the numerical simulations discussed later in this section), and a threshold voltage of 2.7 V (to match the numerical simulations) in the above equation for the power GD-MOSFET design with a cell width of 1.5 μm, the specific resistance contributed by the channel at a gate bias of 4.5 V is found to be 0.0326 mΩ $\text{cm}^2$. The specific on-resistance for the channel in the power CC-MOSFET structure is reduced to 0.008 mΩ cm$^2$ when the gate bias is increased to 10 V. These values are an order of magnitude smaller than for the power D-MOSFET and U-MOSFET structures due to the combination of a very short channel length and a small cell pitch.

## *6.3.2 Accumulation Resistance for Current Spreading Region*

In the power GD-MOSFET structure, the current flowing through the inversion channel enters the drift region at the edge of the P-base junction. The current then spreads from the edge of the P-base junction into the drift region. The current flow is strongly aided by the formation of an accumulation layer in the semiconductor at the trench sidewalls adjacent to the gate electrode due to the positive gate bias applied to turn-on the device. The specific on-resistance contributed by current spreading resistance in the drift region can be neglected because the accumulation layer resistance ($R_{A1}$) for this portion in the power GD-MOSFET structure is much smaller. The accumulation resistance for this portion is given by the same formulation derived for the power MOSFET structures in the textbook [5]:

$$R_{A,SP} = \frac{(L_G - x_{JP})W_{Cell}}{2\mu_{nA}C_{OX}(V_G - V_{THA})} \tag{6.17}$$

The gate oxide thickness must be used for computation of the oxide capacitance in this case. The threshold voltage ($V_{THA}$) in the expression is for the on-set of formation of the accumulation layer. A zero threshold voltage will be assumed here when performing the analytical computations. The gate length ($L_G$) will be assumed to 0.4 μm to match the simulations.

For the 60-V power GD-MOSFET design with a cell width of 1.5 μm and trench width of 1.0 μm, the specific resistance contributed by the accumulation layer in the current spreading portion at a gate bias of 4.5 V is 0.00196 mΩ cm$^2$ for a gate oxide thickness of 500 Å. An accumulation layer mobility of 1,000 cm$^2$/V s was used in this calculation to match the mobility used in the numerical simulations (discussed later in this section). When the gate bias is increased to 10 V, the specific resistance contributed by the accumulation layer in the current spreading portion is reduced to 0.00088 mΩ cm$^2$.

## *6.3.3 Drift Region Resistance*

The resistance contributed by the drift region in the power GD-MOSFET structure is reduced well below that for the ideal drift region due to the high doping concentration in the drift region. The specific on-resistance contributed by the drift region in the power GD-MOSFET structure can be computed by analysis of the two resistances shown in Fig. 6.26. The analytical modeling of the drift region resistance is complicated by the non-uniform doping profile with an electron mobility that also varies with doping concentration.

In the mesa portion of the structure, the current density is uniform with a current density enhanced by the ratio of the cell pitch to the mesa width. The drift region resistance contribution from the mesa region can be computed by considering a small segment (dy) of the drift region at a depth y from the bottom of the gate electrode. The specific resistance of the drift region for the mesa portion is given by:

$$R_{D1,SP} = \left(\frac{W_{Cell}}{W_M}\right) \int_0^{L_D} \rho_D(y)\, dy \tag{6.18}$$

where the resistivity $\rho_D$ is a function of the position in the drift region due to the graded doping profile. The resistivity of the drift region is given by:

$$\rho_D(y) = \frac{1}{q\mu_n(y)N_D(y)} \tag{6.19}$$

The dependence of the electron mobility on the doping concentration must be taken into account when analyzing the drift region resistance in the power GD-MOSFET structure because the doping concentration exceeds $1 \times 10^{16}/\text{cm}^3$. The electron mobility is given by [5]:

$$\mu_n = \frac{5.1 \times 10^{18} + 92\, N_D^{0.91}}{3.75 \times 10^{15} + N_D^{0.91}} \tag{6.20}$$

In order to simplify the analysis, the following approximation for the dependence of the electron mobility on doping concentration is adequate for the doping levels ranging between $1 \times 10^{16}/\text{cm}^3$ and $3 \times 10^{17}/\text{cm}^3$:

$$\mu_n = \frac{5.1 \times 10^{18}}{3.75 \times 10^{15} + N_D^{0.91}} \tag{6.21}$$

Using (6.19) and (6.21) in (6.18):

$$R_{D1,SP} = \left(\frac{W_{Cell}}{W_M}\right) \int_0^{L_D} \left[\frac{4.6 \times 10^{15}}{N_D(y)} + \frac{1.225}{N_D(y)^{0.09}}\right] dy \tag{6.22}$$

For a linearly graded doping profile with a gradient G:

$$N_D(y) = N_0 + G\,y \tag{6.23}$$

where $N_0$ is the initial doping concentration in the mesa region at the depth of the gate electrode. Integrating (6.22) with (6.23) yields:

$$R_{D1,SP} = \left(\frac{W_{Cell}}{W_M}\right)\left\{\begin{array}{l}\dfrac{4.6 \times 10^{15}}{G} \ln\left(\dfrac{N_0 + GL_D}{N_0}\right) + \\ \dfrac{1.346}{G}\left[(N_0 + GL_D)^{0.91} - N_0^{0.91}\right]\end{array}\right\} \quad (6.24)$$

An additional resistance contribution ($R_{D2}$) in the power GD-MOSFET structure is associated with the buffer layer located below the bottom of the trenches. In this portion of the structure, the current density is uniform and current flow occurs across the entire cell width. Using the above approach, the specific resistance contributed by the portion of the drift region below the trench in the power GD-MOSFET structure can be derived:

$$R_{D2,SP} = \left\{\begin{array}{l}\dfrac{4.6 \times 10^{15}}{G} \ln\left[\dfrac{N_0 + G(L_D + L_B)}{N_0 + GL_D}\right] + \\ \dfrac{1.346}{G}\left[[N_0 + G(L_D + L_B)]^{0.91} - (N_0 + GL_D)^{0.91}\right]\end{array}\right\} \quad (6.25)$$

In the case of the 60-V GD-MOSFET structure, the optimum doping gradient has been shown to be $8 \times 10^{20}/\text{cm}^4$ with a drift region length of 2 μm to support the voltage. The doping concentration for the drift region then increases from an initial value of $1 \times 10^{16}$ to $2.1 \times 10^{17}/\text{cm}^3$ at a depth of 2.5 μm below the bottom of the gate electrode. The doping concentration at the bottom of the source electrode in the trench (i.e. at a depth of $L_D$ below the gate electrode) in this case is $1.7 \times 10^{17}/\text{cm}^3$. For a device structure with cell pitch of 1.5 μm and a mesa width of 0.5 μm, the specific resistance of the drift region in the mesa portion ($R_{D1,SP}$) is found to be 0.0712 mΩ cm² using (6.24). For this doping profile, the specific resistance of the drift region in the buffer layer below the trenches ($R_{D2,SP}$) is found to be 0.0029 mΩ cm² using (6.25). The total drift region specific resistance for this power GD-MOSFET structure with an optimum doping gradient of $8 \times 10^{20}/\text{cm}^4$ is then 0.0741 mΩ cm².

### *6.3.4 Total On-Resistance*

The total specific on-resistance for the power GD-MOSFET structure can be computed by combining the above components for the on-resistance as defined by (6.15). In the case of the 60-V power GD-MOSFET design with a cell pitch ($W_{Cell}$) of 1.5 μm and mesa width of 0.5 μm, the total specific on-resistance is found to be 0.109 mΩ cm² at a gate bias of 4.5 V and 0.083 mΩ cm² at a gate bias of 10 V by using the analytical models. The contributions from each of the components of the on-resistance are summarized in Fig. 6.27. The specific on-resistance for the power GD-MOSFET

| Resistance | $V_G = 4.5$ v (mΩ-cm$^2$) | $V_G = 10$ v (mΩ-cm$^2$) |
|---|---|---|
| Channel ($R_{CH, SP}$) | 0.0326 | 0.00803 |
| Accumulation ($R_{A, SP}$) | 0.00196 | 0.00088 |
| Drift ($R_{D1, SP}$) | 0.0712 | 0.0712 |
| Drift ($R_{D2, SP}$) | 0.00294 | 0.00294 |
| Total ($R_{T, SP}$) | 0.1087 | 0.0830 |

**Fig. 6.27** Resistance components in the 60-V power GD-MOSFET structure with source electrode in the trenches

structure is found to be an order of magnitude smaller than that for the power D-MOSFET and the U-MOSFET structures at a gate bias of 4.5 V. It is also 50% smaller than the specific on-resistance for the 60-V power CC-MOSFET structure.

From the values provided in Fig. 6.27, it can be seen that the drift region resistance for the power GD-MOSFET structure is far smaller than that for the power D-MOSFET and U-MOSFET structures. The very low specific resistance of the drift region is associated with the high doping concentration of the drift region enabled by the charge-coupling phenomenon. Due to this reduction in drift region resistance, the channel resistance contribution in the power GD-MOSFET structure remains a significant contributor to the total specific on-resistance. In addition, it is important to reduce the resistance contributed by the $N^+$ substrate by wafer thinning technology to realize the benefits of the charge-coupling concept.

The ideal specific on-resistance for a drift region is given by:

$$R_{IDEAL,SP} = \frac{W_{PP}}{q\mu_n N_D} \tag{6.26}$$

where $W_{PP}$ is the parallel-plane depletion width at breakdown, $N_D$ is the doping concentration of the drift region to sustain the blocking voltage, and $\mu_n$ is the mobility for electrons corresponding to this doping concentration. For the case of a blocking voltage of 60 V, the depletion width and doping concentration are found to be $6.7 \times 10^{15}/\text{cm}^3$ and 3.4 μm, respectively. Using the electron mobility for this doping level, the ideal specific on-resistance is found to be 0.247 mΩ cm$^2$. Since the conventional device is constrained by the impact of an 80% reduction of breakdown voltage due to the edge termination, it is worth computing the ideal specific on-resistance for this case for comparison with the device. For the case of a blocking voltage of 75 V, the depletion width and doping concentration are found to be $5.0 \times 10^{15}/\text{cm}^3$ and 4.4 μm, respectively. Using the electron mobility for this doping level, the ideal specific on-resistance is found to be 0.426 mΩ cm$^2$. The specific on-resistance for the 60-V GD-MOSFET structure is smaller than these ideal specific on-resistances for a gate bias of 10 V by a factor of three times and five times. Consequently, the charge-coupling concept with the optimized doping profile provides the opportunity to overcome the ideal specific resistance barrier using silicon device structures.

#### 6.3.4.1 Simulation Results

The transfer characteristics for the 60-V GD-MOSFET structure were obtained using numerical simulations with a drain bias of 0.1 V at 300 and 400°K. The resulting transfer characteristics are shown in Fig. 6.28. From this graph, a threshold voltage of 2.7 and 2.3 V can be extracted at 300 and 400°K, respectively. The specific on-resistance can be obtained from the transfer characteristics at any gate bias voltage. For the case of a gate bias of 4.5 V and 300°K, the specific in-resistance is found to be 0.104 mΩ cm$^2$. For the case of a gate bias of 10 V and 300°K, the specific in-resistance is found to be 0.080 mΩ cm$^2$. The close agreement of these values with the analytical model provides validity for the analytical approach.

The on-state current flow pattern within the GD-MOSFET structure at a small drain bias of 0.1 V and a gate bias of 4.5 V is shown in Fig. 6.29. In the figure, the depletion layer boundary is shown by the dotted lines and the junction boundary is delineated by the dashed line. From the figure, it can be seen that the current spreads from the channel to the drift region at a 45° angle. The current distribution is

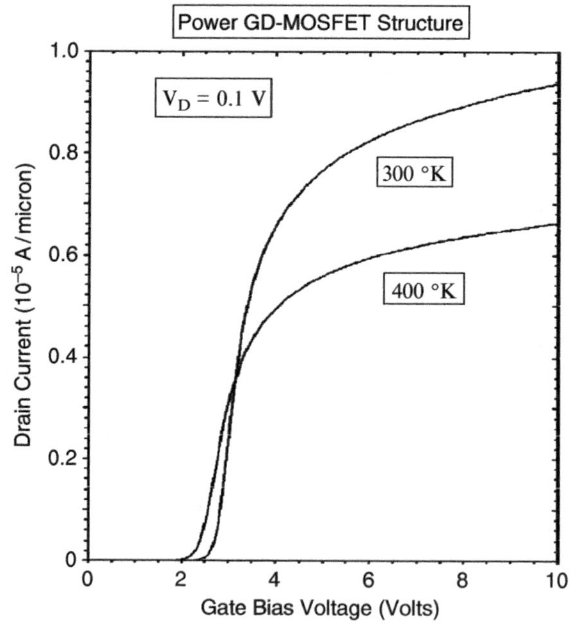


**Fig. 6.28** Transfer characteristics of the 60-V GD-MOSFET structure

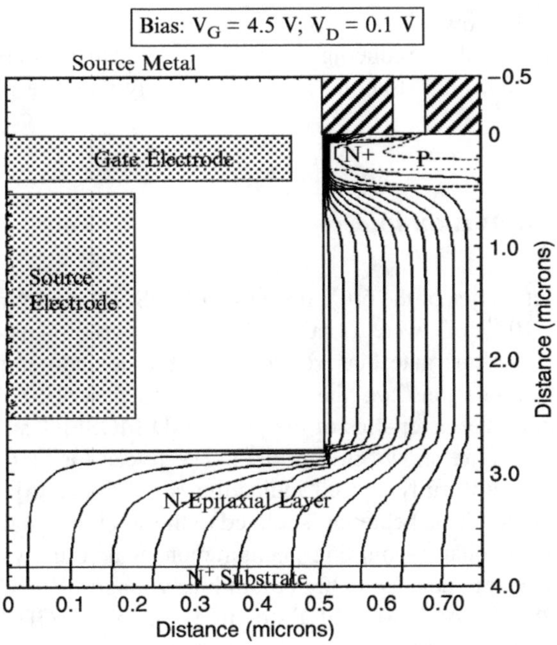


**Fig. 6.29** Current distribution in the GD-MOSFET structure

uniform in the mesa region at depths below the bottom of the gate electrode justifying the assumptions made in the analytical model.

## 6.4 Output Characteristics

The output characteristics of the power GD-MOSFET structure are important to the loci for the switching waveforms when it is operating in power circuits. Due to shielding of the P-base region, the power CC-MOSFET structure exhibits super-linear transfer characteristics [5]. The saturated drain current for the power GD-MOSFET structure is then given by:

$$I_{Dsat} = C_{ox}(V_G - V_T)v_{sat,n}Z \tag{6.27}$$

With sufficiently high doping concentration of the P-base region, the modulation of channel length can be made sufficiently small to ensure a high output resistance. In the power GD-MOSFET structure, the charge coupling phenomenon suppresses the depletion of the P-base region with increasing drain bias voltage because the voltage is spread into the drift region. This is beneficial for obtaining a reasonable output resistance despite the very short channel length in the device. The short channel length in the power GD-MOSFET structure produces a high transconductance which is beneficial for reducing switching losses. The saturated drain current in the power GD-MOSFET structure increases linearly with the gate bias voltage for this model.

### *6.4.1 Simulation Example*

The output characteristics of the 60-V power GD-MOSFET structure with a doping gradient of $8 \times 10^{20}/\text{cm}^4$ were obtained by using two-dimensional numerical simulations using various gate bias voltages. All the device parameters used for these numerical simulations are the same as those described in the previous sections. The output characteristics of the power GD-MOSFET structure obtained using the simulations are shown in Fig. 6.30. The traces for increasing gate bias voltages are nearly uniformly spaced indicating super-linear [5] behavior of the transfer characteristics. This behavior is related to the shielding of the channel from the drain bias by the charge-coupling phenomenon, which allows the channel to operate in the linear regime even at high drain bias voltages. When compared with the power D-MOSFET and U-MOSFET structures, the power GD-MOSFET structure exhibits a gradual transition from the on-state to the current saturation regime due to the on-set of the charge-coupling effect at a drain bias above 10 V as described earlier with the aid of Figs. 6.15–6.17.

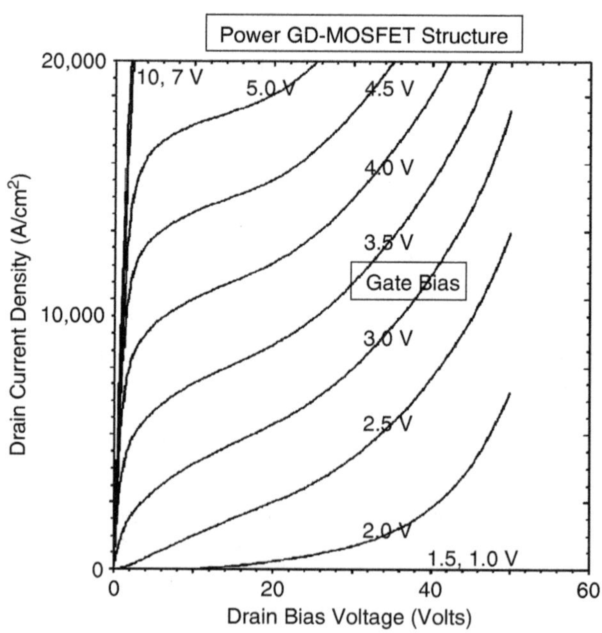


**Fig. 6.30** Output characteristics for the power GD-MOSFET structure

## 6.5 Device Capacitances

For the power MOSFET structures, the switching speed is limited by the device capacitances in practical applications as previously discussed in Chap. 2. The rate at which the power MOSFET structure can be switched between the on- and off-states is determined by the rate at which the input capacitance can be charged or discharged. In addition, the capacitance between the drain and the gate electrodes has been found to play an important role in determining the drain current and voltage transitions during the switching event.

The capacitances within the power GD-MOSFET structure can be analyzed using the same approach used in the textbook [5] for the power U-MOSFET structure. The specific input capacitance for the power GD-MOSFET structure with source electrode in the trench is the same as that for the power CC-MOSFET structure with source electrode in the trench region. In addition to the capacitance associated with the gate electrode overlap with the P-base and $N^+$ source regions on the trench sidewall, it is necessary to account for the capacitances due to overlap of the source metal electrode and the overlap of the gate electrode with the source electrode embedded within the trench. The specific input (or gate) capacitance for the power GD-MOSFET structure with source electrode in the trench is given by:

$$C_{IN,SP} = C_{N+} + C_P + C_{SM} + C_{SG} = \frac{2x_{JP}}{W_{Cell}}\left(\frac{\varepsilon_{OX}}{t_{GOX}}\right) + \frac{W_T}{W_{Cell}}\left(\frac{\varepsilon_{OX}}{t_{IEOX}}\right) + \frac{(W_T - 2t_{TOX})}{W_{Cell}}\left(\frac{\varepsilon_{OX}}{t_{IGOX}}\right) \quad (6.28)$$

where $t_{GOX}$, $t_{IEOX}$, and $t_{IGOX}$ are the thicknesses of the gate oxide and inter-electrode oxide, and the oxide between the gate electrode and source electrode in the trench, respectively. The junction depth for the P-base region is smaller in the power GD-MOSFET structure when compared with the power U-MOSFET structure resulting in a smaller input capacitance. For a 60-V power GD-MOSFET structure with a cell pitch of 1.5 μm and mesa width of 0.5 μm, the specific input capacitance is found to be 42.8 nF/cm$^2$ for a gate oxide thickness of 500 Å, an inter-metal dielectric thickness of 5,000 Å, and a oxide thickness of 1,000 Å between the gate electrode and the source electrode within the trenches. This value is larger than 22 nF/cm$^2$ for the power D-MOSFET structure and approximately equal to 47 nF/cm$^2$ for the power U-MOSFET structure.

The capacitance between the gate and drain electrodes (also called the reverse transfer capacitance) for the power GD-MOSFET structure can be obtained using the same approach as used for the power CC-MOSFET structures with source electrode in the trenches. The gate electrode for power GD-MOSFET structure with source electrode in the trenches gets screened from the drain by the presence of the source electrode in the trenches.

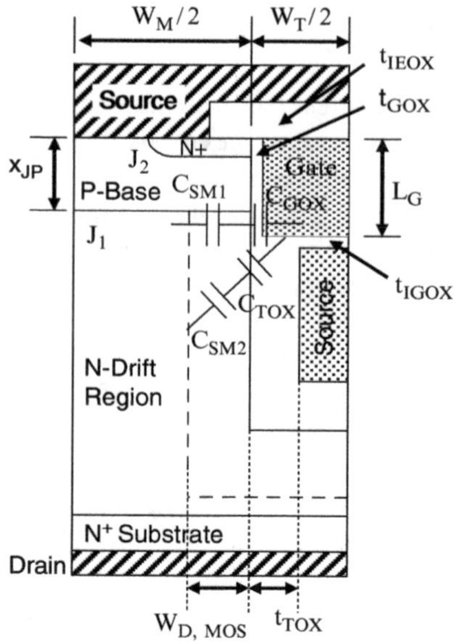


**Fig. 6.31** Depletion boundaries and capacitances within the power GD-MOSFET structure with source electrode in the trench region

The gate transfer capacitance for the power GD-MOSFET structure with source electrode in the trenches can be modeled using the depletion layers extending from the gate electrode into the drift region as illustrated in Fig. 6.31. Two sets of capacitances are indicated in the figure. The first set pertains to the vertical extension of the gate electrode ($L_G - x_{JP}$) into the drift region below the junction ($J_1$) between the P-base region and the drift region. The second set pertains to the bottom surface of the gate electrode ($t_{TOX} - t_{GOX}$) that does not overlap the source electrode within the trenches. Without taking the screening by the source electrode into account, the gate transfer capacitance for the power GD-MOSFET structure with source electrode in the trenches is given by:

$$C_{GD,SP} = \frac{2(L_G - x_{JP})}{W_{Cell}} \left( \frac{C_{GOX} C_{S,M1}}{C_{GOX} + C_{S,M1}} \right) + 2 \left( \frac{t_{TOX} - t_{GOX}}{W_{Cell}} \right) \left( \frac{C_{TOX} C_{S,M2}}{C_{TOX} + C_{S,M2}} \right) \quad (6.29)$$

where $C_{S,M1}$ is the semiconductor capacitance corresponding to the gate oxide thickness; $C_{GOX}$ is the gate oxide specific capacitance; $C_{S,M2}$ is the semiconductor capacitance corresponding to the trench oxide thickness; and $C_{TOX}$ is the trench oxide specific capacitance.

It is convenient to express the capacitances for the power GD-MOSFET structure with source electrode in the trenches in terms of a gate geometry factor [5]:

$$C_{GD,SP} = K_{G4} \left( \frac{C_{GOX} C_{S,M1}}{C_{GOX} + C_{S,M1}} \right) + K_{G5} \left( \frac{C_{TOX} C_{S,M2}}{C_{TOX} + C_{S,M2}} \right) \quad (6.30)$$

where

$$K_{GD4} = \frac{2(L_G - x_{JP})}{W_{Cell}} \tag{6.31}$$

and

$$K_{GD5} = \frac{2(t_{TOX} - t_{GOX})}{W_{Cell}} \tag{6.32}$$

The screening of the gate electrode by the source electrode in the trenches can be accounted for by using a screening factor that takes into account the depletion of the mesa region by the extension of a depletion layer for the MOS structure comprising the source electrode, the thick trench oxide, and the drift region. The screening factor is then given by:

$$K_{S,GD} = \frac{W_M - 2W_{D,MOS2}}{W_M} \tag{6.33}$$

The screening factor as defined above decreases with increasing drain bias voltage and becomes equal to zero at the drain bias required to completely deplete the mesa region. The gate transfer capacitance for the power GD-MOSFET structure with source electrode in the trenches with screening taken into account can be computed by multiplying the gate transfer capacitance given by (6.30) by the screening factor given by (6.33).

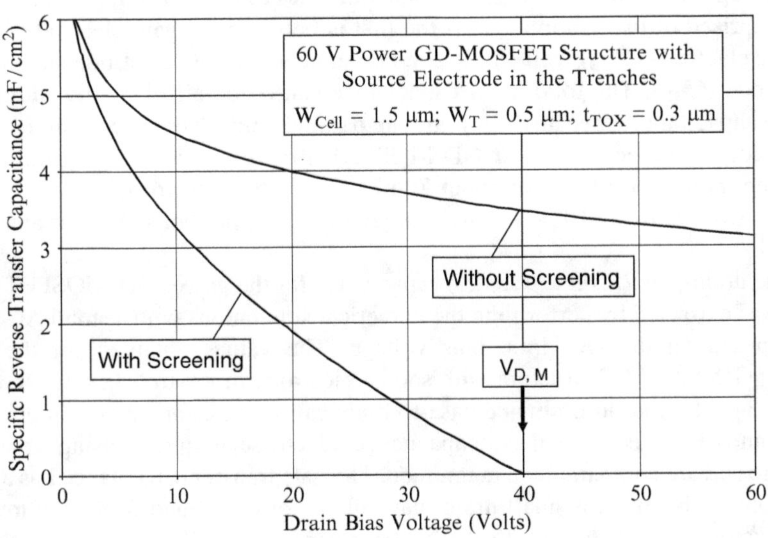


**Fig. 6.32** Gate-drain capacitance for the power GD-MOSFET structure with source electrode in the trenches

The specific gate transfer capacitance obtained by using the above analytical formulae is shown in Fig. 6.32 for the 60-V power GD-MOSFET structure with source electrode in the trenches. This structure has a cell pitch of 1.5 μm, a mesa width of 0.5 μm, gate oxide thickness of 500 Å, and trench oxide thickness of 3,000 Å. The gate was assumed to extend to a depth of 0.4 μm which is 0.08 μm below the junction between the P-base region and the drift region. Without taking into account the screening by the source electrode, the specific gate transfer capacitance reduces from 6.6 to 3.1 nF/cm$^2$ at a drain bias of 60 V. In contrast to this, when screening by the source electrode is taken into account, the specific gate transfer capacitance has about the same value of 6.4 nF/cm$^2$ without the screening at a drain bias of 0.5 V but reduces much more rapidly and becomes equal to 0 nF/cm$^2$ at a drain bias of 40 V when the mesa region becomes completely depleted under the assumption of an effective doping concentration of 1 × 10$^{17}$/cm$^3$ in the drift region.

### *6.5.1 Simulation Example*

The capacitances of the 60-V power GD-MOSFET structure with doping gradient of 8 × 10$^{20}$/cm$^4$ were extracted using two-dimensional numerical simulations on the structure with device parameters described in the previous sections. The input capacitance was extracted by performing the numerical simulations with a small AC signal superposed on the DC gate bias voltage. The input capacitances obtained for the power GD-MOSFET structure with source electrode within the trench region are shown in Fig. 6.33 at a drain bias of 40 V. The input capacitance is comprised of two components – the first is between the gate electrode and the source electrode ($C_{GS}$) while the second is between the gate electrode and the base electrode ($C_{GB}$). The total input capacitance can be obtained by the addition of these capacitances because they are in parallel and share a common contact electrode in the actual power GD-MOSFET structure. From the figure, a total specific input capacitance of about 70 nF/cm$^2$ is observed for gate bias voltages raging from zero to 10 V – much larger than that predicted by the analytical model.

The drain-gate (reverse transfer) capacitance for the power GD-MOSFET structure was extracted by performing the numerical simulations with a small AC signal superposed on the DC drain bias voltage. The values obtained for the 60-V power GD-MOSFET structure with source electrode in the trenches are shown in Fig. 6.34. The gate-to-drain and base-to-drain capacitances are shown in the figure for comparison. Both of these capacitances decrease with increasing drain bias voltage and are comparable in magnitude. The gate transfer capacitance has a value of about 6 nF/cm$^2$ at a small drain bias voltage of 0.5 V and decreases to about 0.7 nF/cm$^2$ at a drain bias of 30 V. This behavior is in excellent agreement with the values predicted by the analytical model (see Fig. 6.32) when screening by the source electrode is included.

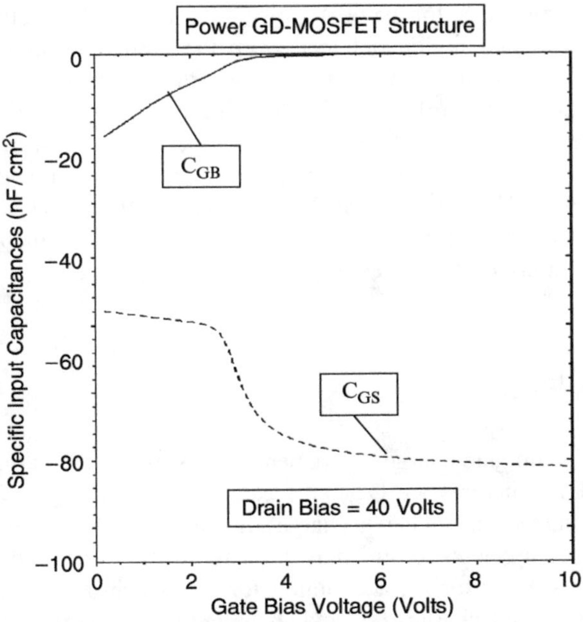


**Fig. 6.33** Input capacitances for the GD-MOSFET structure with source electrode in the trench region

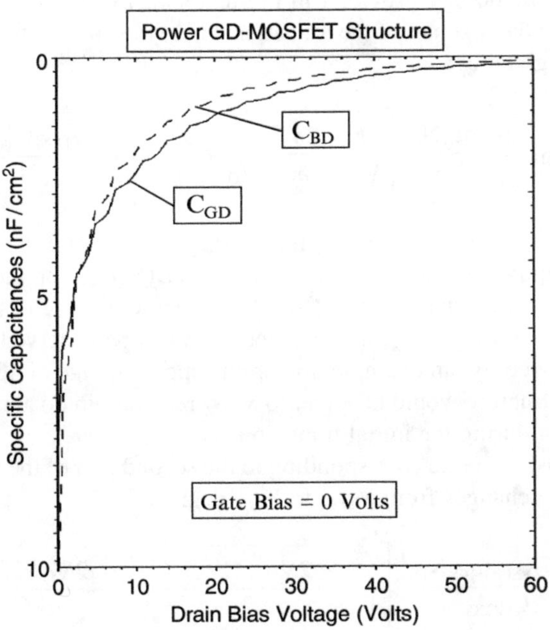


**Fig. 6.34** Reverse transfer and output capacitances for the GD-MOSFET structure with source electrode in the trench region

The output capacitance for the 60-V power GD-MOSFET structure with source electrode in the trench region obtained from the numerical simulations is also provided in Fig. 6.34. It can be observed that the output capacitance is about 6 nF/cm$^2$ at a drain bias of 0.5 V and reduces to about 0.5 nF/cm$^2$ at a drain bias of 30 V. Note that the gate transfer capacitance for the power GD-MOSFET structure with source electrode in the trench region is approximately equal to its output capacitance. The low gate transfer capacitance for the power GD-MOSFET structure with source electrode in the trench region makes it suitable for applications operating at high frequencies.

## 6.6 Gate Charge

The most significant gate charge components for assessing the performance of the power MOSFET structures are $Q_{SW}$ (the gate switching charge), $Q_{GD}$ (the gate-drain charge), and $Q_G$ (the total gate charge). In the case of the power GD-MOSFET structure, these components are given by the same equations derived in the textbook with adjustments made for the accounting for the trench oxide capacitance. The expressions for the gate charge $Q_{GS1}$ and $Q_{GS2}$ are the same as those provided in the textbook. However, in the case of the gate transfer charge, the gate transfer capacitance must be modeled by taking into account the presence of the source electrode in the trenches as done for the power CC-MOSFET structure.

Using the methodology provided in the textbook [5], the gate transfer charge corresponding to the first part of the transition where the drain voltage changes from $V_{DS}$ to $V_{D,M}$ is given by:

$$Q_{GD1} = \frac{2q\varepsilon_S N_D}{C_{TOX}}\left[\sqrt{1+\frac{2V_{DS}C_{TOX}^2}{q\varepsilon_S N_D}} - \sqrt{1+\frac{2V_{D,M}C_{TOX}^2}{q\varepsilon_S N_D}}\right] \tag{6.34}$$

Note that the gate geometry factor is unity during this phase of the turn-on process. The voltage required to deplete the mesa region ($V_{D,M}$) for the GD-MOSFET structure is larger than that for the CC-MOSFET structure due to the larger trench oxide thickness. Consequently, this transient may not be observed during applications for this device. As an example, the drain supply voltage of 40 V for the 60-V GD-MOSFET structure would be equal to $V_{D,M}$ resulting in no abrupt reduction of the drain voltage during the initial transition.

The gate transfer charge corresponding to the second part of the transition where the drain voltage changes from $V_{D,M}$ to $V_{ON}$ is given by:

$$\begin{aligned} Q_{GD2} = {} & \frac{2K_{G4}K_{SGD}q\varepsilon_S N_{DE}}{C_{GOX}}\left[\sqrt{1+\frac{2V_{D,M}C_{GOX}^2}{q\varepsilon_S N_{DE}}} - \sqrt{1+\frac{2V_{ON}C_{GOX}^2}{q\varepsilon_S N_{DE}}}\right] \\ & + \frac{2K_{G5}K_{SGD}q\varepsilon_S N_{DE}}{C_{TOX}}\left[\sqrt{1+\frac{2V_{D,M}C_{TOX}^2}{q\varepsilon_S N_{DE}}} - \sqrt{1+\frac{2V_{ON}C_{TOX}^2}{q\varepsilon_S N_{DE}}}\right] \end{aligned} \tag{6.35}$$

In order to simplify the analysis, the screening factor was assumed to have a constant value at a drain bias of half the supply voltage and an effective doping concentration was utilized. The voltage transition in the power GD-MOSFET structure occurs in two steps – first with a lower capacitance when the mesa region is entirely depleted at the larger drain bias voltages, and later with a larger capacitance when the entire mesa region is no longer depleted. The first transition will be defined to occur until a time $t_3$ with the second part completed at a time $t_4$. As discussed above, the first abrupt transition may not be usually observed in the GD-MOSFET structure due to a large value for $V_{D,M}$. The other components of the gate charge are similar to those already provided in the textbook:

$$Q_{SW} = Q_{GS2} + Q_{GD} \tag{6.36}$$

$$Q_G = [C_{GS} + C_{GD}(V_{DS})]V_{GP} + Q_{GD} + [C_{GS} + C_{GD}(V_{ON})](V_G - V_{GP}) \tag{6.37}$$

The gate charge values obtained for the 60-V power GD-MOSFET structure with source electrode in the trenches by using the above analytical equations are: $Q_{GD}$ = 218 nC/cm$^2$; $Q_{SW}$ = 229 nC/cm$^2$; and $Q_G$ = 1,060 nC/cm$^2$. It can be concluded that the power GD-MOSFET structure with source electrode in the trenches is suitable for high frequency applications where both the specific on-resistance and switching losses must be optimized.

Equations for the gate voltage, drain current, and drain voltage waveforms obtained by using the analytical model are provided in the textbook [5]. However, the drain voltage waveform for the power GD-MOSFET structure with source electrode in the trenches must be reformulated because of the more complex nature of the gate transfer capacitance. Due to the very small (or ideally zero) reverse transfer capacitance for the power GD-MOSFET structure with source electrode in the trenches at larger drain bias voltages, the drain voltage first reduces abruptly from $V_{DS}$ to $V_{D,M}$. After this, the drain voltage is determined by the gate transfer capacitance given by (6.30):

$$v_D(t) = \left[\frac{J_G(t-t_2)}{(K_{G1}+K_{G2})K_{SCC}\sqrt{2q\,\varepsilon_S N_D}} - \sqrt{V_{D,M}}\right]^2 \tag{6.38}$$

from $t = t_2$ to $t = t_3$.

The waveforms obtained for the 60-V power GD-MOSFET structure with source electrode in the trenches – using 1.5 μm cell pitch and a trench width of 1.0 μm with a gate oxide thickness of 500 Å and trench oxide thickness of 3,000 Å – using these equations are provided in Fig. 6.35. A gate drive current density of 1.33 A/cm$^2$ was used to turn on the device from a steady-state blocking voltage of 40 V to match the results of two dimensional numerical simulations discussed below. The specific input capacitance was assumed to be 75 nF/cm$^2$.

The gate voltage initially increases linearly with time. After reaching the threshold voltage, the drain current can be observed to increase very quickly because of the large transconductance for this device structure. The drain current density

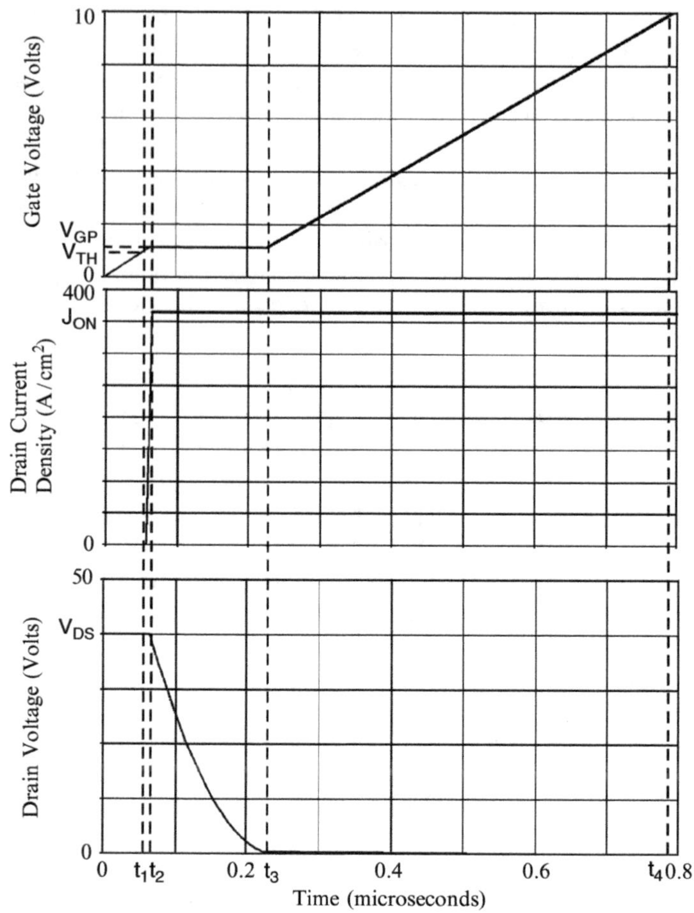


**Fig. 6.35** Analytically computed waveforms for the 30-V power CC-MOSFET structure with source electrode in the trenches

increases until it reaches an on-state current density of 365 A/cm$^2$. The on-state current density determines the gate plateau voltage which has a value of 1.15 V for a threshold voltage of 1.0 V at this drain bias. During the gate voltage plateau phase, the drain voltage decreases in a non-linear manner until it reaches the on-state voltage drop. After this time, the gate voltage again increases but at a slower rate than during the initial turn-on phase due to the larger gate transfer capacitance.

### *6.6.1 Simulation Example*

The gate charges for the 60-V power GD-MOSFET structures were extracted by using the results of two-dimensional numerical simulations of the cell structure described in the previous sections with a doping gradient of $8 \times 10^{20}$/cm$^4$.

The devices were turned-on from blocking state with a drain bias of 40 V by using a gate current of $1 \times 10^{-8}$ A/μm (equivalent to 1.33 A/cm$^2$ for the area of 0.75 $\times$ $10^{-8}$ cm$^2$). Once the drain current density reached the on-state value (365 A/cm$^2$), the drain current was held constant resulting in a reduction of the drain voltage. Once the drain voltage reached the on-state value corresponding to the gate plateau voltage, the gate voltage increased to the steady-state value of 10 V.

The gate charge waveforms obtained by using an input gate current density of 1.33 A/cm$^2$ when turning on the power GD-MOSFET structure with source electrode in the trench region from a blocking state with drain bias of 40 V are shown in Fig. 6.36. The on-state current density in this case is 365 A/cm$^2$ at a DC gate bias of

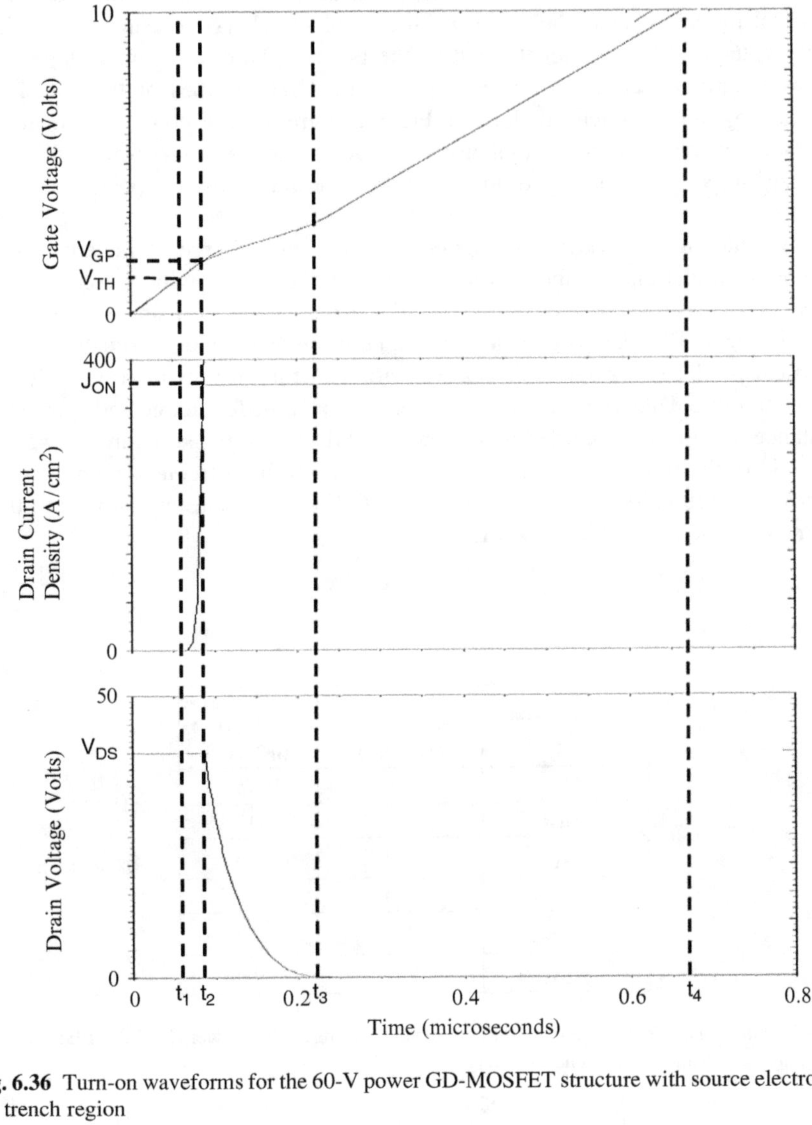


**Fig. 6.36** Turn-on waveforms for the 60-V power GD-MOSFET structure with source electrode in the trench region

10 V at the end of the turn-on transient. The gate voltage increases at a constant rate at the beginning of the turn-on process as predicted by the analytical model. When the gate voltage reaches the threshold voltage, the drain current begins to increase. The drain current increases very rapidly until it reaches the on-state current density of 365 A/cm$^2$.

Once the drain current reaches the on-state value, the gate voltage remains approximately constant at the plateau voltage ($V_{GP}$). The plateau voltage for this structure is 1.8 V for the drain current density of 365 A/cm$^2$ as governed by the transconductance of the device. The drain voltage decreases in a non-linear manner during the plateau phase as predicted by the analytical model. There is no abrupt reduction of the drain voltage at the beginning of the transient as predicted by the analytical model because the mesa pinch-off voltage ($V_{D,M}$) is equal to the drain supply voltage (40 V). After the end of the plateau phase, the gate voltage again increases until it reaches the gate supply voltage. The shapes of the waveforms predicted by the analytical model (see Fig. 6.35) are in very good agreement with the results of the numerical simulations indicating that the waveforms can be predicted by properly accounting for the variation of the gate transfer capacitance for this structure.

The values for the various components of the gate charge extracted from the numerical simulations for the power GD-MOSFET structure with source electrode in the trench region are compared with those calculated by using the analytical model in Fig. 6.37. There is good agreement between these values indicating that the analytical model is a reasonable representation of the physics of turn-on for this device structure. Due to the small values for the gate transfer charge and specific on-resistance, it can be concluded that the power GD-MOSFET structure with source electrode in the trench region is suitable for low and high frequency applications. The gate charge values for the power GD-MOSFET structure are similar to those obtained for the power CC-MOSFET structure.

| **Specific Gate Charge** | **Numerical Simulation (nC/cm$^2$)** | **Analytical Model (nC/cm$^2$)** |
|---|---|---|
| $Q_{GS1}$ | 80 | 75 |
| $Q_{GS2}$ | 40 | 11 |
| $Q_{GS}$ | 120 | 86 |
| $Q_{GD}$ | 180 | 218 |
| $Q_{SW}$ | 220 | 229 |
| $Q_G$ | 900 | 1060 |

**Fig. 6.37** Gate charge extracted from numerical simulations for the power GD-MOSFET structure with source electrode in the trench region

## 6.7 Device Figures of Merit

Significant power switching losses can arise from the charging and discharging of the large input capacitance in power MOSFET devices at high frequencies. The input capacitance ($C_{IN}$) of the power MOSFET structure must be charged to the gate supply voltage ($V_{GS}$) when turning on the device and then discharged to 0 V when turning off the device during each period of the operating cycle. The total power loss can be obtained by summing the on-state power dissipation for a duty cycle $\delta = t_{ON}/T$ and the switching power losses:

$$P_T = P_{ON} + P_{SW} = \delta R_{ON} I_{ON}^2 + C_{IN} V_{GS}^2 f \tag{6.39}$$

where $R_{ON}$ is the on-resistance of the power MOSFET structure, $I_{ON}$ is the on-state current, and f is the operating frequency. In writing this equation, the switching power losses due to the drain current and voltage transitions has been neglected. A minimum total power loss occurs for each power MOSFET structure at an optimum active area as shown in the textbook [5]. The on-state and switching power losses are equal at the optimum active area. The optimum active area at which the power dissipation is minimized is given by:

$$A_{OPT} = \sqrt{\frac{R_{ON,sp}}{C_{IN,sp}}}\left(\frac{I_{ON}}{V_{GS}}\right)\left(\sqrt{\frac{\delta}{f}}\right) \tag{6.40}$$

From the first term in this expression, a useful technology figure-of-merit can be defined:

$$FOM(A) = \frac{R_{ON,sp}}{C_{IN,sp}} \tag{6.41}$$

In the power electronics community, there is trend towards increasing the operating frequency for circuits in order to reduce the size and weight of the magnetic components. The ability to migrate to higher operating frequencies in power conversion circuits is dependent on making enhancements to the power MOSFET technology. From the above equations, an expression for the minimum total power dissipation can be obtained [5]:

$$P_T(\min) = 2 I_{ON} V_{GS} \sqrt{\delta R_{ON,sp} C_{IN,sp} f} \tag{6.42}$$

A second technology figure of merit related to the minimum power dissipation can be defined as:

$$FOM(B) = R_{ON,sp} C_{IN,sp} \tag{6.43}$$

In most applications for power MOSFET structures with high operating frequency, the switching losses associated with the drain current and voltage transitions become a dominant portion of the total power loss. The time period associated with the increase of the drain current and decrease of the drain voltage is determined by the charging of the device capacitances. It is therefore common practice in the industry to use the following figures-of-merit to compare the performance of power MOSFET products [5]:

$$FOM(C) = R_{ON,sp}Q_{GD,sp} \tag{6.44}$$

and

$$FOM(D) = R_{ON,sp}Q_{SW,sp} \tag{6.45}$$

Although FOM(D) encompasses both the drain current and voltage transitions, it is customary to use FOM(C) because the gate-drain charge tends to dominate in the switching gate charge. One advantage of using these expressions is that the figure-of-merit becomes independent of the active area of the power MOSFET device.

The figures of merit computed for the power GD-MOSFET structure with source electrode in the trench are provided in Fig. 6.38. The figure of merit usually used for comparison of device technologies in the literature is FOM(C). Most often, the value for this figure of merit at a gate bias of 4.5 V is utilized for selection of devices. In comparison with the 60-V power D-MOSFET structure, the 60-V power GD-MOSFET structure with source electrode in the trench has a FOM(C) that is 24-times smaller. In comparison with the 60-V power U-MOSFET structure with thick oxide at the trench bottom surface, the 60-V power GD-MOSFET structure with source electrode in the trench has a FOM(C) that is 6.5-times smaller. Consequently, the power GD-MOSFET structure with source electrode in the trench offers significant improvement in circuit performance when compared with these structures because of its very low specific on-resistance and small gate transfer charge.

| Figures of Merit | $V_G = 4.5$ V | $V_G = 10$ V |
|---|---|---|
| FOM(A) ($\Omega^2cm^4s^{-1}$) | 1450 | 1110 |
| FOM(B) (ps) | 8.2 | 6.2 |
| FOM(C) (mΩ*nC) | 23.8 | 18.1 |
| FOM(D) (mΩ*nC) | 25.0 | 19.0 |

**Fig. 6.38** Figures of merit for the 60-V power GD-MOSFET structure with source electrode in the trench region

## 6.8 Edge Termination

A critical challenge for the development of the power GD-MOSFET structure, which operates with a breakdown voltage larger than the parallel-plane breakdown voltage of its drift region, is the formulation of an edge termination that can also support the blocking voltage. Fortunately, an elegant edge termination has been proposed [2] and demonstrated that can be implemented in a manner compatible with the process for the fabrication of the power GD-MOSFET structure.

Two edge terminations that allow the power CC-MOSFET structure to achieve the same breakdown voltage as the cell structure were discussed in Chap. 5. These terminations can also be utilized for the power GD-MOSFET structure to achieve the same breakdown voltage as its cell structure. The preferred edge termination for the power GD-MOSFET structure is illustrated in Fig. 6.39 including a portion of the device with a mesa region located between two trenches as in the case of the device cell structure. The $N^+$ source region is not formed on this mesa region. The trench at the termination contains an electrode that overlaps the trench sidewalls with only the thick trench oxide. The electric field in the oxide is reduced to levels suitable for reliable operation of the power GD-MOSFET structure as demonstrated in Chap. 5 for the power CC-MOSFET structure.

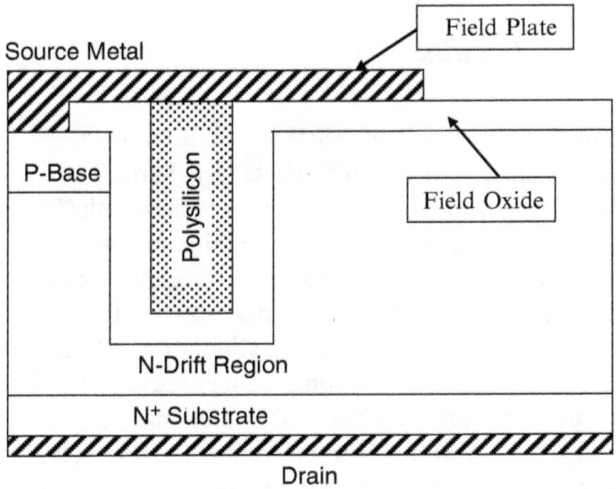


**Fig. 6.39** Edge termination for the GD-MOSFET structure

## 6.9 High Voltage Devices

As demonstrated in the previous sections of this chapter, the charge-coupling concept allows increasing the doping concentration of the drift region well above that dictated by ideal parallel-plane breakdown considerations. This approach has utility in the development of power MOSFET structures with larger breakdown voltages. The characteristics of 120 and 200 V devices are first described in this section. Under the assumption of approximately a uniform electric field in the vertical (or y) direction, the higher blocking voltages can be achieved by increasing the length ($L_D$) of the source electrode inside the trenches to 6 and 8 μm, respectively. It is also necessary to increase the trench oxide thickness and reduce the doping concentration of the drift region when the blocking voltage is made larger. These structures have dimensions that are practical for manufacturing the devices with current technology. In addition, the characteristics of devices with blocking voltages of 600 and 1,000 V are also described in this section for the purpose of comparison with the power SJ-MOSFET structure discussed in the next chapter. The dimensions for these higher voltage power GD-MOSFET structures may be challenging for current manufacturing technology but may be feasible in the future.

The analytical formulations presented in the previous sections are applicable for any breakdown voltage design when appropriate values for the device parameters are used during the computations. This section will focus on the results of two-dimensional numerical simulations for the four higher blocking voltage ratings for the power GD-MOSFET structure. The values computed with the analytical models, provided in the last section of this chapter for devices with various blocking voltages, are in excellent agreement with the simulation results.

### *6.9.1 Simulation Results*

The results of two-dimensional numerical simulations on the 120-V and 200-V power GD-MOSFET structures are first described here. The 120-V power GD-MOSFET structure used for the simulations had a mesa width ($W_M$) of 0.5 μm and a gate oxide thickness of 500 Å. The trench oxide thickness was increased to 6,000 Å due to the larger blocking voltage capability. In order to accommodate this larger trench oxide thickness, it is necessary to increase the trench width to 1.6 μm, leading to a half-cell width of 1.05 μm, in the simulations. The drift region consists of a total thickness of 6.5 μm with a doping concentration of $1 \times 10^{16}$/cm$^3$ for the upper 0.4 μm where the channel is formed. The linear doping gradient for the rest of the drift region was varied to study its impact on the blocking characteristics. The P-base region for this structure was formed by using ion-implantation to create a peak doping concentration at 0.15 μm below the surface with a depth of 0.32 μm. The $N^+$ source region has a depth of about 0.08 μm. The narrow width for the P-base region is possible because of the unique distribution of the electric field within the power GD-MOSFET structure. A heavily doped $P^+$ region is also included in the structure in the center of the mesa at the upper surface to allow fabrication of ohmic contacts to the P-base region.

The vertical doping profile taken along the surface of the trench is shown in Fig. 6.40 for the case of the 120-V power GD-MOSFET structure with a doping gradient of $3 \times 10^{20}$/cm$^4$. From the profile, it can be observed that the P-base region has a retrograde doping profile with a peak doping concentration of

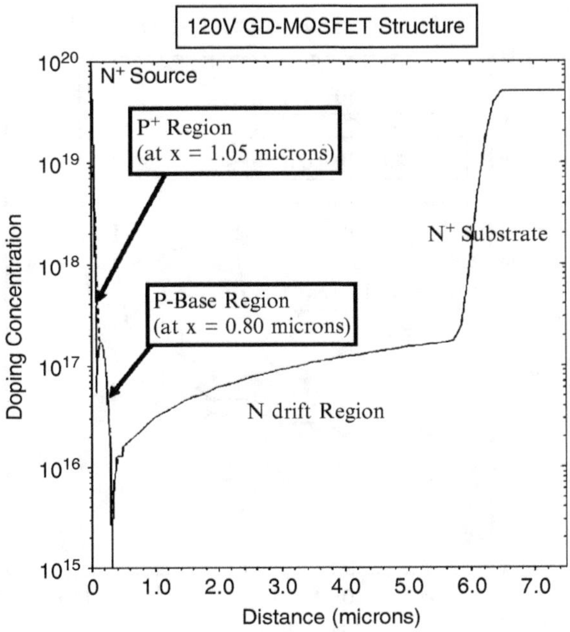


**Fig. 6.40** Vertical doping profiles in the 120-V GD-MOSFET structure

$1.5 \times 10^{17}/\text{cm}^3$ to obtain the desired threshold voltage. The vertical depths of the P-base and $N^+$ source regions are 0.32 and 0.08 μm leading to a channel length of only 0.24 μm. The drift region doping concentration increases linearly from $1 \times 10^{16}/\text{cm}^3$ to $1.5 \times 10^{17}/\text{cm}^3$ at a depth of 6 μm.

The blocking characteristics for the 120-V power GD-MOSFET cell structure are shown in Fig. 6.41 at 300°K for four cases of doping gradient in the mesa region. It can be observed that the blocking voltage increases from 109 V for a doping gradient of $2.0 \times 10^{20}/\text{cm}^4$ to 120 V at a doping gradient of $3.0 \times 10^{20}/\text{cm}^4$ and then decreases to 74 V when doping gradient is increased to $4.0 \times 10^{20}/\text{cm}^4$. This demonstrates that the power GD-MOSFET structure is capable of supporting 120 V with an optimum doping gradient of $3.0 \times 10^{20}/\text{cm}^4$. This also demonstrates the benefit of the two-dimensional charge-coupling phenomenon in obtaining a high breakdown voltage with large doping concentrations in the drift region. The optimum doping gradient for the 120-V power GD-MOSFET cell structure is smaller than for the case of the device with 60-V blocking capability as predicted by the analytical model (see Fig. 6.5). Unlike the power CC-MOSFET structure, the leakage current before breakdown is independent of the doping gradient in the drift region for the power GD-MOSFET structure.

In order to demonstrate the improved two-dimensional charge coupling phenomenon obtained by using a linearly graded doping profile, it is instructive to examine the potential contours inside the power 120-V GD-MOSFET structure

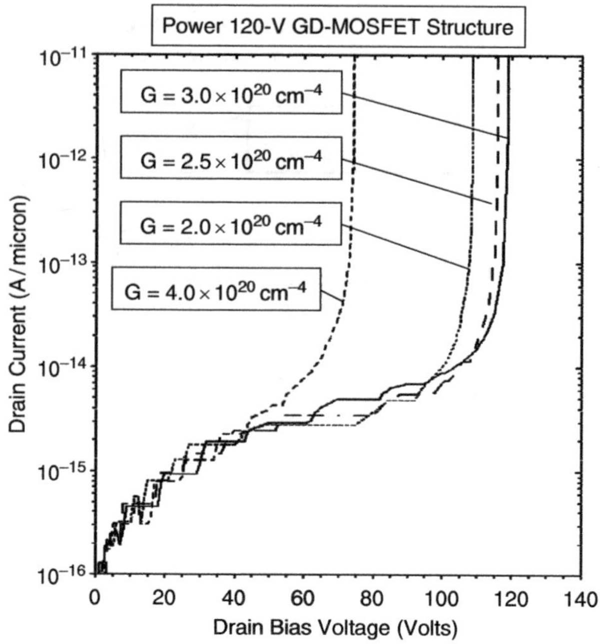


**Fig. 6.41** Blocking characteristics for the 120-V GD-MOSFET structures

when it is operating in the blocking mode. This also allows determination of the voltage distribution within the structure and the penetration of the depletion region in the P-base region. The potential contours obtained using the numerical simulations with zero gate bias and a drain bias voltage of 120 V for the 120-V GD-MOSFET structure with optimum doping gradient of $3 \times 10^{20}/\text{cm}^4$ are shown in Fig. 6.42. It can be observed that the drain bias voltage is supported in the drift region below the P-base/N-drift junction. This suppresses the depletion of the P-base region by the drain bias allowing a very short channel length for the power 120-V GD-MOSFET structure. Since the potential contours are uniformly spaced in the power 120-V GD-MOSFET structure due to the linearly graded doping profile, the breakdown voltage can be scaled with increasing drift region length ($L_D$).

The electric field profiles along the vertical direction through the center of the mesa region are shown in Fig. 6.43 for the 120-V power GD-MOSFET structure with optimum doping gradient of $3 \times 10^{20}/\text{cm}^4$. At a drain bias of 10 V, the electric field has a triangular shape representative of a one-dimensional junction. This shape prevails until a drain bias of about 40 V. At drain bias voltages above 40 V, a uniform electric field is observed throughout the length of the drift region (unlike in the power 60-V CC-MOSFET structure) due to the improved two-dimensional charge coupling phenomenon achieved with the linearly graded doping profile. The electric field at the vicinity of the P-N junction remains at $2.5 \times 10^5$ V/cm which suppresses reach-through in the P-base region. This magnitude of the electric

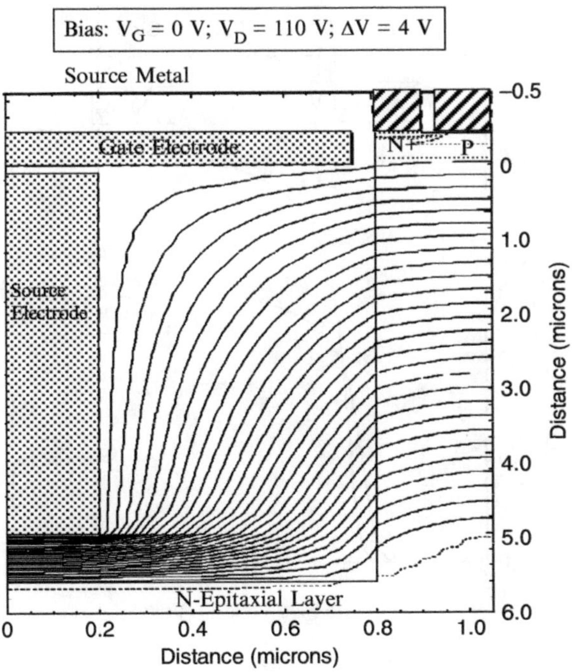


**Fig. 6.42** Potential contours in the 120-V GD-MOSFET structure

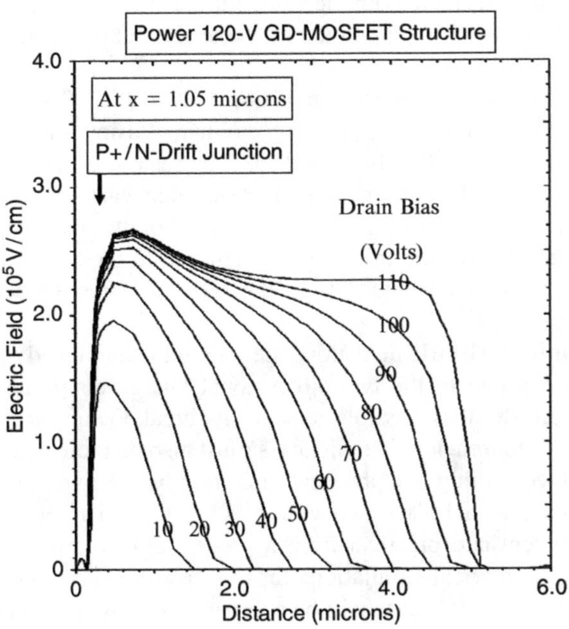


**Fig. 6.43** Electric field profiles in the 120-V GD-MOSFET structure

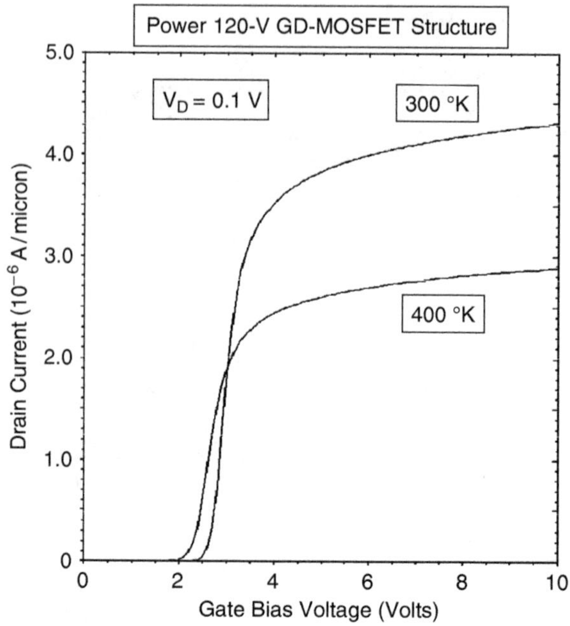


**Fig. 6.44** Transfer characteristics of the 120-V GD-MOSFET structure

field is in excellent agreement with the value for the critical electric field in the case of a uniform field distribution predicted by the analytical model (see Fig. 6.4) for a 120-V device.

The transfer characteristics for the 120-V GD-MOSFET structure with optimum doping gradient of $3 \times 10^{20}/\text{cm}^4$ were obtained using numerical simulations with a drain bias of 0.1 V at 300 and 400°K. The resulting transfer characteristics are shown in Fig. 6.44. The specific on-resistance can be obtained from the transfer characteristics at any gate bias voltage. For the case of a gate bias of 4.5 V and 300°K, the specific in-resistance is found to be 0.283 mΩ cm$^2$. For the case of a gate bias of 10 V and 300°K, the specific in-resistance is found to be 0.244 mΩ cm$^2$.

Due to the uniform electric field along the y-direction achieved in the power GD-MOSFET structure due to the two-dimensional charge coupling with a linearly graded doping profile, it is possible to scale its breakdown voltage to more than 120 V. However, the trench oxide thickness must also be increased with increasing breakdown voltage leading to a practical limitation based up on process considerations. For the case of a breakdown voltage of 200 V, a trench oxide thickness of 1 μm is sufficient and practical from a processing stand point. The results of two-dimensional numerical simulations for this case are therefore described here. The results obtained for this case will be compared to those obtained for the super-junction devices discussed in the next chapter.

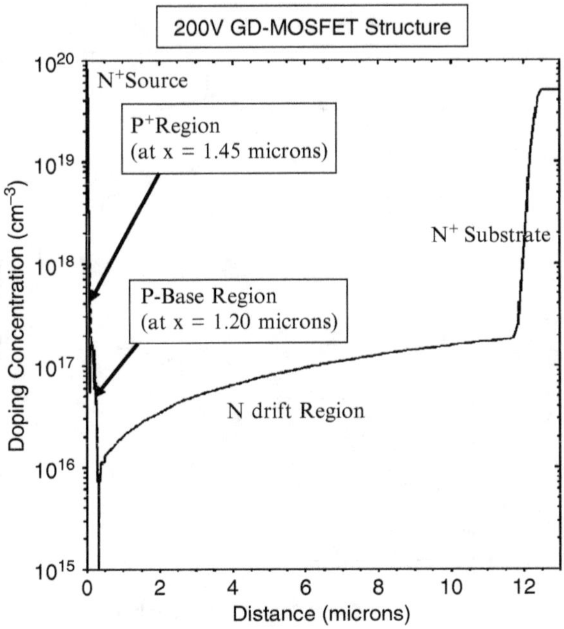


**Fig. 6.45** Vertical doping profiles in the 200-V GD-MOSFET structure

The vertical doping profile taken along the surface of the trench is shown in Fig. 6.45 for the case of the 200-V power GD-MOSFET structure with a doping gradient of $1.5 \times 10^{20}/\text{cm}^4$. From the profile, it can be observed that the P-base region has a retrograde doping profile with a peak doping concentration of $1.5 \times 10^{17}/\text{cm}^3$ to obtain the desired threshold voltage. The vertical depths of the P-base and $N^+$ source regions are 0.32 and 0.08 μm leading to a channel length of only 0.24 μm. The drift region doping concentration increases linearly from $1 \times 10^{16}/\text{cm}^3$ to $1.75 \times 10^{17}/\text{cm}^3$ at a depth of 11 μm. The trench depth for this structure is 11 μm with the source electrode in the trench extending to a depth of 10 μm.

The blocking characteristics for the 200-V power GD-MOSFET cell structure are shown in Fig. 6.46 at 300°K for three cases of doping gradient in the mesa region. It can be observed that the blocking voltage increases from 173 V for a doping gradient of $1.0 \times 10^{20}/\text{cm}^4$ to 204 V at a doping gradient of $1.5 \times 10^{20}/\text{cm}^4$ and then decreases to 123 V when doping gradient is increased to $2.0 \times 10^{20}/\text{cm}^4$. This demonstrates that the power GD-MOSFET structure is capable of supporting 200 V with an optimum doping gradient of $1.5 \times 10^{20}/\text{cm}^4$. This further demonstrates the benefit of the two-dimensional charge-coupling phenomenon in obtaining a high breakdown voltage with large doping concentrations in the drift region. The optimum doping gradient for the 200-V power GD-MOSFET cell structure is smaller than for the case of the device with 120-V blocking capability

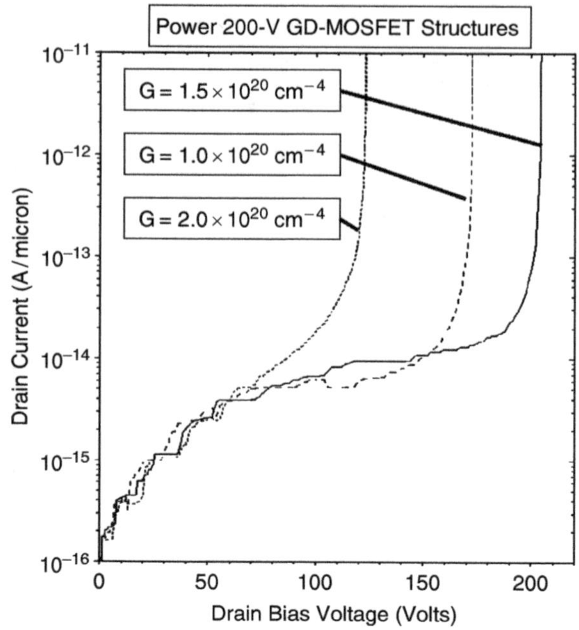


**Fig. 6.46** Blocking characteristics for the 200-V GD-MOSFET structures

as predicted by the analytical model (see Fig. 6.5). Unlike the power CC-MOSFET structure, the leakage current before breakdown is independent of the doping gradient in the drift region for the 200-V power GD-MOSFET structure.

In order to demonstrate the improved two-dimensional charge coupling phenomenon obtained by using a linearly graded doping profile, it is instructive to examine the potential contours inside the power 200-V GD-MOSFET structure when it is operating in the blocking mode. This also allows determination of the voltage distribution within the structure and the penetration of the depletion region in the P-base region. The potential contours obtained using the numerical simulations with zero gate bias and a drain bias voltage of 200 V for the 200-V GD-MOSFET structure with optimum doping gradient of $1.5 \times 10^{20}$/cm$^4$ are shown in Fig. 6.47. It can be observed that the drain bias voltage is supported in the drift region below the P-base/N-drift junction. This suppresses the depletion of the P-base region by the drain bias allowing a very short channel length for the power 200-V GD-MOSFET structure. Since the potential contours are uniformly spaced in the power 200-V GD-MOSFET structure due to the linearly graded doping profile, the breakdown voltage can be scaled to 200 V by increasing drift region length ($L_D$) to 9.5 μm.

The electric field profiles along the vertical direction through the center of the mesa region are shown in Fig. 6.48 for the 200-V power GD-MOSFET structure with optimum doping gradient of $1.5 \times 10^{20}$/cm$^4$. At a drain bias of 10 V, the electric field has a triangular shape representative of a one-dimensional junction.

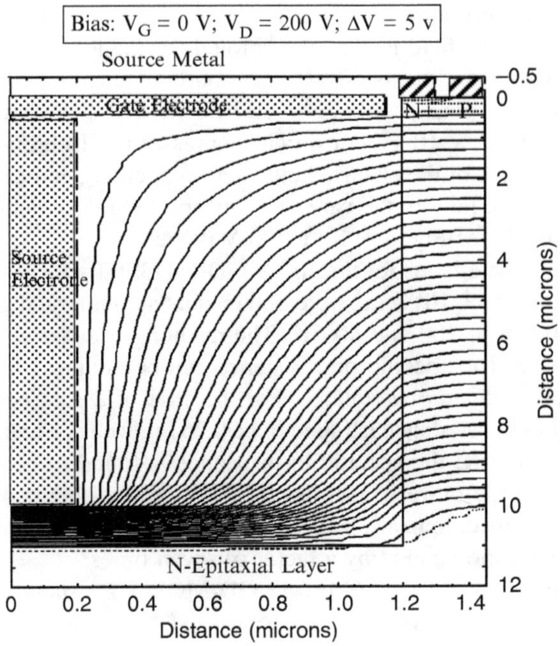


**Fig. 6.47** Potential contours in the 200-V GD-MOSFET structure

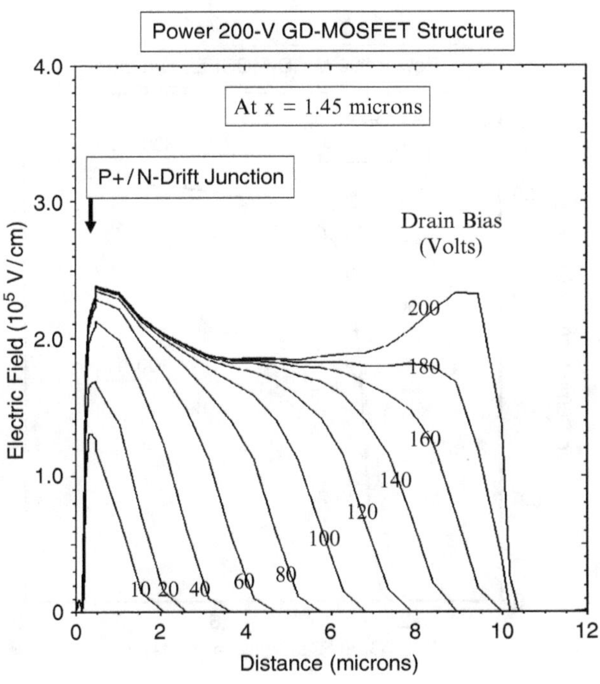


**Fig. 6.48** Electric field profiles in the 200-V GD-MOSFET structure

This shape prevails until a drain bias of about 40 V. At drain bias voltages above 40 V, a uniform electric field is observed throughout the length of the drift region due to the improved two-dimensional charge coupling phenomenon achieved with the linearly graded doping profile. The electric field at the vicinity of the P-N junction remains at $2.2 \times 10^5$ V/cm which suppresses reach-through in the P-base region. This magnitude of the electric field is in excellent agreement with the value for the critical electric field in the case of a uniform field distribution predicted by the analytical model (see Fig. 6.4) for a 200-V device.

The transfer characteristics for the 200-V GD-MOSFET structure with optimum doping gradient of $1.5 \times 10^{20}/\mathrm{cm}^4$ were obtained using numerical simulations with a drain bias of 0.1 V at 300 and 400°K. The resulting transfer characteristics are shown in Fig. 6.49. The specific on-resistance can be obtained from the transfer characteristics at any gate bias voltage. For the case of a gate bias of 4.5 V and 300°K, the specific in-resistance is found to be 0.751 mΩ cm$^2$. For the case of a gate bias of 10 V and 300°K, the specific in-resistance is found to be 0.690 mΩ cm$^2$. Since the ideal specific on-resistance for a 200-V device obtained using Baliga's power law for the impact ionization coefficients is 4.79 mΩ cm$^2$, the GD-MOSFET structure offers an improvement by a factor of seven times.

The gate charges for the 200-V power GD-MOSFET structure with optimum doping gradient of $1.5 \times 10^{20}/\mathrm{cm}^4$ was extracted by using the results of two-dimensional numerical simulations. The device was turned-on from blocking state with a drain bias of 150 V by using a gate current of $1 \times 10^{-8}$ A/μm

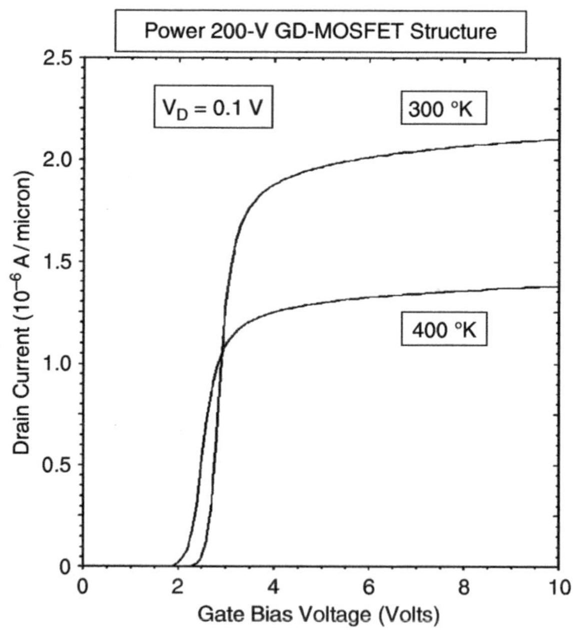


**Fig. 6.49** Transfer characteristics of the 200-V GD-MOSFET structure

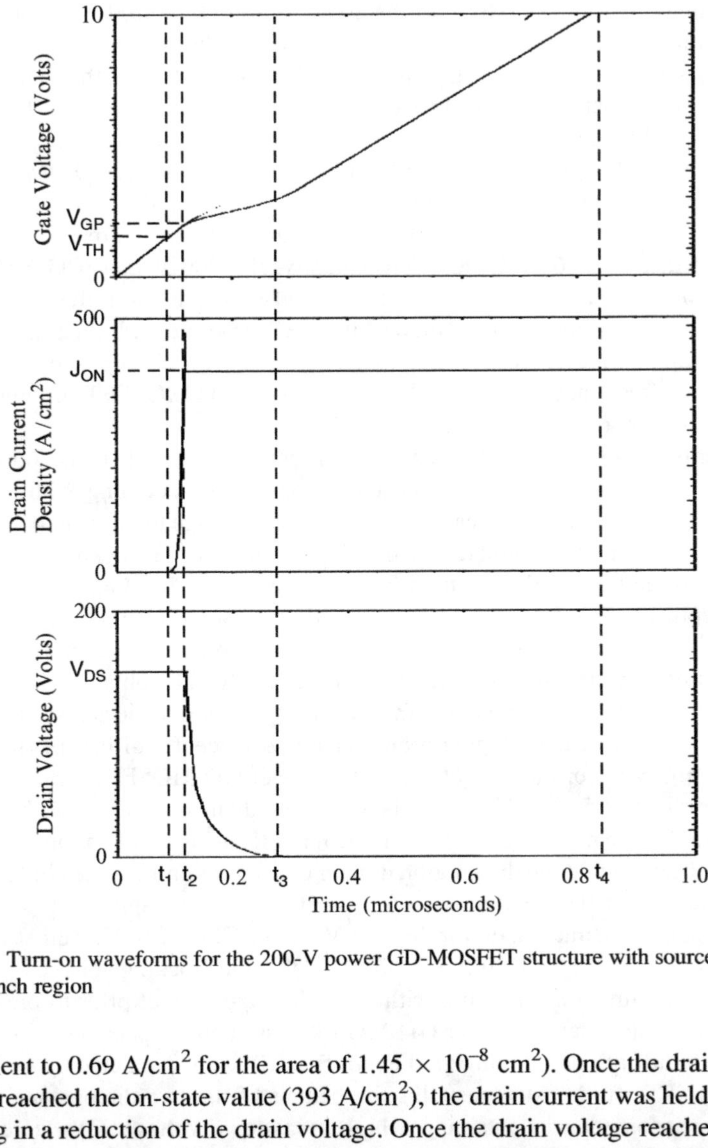


**Fig. 6.50** Turn-on waveforms for the 200-V power GD-MOSFET structure with source electrode in the trench region

(equivalent to 0.69 A/cm$^2$ for the area of 1.45 $\times$ 10$^{-8}$ cm$^2$). Once the drain current density reached the on-state value (393 A/cm$^2$), the drain current was held constant resulting in a reduction of the drain voltage. Once the drain voltage reached the on-state value corresponding to the gate plateau voltage, the gate voltage increased to the steady-state value of 10 V. The gate charge waveforms obtained by using the numerical simulations are shown in Fig. 6.50.

The values for the various components of the gate charge extracted from the numerical simulations for the 200-V power GD-MOSFET structure with source electrode in the trench region are: $Q_{GS1}$ = 60 nC/cm$^2$; $Q_{GS2}$ = 20 nC/cm$^2$; $Q_{GD}$ = 110 nC/cm$^2$; $Q_{GW}$ = 130 nC/cm$^2$; and $Q_G$ = 580 nC/cm$^2$. These values are smaller than those obtained for the 60-V power GD-MOSFET structure because of the larger cell pitch for the 200-V device structure.

The Figure-of-Merit FOM(C) for the 200-V power GD-MOSFET structure with an optimum doping gradient can be obtained using the results of the numerical simulations provided above. The Figure-of-Merit FOM(C) for the 200-V power GD-MOSFET structure at a gate bias of 4.5 V is found to be 83 mΩ nC. The Figure-of-Merit FOM(C) for the 200-V power GD-MOSFET structure at a gate bias of 10 V is found to be 76 mΩ nC. In comparison, the Figure-of-Merit FOM(C) for the 200-V power D-MOSFET structure is 2,590 mΩ nC and the Figure-of-Merit FOM(C) for the 200-V power U-MOSFET structure with thicker oxide at the trench bottom surface is 1,560 mΩ nC. Consequently, the 200-V power GD-MOSFET structure has a performance metric that is 34-times that for the D-MOSFET structure and 20.5-times that for the U-MOSFET structure with thick oxide at the trench bottom surface. These results indicate very favorable performance in both low and high frequency circuits for the 200-V power GD-MOSFET structure due to a low specific on-resistance and low gate charge.

The uniform electric field along the y-direction achieved in the power GD-MOSFET structure, due to the two-dimensional charge coupling with a linearly graded doping profile, allows scaling its breakdown voltage to more than 200 V. For the case of a breakdown voltage of 600 V, a trench oxide thickness of 3 μm is sufficient to reduce the electric field in the oxide to $2 \times 10^6$ V/cm at 600 V. The trench depth for this structure is 38 μm with the source electrode in the trench extending to a depth of 35 μm. The results of two-dimensional numerical simulations for this case are therefore described here. The results obtained for this case will be compared to those obtained for the super-junction devices discussed in the next chapter. The vertical doping profile taken at the center of the mesa region is shown in Fig. 6.51 for the case of the 600-V power GD-MOSFET structure with a doping gradient of $5 \times 10^{19}/\text{cm}^4$. This doping gradient was chosen to be close to the optimum value predicted by the analytical model using (6.13). From the profile, it can be observed that the drift region doping concentration increases linearly from $2 \times 10^{15}/\text{cm}^3$ near the P-base region to $2 \times 10^{17}/\text{cm}^3$ at a depth of 40 μm.

The blocking characteristic for the 600-V power GD-MOSFET cell structure is shown in Fig. 6.52 at 300°K. It can be observed that the blocking voltage is just over 600 V for the simulated structure with a low leakage current prior to breakdown. This demonstrates that the power GD-MOSFET structure is capable of supporting 600 V with the optimum doping gradient of $5 \times 10^{19}/\text{cm}^4$ predicted by the analytical model. This further demonstrates the benefit of the two-dimensional charge-coupling phenomenon in obtaining a high breakdown voltage with large doping concentrations in the drift region.

In order to demonstrate the improved two-dimensional charge coupling phenomenon obtained by using a linearly graded doping profile, it is instructive to examine the potential contours inside the power 600-V GD-MOSFET structure when it is operating in the blocking mode. The potential contours obtained using the numerical simulations with zero gate bias and a drain bias voltage of 600 V for the 600-V GD-MOSFET structure are shown in Fig. 6.53. Note that most of the cell structure is consumed by the trench region filled with the thick oxide and the mesa region extends only from 3.15 to 3.4 μm. The potential contours are uniformly

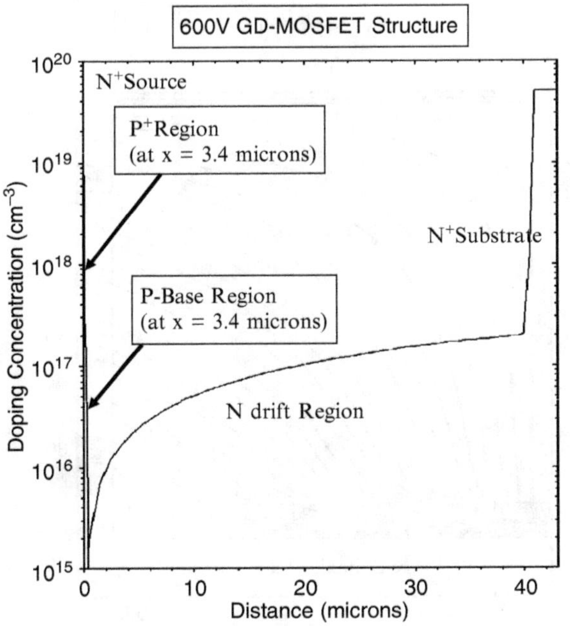


**Fig. 6.51** Vertical doping profile in the 600-V GD-MOSFET structure

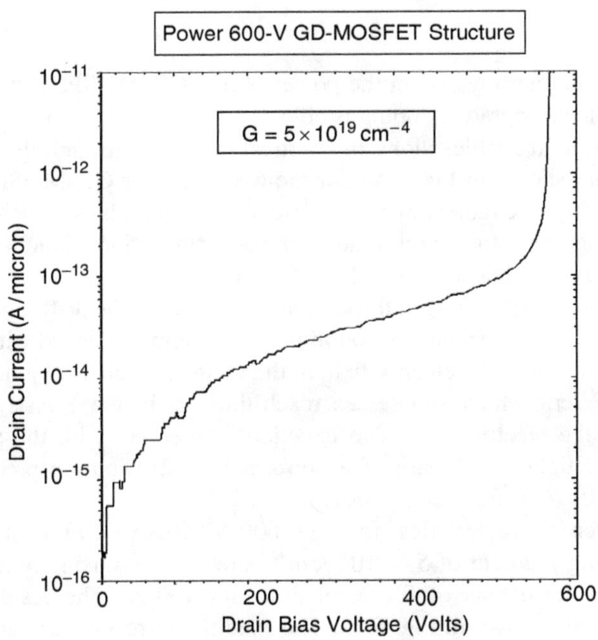


**Fig. 6.52** Blocking characteristics for the 600-V GD-MOSFET structure

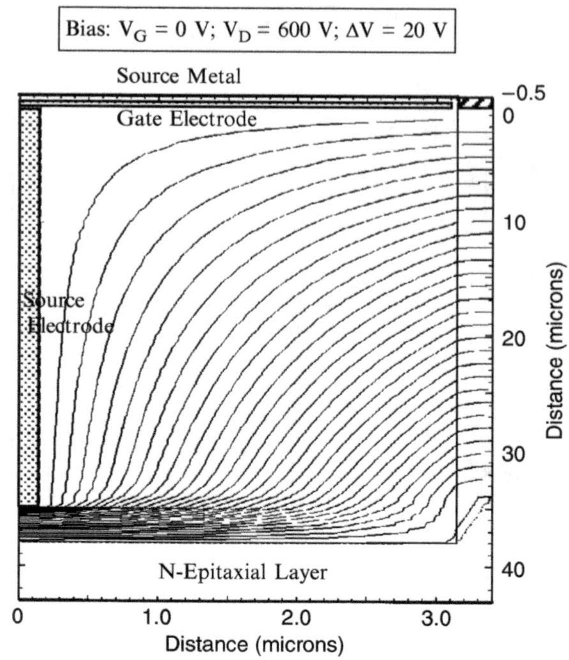


**Fig. 6.53** Potential contours in the 600-V GD-MOSFET structure

spaced within the mesa region in the power 600-V GD-MOSFET structure due to the optimum linearly graded doping profile.

The electric field profiles along the vertical direction through the center of the mesa region are shown in Fig. 6.54 for the 600-V power GD-MOSFET structure with optimum doping gradient of $5 \times 10^{19}/\text{cm}^4$. At a drain bias of 10 V, the electric field has a triangular shape representative of a one-dimensional junction. This shape prevails until a drain bias of about 100 V. At drain bias voltages above 100 V, a uniform electric field is observed throughout the length of the drift region due to the improved two-dimensional charge coupling phenomenon achieved with the linearly graded doping profile. The electric field at the vicinity of the P-N junction remains at $1.8 \times 10^5$ V/cm which suppresses reach-through in the P-base region. This magnitude of the electric field is in excellent agreement with the value for the critical electric field in the case of a uniform field distribution predicted by the analytical model (6.5) for a 600-V device.

The transfer characteristics for the 600-V GD-MOSFET structure with optimum doping gradient of $5 \times 10^{19}/\text{cm}^4$ were obtained using numerical simulations with a drain bias of 0.1 V at 300 and 400°K. The resulting transfer characteristics are shown in Fig. 6.55. The specific on-resistance can be obtained from the transfer characteristics at any gate bias voltage. For the case of a gate bias of 4.5 V and 300°K, the specific in-resistance is found to be 6.88 mΩ cm$^2$. For

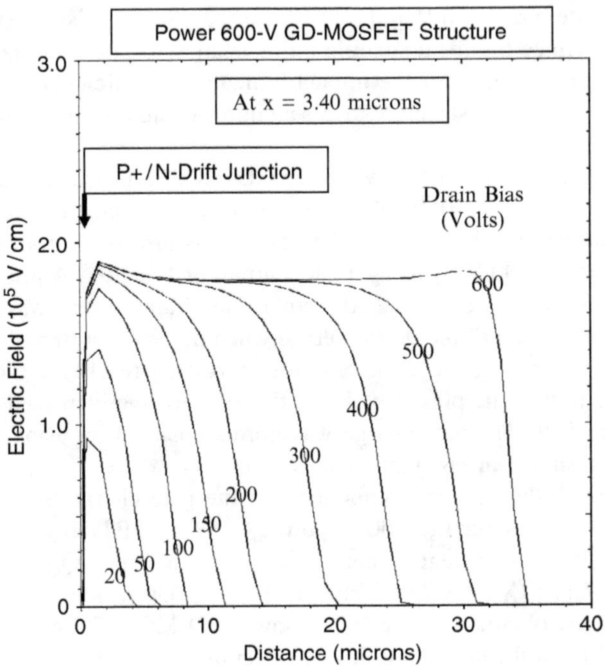


**Fig. 6.54** Electric field profiles in the 600-V GD-MOSFET structure

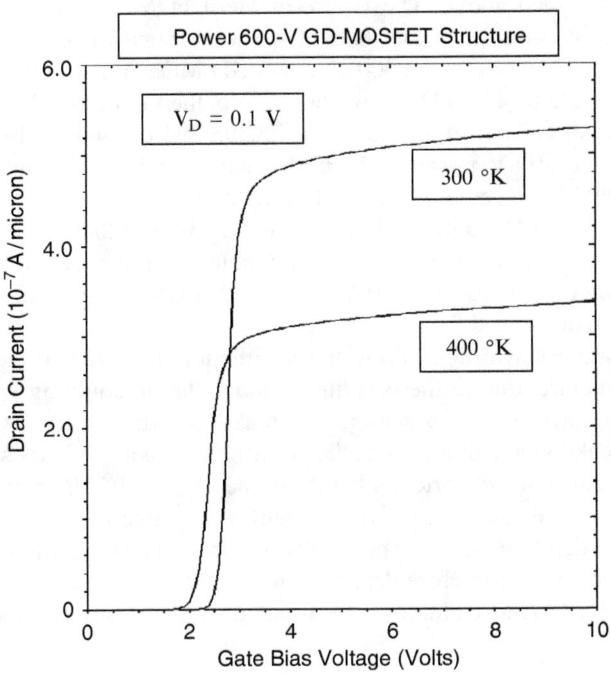


**Fig. 6.55** Transfer characteristics of the 600-V GD-MOSFET structure

the case of a gate bias of 10 V and 300°K, the specific in-resistance is found to be 6.42 mΩ cm$^2$. Since the ideal specific on-resistance for a 600-V device obtained using Baliga's power law for the impact ionization coefficients is 73.9 mΩ cm$^2$, the power GD-MOSFET structure offers an improvement by a factor of more than ten-times.

The gate charges for the 600-V power GD-MOSFET structure with optimum doping gradient of $5 \times 10^{19}$/cm$^4$ were extracted by using the results of two-dimensional numerical simulations. The device was turned-on from blocking state with a drain bias of 400 V by using a gate current of $1 \times 10^{-8}$ A/μm (equivalent to 0.29 A/cm$^2$ for the area of $3.4 \times 10^{-8}$ cm$^2$). The drain current was held constant resulting in a reduction of the drain voltage when the drain current density reached the on-state value (67 A/cm$^2$). Once the drain voltage reached the on-state value corresponding to the gate plateau voltage, the gate voltage increased to the steady-state value of 10 V. The gate charge waveforms obtained by using the numerical simulations are shown in Fig. 6.56.

The values for the various components of the gate charge extracted from the numerical simulations for the 600-V power GD-MOSFET structure with source electrode in the trench region are: $Q_{GS1} = 45$ nC/cm$^2$; $Q_{GS2} = 10$ nC/cm$^2$; $Q_{GD} = 65$ nC/cm$^2$; $Q_{GW} = 75$ nC/cm$^2$; and $Q_G = 330$ nC/cm$^2$. These values are smaller than those obtained for the 200-V power GD-MOSFET structure because of the larger cell pitch for the 600-V device structure.

The Figure-of-Merit FOM(C) for the 600-V power GD-MOSFET structure with an optimum doping gradient can be obtained using the results of the numerical simulations provided above. The Figure-of-Merit FOM(C) for the 600-V power GD-MOSFET structure at a gate bias of 4.5 V is found to be 447 mΩ nC. The Figure-of-Merit FOM(C) for the 600-V power GD-MOSFET structure at a gate bias of 10 V is found to be 417 mΩ nC. In comparison, the Figure-of-Merit FOM(C) for the 600-V power D-MOSFET structure is 23,395 mΩ nC and the Figure-of-Merit FOM(C) for the 600-V power U-MOSFET structure is 11,529 mΩ nC. Consequently, the 600-V power GD-MOSFET structure has a performance metric that is 56-times that for the D-MOSFET structure and 27.5-times that for the U-MOSFET structure. These results indicate very favorable performance in both low and high frequency circuits for the 600-V power GD-MOSFET structure due to a low specific on-resistance and low gate charge.

The uniform electric field along the y-direction achieved in the power GD-MOSFET structure, due to the two-dimensional charge coupling with a linearly graded doping profile, allows scaling its breakdown voltage to 1,000 V. For the case of a breakdown voltage of 1,000 V, a trench oxide thickness of 5 μm is sufficient to reduce the electric field in the oxide to $2 \times 10^6$ V/cm at 1,000 V. The trench depth for this structure is 65 μm with the source electrode in the trench extending to a depth of 60 μm. The results of two-dimensional numerical simulations for this case are therefore described here. The results obtained for this case will be compared to those obtained for the super-junction devices discussed in the next chapter.

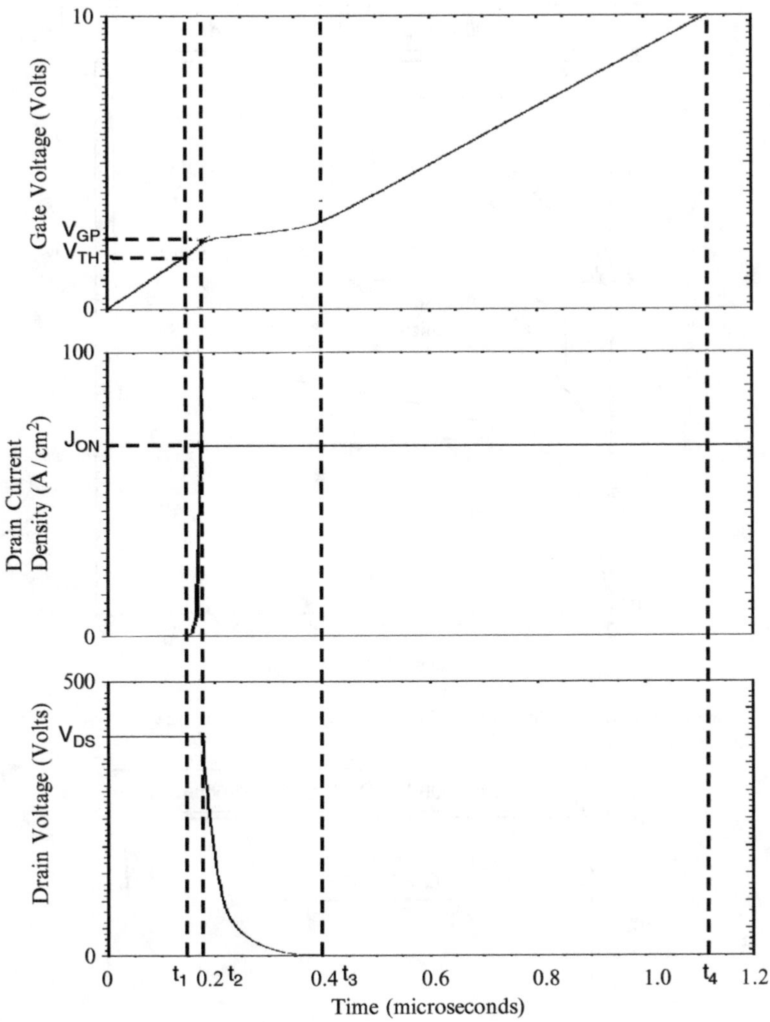


**Fig. 6.56** Turn-on waveforms for the 600-V power GD-MOSFET structure

The vertical doping profile taken at the center of the mesa region is shown in Fig. 6.57 for the case of the 1,000-V power GD-MOSFET structure with a doping gradient of $3 \times 10^{19}/\text{cm}^4$. This doping gradient was chosen to be close to the optimum value predicted by the analytical model using (6.13). From the profile, it can be observed that the drift region doping concentration increases linearly from $1 \times 10^{15}/\text{cm}^3$ near the P-base region to $2 \times 10^{17}/\text{cm}^3$ at a depth of 70 μm.

The blocking characteristic for the 1,000-V power GD-MOSFET cell structure is shown in Fig. 6.58 at 300°K. It can be observed that the blocking voltage is just below 1,000 V for the simulated structure with a low leakage current prior to

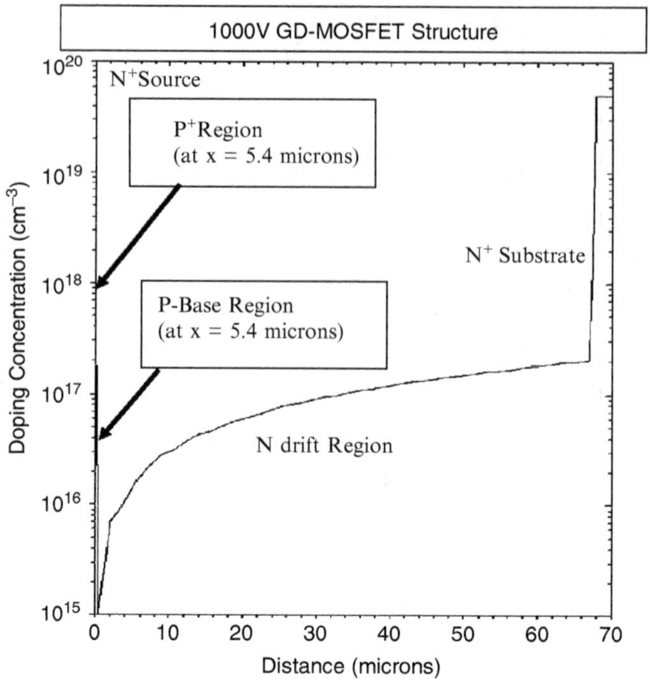


**Fig. 6.57** Vertical doping profile in the 1,000-V GD-MOSFET structure

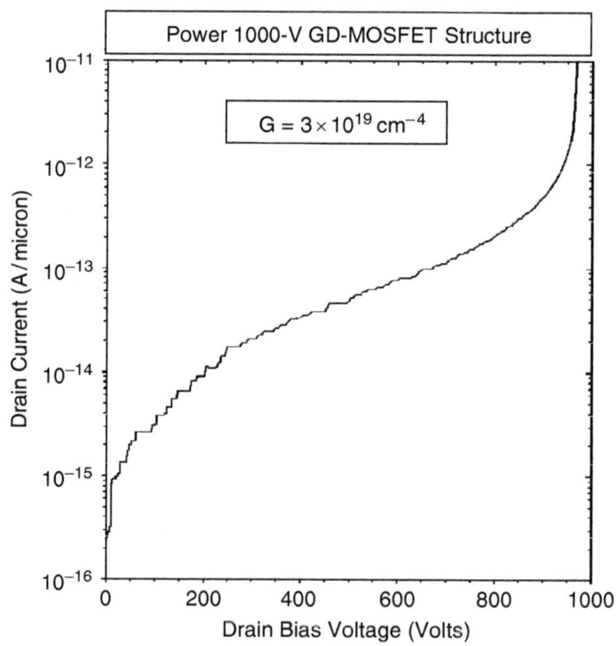


**Fig. 6.58** Blocking characteristics for the 1,000-V GD-MOSFET structure

breakdown. This demonstrates that the power GD-MOSFET structure is capable of supporting 1,000 V with the optimum doping gradient of $3 \times 10^{19}/\text{cm}^4$ predicted by the analytical model. This further demonstrates the benefit of the two-dimensional charge-coupling phenomenon in obtaining a high breakdown voltage with large doping concentrations in the drift region.

It is instructive to examine the potential contours inside the power 1,000-V GD-MOSFET structure when it is operating in the blocking mode. The potential contours obtained using the numerical simulations with zero gate bias and a drain bias voltage of 900 V are shown in Fig. 6.59. Note that most of the cell structure is consumed by the trench region filled with the thick oxide and the mesa region extends only from 5.15 to 5.4 μm. The potential contours are uniformly spaced within the mesa region in the power 1,000-V GD-MOSFET structure due to the optimum linearly graded doping profile.

The electric field profiles along the vertical direction through the center of the mesa region are shown in Fig. 6.60 for the 1,000-V power GD-MOSFET structure with optimum doping gradient of $3 \times 10^{19}/\text{cm}^4$. At a drain bias of 10 V, the electric field has a triangular shape representative of a one-dimensional junction. This shape prevails until a drain bias of about 100 V. At drain bias voltages above 100 V, a uniform electric field is observed throughout the length of the drift region due to the improved two-dimensional charge coupling phenomenon achieved with the linearly

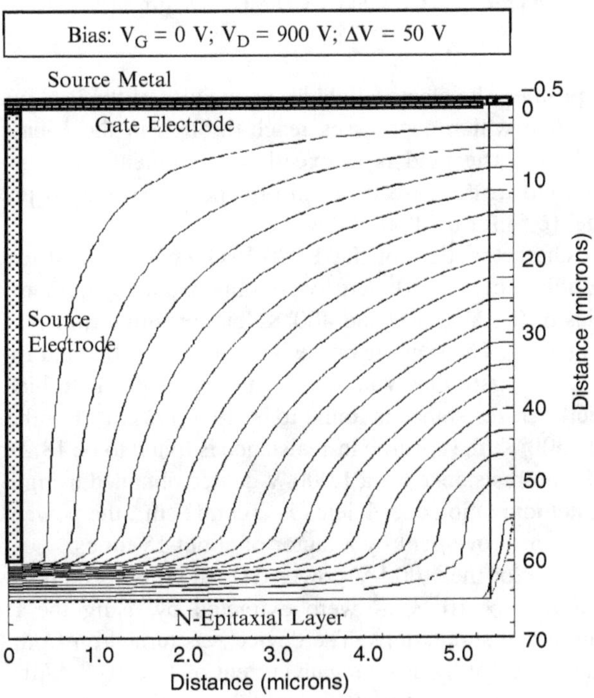


**Fig. 6.59** Potential contours in the 1,000-V GD-MOSFET structure

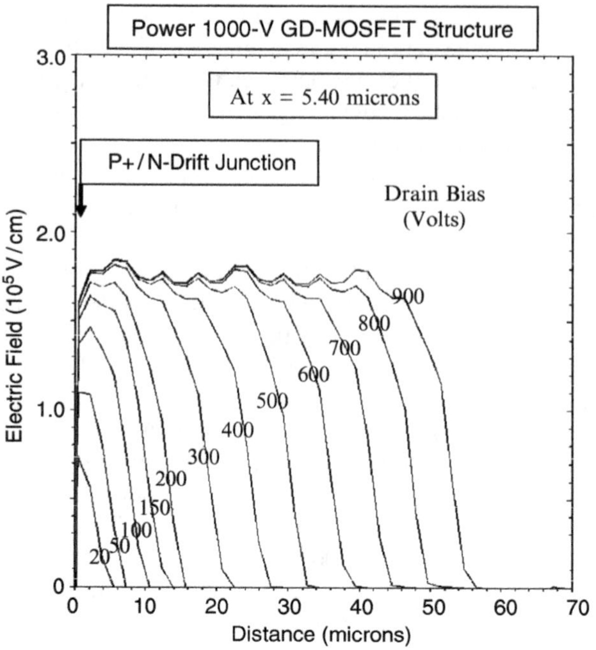


**Fig. 6.60** Electric field profiles in the 1,000-V GD-MOSFET structure

graded doping profile. The electric field at the vicinity of the P-N junction remains at $1.75 \times 10^5$ V/cm which suppresses reach-through in the P-base region. This magnitude of the electric field is in excellent agreement with the value for the critical electric field in the case of a uniform field distribution predicted by the analytical model (6.5) for a 1,000-V device.

The transfer characteristics for the 1,000-V GD-MOSFET structure with optimum doping gradient of $3 \times 10^{19}$/cm$^4$ were obtained using numerical simulations with a drain bias of 0.1 V at 300 and 400°K. The resulting transfer characteristics are shown in Fig. 6.61. The specific on-resistance can be obtained from the transfer characteristics at any gate bias voltage. For the case of a gate bias of 4.5 V and 300°K, the specific in-resistance is found to be 19.3 mΩ cm$^2$. For the case of a gate bias of 10 V and 300°K, the specific in-resistance is found to be 18.2 mΩ cm$^2$. Since the ideal specific on-resistance for a 1,000-V device obtained using Baliga's power law for the impact ionization coefficients is 265 mΩ cm$^2$, the power GD-MOSFET structure offers an improvement by a factor of about 15-times.

The gate charges for the 1,000-V power GD-MOSFET structure with optimum doping gradient of $3 \times 10^{19}$/cm$^4$ were extracted by using the results of two-dimensional numerical simulations. The device was turned-on from blocking state with a drain bias of 700 V by using a gate current of $1 \times 10^{-8}$ A/μm (equivalent to 0.185 A/cm$^2$ for the area of $5.4 \times 10^{-8}$ cm$^2$). The drain current was held constant resulting in a reduction of the drain voltage when the drain current density reached

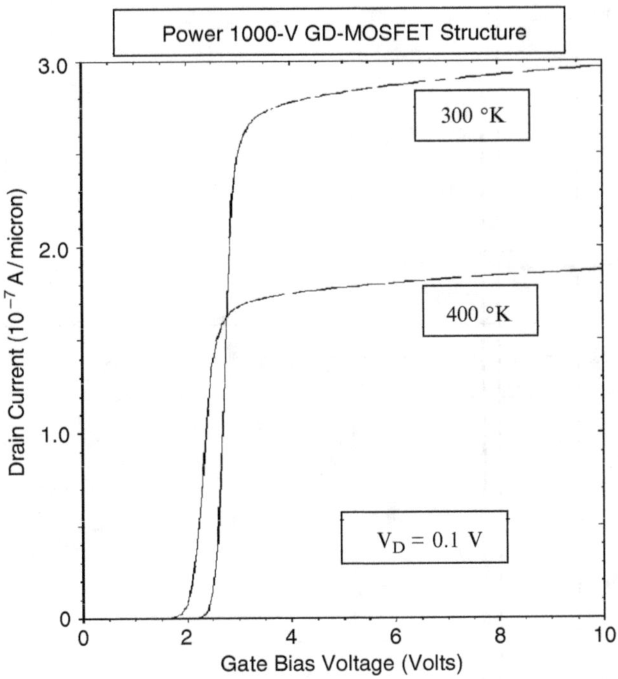


**Fig. 6.61** Transfer characteristics of the 1,000-V GD-MOSFET structure

the on-state value (85 A/cm$^2$). Once the drain voltage reached the on-state value corresponding to the gate plateau voltage, the gate voltage increased to the steady-state value of 10 V. The gate charge waveforms obtained by using the numerical simulations are shown in Fig. 6.62.

The values for the various components of the gate charge extracted from the numerical simulations for the 1,000-V power GD-MOSFET structure with source electrode in the trench region are: $Q_{GS1} = 33$ nC/cm$^2$; $Q_{GS2} = 7.4$ nC/cm$^2$; $Q_{GD} = 56$ nC/cm$^2$; $Q_{GW} = 63.4$ nC/cm$^2$; and $Q_G = 267$ nC/cm$^2$. These values are smaller than those obtained for the 600-V power GD-MOSFET structure because of the larger cell pitch for the 1,000-V device structure.

The Figure-of-Merit FOM(C) for the 1,000-V power GD-MOSFET structure with an optimum doping gradient can be obtained using the results of the numerical simulations provided above. The Figure-of-Merit FOM(C) for the 1,000-V power GD-MOSFET structure at a gate bias of 4.5 V is found to be 1,078 mΩ nC. The Figure-of-Merit FOM(C) for the 1,000-V power GD-MOSFET structure at a gate bias of 10 V is found to be 1,016 mΩ nC. In comparison, the Figure-of-Merit FOM (C) for the 1,000-V power D-MOSFET structure is 82,453 mΩ nC and the Figure-of-Merit FOM(C) for the 1,000-V power U-MOSFET structure is 37,553 mΩ nC. Consequently, the 1,000-V power GD-MOSFET structure has a performance metric that is 80-times that for the D-MOSFET structure and 37-times that for the U-MOSFET structure. These results indicate very favorable performance in both low

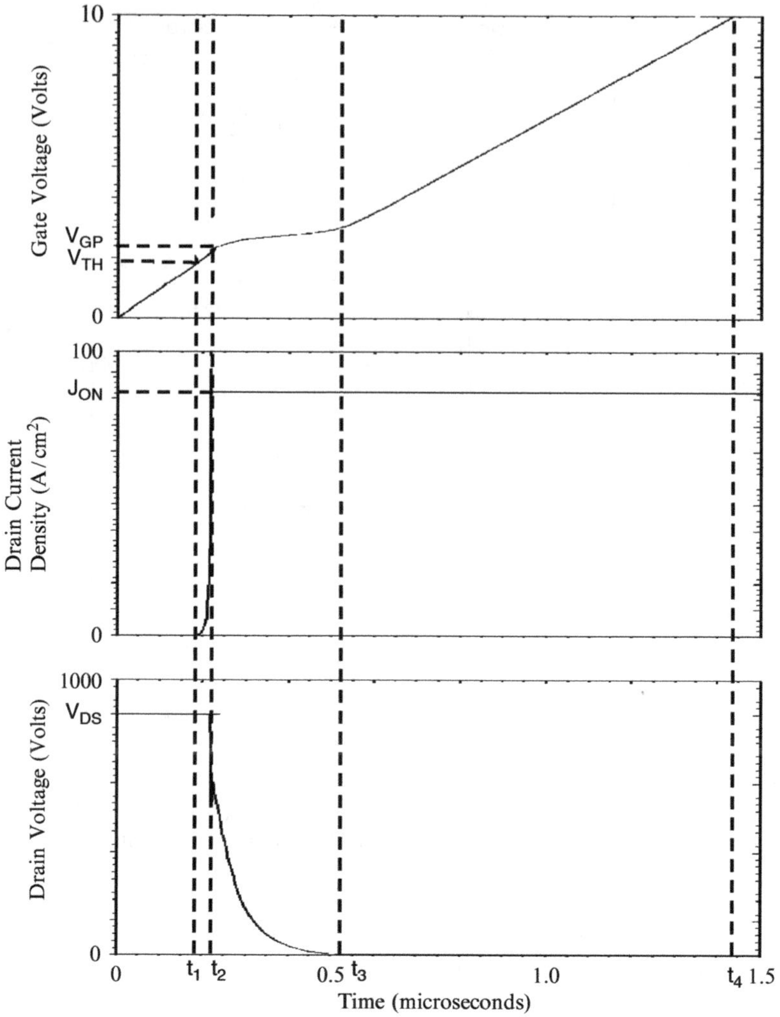


**Fig. 6.62** Turn-on waveforms for the 1,000-V power GD-MOSFET structure

and high frequency circuits for the 1,000-V power GD-MOSFET structure due to a low specific on-resistance and low gate charge.

## 6.10 Process Sensitivity Analysis

As described in the previous sections, the gradient of the linear doping profile in the drift region of the power GD-MOSFET structure must be optimized in order to achieve the desired breakdown voltage (together with the proper length for the drift region). In addition, the breakdown voltage has been demonstrated to be dependent on the trench oxide thickness. In this section, the impact of these parameters is described for the case of the 60-V GD-MOSFET structure based up on the results obtained using the two-dimensional numerical simulations.

The blocking voltage for the power GD-MOSFET structures is plotted in Fig. 6.63 as a function of the doping gradient in the drift region for the case of a trench oxide thickness of 3,000 Å and a drift region length ($L_D$) of 2.0 μm. When making this plot, the blocking voltage was defined at a leakage current density of 1 mA/cm$^2$. The blocking voltage is observed to reduce rapidly when the doping gradient exceeds $8 \times 10^{20}$/cm$^4$. More significantly, from the point of view of process margin, it can be observed that a breakdown voltage above 60 V is obtained for a wide range of doping gradients ranging from $4 \times 10^{20}$/cm$^4$ to $8 \times 10^{20}$/cm$^4$ in the drift region.

The specific on-resistance for the 60-V power GD-MOSFET structure is plotted in Fig. 6.64 as a function of the doping gradient in the drift region for two gate bias voltages for the case of a trench oxide thickness of 3,000 Å and a drift region length ($L_D$) of 2.0 μm. As expected, the specific on-resistance reduces monotonically with increasing doping gradient of the drift region. Based up on a optimum doping concentration of $8 \times 10^{20}$/cm$^4$ in the drift region to achieve the desired 60-V blocking voltage capability, the smallest specific on-resistance for the 60-V power GD-MOSFET structure is found to be 0.104 mΩ cm$^2$ at a gate bias of 4.5 V and 0.080 mΩ cm$^2$ at a gate bias of 10 V.

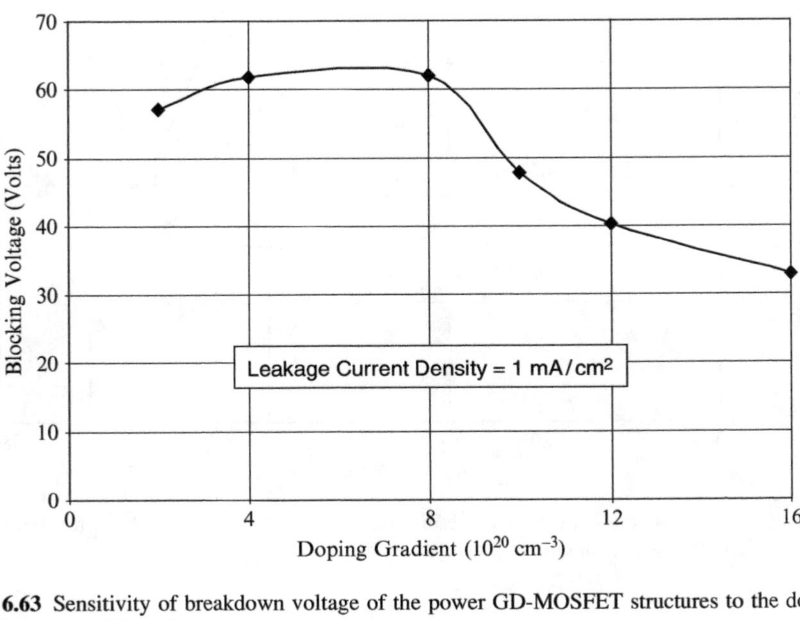


**Fig. 6.63** Sensitivity of breakdown voltage of the power GD-MOSFET structures to the doping gradient in the drift region

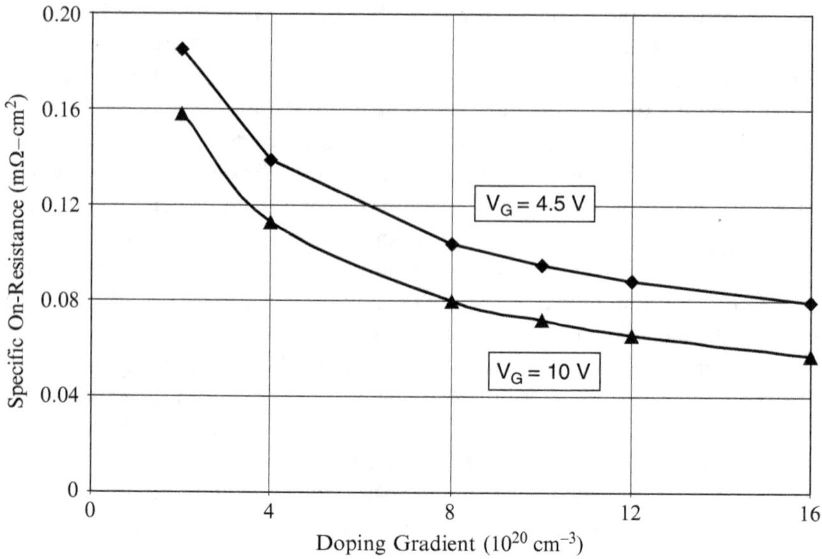


**Fig. 6.64** Sensitivity of specific on-resistance of the power GD-MOSFET structures to the doping gradient in the drift region

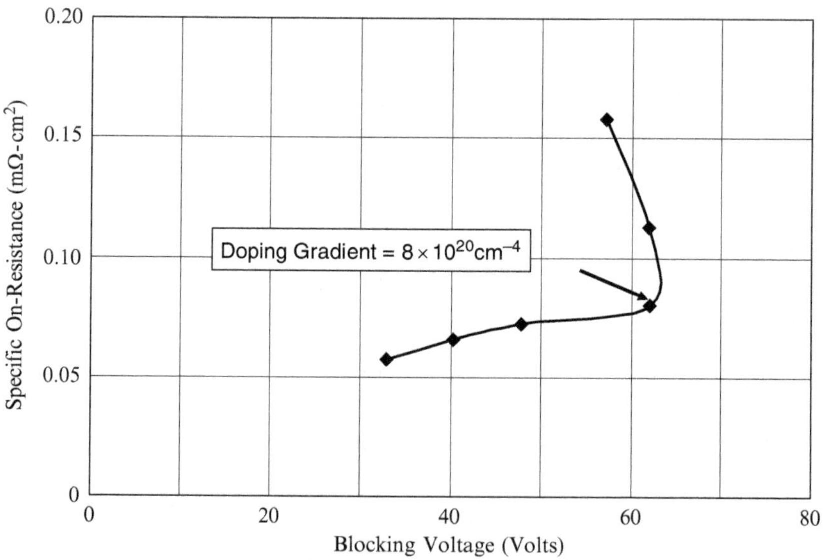


**Fig. 6.65** Trade-off curve between specific on-resistance and breakdown voltage for the power GD-MOSFET structures with doping gradient as a parameter

An interesting trade-off curve between the specific on-resistance and the blocking voltage capability can be created by using the doping gradient in the drift region as a parametric variable. From the plot shown in Fig. 6.65, it can be observed that the blocking voltage degrades significantly while producing only a small improvement in the specific on-resistance once the doping gradient exceed $8 \times 10^{20}/cm^4$ in the drift region. It is therefore appropriate to use a doping gradient of $8 \times 10^{20}/cm^4$ for the 60-V GD-MOSFET structure.

The blocking voltage for the 60-V power GD-MOSFET structures is plotted in Fig. 6.66 as a function of the trench oxide thickness for the case of a drift region doping gradient of $8 \times 10^{20}/cm^4$. A blocking voltage remains above 60 V for a range of trench oxide thickness between 2,300 and 3,100 Å. This is consistent with the optimum trench oxide thickness predicted by the analytical model (see Fig. 6.6). Since the trench oxide thickness can be precisely controlled by thermal oxidation, the broad maximum in blocking voltage capability is indicative of good process tolerance for the power GD-MOSFET structure.

The specific on-resistance for the power GD-MOSFET structures is independent of the trench oxide thickness when the cell pitch and mesa widths are held constant with the same doping concentration of $8 \times 10^{20}/cm^4$ in the drift region. It is therefore only necessary to optimize the trench oxide thickness to achieve the desired breakdown voltage.

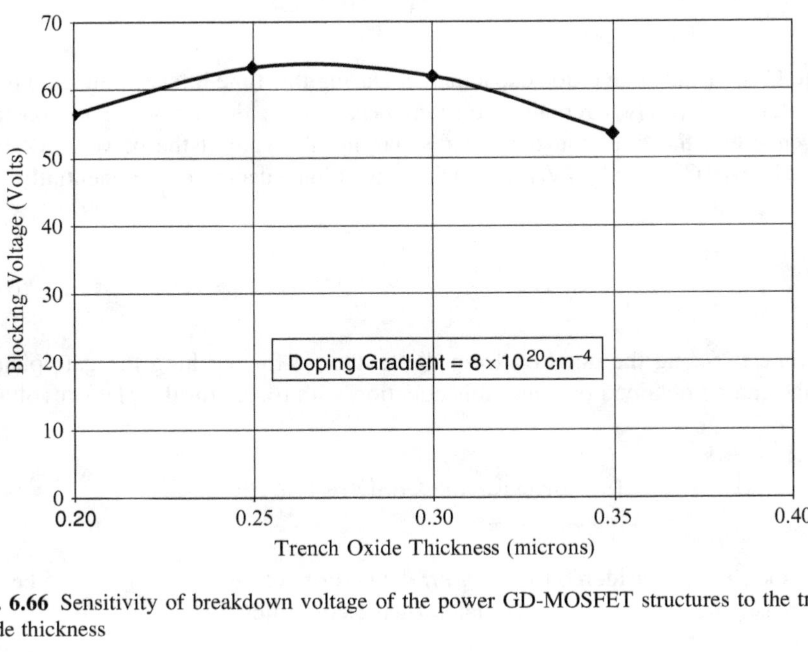


**Fig. 6.66** Sensitivity of breakdown voltage of the power GD-MOSFET structures to the trench oxide thickness

## 6.11 Inductive Load Turn-Off Characteristics

High voltage power MOSFET devices are often used for adjustable speed motor drives which behave as inductive loads. The basic half-bridge circuit utilized in these applications is shown in Fig. 10.1 of the textbook [5]. The waveforms for the current and voltage in the power switch is shown in Fig. 10.2 of the textbook. A high power dissipation occurs during each turn-off event due to the simultaneous high current and voltage during the transient. This power loss is usually characterized as an energy loss per cycle. The operation of a power MOSFET device in an inductive load circuit was illustrated in Fig. 3.42 together with a description of its operation.

Prior to the turn-off transient, the device is operating in its on-state because switch $S_1$ is closed and switch $S_2$ is open. These initial conditions are defined by: $v_G = V_{GS}$; $i_D = I_L$; and $v_D = V_{ON}(V_{GS})$. In order to initiate the turn-off process, switch $S_1$ is opened and switch $S_2$ is subsequently closed by the control circuit. The gate electrode of the power MOSFET device is then connected to the source via the gate resistance to discharge its capacitances. However, no changes in the drain current or voltage can occur until the gate voltage reaches the magnitude required to operate the power MOSFET device at a saturated drain current equal to the load current. (The small increase in the drain voltage, due to the increase in on-resistance resulting from the reduction of the gate bias voltage, has been neglected here). This gate plateau voltage is given by:

$$V_{GP} = V_{TH} + \sqrt{\frac{J_{D,ON} W_{Cell} L_{CH}}{\mu_{ni} C_{GOX}}} \tag{6.46}$$

where $C_{GOX}$ is the gate oxide capacitance. During this time interval, the gate-drain capacitance $C_{GD}(V_{ON})$ remains constant because the drain voltage is constant. Consequently, the time constant for discharging the gate of the power MOSFET device is $R_G$*$[C_{GS} + C_{GD}(V_{ON})]$ and the gate voltage decreases exponentially with time as given by:

$$v_G(t) = V_{GS} e^{-t/R_{G,SP}[C_{GS}+C_{GD}(V_{ON})]} \tag{6.47}$$

The time $t_4$ (using the notation from the textbook) for reaching the gate plateau voltage can be obtained by using this equation with (6.46) for the plateau voltage:

$$t_4 = R_{G,SP}[C_{GS} + C_{GD}(V_{ON})] \ln\left[\frac{V_{GS}}{V_{GP}}\right] \tag{6.48}$$

This time can be considered to a *turn-off delay time* before the drain voltage begins to increase after the turn-off is initiated by the control circuit.

The drain voltage begins to increase at time $t_4$ but the drain current remains constant at the load current $I_L$ because the current cannot be transferred to the diode until the voltage at the drain of the MOSFET device exceeds the supply voltage $V_{DS}$ by one diode drop to forward bias the diode. Since the drain current density is constant, the gate voltage also remains constant at the gate plateau voltage. Consequently:

$$J_{GP} = \frac{V_{GP}}{R_{G,SP}} \tag{6.49}$$

Since all of the gate current is used to discharge the gate-drain capacitance during the plateau phase because there is no change in the voltage across the gate-source capacitance:

$$J_{GP} = C_{GD,SP}\frac{dv_D}{dt} \tag{6.50}$$

where $C_{GD,SP}$ is the specific gate transfer capacitance of the power MOSFET structure which is a function of the drain voltage. This voltage dependence of the gate transfer capacitance was not taken into account in the derivation provided in the textbook but is important to include here to allow comparison of the behavior of various power MOSFET structures.

The gate transfer capacitance for the power GD-MOSFET structure with source electrode in the trenches was previously shown to be given by:

$$C_{GD,SP} = K_{G4}\left(\frac{C_{GOX}C_{S,M1}}{C_{GOX} + C_{S,M1}}\right) + K_{G5}\left(\frac{C_{TOX}C_{S,M2}}{C_{TOX} + C_{S,M2}}\right) \tag{6.51}$$

where

$$K_{GD4} = \frac{2(L_G - x_{JP})}{W_{Cell}} \tag{6.52}$$

and

$$K_{GD5} = \frac{2(t_{TOX} - t_{GOX})}{W_{Cell}} \tag{6.53}$$

The screening of the gate electrode by the source electrode in the trenches can be accounted for by using a screening factor that takes into account the depletion of the mesa region by the extension of a depletion layer for the MOS structure comprising the source electrode, the thick trench oxide, and the drift region. The screening factor is then given by:

$$K_{S,GD} = \frac{W_M - 2W_{D,MOS2}}{W_M} \tag{6.54}$$

The screening factor as defined above decreases with increasing drain bias voltage and becomes equal to zero at the drain bias required to completely deplete the mesa region. The gate transfer capacitance for the power GD-MOSFET structure with source electrode in the trenches with screening taken into account can be computed by multiplying the gate transfer capacitance given by (6.51) by the screening factor given by (6.54).

In (6.51), the specific semiconductor capacitances are given by:

$$C_{S,M1} = \frac{\varepsilon_S}{W_{D,M1}} \tag{6.55}$$

with the depletion width under the gate oxide region:

$$W_{D,M1} = \frac{\varepsilon_S}{C_{GOX}}\left\{\sqrt{1+\frac{2v_D(t)C_{GOX}^2}{q\varepsilon_S N_D}}-1\right\} \tag{6.56}$$

and

$$C_{S,M2} = \frac{\varepsilon_S}{W_{D,M2}} \tag{6.57}$$

with the depletion width under the trench oxide region:

$$W_{D,M2} = \frac{\varepsilon_S}{C_{TOX}}\left\{\sqrt{1+\frac{2v_D(t)C_{TOX}^2}{q\varepsilon_S N_D}}-1\right\} \tag{6.58}$$

The screening factor $K_{S,GD}$ is also a function of the drain voltage but in order to simplify the analysis, it will be assumed to have a constant value at a drain bias equal to half the supply voltage with an effective doping concentration (see Sect. 6.6).

Combining the above relationships yields the following differential equation for the voltage increase phase of the turn-off transient:

$$dt = \frac{K_{S,GD}}{J_{GP}}\left[\frac{K_{G4}C_{GOX}}{\sqrt{1+\frac{2v_D(t)C_{GOX}^2}{q\varepsilon_S N_D}}}+\frac{K_{G5}C_{TOX}}{\sqrt{1+\frac{2v_D(t)C_{TOX}^2}{q\varepsilon_S N_D}}}\right]dv_D \tag{6.59}$$

Integration of this equation yields:

$$(t-t_4) = \frac{4K_{S,GD}q\varepsilon_S N_D}{J_{GP}}\left\{\begin{aligned}&\left[\frac{K_{G4}}{C_{GOX}}\left(\sqrt{1+\frac{2v_D(t)C_{GOX}^2}{q\varepsilon_S N_D}}-\sqrt{1+\frac{2V_{ON}C_{GOX}^2}{q\varepsilon_S N_D}}\right)\right]\\&+\left[\frac{K_{G5}}{C_{TOX}}\right]\left(\sqrt{1+\frac{2v_D(t)C_{TOX}^2}{q\varepsilon_S N_D}}-\sqrt{1+\frac{2V_{ON}C_{TOX}^2}{q\varepsilon_S N_D}}\right)\end{aligned}\right\} \tag{6.60}$$

The voltage rise-time, i.e. the time taken for the voltage to increase from the on-state voltage drop ($V_{ON}$) to the drain supply voltage ($V_{DS}$) can be derived from the above expression:

$$t_{V,OFF} = \frac{4K_{S,GD}q\varepsilon_S N_D}{J_{GP}}\left\{\begin{array}{l}\left[\frac{K_{G4}}{C_{GOX}}\left(\sqrt{1+\frac{2V_{DS}C_{GOX}^2}{q\varepsilon_S N_D}}-\sqrt{1+\frac{2V_{ON}C_{GOX}^2}{q\varepsilon_S N_D}}\right)\right]\\ +\left[\frac{K_{G5}}{C_{TOX}}\right]\left(\sqrt{1+\frac{2V_{DS}C_{TOX}^2}{q\varepsilon_S N_D}}-\sqrt{1+\frac{2V_{ON}C_{TOX}^2}{q\varepsilon_S N_D}}\right)\end{array}\right\} \tag{6.61}$$

When deriving this relationship, it has been assumed that the voltage required to deplete the mesa ($V_{D,M}$) is larger than the drain supply voltage. A simple closed form equation for the drain voltage waveform cannot be derived from (6.60). However, the drain voltage waveform during the rise-time can be obtained by using (6.60) with an iterative process.

At the end of the plateau phase (at time $t_5$), the load current begins to transfer from the power MOSFET device to the free wheeling diode. Since the drain voltage remains constant, the gate-drain capacitance can also be assumed to remain constant during this phase. The current flowing through the gate resistance ($R_G$) discharges both the gate-drain and gate-source capacitances leading to an exponential fall in gate voltage from the plateau voltage:

$$v_G(t) = V_{GP}e^{-(t-t_5)/R_{G,SP}[C_{GS}+C_{GD}(V_{DS})]} \tag{6.62}$$

The drain current follows the gate voltage as given by:

$$J_D(t) = g_m[v_G(t) - V_{TH}] = \frac{\mu_{ni}C_{OX}}{L_{CH}W_{Cell}}[v_G(t) - V_{TH}]^2 \tag{6.63}$$

The drain current decreases rapidly with time due to the exponential reduction of the gate voltage, as given by (6.62), during the current fall phase. The drain current becomes equal to zero when the gate voltage reaches the threshold voltage. The current fall time can therefore be obtained from (6.62):

$$t_{I,OFF} = R_{G,SP}[C_{GS} + C_{GD}(V_{DS})]\ \ln\left(\frac{V_{GP}}{V_{TH}}\right) \tag{6.64}$$

Specific capacitances should be used in this expression for computation of the current fall time. Beyond this point in time, the gate voltage decreases exponentially until it reaches zero. The time constant for this exponential decay is different from the initial phase due to the smaller gate-drain capacitance.

The largest power dissipation during the turn-off transient occurs during the drain voltage rise-time time interval ($t_{V,OFF}$). The turn-off energy loss per cycle can be obtained using:

$$E_{OFF} = \frac{1}{2} J_{ON} \left( V_{D,M} t_{V,OFF} + V_{DS} t_{I,OFF} \right) \tag{6.65}$$

under the assumption that the drain current and voltage excursions are approximately linear with time.

In the case of the 600-V power GD-MOSFET structure, the typical drain supply voltage is 400 V. Using the specific on-resistance for this device of 6.4 mΩ cm$^2$

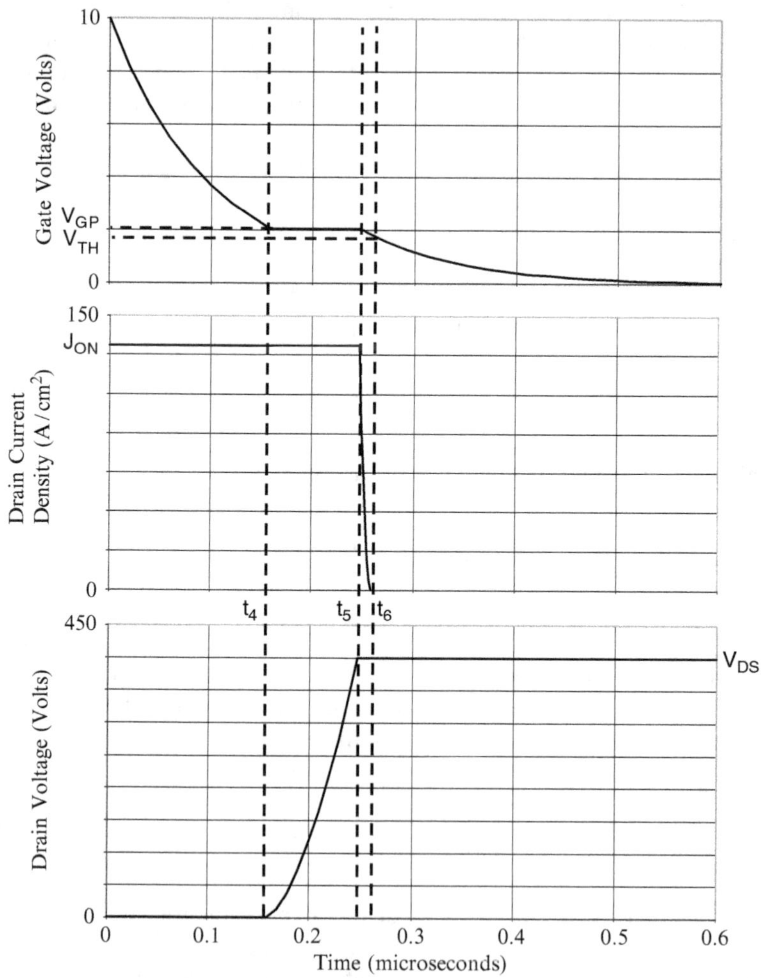


**Fig. 6.67** Analytically computed turn-off waveforms for the 600-V power GD-MOSFET structure

with an on-state power dissipation of 100 W/cm$^2$, the on-state current density is found to be 125 A/cm$^2$. The on-state voltage drop at this current density is 0.8 V. The device structure has a cell pitch of 6.8 μm with a mesa width of 0.5 μm, trench depth of 38 μm, trench oxide thickness of 3 μm and a gate oxide thickness of 500 Å. The drift region has a doping gradient of 5 × 10$^{19}$/cm$^4$ and thickness of 40 μm. The specific gate resistance used in the turn-off circuit was assumed to have a value of 3.4 Ω cm$^2$.

Using the above parameters, the specific input capacitance for the 600-V power GD-MOSFET structure computed using (6.28) is found to be 28 nF/cm$^2$. The specific gate reverse transfer capacitance computed by using (6.29) at the on-state voltage drop is 1.5 nF/cm$^2$. The pinch-off voltage ($V_{D,M}$) for the 600-V power GD-MOSFET structure computed using (5.55) is found to be 350 V. Consequently, the specific gate reverse transfer capacitance is zero at the drain supply voltage. Using these values in (6.48), the time ($t_4$) to reach the gate plateau voltage is found to be 0.16 μs. Using these parameters in (6.61), the voltage rise-time is computed as 0.088 μs. Using these parameters in (6.64), the current fall-time is computed as 0.013 μs. It can be observed that the current fall time is much smaller than the voltage rise-time. The energy loss per cycle obtained by using these values for the voltage rise-time and current fall-time in (6.65) is 2.52 mJ/cm$^2$. Although this is twice that observed for the power U-MOSFET structure (see Sect. 3.10), the power GD-MOSFET structure is operating at 3.8-times larger output power. Consequently, the power GD-MOSFET structure has a superior energy loss per cycle when compared with the power U-MOSFET structure. The waveforms that can be generated by using the above equations are shown in Fig. 6.67.

### *6.11.1 Simulation Results*

The results of two-dimensional numerical simulations on the turn-off of the 600-V power GD-MOSFET structure are described below. The drain supply voltage was chosen as 400 V for the turn-off analysis. During the turn-off simulations, the gate voltage was reduced to zero with a gate resistance of 1 × 10$^8$ Ω μm for the 3.4 μm half-cell structure, which is equivalent to a specific gate resistance of 3.4 Ω cm$^2$. The current density was initially held constant at an on-state current density of 125 A/cm$^2$ allowing the drain voltage to rise to the drain supply voltage. The drain supply voltage was then held constant allowing the drain current density to reduce to zero.

The turn-off waveforms obtained for the 600-V power GD-MOSFET structure by using the numerical simulations are shown in Fig. 6.68. The gate voltage initially reduces to the gate plateau voltage corresponding to the on-state current density. The drain voltage then increases from the on-state voltage drop to the drain supply voltage. After this, the drain current rapidly falls to zero. The drain voltage

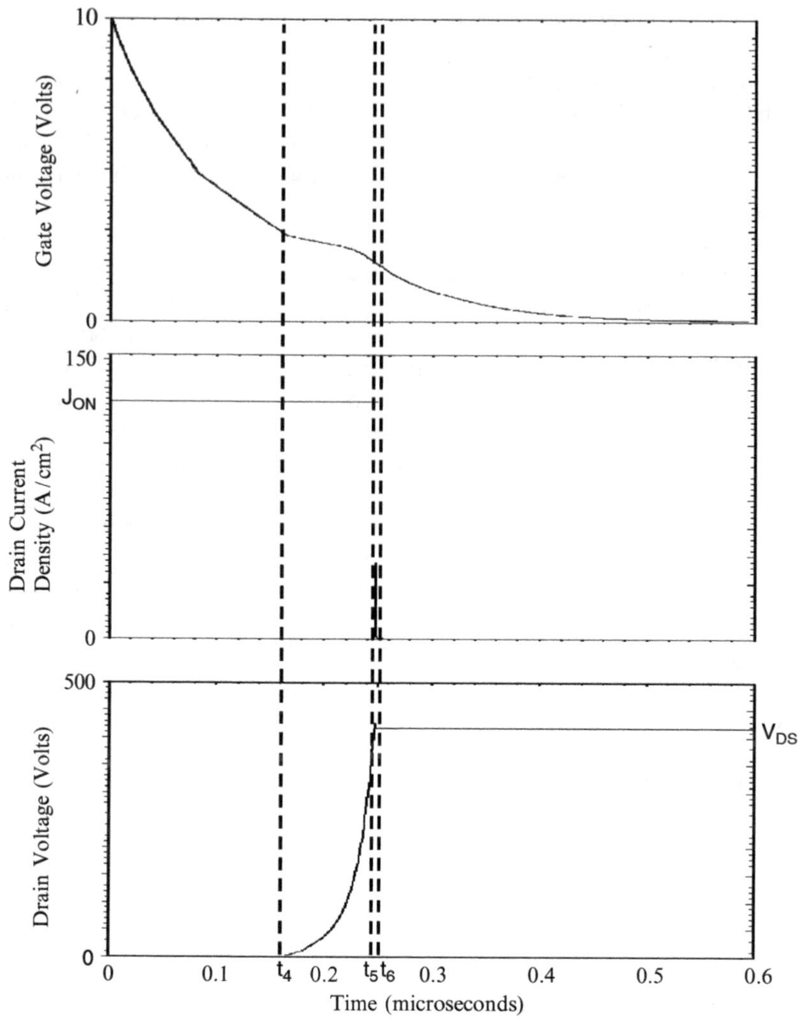


**Fig. 6.68** Turn-off waveforms for the 600-V power GD-MOSFET structure

rise-time ($t_5 - t_4$) can be observed to be much greater than the drain current fall time ($t_6 - t_5$). The drain voltage rise-time obtained from the simulations of the power GD-MOSFET structure is 0.16 μs and the drain current fall-time obtained from the simulations is 0.005 μs. The shape of the waveforms predicted by the analytical model (see Fig. 6.67) for the gate voltage, drain voltage and drain current are in good agreement with the simulation results. The numerical values predicted for the analytical model for the voltage rise-time and the drain fall-time are also in good agreement with those observed in the numerical simulations allowing an accurate computation of the energy loss per cycle for the power GD-MOSFET structure.

## 6.12 Discussion

The physics of operation and resulting electrical characteristics of the power GD-MOSFET structure have been described in this chapter. An improved two-dimensional charge-coupling is achieved in this device structure by utilizing a linearly graded doping profile. Analytical formulations for the optimum doping gradient have been derived in the chapter to aid in the design of the structures. With the optimum doping gradient, it has been demonstrated that the electric field in the center of the mesa region becomes nearly constant with depth (i.e. along the y-direction). The uniform electric field along the drift region in the power GD-MOSFET structure allows scaling of its breakdown voltage with increasing length of the drift region. However, the thickness of the oxide in the trenches must also be increased with increasing breakdown voltage. This may limit its maximum blocking voltage capability to about 200 V with a trench oxide thickness of 1 μm and a trench depth of 11 μm. Device with larger breakdown voltages may be feasible as technology evolves for making deeper trenches with thicker oxide coatings.

Due to the relatively high doping levels in the drift region, the specific on-resistance for the power GD-MOSFET structure is reduced to below that for the ideal parallel-plane junction case. With this device structure, it is possible to reduce the specific on-resistance to five times smaller than the ideal specific on-resistance at a breakdown voltage of 60 V. An even greater improvement over the ideal specific on-resistance is obtained for devices with larger breakdown voltages: a factor of 12-times for a 200-V device; a factor of 20-times for a 600-V device; and factor of 25-times for a 1,000-V device. The commercially developed power GD-MOSFET devices [7] have been found to exhibit the characteristics described in this chapter.

For purposes of comparison with the power MOSFET structures discussed in subsequent chapters, the analysis of the power GD-MOSFET structure is provided here for blocking voltages ranging from 30 to 1,000 V. As demonstrated in the chapter, it is feasible to extend the breakdown voltage for the power GD-MOSFET structure beyond 100 V due to the uniform electric field distribution along the drift region. In this analysis, the power GD-MOSFET structure was assumed to have the following parameters: (a) $N^+$ source junction depth of 0.08 μm; (b) P-base junction depth of 0.32 μm; (c) gate oxide thickness of 500 Å; (d) mesa width of 0.5 μm; (e) threshold voltage of 2 V; (f) gate drive voltage of 10 V; (g) inversion mobility of 450 $cm^2/V$ s; (h) accumulation mobility of 1,000 $cm^2/V$ s. The contributions from the contacts and the $N^+$ substrate were neglected during the analysis. The doping gradient and thickness of the drift region were determined under the assumption that the edge termination limits the breakdown voltage to the breakdown voltage of the cells, which exceeds the parallel-plane breakdown voltage. The device parameters pertinent to each blocking voltage are provided in Fig. 6.69. It can be observed that the doping

| Blocking Voltage (V) | Drift Doping Gradient ($cm^{-4}$) | Trench Oxide Thickness (microns) | Drift Region Thickness (microns) |
|---|---|---|---|
| 30 | $1.4 \times 10^{21}$ | 0.15 | 1.0 |
| 60 | $8.0 \times 10^{20}$ | 0.30 | 2.0 |
| 120 | $3.0 \times 10^{20}$ | 0.60 | 4.5 |
| 200 | $1.5 \times 10^{20}$ | 1.00 | 9.5 |
| 300 | $1.2 \times 10^{20}$ | 1.50 | 15 |
| 600 | $5.0 \times 10^{19}$ | 3.00 | 35 |
| 1000 | $3.0 \times 10^{19}$ | 5.00 | 60 |

**Fig. 6.69** Device parameters for the power GD-MOSFET structures

| Blocking Voltage (V) | Cell Pitch (microns) | Specific On-Resistance ($m\Omega$–$cm^2$) |
|---|---|---|
| 30 | 1.0 | 0.0337 |
| 60 | 1.4 | 0.0816 |
| 120 | 2.0 | 0.235 |
| 200 | 2.8 | 0.618 |
| 300 | 3.8 | 1.31 |
| 600 | 6.8 | 5.29 |
| 1000 | 10.8 | 14.7 |

**Fig. 6.70** Specific on-resistances for the power GD-MOSFET structures

concentration of the drift region is much larger than that for the previous power high voltage MOSFET structures especially near the drain.

The trench width in the power GD-MOSFET structure was chosen for each breakdown voltage to accommodate the thickness of the trench oxide with enough room for the polysilicon refill to form the source connected electrode in the trench. The trench oxide thickness is provided in Fig. 6.69. The cell pitch for the power GD-MOSFET structures is provided in Fig. 6.70 for the various blocking voltages. The analytical model described in Sect. 6.3 for computing the specific on-resistance was used for all of the power GD-MOSFET structures. These values are also provided in Fig. 6.70.

The specific on-resistance for the power GD-MOSFET structure can be compared with the ideal specific on-resistance obtained by using Baliga's power law for the impact ionization coefficients in Fig. 6.71. From this figure, it can be concluded that the specific on-resistance for the GD-MOSFET structure is less than the ideal

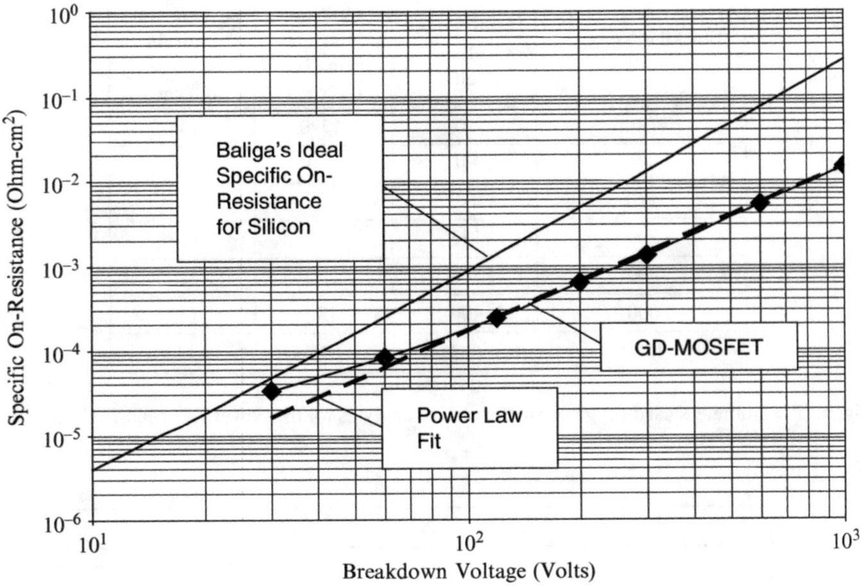


**Fig. 6.71** Specific on-resistances for the power GD-MOSFET structures

specific on-resistance. The reduction of the specific on-resistance for the power GD-MOSFET structure below the ideal specific on-resistance becomes larger with increasing breakdown voltage. The figure clearly demonstrates that extremely low specific on-resistance, well below the ideal specific on-resistance of one-dimensional device structures, can be achieved by using the power GD-MOSFET structure at high blocking voltages. The specific on-resistance increases in a linear fashion on this log–log graph for breakdown voltages ranging from 120 to 1,000 V indicating a power law relationship. The power law relationship that fits the data is shown in the figure by the dashed line. The equation for this line is:

$$R_{on,sp} = 2.08 \times 10^{-8}\ BV^{1.95} \tag{6.66}$$

where the specific on-resistance has units of $\Omega$ cm$^2$.

At a blocking voltage of 600 V, the specific on-resistance for the power GD-MOSFET structure is 14-times smaller than the ideal specific on-resistance while at a blocking voltage of 1,000 V, the specific on-resistance for the power GD-MOSFET structure is 18-times smaller than the ideal specific on-resistance. The ability to achieve this superlative performance is constrained by the technology required to make very deep trenches with thick sidewall oxide coatings.

The specific gate transfer charge for the power GD-MODSFET structures was obtained by using the analytical model (see 6.35). During this analysis, the devices were assumed to be operated at an on-state current density that results in a power

| Blocking Voltage (V) | Drain Supply Voltage (Volts) | On-State Current Density (A/cm²) | Specific Gate Transfer Charge (nC/cm²) |
|---|---|---|---|
| 30 | 20 | 1720 | 206 |
| 60 | 40 | 1110 | 190 |
| 120 | 80 | 652 | 150 |
| 200 | 133 | 402 | 110 |
| 300 | 200 | 276 | 88 |
| 600 | 400 | 137 | 65 |
| 1000 | 667 | 82.5 | 44 |

**Fig. 6.72** Parameters used for gate transfer charge analysis for the power GD-MOSFET structures

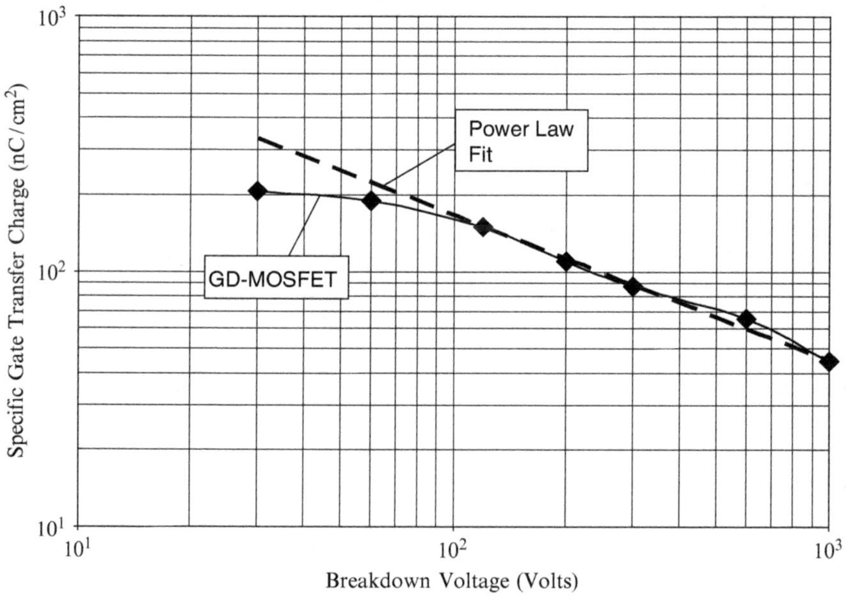


**Fig. 6.73** Specific gate transfer charge for the power GD-MOSFET structures

dissipation of 100 W/cm$^2$ as determined by the specific on-resistance for each device. The on-state current density values are provided in Fig. 6.72. The drain supply voltage used for this analysis was chosen to be two-thirds of the breakdown voltage. The drain supply voltage values are also provided in Fig. 6.72. The specific gate transfer charge values calculated by using the analytical model for the power GD-MOSFET structure are given in Fig. 6.72 as a function of the breakdown voltage. These values are also plotted in Fig. 6.73 as a function of the breakdown

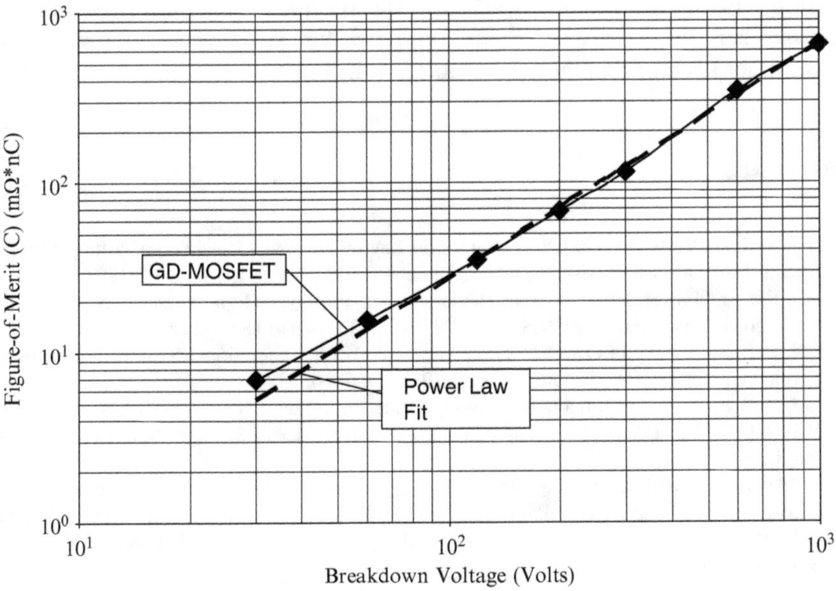


**Fig. 6.74** Figure-of-merit (C) for the power GD-MOSFET structures

voltage. It can be observed that the gate transfer charge decreases with increasing breakdown voltage in spite of the larger drain supply voltage during the transient for higher voltage devices. This reduction of specific gate transfer charge is related to the increase in the device cell pitch with increasing breakdown voltage. The specific gate transfer charge reduces in a linear fashion on this log-log graph for breakdown voltages ranging from 120 to 1,000 V indicating a power law relationship. The power law relationship that fits the data is shown in the figure by the dashed line. The equation for this line is:

$$Q_{GD} = 2,350\ BV^{-0.575} \tag{6.67}$$

where the specific gate transfer charge has units of nC/cm$^2$.

The figure-of merit (C) – product of the specific on-resistance and the specific gate transfer charge – for the power GD-MODSFET structures was obtained by using the specific on-resistance and specific gate transfer charge values calculated using the analytical model as described above. The resulting values for the FOM(C) are plotted in Fig. 6.74 as a function of the breakdown voltage of the power GD-MOSFET structure. It can be observed that the FOM(C) increases in a linear fashion on this log–log graph indicating a power law relationship. The power law relationship that fits the data is shown in the figure by the dashed line. The equation for this line is:

$$FOM(C) = R_{ON,sp}Q_{GD,sp} = 0.0878\ BV^{1.376} \tag{6.68}$$

where the FOM(C) has units of mΩ nC. The specific on-resistance and FOM(C) discussed here for the power GD-MOSFET structure will be compared with those for the other structures in the final chapter of the book.

## References

1. B.J. Baliga, "Vertical Field Effect Transistors having improved Breakdown Voltage Capability and Low On-state Resistance", U.S. Patent # 5,637,898, Issued June 10, 1997.
2. B.J. Baliga, "Power Semiconductor Devices having improved High Frequency Switching and Breakdown Characteristics", U.S. Patent # 5,998,833, Issued December 7, 1999.
3. B.J. Baliga, "Trends in Power Discrete Devices", IEEE International Symposium on Power Semiconductor Devices and ICs, Abstract P-2, pp. 5–10, 1997.
4. B.J. Baliga, "Epitaxial Silicon Technology", Academic Press, New York, 1986.
5. B.J. Baliga, "Fundamentals of Power Semiconductor Devices", Springer-Science, New York, 2008.
6. S. Mahalingam and B.J. Baliga, "The Graded Doped Trench MOS Barrier Schottky Rectifier: a Low Forward Drop High Voltage Rectifier", Solid State Electronics, Vol. 43, pp. 1–9, 1999.
7. B.J. Baliga, "Silicon RF Power MOSFETs", World Scientific Publishing Company, Singapore, 2005.

# Chapter 7
# SJ-MOSFET Structure

The power MOSFET structures discussed in the last two chapters utilize two-dimensional charge coupling for supporting the drain voltage when operating in the blocking mode. The charge coupling is achieved in these devices by utilizing a source electrode located within an oxide coated trench oriented orthogonal to the wafer surface. In these structures, a depletion layer is simultaneously formed across a horizontal P-N junction and a vertical MOS interface at the trench sidewalls. The simultaneous depletion in the x- and y-directions produces the desired two-dimensional charge coupling which improves the electric field distribution and allows using very high doping levels for the drift region. The high doping concentrations in the drift region result in a very low specific on-resistance which has been demonstrated to be well below the ideal specific on-resistance as given by:

$$R_{on-ideal} = \frac{4BV^2}{\varepsilon_S\ \mu_n E_C^3} \tag{7.1}$$

An alternate approach has been proposed and demonstrated for achieving the desired two-dimensional charge coupling within power MOSFET structures [1]. The resulting power MOSFET structure has been named 'COOLMOS'. The proposed approach is based up on utilizing alternating P-type and N-type columns as the drift region. These columns create vertically oriented P-N junctions. When the drain bias is applied, a depletion layer is simultaneously formed under the planar MOS gate electrode in the MOSFET structure and the vertical P-N junction formed by the P/N columns in the drift region. The simultaneous depletion in the x- and y-directions produces the desired two-dimensional charge coupling which improves the electric field distribution and allows using high doping levels for the drift region. The high doping concentrations in the drift region result in a very low specific on-resistance which has been demonstrated to be well below the specific on-resistance of the conventional D-MOSFET structure.

This approach is particularly well suited for the development of high performance power MOSFET structures with large blocking voltage capability in the

B.J. Baliga, *Advanced Power MOSFET Concepts*,
DOI 10.1007/978-1-4419-5917-1_7, © Springer Science+Business Media, LLC 2010

range of 300–1,200 V. The approach has become generally known as a super-junction concept in the literature [2]. The first super-junction power MOSFET structures were fabricated by using multiple epitaxial layers grown with ion-implantation of boron and phosphorus between epitaxial growth steps to create the alternating P and N columns. This is a relatively cumbersome process because of the large number of process steps to achieve devices with breakdown voltage of 600 V. Another method for creating the super-junction devices that has been proposed and demonstrated is by etching a deep trench in an N-type drift region followed by refill with P-type silicon [3]. Many other structural variations of the super-junction power MOSFET structure have also been proposed and demonstrated in the literature [4–8].

## 7.1 The SJ-MOSFET Structure

A cross-section of the basic cell structure for the power SJ-MOSFET structure is illustrated in Fig. 7.1. Two basic approaches to the fabrication of the power SJ-MOSFET structure have been proposed and demonstrated. The first method that was proposed is based up on multiple ion-implantations of P-type and N-type dopants followed by epitaxial growth [1]. The second method is based up on etching deep trenches into an N-type drift region followed by the refilling of the trenches with the P-type drift region [3]. Other variation on these two basic approaches have also been reported [9,10]. After formation of the alternating vertical P-type and N-type columns, the surface of the wafer is planarized by using chemi-mechanical-polishing (CMP). The DMOS gate structure is then

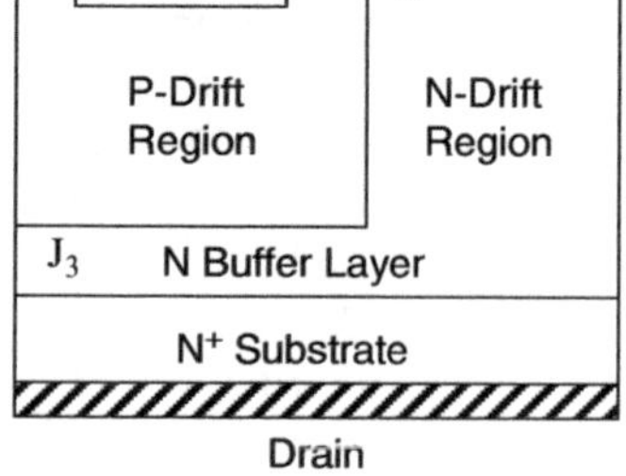


**Fig. 7.1** The SJ-MOSFET structure

formed using a process similar to that described for the power D-MOSFET structure in Chap. 2. A mask is required during the ion-implantation of the $N^+$ source regions to create the short between the $N^+$ source and P-base regions. In the power SJ-MOSFET structure, the gate must be carefully aligned to the P-type columns as shown in Fig. 7.1 so that the channel extends from the $N^+$ source regions into the N-type drift region column. Although the threshold voltage of the power SJ-MOSFET structure could in principle be controlled by the doping concentration of the P-type column, it is preferable to determine the doping level of the P-type drift region by charge-coupling considerations. Consequently, it is better to utilize a P-base region (formed with the DMOS process) to control the threshold voltage as illustrated in Fig. 7.1 by the dashed line.

The physics of operation of the power SJ-MOSFET structure in the blocking mode is different from that for the power MOSFET structures described in the previous chapters. Without the application of a gate bias, a high voltage can be supported in the SJ-MOSFET structure when a positive bias is applied to the drain. In this case, vertical junction $J_2$ formed between the P-type drift region and the N-type drift region becomes reverse biased. Simultaneously, the drain voltage is applied across the horizontal MOS structure formed between the gate electrode on the upper surface of the N-type drift region as well as the horizontal junction $J_3$ formed between the P-type drift region and the N-buffer layer. The MOS structure operates in the deep-depletion mode due to the presence of the reverse bias across junction $J_2$ between the P-type drift region and the N-type drift region. Consequently, depletion regions are formed across the vertical junction $J_2$ and the horizontal MOS interface creating the desired two-dimensional charge-coupling phenomenon. Simultaneously, depletion regions are formed across the vertical junction $J_2$ and the horizontal junction $J_3$ creating a two-dimensional depletion phenomenon. These two-dimensional depletion phenomena alter the electric field distribution along the y-direction from the triangular shape observed in conventional parallel-plane junctions to a rectangular shape. This allows supporting a required blocking voltage over a shorter distance. In addition, the doping concentration in the N-type drift region can be made much greater that predicted by the one-dimensional theory. This allows substantial reduction of the specific on-resistance to below the ideal specific on-resistance at any desired breakdown voltage. However, a precise charge concentration must be achieved for the P-type and the N-type drift region in order for the two-dimensional charge-coupling to produce a high breakdown voltage.

Drain current flow in the power SJ-MOSFET structure is induced by the application of a positive bias to the gate electrode. This produces an inversion layer at the surface of the P-base region at the upper surface. The threshold voltage for the power SJ-MOSFET structure can be controlled by adjusting the dose for the boron ion-implantation for the P-base region. The inversion layer channel provides a path for transport of electrons from the source to the N-type drift region when a positive drain voltage is applied.

After transport from the source region through the channel, the electrons enter the N-type drift region and are then transported vertically down to the $N^+$ substrate.

The resistance of the drift region is very low in the power SJ-MOSFET structure due to the high doping concentration in the N-type drift region. Note that an accumulation layer forms at the upper surface under the gate electrode which reduces the on-resistance by distributing the current into the N-type drift region. The channel resistance in the power SJ-MOSFET structure is small due to the small cell pitch or high channel density.

## 7.2 Charge-Coupling Physics

A much larger doping concentration can be employed in the drift region of the power SJ-MOSFET structure when compared with the D-MOSFET, U-MOSFET or SC-MOSFET structures. This enables a drastic reduction of the drift region resistance ($R_{D1}$ in Fig. 7.2) for devices capable of supporting high voltages. The doping concentration in the drift region is determined by two-dimensional charge coupling for the power SJ-MOSFET structure. In order to achieve good charge coupling, the N-type and P-type drift regions must be completely depleted when the drain bias approaches the breakdown voltage. The electric field then becomes uniform along the y-direction in the N-type and P-type drift regions. The super-junction concept has been found to be particularly effective for devices with breakdown voltages above 300 V.

Based up on the depletion of the N-type and P-type drift regions when the electric field at junction $J_2$ becomes equal to the critical electric field ($E_{CU}$) for breakdown for the uniform electric field distribution case:

$$Q_{Optimum} = qN_D \frac{W_N}{2} = \varepsilon_S E_{CU} = qN_A \frac{W_P}{2} \tag{7.2}$$

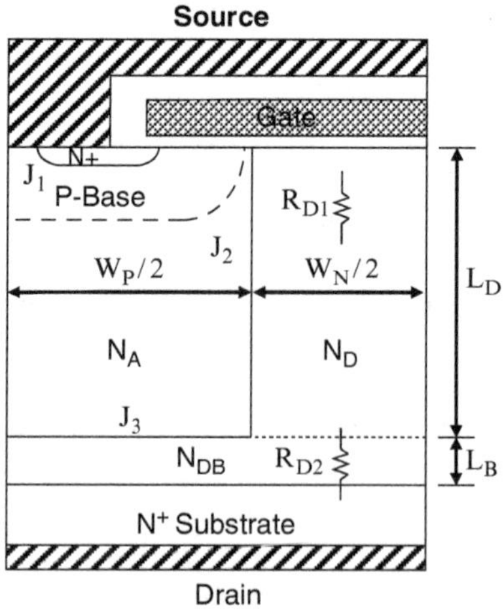


**Fig. 7.2** Resistances in the SJ-MOSFET structure

where $W_N$ and $W_P$ are the widths of the N-type and P-type drift regions, respectively; $N_D$ and $N_A$ are the doping concentration of the N-type and P-type drift regions, respectively. This provides the criterion for choosing the dopant dose (product of doping concentration and thickness) in the N-drift region to achieve the desired two-dimensional charge coupling:

$$N_D W_N = N_A W_P = \frac{2\varepsilon_S E_{CU}}{q} \tag{7.3}$$

Using Baliga's formula for the impact ionization coefficient for silicon, an expression for the critical electric field for the case of a uniform (or constant) electric field is obtained (see Chap. 5):

$$E_{CU} = 8.36 \times 10^4 \; L_D^{-1/7} \tag{7.4}$$

The breakdown voltage for the two dimensional charge coupled device is then given by:

$$BV_U = E_{CU} \; L_D = 8.36 \times 10^4 \; L_D^{6/7} \tag{7.5}$$

by using (7.4).

The breakdown voltage for devices with a uniform electric field can be observed in Fig. 7.3 to increase non-linearly with increasing length of the drift region. For comparison purposes, the dashed line in the figure is included to show a linear trend with a critical electric field of 2.4 × $10^5$ V/cm independent of the

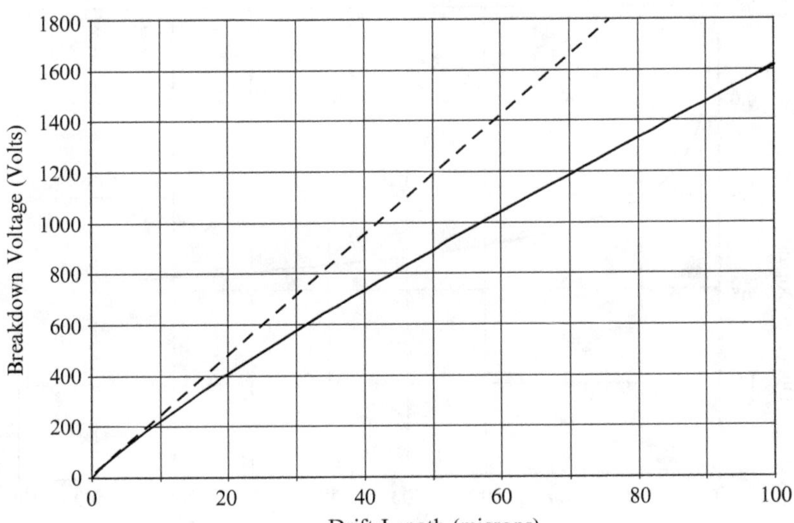


**Fig. 7.3** Breakdown voltages for the uniform electric field case

length of the drift region. For a drift region length of 10 μm, the breakdown voltage predicted by this analytical model is about 200 V. According to the analytical model, a drift region length of 32 μm is required to obtain a breakdown voltage of 600 V.

Combining (7.4) and (7.5), the critical electric field for breakdown in the uniform electric field case can be related to the breakdown voltage:

$$E_{CU} = 5.53 \times 10^5 \, BV^{-1/6} \tag{7.6}$$

Using this equation with (7.3) yields:

$$N_D W_N = 1.106 \times 10^6 \frac{\varepsilon_S}{q} \, BV^{-1/6} \tag{7.7}$$

The optimum dose predicted by this expression is shown in Fig. 7.4 as a function of the breakdown voltage. It can be observed that the optimum dose decreases gradually with increasing breakdown voltage. The optimum dose for the N-type and P-type drift regions predicted by the analytical model is $2.96 \times 10^{12}$, $2.77 \times 10^{12}$, $2.47 \times 10^{12}$, and $2.27 \times 10^{12}/cm^2$ for breakdown voltages of 200, 300, 600 and 1,000 V, respectively.

From (7.7), the optimum doping concentration for the two-dimensional charge coupled case is given by:

$$N_D = 1.106 \times 10^6 \frac{\varepsilon_S}{q W_N} \, BV^{-1/6} \tag{7.8}$$

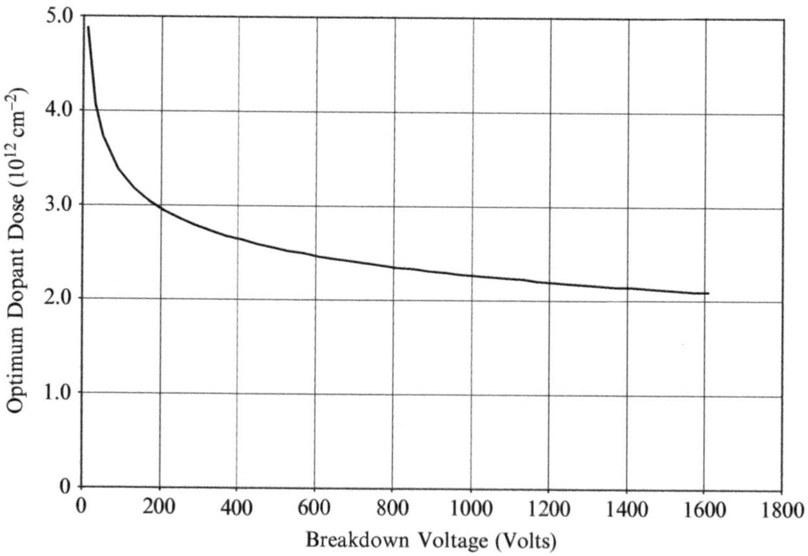


**Fig. 7.4** Optimum dose for two-dimensional charge-coupling

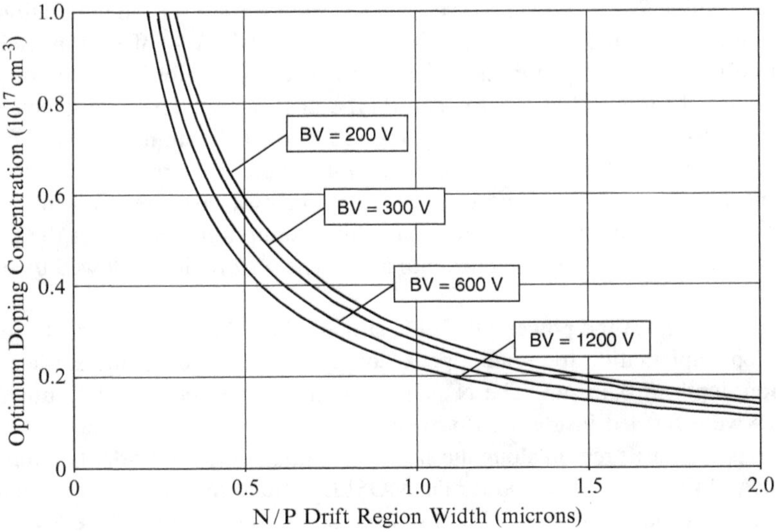


**Fig. 7.5** Optimum doping concentration for two-dimensional charge-coupling

The optimum doping concentration predicted by this expression is shown in Fig. 7.5 as a function of the P/N drift region width using the breakdown voltage as a parameter. It can be observed that the optimum doping concentration decreases with increasing P/N drift region width and is a weak function of the breakdown voltage. For the case of a P/N drift region width of 1.5 μm, the optimum doping concentration is found to be $1.98 \times 10^{16}/\text{cm}^3$ for a breakdown voltage of 200 V and $1.64 \times 10^{16}/\text{cm}^3$ for a breakdown voltage of 600 V. It is worth pointing out that the doping concentration and width of the P-type and N-type drift regions do not have to be the same for the super-junction devices. In order to achieve the desired charge balance for two-dimensional charge-coupling, it is sufficient that the dose for both the regions be equal to the optimum dose shown in Fig. 7.4.

### *7.2.1 Simulation Results*

The results of two-dimensional numerical simulations on the 200-V power SJ-MOSFET structure are first described here to provide a more detailed understanding of the underlying device physics and operation during the blocking mode. This structure was designed to have the same 200-V blocking voltage as the power GD-MOSFET structure in the previous chapter for comparison purposes. However, the super-junction concept is more effective for devices with larger breakdown voltages. For this reason, the performance of a power SJ-MOSFET structure with a breakdown voltage of 600 V is also discussed here.

The baseline 200-V power SJ-MOSFET structure used for the simulations had P-type and N-type drift regions with the same width ($W_N$, $W_P$) of 1.5 μm leading to a half-cell width of 1.5 μm in the simulations. The drift region consisted of a total thickness of 12 μm with a 10-μm deep P-type drift region. The doping concentration in the N-type and P-type drift regions was varied to examine the impact on the breakdown voltage. For the structure used for the numerical simulations, the doping concentration in the N-type and P-type drift regions was assumed to be uniform in both the x-direction and the y-direction. This assumption is valid for the devices fabricated by etching deep trenches into an N-type drift region followed by epitaxial refill.

In order to form the power MOSFET structure, a P-base region was formed by using ion-implantation to create a peak doping concentration at the surface and a junction depth of 0.45 μm. The $N^+$ source region had a depth of 0.07 μm. These regions were formed inside the P-type drift region so that the channel can extend into the N-type drift region along the surface. Although the P-type drift region could be used to form the base region for the MOSFET structure, it is more convenient to create the channel using the D-MOS process so that the doping of the P-type drift region can be independently controlled for optimization of the charge-coupling.

A three dimensional view of the doping distribution in the upper portion of the 200-V SJ-MOSFET structure is shown in Fig. 7.6. The $N^+$ source and P-base regions can be observed on the upper left hand side within the P-type drift region. The N-type drift region can be observed on the right hand side of the structure. The doping concentration for both the P-type and N-type drift regions is $2 \times 10^{16}/cm^3$

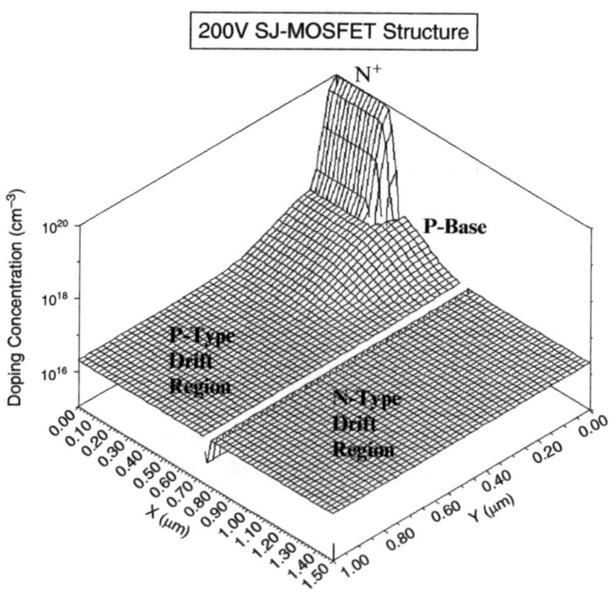


**Fig. 7.6** Doping distribution for the SJ-MOSFET structure

to maintain charge balance. This concentration for the P-type and N-type drift regions corresponds to a dose of $3 \times 10^{12}/cm^2$ – the optimum dose predicted by the analytical model for a 200-V device (see Fig. 7.4).

The lateral doping profile taken along the surface of the semiconductor is shown in Fig. 7.7. From the profile, it can be observed that the P-base region has a doping profile with a peak doping concentration of $1.5 \times 10^{17}/cm^3$ to obtain the desired threshold voltage. The lateral junction depths of the P-base and $N^+$ source regions are 0.75 and 0.37 μm leading to a channel length of 0.38 μm. The N-type drift region has a doping concentration of $2 \times 10^{16}/cm^3$.

The vertical doping profiles taken at two positions within the power SJ-MOSFET structure are provided in Fig. 7.8. From the profile taken at x = 0 μm (solid line), it can be observed that the doping concentration of the P-type drift region is $2 \times 10^{16}/cm^3$ with a higher concentration at the upper 0.5 μm due to the P-Base region. From the profile taken at x = 1.5 μm (dashed line), it can be observed that the N-type drift region also has doping concentration of $2 \times 10^{16}/cm^3$. The P-type and N-type drift region extend to a vertical depth of 11 μm.

The blocking characteristics for the 200-V SJ-MOSFET cell structure are shown in Fig. 7.9 for various doping concentrations in the N-type and P-type drift regions. In all these structures, the doping concentrations for the N-type and P-type drift regions was kept at the same value (provided in the figure) to maintain charge balance. It can be observed that the cell structure is capable of supporting 225 V when the doping concentration has an optimum value of $2 \times 10^{16}/cm^3$.

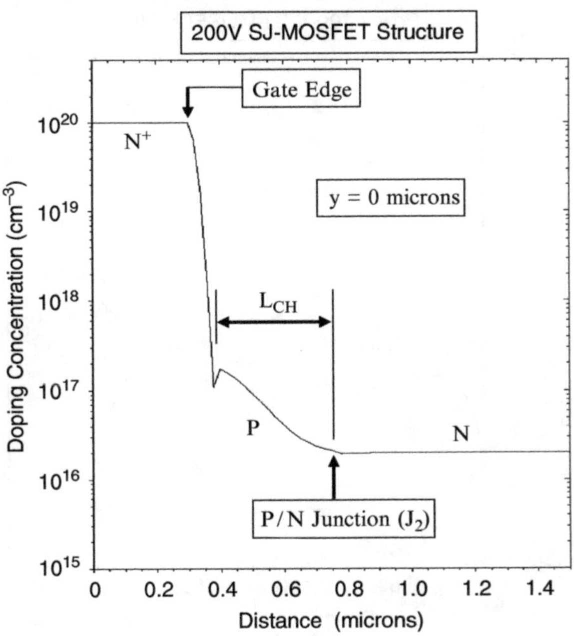


**Fig. 7.7** Channel doping profile for the 200-V SJ-MOSFET structure

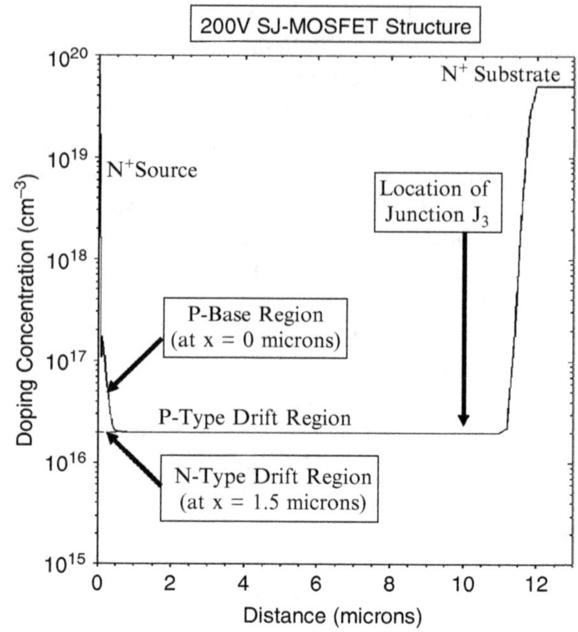


**Fig. 7.8** Vertical doping profiles in the 200-V SJ-MOSFET structure

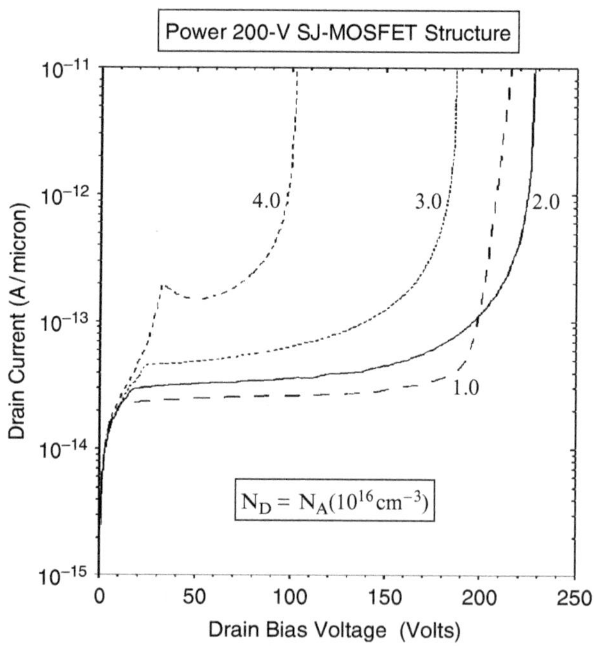


**Fig. 7.9** Blocking characteristics for the 200-V SJ-MOSFET structures

The corresponding optimum charge or dose in the N-type and P-type drift region (using a width of 1.5 μm for these regions) is $3 \times 10^{12}/\text{cm}^2$. The optimum dose obtained by using the analytical model (see Fig. 7.4) is in excellent agreement with the dose obtained using the numerical simulations. The optimum doping concentration for the super-junction structure is far greater than the doping concentration of $1.35 \times 10^{15}/\text{cm}^3$ for the one-dimensional structure to achieve the same breakdown voltage of 200 V. The larger doping concentration in the super-junction structure, achieved by using the two-dimensional charge coupling phenomenon, results in a large reduction of the specific on-resistance despite loss of half the active area for current flow in the drift region.

It can be observed from Fig. 7.9 that the breakdown voltage is reduced when the doping concentration for the N-type and P-type drift regions is either increased or reduced. When the doping concentration in the drift regions is reduced to $1 \times 10^{16}/\text{cm}^3$, the pre-breakdown leakage current is reduced but this is accompanied by a reduction in the breakdown voltage from 228 to 215 V. When the doping concentration in the drift regions is increased to $3 \times 10^{16}/\text{cm}^3$, the pre-breakdown leakage current increases and the breakdown voltage is reduced 190 V. An even greater reduction in the breakdown voltage to 100 V occurs when the doping concentration in the drift regions is increased to $4 \times 10^{16}/\text{cm}^3$.

In order to demonstrate the two-dimensional charge coupling phenomenon in super-junction devices, it is instructive to examine the potential contours inside the 200-V power SJ-MOSFET structure when it is operating in the blocking mode. This also allows determination of the voltage distribution within the structure and the penetration of the depletion region in the P-base region with increasing drain bias voltage. The potential contours for the SJ-MOSFET structure obtained using the numerical simulations with zero gate bias and various drain bias voltages are shown in Figs. 7.10–7.13.

The potential contours at a drain bias of 20 V (see Fig. 7.10) show a one-dimensional distribution across the P-drift/N-drift junction ($J_2$) because the drift regions are not completely depleted at this bias voltage. At the same time, the potential is supported in a one-dimensional configuration across the junction $J_3$ as well as under the gate electrode overlapping the N-type drift region. When the voltage is increased to 50 V (see Fig. 7.11), the two-dimensional charge coupling begins to take effect. At a drain bias of 100 V (see Fig. 7.12), the two-dimensional charge coupling produces a fairly uniform potential distribution within the N-type and P-type drift regions. This behavior is also visible in the case of a drain bias of 200 V in Fig. 7.13.

In the case of the power SJ-MOSFET structure, it is not feasible to utilize the edge termination approaches used for the device structures discussed in the textbook because of the very high doping concentration in the drift region. Special edge terminations compatible with the high doping concentrations in the drift region of power SJ-MOSFET structures are discussed later in the chapter.

It is insightful to examine the electric field profile inside the power SJ-MOSFET structure when it is operating in the blocking mode. A three-dimensional view of the electric field within the 200-V power SJ-MOSFET structure is shown in

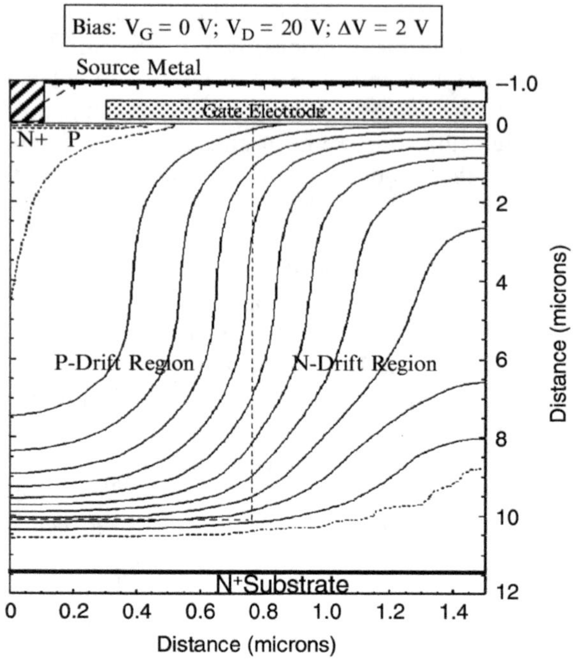


**Fig. 7.10** Potential contours in the 200-V SJ-MOSFET structure

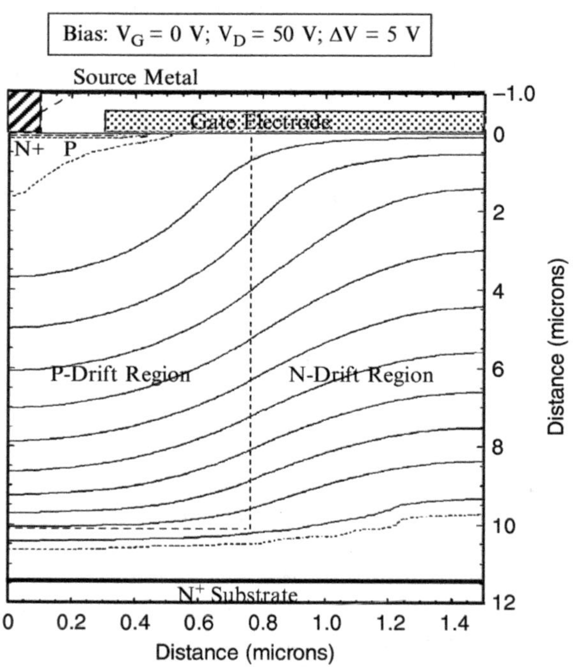


**Fig. 7.11** Potential contours in the 200-V SJ-MOSFET structure

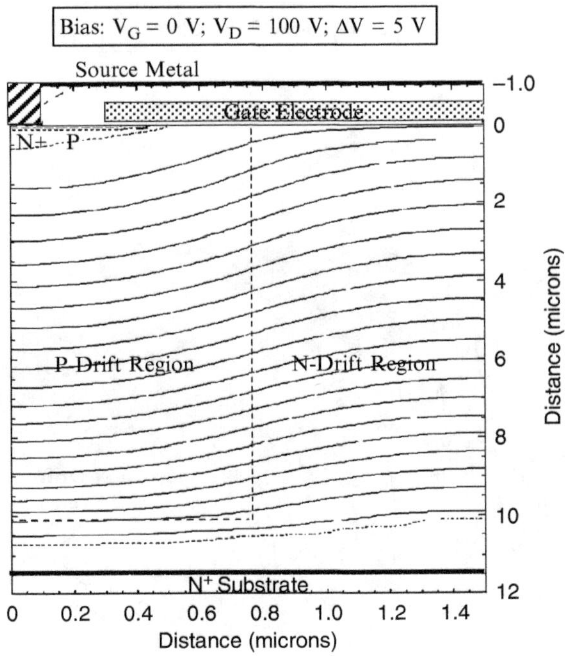


**Fig. 7.12** Potential contours in the 200-V SJ-MOSFET structure

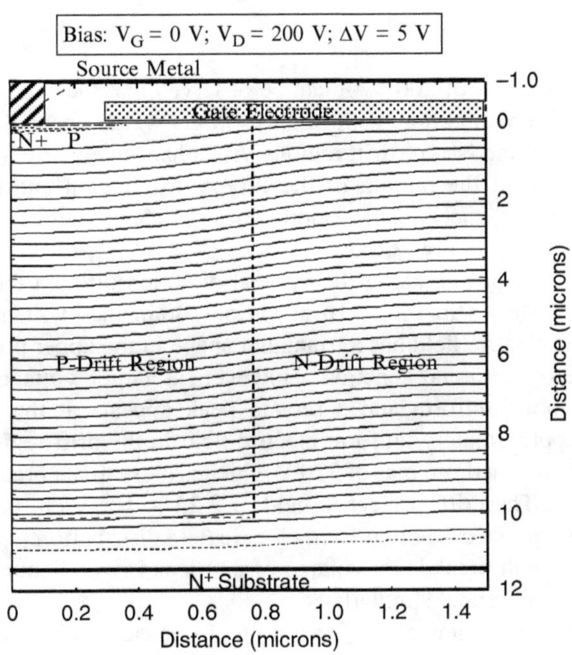


**Fig. 7.13** Potential contours in the 200-V SJ-MOSFET structure

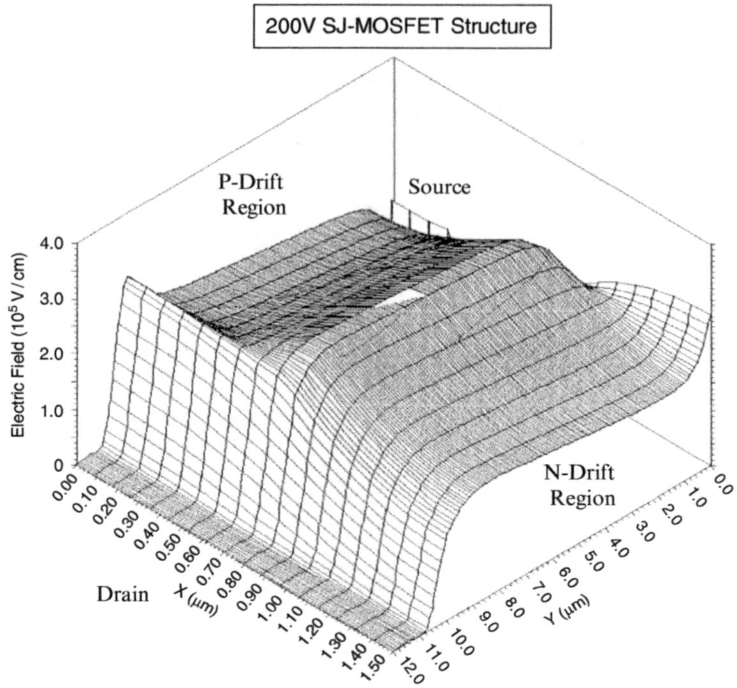


**Fig. 7.14** Three-dimensional electric field distribution in the 200-V SJ-MOSFET structure

Fig. 7.14 at a drain bias of 200 V. It can be observed that the electric field is very uniform along the y-direction with slightly larger values near the junction ($J_2$ between the P-type and N-type drift regions. An enhanced electric field is observed at the junction ($J_3$) at the bottom of the P-type drift region and under the gate electrode overlapping the N-drift region. These enhanced electric fields can be observed more clearly in Fig. 7.15.

The electric field profile along the vertical direction through the center of the N-drift region is shown in Fig. 7.16 at various drain bias voltages. At a drain bias of 20 V, the electric field has a triangular shape in the upper portion representative of a one-dimensional junction. However, at larger voltages, the electric field becomes more uniform and a second peak appears at the bottom of the trenches. The approximately uniform electric field in the drift region has a value of $2.2 \times 10^5$ V/cm at a drain bias of 220 V which is close to the breakdown voltage for this structure. The critical electric field of $2.25 \times 10^5$ V/cm predicted by the analytical model for the case of uniform electric field distribution (see Figs. 6.4 and 6.5) for a device with breakdown voltage of 220 V is in excellent agreement with the results of the numerical simulations.

A three dimensional view of the generation of free carriers by impact ionization is shown in Fig. 7.17 for the 200-V power SJ-MOSFET structure at a drain bias

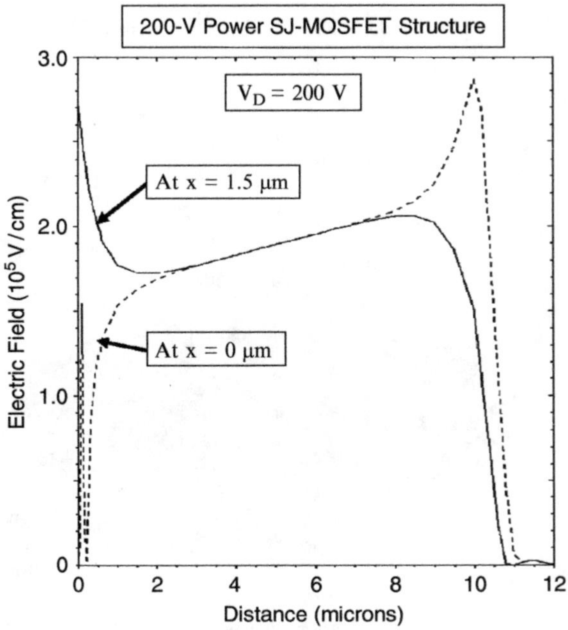


**Fig. 7.15** Electric field profiles in the 200-V SJ-MOSFET structure

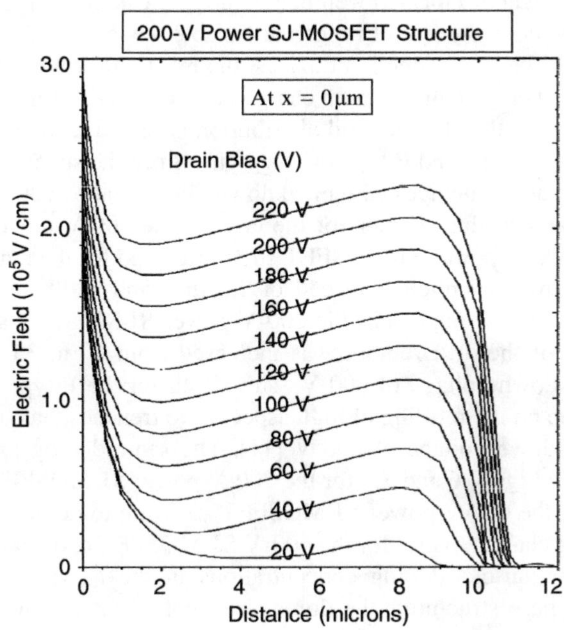


**Fig. 7.16** Electric field profiles in the 200-V SJ-MOSFET structure at various drain bias voltages

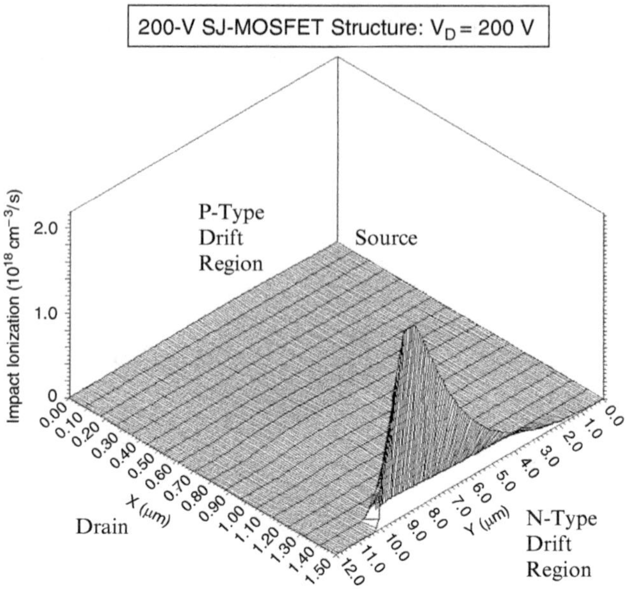


**Fig. 7.17** Three-dimensional impact ionization distribution in the 200-V SJ-MOSFET structure

of 200 V. It can be observed that the impact ionization occurs at the middle of the N-type drift region and is located well below the gate electrode. This distribution is favorable for the development of rugged and stable devices.

In the case of the super-junction devices, the most uniform electric field distribution is obtained only at an optimum charge or dose in the N-type and P-type drift regions. In addition, the electric field distribution is sensitive to any charge imbalance between the N-type and P-type drift regions. Since the super-junction concept has the most impact on devices with breakdown voltages above 200 V, these issues are discussed later in this chapter for the case of the 600-V power SJ-MOSFET structure. The 600-V power SJ-MOSFET structures used for the numerical simulations had the same cell pitch (half-cell of 1.5 μm) and widths (1.5 μm) for the N-type and P-type drift regions as the 200-V power SJ-MOSFET structure. However, the depth of the drift regions was increased from 10 to 35 μm in order to achieve a breakdown voltage of 600 V. This depth for the P-type drift region, in devices based up on the refilling of high aspect ratio trenches, has been reported to achieve a breakdown voltage of 600 V [11]. The same doping profile along the surface shown in Figs. 7.6 and 7.7 for the 200-V power SJ-MOSFET structure was also utilized for the 600-V power SJ-MOSFET structure to define the channel.

The blocking characteristics for the 600-V SJ-MOSFET cell structure are shown in Fig. 7.18 for various doping concentrations in the N-type and P-type drift regions. In all these structures, the doping concentrations for the N-type and P-type drift regions was kept equal (as provided in the figure) to maintain charge balance. It can be observed that the cell structure is capable of supporting 570 V

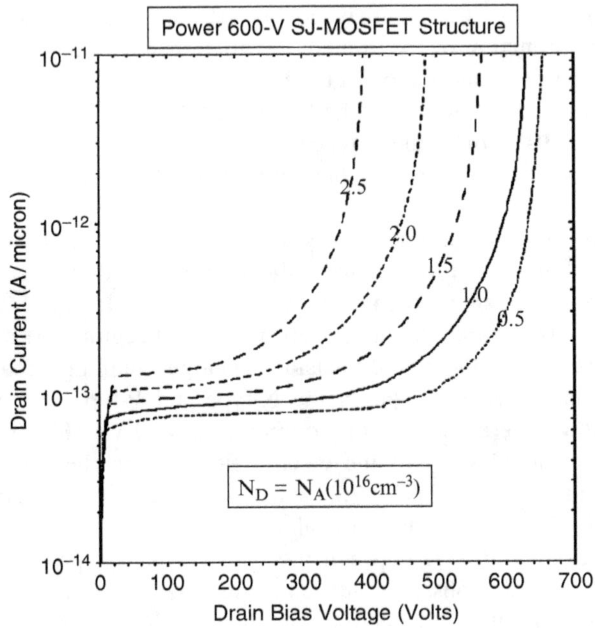


**Fig. 7.18** Blocking characteristics for the 600-V SJ-MOSFET structures

when the doping concentration has an optimum value of $1.5 \times 10^{16}/\text{cm}^3$. The corresponding optimum charge or dose in the N-type and P-type drift region (using a width of 1.5 μm for these regions) is $2.25 \times 10^{12}/\text{cm}^2$. The optimum dose obtained by using the analytical model (see Fig. 7.4) is in excellent agreement with the dose obtained using the numerical simulations. The optimum doping concentration of the N-type drift region in the super-junction structure is far greater than the doping concentration of $3.12 \times 10^{14}/\text{cm}^3$ for the one-dimensional structure to achieve the same breakdown voltage of 600 V. The larger doping concentration in the super-junction structure, achieved by using the two-dimensional charge coupling phenomenon, results in a large reduction of the specific on-resistance despite loss of half the active area for current flow in the drift region.

It can be observed from Fig. 7.18 that the breakdown voltage is reduced when the doping concentration for the N-type and P-type drift regions is increased. When the doping concentration in the drift regions is reduced to $1.0 \times 10^{16}/\text{cm}^3$, the pre-breakdown leakage current is reduced and the breakdown voltage increases from 570 to 635 V. When the doping concentration in the drift regions is increased to $2.0 \times 10^{16}/\text{cm}^3$, the pre-breakdown leakage current increases and the breakdown voltage is reduced 485 V. An even greater reduction in the breakdown voltage to 395 V occurs when the doping concentration in the drift regions is increased to $2.5 \times 10^{16}/\text{cm}^3$.

In order to demonstrate the two-dimensional charge coupling phenomenon in super-junction devices, it is instructive to examine the potential contours inside the

600-V power SJ-MOSFET structure when it is operating in the blocking mode. This also allows determination of the voltage distribution within the structure and the penetration of the depletion region in the P-base region with increasing drain bias voltage. The potential contours for the SJ-MOSFET structure with a doping concentration of $1 \times 10^{16}/\text{cm}^3$ in the N-type and P-type drift regions obtained using the numerical simulations with zero gate bias and various drain bias voltages are shown in Figs. 7.19–7.22.

The potential contours at a drain bias of 50 V (see Fig. 7.19) show a potential crowding at the bottom of the structure indicating that the doping concentration is less than the optimum value. When the voltage is increased to 100 V (see Fig. 7.20), the voltage spreads towards the upper part of the drift regions. At a drain bias of 300 V (see Fig. 7.21), the two-dimensional charge coupling produces a fairly uniform potential distribution within the N-type and P-type drift regions. This behavior is also visible in the case of a drain bias of 600 V in Fig. 7.22.

A three-dimensional view of the electric field within the 600-V power SJ-MOSFET structure is shown in Fig. 7.23 at a drain bias of 600 V. It can be observed that the electric field is very uniform along the y-direction with slightly larger values near the junction ($J_2$) between the P-type and N-type drift regions. An enhanced electric field is observed at the junction ($J_3$) at the bottom of the P-type drift region and under the gate electrode overlapping the N-drift region.

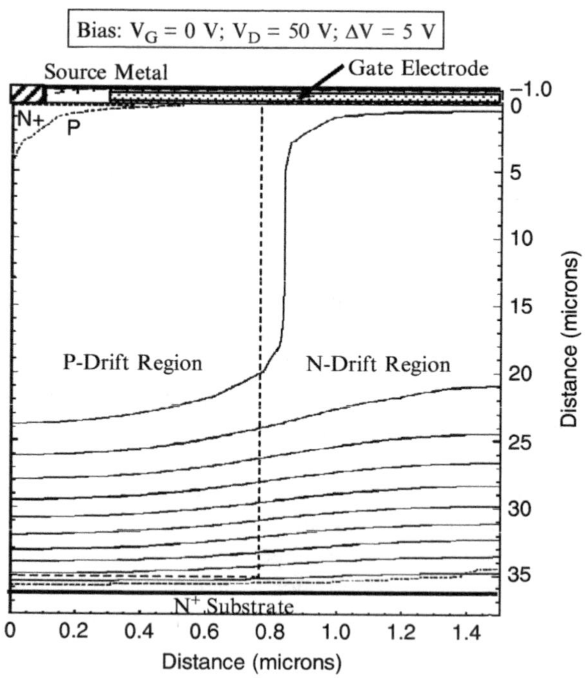


**Fig. 7.19** Potential contours in the 600-V SJ-MOSFET structure

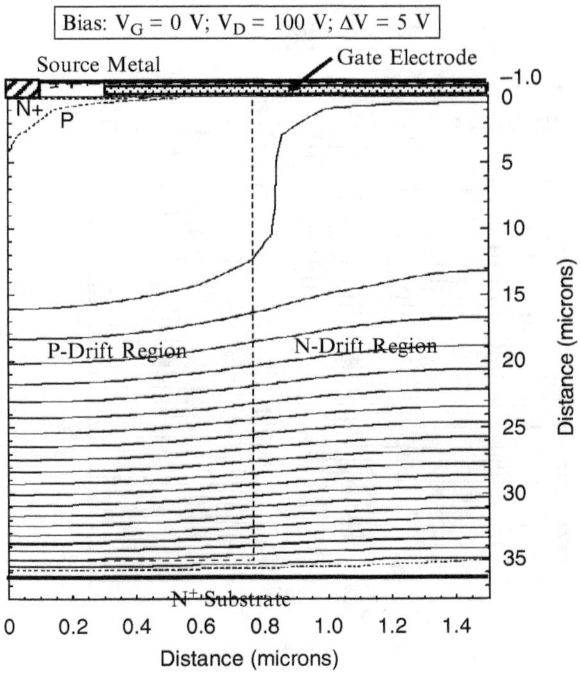


**Fig. 7.20** Potential contours in the 600-V SJ-MOSFET structure

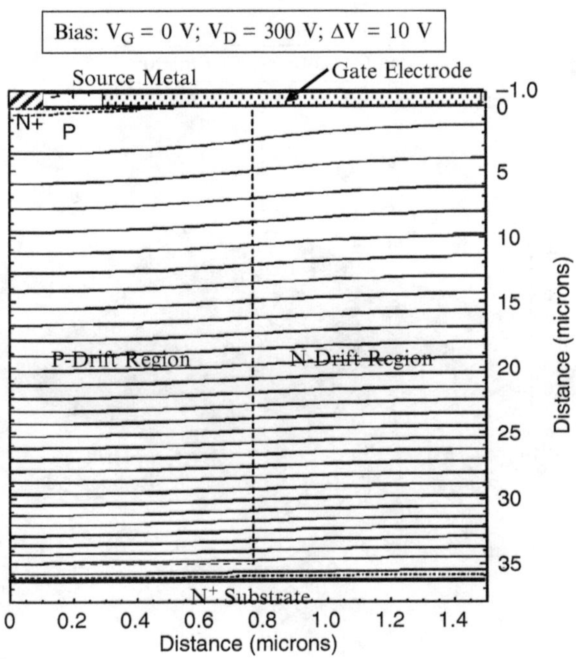


**Fig. 7.21** Potential contours in the 600-V SJ-MOSFET structure

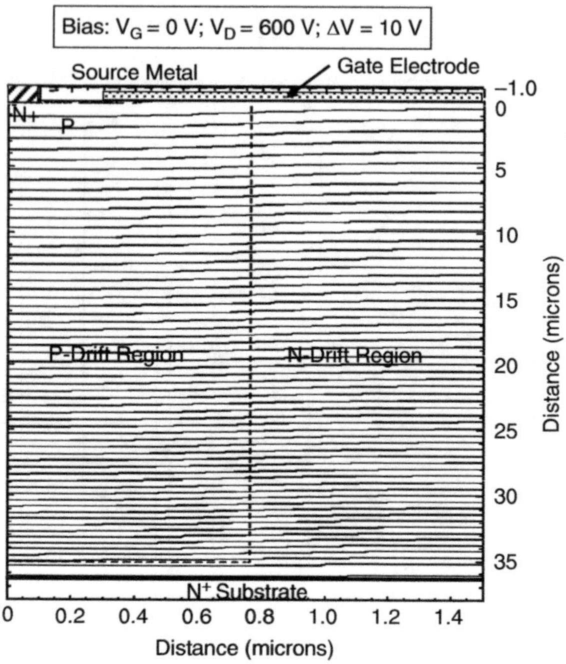


**Fig. 7.22** Potential contours in the 600-V SJ-MOSFET structure

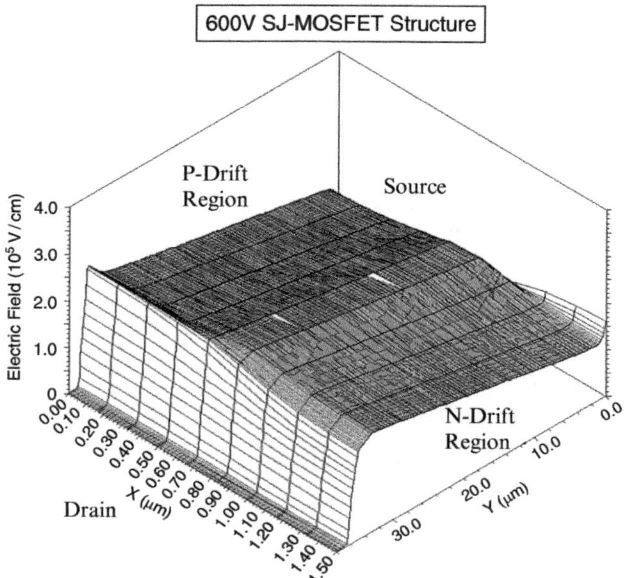


**Fig. 7.23** Three-dimensional electric field distribution in the 600-V SJ-MOSFET structure

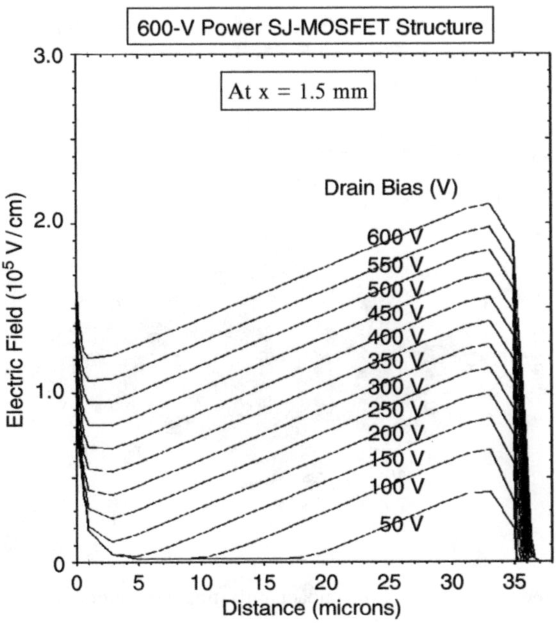


**Fig. 7.24** Electric field profiles in the 600-V SJ-MOSFET structure at various drain bias voltages

The electric field profile along the vertical direction through the center of the N-drift region is shown in Fig. 7.24 at various drain bias voltages. At a drain bias of 50 V, the electric field has a triangular shape in the lower portion of the drift region representative of a one-dimensional junction. However, at larger voltages, the electric field becomes more uniform. The highest electric field in the drift region has a value of $2.0 \times 10^5$ V/cm at a drain bias of 600 V which is close to the breakdown voltage for this structure. The critical electric field of $1.90 \times 10^5$ V/cm predicted by the analytical model for the case of uniform electric field distribution (see 7.6) for a device with breakdown voltage of 600 V is in good agreement with the results of the numerical simulations.

A three dimensional view of the generation of free carriers by impact ionization is shown in Fig. 7.25 for the 600-V power SJ-MOSFET structure at a drain bias of 600 V. It can be observed that the impact ionization occurs at the middle of the N-type drift region and is located well below the gate electrode. This distribution is favorable for the development of rugged and stable devices.

In the case of the super-junction devices, the highest breakdown voltage occurs at the optimum dose in the N-type and P-type drift region. The electric field profiles along the y-direction in the middle of the N-type drift region of the 600-V SJ-MOSFET structures are shown in Fig. 7.26 at a drain bias of 300 V. It can be observed that the electric field becomes flatter along the y-direction when the doping concentration is reduced. This leads to an increase in the breakdown voltage as previous shown in Fig. 7.18. However, a reduced doping concentration in the N-type

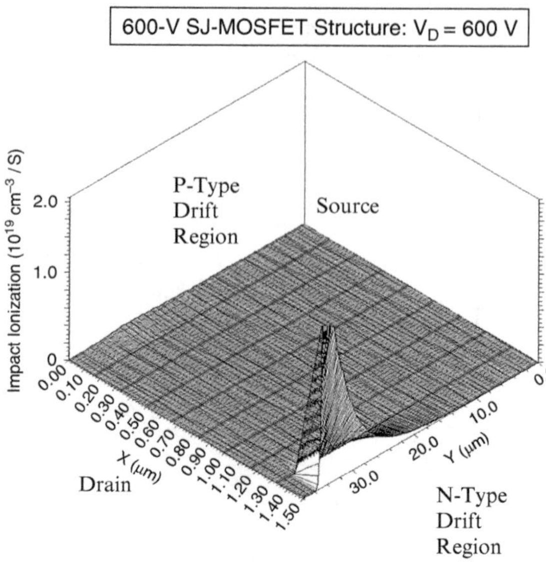


**Fig. 7.25** Three-dimensional view of the impact ionization distribution in the 600-V SJ-MOSFET structure

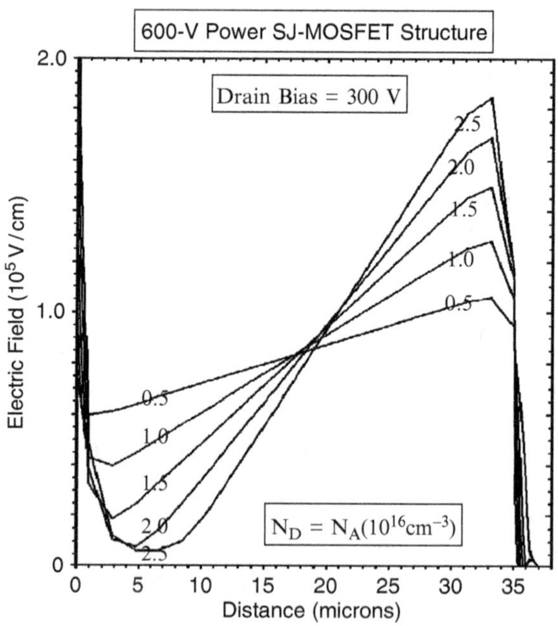


**Fig. 7.26** Electric field profiles in the 600-V SJ-MOSFET structures

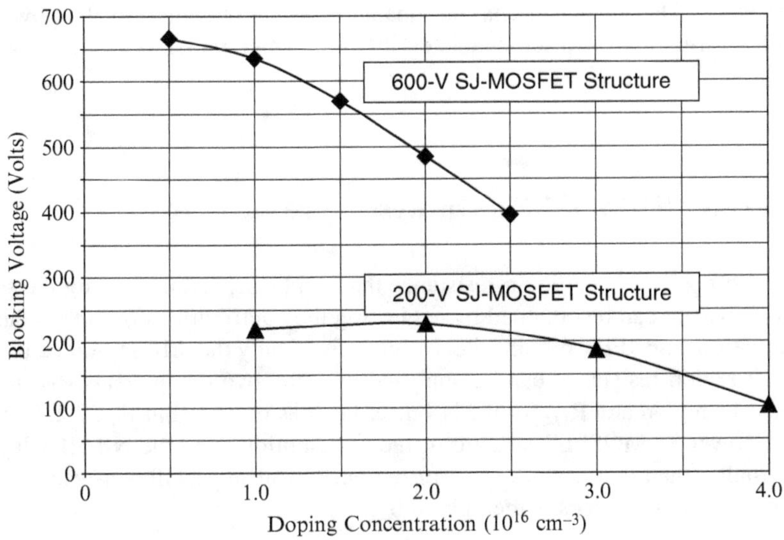


**Fig. 7.27** Impact of doping concentration in the drift regions on the breakdown voltage of SJ-MOSFET structures

drift region leads to a large specific on-resistance. It is therefore better to utilize a large doping concentration even though the electric field profile is not flat in nature.

As discussed previously, the breakdown voltage for the super-junction device structures is dependent on the doping concentration of the N-type and P-type drift regions. The variation of the breakdown voltage with doping concentration obtained using the two-dimensional numerical simulations is shown in Fig. 7.27 for the case of the 200-V and 600-V power SJ-MOSFET structures. In the case of the 200-V structure, it can be concluded that the maximum doping concentration of the drift region should be $2 \times 10^{16}/\text{cm}^3$. In contrast, it can be concluded that the maximum doping concentration of the drift region is $1.3 \times 10^{16}/\text{cm}^3$ in the case of the 600-V structure. The maximum doping concentration in the drift region for the power SJ-MOSFET structure, therefore, reduces with increasing breakdown voltage. The predictions of the analytical model (see Figs. 7.5 and 7.8) are consistent with these results obtained using the two-dimensional numerical simulations.

The breakdown voltage of the power SJ-MOSFET structure is sensitive to the charge balance between the N-type and P-type drift regions. For the structures discussed above, it was assumed that the doping concentration and width of the N-type and P-type regions are exactly equal. The charge imbalance between these regions can be analyzed by varying the doping concentration of either the N-type drift region or the P-type drift region. When the doping concentration of the N-type drift region is changed, the specific on-resistance is also altered. In contrast, varying the doping concentration of the P-type drift region does not alter the specific on-resistance. The impact of charge imbalance on the operation of the power SJ-MOSFET structures is discussed later in the chapter under process sensitivity analysis.

## 7.3 Power SJ-MOSFET On-Resistance

The lowest possible specific on-resistance that can be achieved by using the super-junction concept can be obtained by neglecting the contributions from the channel, accumulation, and JFET regions, i.e. by analysis of only the drift region resistance. The drift region resistance can be analyzed with the two components $R_{D1}$ for the N-type drift region and $R_{D2}$ for the N-buffer layer as indicated in Fig. 7.2. For high voltage power SJ-MOSFET structures, the contribution from the N-buffer layer is much smaller than that from the N-drift region. The ideal specific on-resistance for the super-junction devices is then given by:

$$R_{ON,sp}(\text{Ideal SJ}) = R_{D1,sp} = \rho_{ND} L_D \left(\frac{W_{Cell}}{W_N}\right) = \frac{L_D}{q\,\mu_N N_D}\left(\frac{W_N + W_P}{W_N}\right) \quad (7.9)$$

where $\rho_{ND}$ is the resistivity of the N-type drift region. The length of the drift region can be related to the breakdown voltage of the super-junction device:

$$L_D = \frac{BV}{E_{CU}} \quad (7.10)$$

because the electric field along the y-direction has a constant value equal to the critical electric for breakdown with uniform electric field in the case of super-junction devices. Using (7.3), the optimum doping concentration for super-junction devices is given by:

$$N_D = \frac{2\varepsilon_S E_{CU}}{qW_N} \quad (7.11)$$

Combining the above relationships:

$$R_{ON,sp}(\text{Ideal SJ}) = \frac{BV}{\varepsilon_S \mu_N E_{CU}^2}\left(\frac{W_N + W_P}{2}\right) \quad (7.12)$$

Since the critical electric field for breakdown with uniform electric field is related to the breakdown voltage by (7.6):

$$R_{ON,sp}(\text{Ideal SJ}) = \frac{1.635 \times 10^{-12}\ BV^{4/3}(W_N + W_P)}{\varepsilon_S \mu_N} \quad (7.13)$$

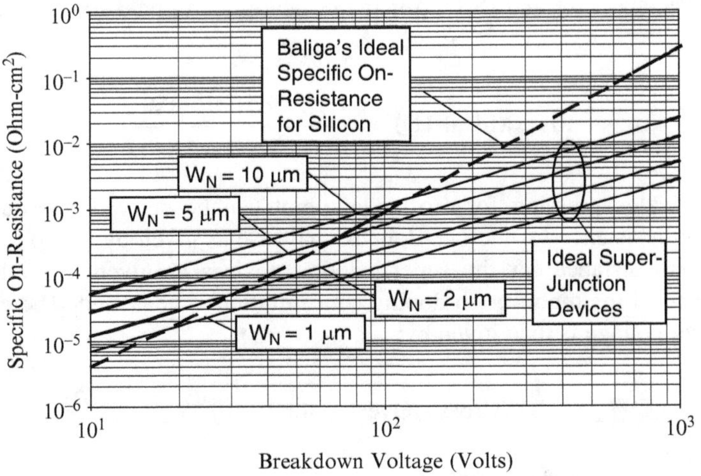


**Fig. 7.28** Ideal specific on-resistance for the super-junction devices

Since it is typical to use the same width for the P-type and N-type drift regions in super-junction devices, the ideal specific on-resistance for super-junction devices can be computed using:

$$R_{ON,sp}(\text{Ideal SJ}) = \frac{3.27 \times 10^{-12}\ BV^{4/3} W_N}{\varepsilon_S \mu_N} \tag{7.14}$$

The ideal specific on-resistance for super-junction devices computed by using (7.14) is provided in Fig. 7.28 for the case of four widths for the N-type drift region under the assumption that the width of the P-type drift region has the same value. Since the optimum doping concentration is relatively high for these devices, the dependence of the mobility on doping concentration was included during this analysis. The ideal specific on-resistance for the drift region in one-dimensional devices as computed by using Baliga's power law for the impact ionization coefficients is also shown in Fig. 7.28 for comparison purposes. It can be observed from this figure that the specific on-resistance for the super-junction devices can be less than the ideal specific on-resistance for silicon devices. At high (>500 V) breakdown voltages, an improvement in the specific on-resistance by more a one order of magnitude can be obtained using the super-junction structure. The range of breakdown voltages for which the performance of the super-junction devices is superior to that of the ideal one-dimensional structure becomes larger when the width of the N-drift region is reduced.

From Fig. 7.28, it can be observed that there is a cross-over point (breakdown voltage) between the specific on-resistance for the super-junction and the ideal one-dimensional devices. Using (7.1) for the ideal specific on-resistance for the one-dimensional case and (5.15) for the critical electric field for the one-dimensional case:

$$R_{ON,sp}(\text{Ideal 1D}) = \frac{1.181 \times 10^{-17}\ BV^{5/2}}{\varepsilon_S \mu_N} \tag{7.15}$$

An analytical expression for cross-over point can be derived by equating the specific on-resistance for the one-dimensional case as given by (7.15) to the ideal specific on-resistance for the super-junction devices as given by (7.14) if the dependence of the mobility on doping concentration is neglected. The cross-over breakdown voltage is then given by:

$$BV(\text{Cross - Over}) = 4.62 \times 10^4 W_N^{6/7} \tag{7.16}$$

The cross-over breakdown voltage computed using this equation is provided in Fig. 7.29. The cross-over breakdown voltage predicted by the above equation is in good agreement with the cross-over points in Fig. 7.28 indicating the assumption of a constant mobility is reasonable. It can be observed that the cross-over breakdown voltage increases with increasing width of the N-drift region. It is therefore advantageous to utilize small widths for the drift region to maximize the performance of the super-junction devices. However, the presence of a depletion region across the vertical junction ($J_3$) can create an increase in the on-resistance for super-junction devices [12]. This has an adverse impact on the specific on-resistance.

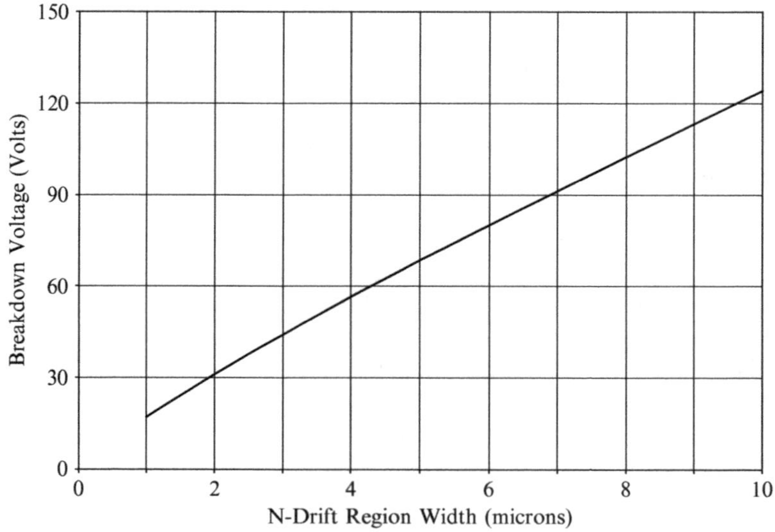


**Fig. 7.29** Cross-over breakdown voltage for the super-junction devices

The components of the on-resistance for the power SJ-MOSFET structure are similar as those already described for the power D-MOSFET structure. However, in the power SJ-MOSFET structure, the JFET effect occurs along the entire length of the drift region. Consequently, the on-resistance for the power SJ-MOSFET structure must be modeled as current flowing from the channel into the N-drift region via the accumulation layer with the width of the current carrying area in the N-drift region reduced by the presence of the depletion region formed at P/N junction $J_2$.

The total on-resistance for the power SJ-MOSFET structure can be obtained by the addition of eight resistances which are considered to be in series in the current path between the source and the drain electrodes:

$$R_{ON} = R_{CS} + R_{N+} + R_{CH} + R_A + R_{D1} + R_{D2} + R_{SUB} + R_{CD} \tag{7.17}$$

Each of the resistances within the power SJ-MOSFET structure is analyzed below by using the procedure described in the textbook [13]. In the textbook, it was demonstrated that the contributions from the source contact resistance ($R_{CS}$), the source resistance ($R_{N+}$), and the drain contact resistance ($R_{CD}$) are very small and will therefore be neglected in this chapter. As in the case of the other power MOSFET structures, the substrate contribution will also be excluded for the comparison of devices.

The power SJ-MOSFET structure is illustrated in Fig. 7.30 with the current flow path shown as the shaded area. The resistances that must be analyzed in the power

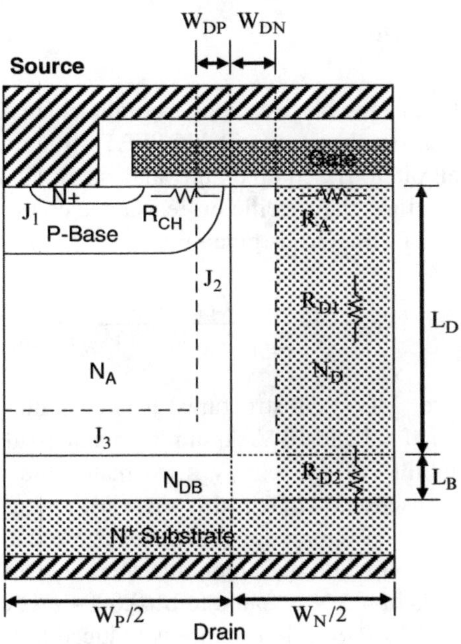


**Fig. 7.30** On-resistance components in the power SJ-MOSFET structure

SJ-MOSFET structure are also indicated in the figure. The figure provides the various dimensions that are used for the analysis of the on-resistance components. Here, $W_P$ is the width of the P-type drift region; $W_N$ is the width of the N-type drift region; $L_D$ is the length of the N-type drift region; $L_B$ is the thickness of the N-type buffer layer; $W_{DN}$ is the depletion width in the N-type drift region; and $W_{DP}$ is the depletion width in the P-type drift region. The cell width ($W_{Cell}$) is the sum of the widths of the N-type and P-type drift regions. The lateral junction depths of the P-base region and the $N^+$ source regions are $x_{JP}$ and $x_{N+}$, respectively.

In this monograph, the characteristics of power SJ-MOSFET structure with 200-V blocking capability will first be analyzed for comparison with the power GD-MOSFET structure. For this voltage rating, the power SJ-MOSFET structure has a cell pitch ($W_{CELL}$) of 3.0 μm with widths of 1.5 μm for the P-type and N-type drift regions. Typical lateral junction depths for the $N^+$ source region and P-base region are 0.37 and 0.75 μm, respectively, leading to a channel length of only 0.38 μm. This short channel length is possible due to the shielding of the channel region in the power SJ-MOSFET structure due to the charge coupling phenomenon which spreads the electric field away from the channel. Based up on the discussion in the previous section, the optimum doping concentration of the N-drift region required to achieve a 200-V blocking voltage capability is $2.0 \times 10^{16}/\text{cm}^3$ for a N-type drift region width of 1.5 μm. The length of the trench region ($L_D$) required to achieve a breakdown voltage of 200 V is 10 μm.

### 7.3.1 Channel Resistance

The contribution to the specific on-resistance from the channel in the SJ-MOSFET structure is smaller than in the power D-MOSFET structure due to the shorter channel length and cell pitch. Based up on the analysis in the textbook [13] for the power D-MOSFET structure, the specific on-resistance contributed by the channel in the power SJ-MOSFET structure is given by:

$$R_{CH,SP} = \frac{L_{CH} W_{Cell}}{2\mu_{ni} C_{OX}(V_G - V_{TH})} \tag{7.18}$$

In the case of the power MOSFET structures, it is customary to provide the on-resistance at a gate bias of 4.5 and 10 V. Assuming a gate oxide thickness is 500 Å, an inversion layer mobility of 450 cm$^2$/V s (to match the mobility used in the numerical simulations discussed later in this section), and a threshold voltage of 2.4 V (to match the numerical simulations) in the above equation for the power SJ-MOSFET design with a cell width of 3.0 μm, the specific resistance contributed by the channel at a gate bias of 4.5 V is found to be 0.0884 mΩ cm$^2$. The specific on-resistance of the power SJ-MOSFET structure is reduced to 0.0244 mΩ cm$^2$ when the gate bias is increased to 10 V. These values are an order of magnitude smaller than for the power D-MOSFET due to the combination of a very short channel length and a small cell pitch.

### 7.3.2 Accumulation Resistance for Current Spreading Region

In the power SJ-MOSFET structure, the current flowing through the inversion channel enters the N-type drift region at the edge of the P/N junction between the P-type and N-type drift regions. The current then spreads from the P/N junction into the N-type drift region. The current flow is strongly aided by the formation of an accumulation layer in the semiconductor at the surface of the N-type drift region due to the positive gate bias applied to turn-on the device. The accumulation resistance for this portion is given by the same formulation derived for the power D-MOSFET structure in the textbook [13]:

$$R_{A,SP} = K_A \frac{W_N W_{Cell}}{4\mu_{nA} C_{OX}(V_G - V_{THA})} \tag{7.19}$$

The gate oxide thickness must be used for computation of the oxide capacitance in this case. The threshold voltage ($V_{THA}$) in the expression is for the onset of formation of the accumulation layer. A zero threshold voltage will be assumed here when performing the analytical computations. The current spreading coefficient ($K_A$) will be assumed to be 0.6 as in the case of the D-MOSFET structure.

For the 200-V power SJ-MOSFET design with a cell width of 3 μm and N-drift region width of 1.5 μm, the specific resistance contributed by the accumulation layer in the current spreading portion at a gate bias of 4.5 V is 0.022 mΩ cm$^2$ for a gate oxide thickness of 500 Å. An accumulation layer mobility of 1,000 cm$^2$/V s was used in this calculation to match the mobility used in the numerical simulations (discussed later in this section). When the gate bias is increased to 10 V, the specific resistance contributed by the accumulation layer in the current spreading portion is reduced to 0.010 mΩ cm$^2$.

### 7.3.3 Drift Region Resistance

The resistance contributed by the drift region in the power SJ-MOSFET structure is reduced well below that for the ideal drift region due to the high doping concentration in the drift region. The specific on-resistance contributed by the drift region in the power SJ-MOSFET structure can be computed by analysis of the two resistances shown in Fig. 7.30. The resistance of the N-drift region ($R_{D1}$) is given by:

$$R_{D1,SP} = \frac{L_D W_{Cell}}{q\mu_n N_D (W_N - 2W_{DN})} \tag{7.20}$$

where $W_{DN}$ is the width of the depletion region at the vertical junction $J_2$. Due to the relatively small width of the N-drift region, it is important to include the impact of the depletion region at junction $J_2$ during the analysis [12]. Since the doping concentration of the P-type and N-type drift regions are comparable in the power SJ-MOSFET structure, the voltage supported by the junction is shared by depletion regions across both sides of the junction. The voltage supported by the junction is equal to the sum of the built-in potential and the on-state voltage drop of the device. For the analysis in this section, it will be assumed that this voltage is 1.5 V. The depletion region width in the N-type drift region is then given by:

$$W_{DN} = \sqrt{\frac{2\varepsilon_S(V_{bi} + V_{ON})}{q}\left[\frac{N_A}{N_D(N_A + N_D)}\right]} \tag{7.21}$$

In the case of most power SJ-MOSFET structures, the doping concentrations in the N-type and P-type drift regions can be assumed to be equal. In these cases:

$$W_{DN} = \sqrt{\frac{\varepsilon_S(V_{bi} + V_{ON})}{qN_D}} \tag{7.22}$$

The current density can be assumed to be uniform in the N-type drift region. However, the N-drift type region contribution is enhanced because part of the device area is occupied by the P-type drift region. When this is taken into account, the specific resistance of the drift region is given by:

$$R_{D1,SP} = \rho_D L_D \left(\frac{W_{Cell}}{W_N - 2W_{DN}}\right) \tag{7.23}$$

In the case of an N-type drift region with width of 1.5 μm, the optimum doping concentration is $1.98 \times 10^{16}/\text{cm}^3$ for a device with breakdown voltage of 200 V. The depletion region width for this doping concentration is found to be 0.2215 μm while the resistivity of the drift region for a doping concentration of $1.98 \times 10^{16}/\text{cm}^3$ is 0.271 Ω cm. Using this value of resistivity in (7.23) with the previously provided structural parameters yields a specific drift region resistance of 0.728 mΩ $\text{cm}^2$.

An additional resistance contribution ($R_{D2}$) in the power SJ-MOSFET structure is associated with the buffer layer located below the bottom of the trenches. A small buffer layer thickness is adequate in the power SJ-MOSFET structure because the depletion extends vertically in the P-type drift region because of the charge coupling phenomenon. Due to the relatively small thickness of the buffer layer, it will be assumed that the current does not spread in the buffer layer until it enters the $N^+$ substrate. Consequently, the specific resistance for the buffer layer is given by:

$$R_{D2,SP} = \rho_D L_B \left(\frac{W_{Cell}}{W_N - 2W_{DN}}\right) \tag{7.24}$$

For a typical buffer layer thickness of 1 μm, the specific resistance is found to be 0.0767 mΩ cm$^2$. The total specific resistance for the drift region in the power SJ-MOSFET structure is then found to be 0.8047 mΩ cm$^2$.

### 7.3.4 Total On-Resistance

The total specific on-resistance for the power SJ-MOSFET structure can be computed by combining the above components for the on-resistance. The total specific on-resistance for the power SJ-MOSFET structure is given by:

$$R_{T,SP} = R_{CH,SP} + R_{A,SP} + R_{D1,SP} + R_{D2,SP} \tag{7.25}$$

For the case of the 200-V power SJ-MOSFET design with a cell pitch ($W_{Cell}$) of 3 μm and N-drift region width of 1.5 μm, the total specific on-resistance is found to be 0.911 mΩ cm$^2$ at a gate bias of 4.5 V and 0.839 mΩ cm$^2$ at a gate bias of 10 V by using the analytical model. The contributions from each of the components of the on-resistance are summarized in Fig. 7.31. The specific on-resistance for the power SJ-MOSFET structure is found to be an order of magnitude smaller than that for the power D-MOSFET and the U-MOSFET structures at a gate bias of 4.5 V.

From the values provided in Fig. 7.31, it can be seen that the drift region resistance for the power SJ-MOSFET structure is far smaller than that for the conventional MOSFET structures. The very low specific resistance of the drift region is associated with the high doping concentration of the drift region enabled by the charge-coupling phenomenon. Due to this reduction in drift region resistance, the channel resistance contribution in the power SJ-MOSFET structure remains a significant contributor to the total specific on-resistance. In addition, it is important to reduce the resistance contributed by the $N^+$ substrate by wafer thinning technology to realize the benefits of the charge-coupling concept.

The ideal specific on-resistance for a drift region is given by:

$$R_{IDEAL,SP} = \frac{W_{PP}}{q\,\mu_n N_D} \tag{7.26}$$

| Resistance | $V_G$ = 4.5 V (mΩ-cm$^2$) | $V_G$ = 10 V (mΩ-cm$^2$) |
|---|---|---|
| Channel ($R_{CH,SP}$) | 0.0844 | 0.0241 |
| Accumulation ($R_{A,SP}$) | 0.0220 | 0.0100 |
| Drift ($R_{D1,SP}$) | 0.728 | 0.728 |
| Drift ($R_{D2,SP}$) | 0.077 | 0.077 |
| Total ($R_{T,SP}$) | 0.9111 | 0.8388 |

**Fig. 7.31** Resistance components in the 200-V power SJ-MOSFET structure

where $W_{PP}$ is the parallel-plane depletion width at breakdown, $N_D$ is the doping concentration of the drift region to sustain the blocking voltage, and $\mu_n$ is the mobility for electrons corresponding to this doping concentration. For the case of a blocking voltage of 200 V, the depletion width and doping concentration are found to be 14 μm and $1.35 \times 10^{15}/cm^3$, respectively. Using the electron mobility for this doping level, the ideal specific on-resistance is found to be 4.79 mΩ cm$^2$. Since the conventional device is constrained by the impact of an 80% reduction of breakdown voltage due to the edge termination, it is worth computing the ideal specific on-resistance for this case for comparison with the device. For the case of a blocking voltage of 250 V, the depletion width and doping concentration are found to be 18 μm and $1.0 \times 10^{15}/cm^3$, respectively. Using the electron mobility for this doping level, the ideal specific on-resistance is found to be 8.34 mΩ cm$^2$. The specific on-resistance for the 200-V SJ-MOSFET structure is about five to ten-times smaller than these ideal specific on-resistances for a gate bias of 10 V. Consequently, the charge-coupling concept provides the opportunity to overcome the ideal specific resistance barrier using silicon device structures.

#### 7.3.4.1 Simulation Results

The transfer characteristics for the 200-V SJ-MOSFET structure were obtained using numerical simulations with a drain bias of 0.1 V at 300 and 400°K. The resulting transfer characteristics are shown in Fig. 7.32. From this graph, a

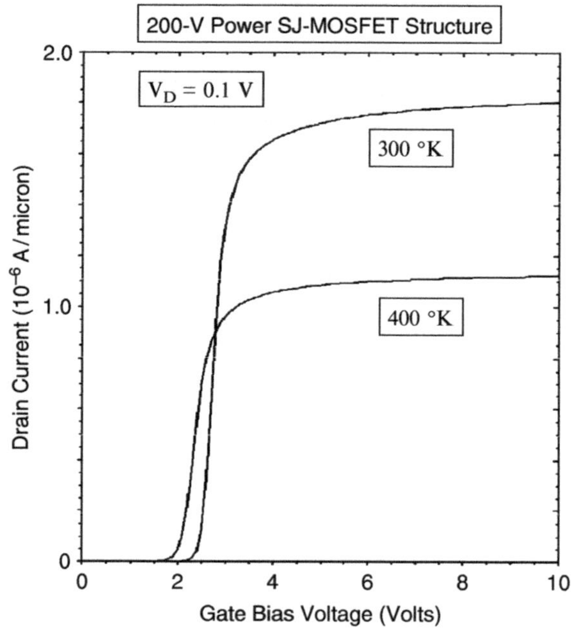


Fig. 7.32 Transfer characteristics of the 200-V SJ-MOSFET structure

threshold voltage of 2.3 and 2.0 V can be extracted at 300 and 400°K, respectively. The threshold voltage decreases by 15% when the temperature increases similar that observed for the power D-MOSFET structure. The specific on-resistance can be obtained from the transfer characteristics at any gate bias voltage. For the case of a gate bias of 4.5 V and 300°K, the specific in-resistance is found to be 0.882 mΩ cm$^2$. For the case of a gate bias of 10 V and 300°K, the specific in-resistance is found to be 0.833 mΩ cm$^2$. These values are in very close agreement with the results obtained from the analytical model providing validation for the model.

The on-state current flow pattern within the 200-V power SJ-MOSFET structure at a small drain bias of 0.1 V and a gate bias of 4.5 V is shown in Fig. 7.33. In the figure, the depletion layer boundary is shown by the dotted lines and the junction boundary is delineated by the dashed line. From the figure, it can be seen that the current spreads from the channel to the drift region very rapidly. The current distribution is uniform in the drift region. The area for current flow in the N-drift region is reduced by the presence of a depletion region of about 0.2 μm in width. This is consistent with the assumptions made in the analytical model. The current spreads within the buffer layer at a 45° angle. Although this was neglected in the analytical model, the impact is small because the buffer layer thickness is a small fraction of the drift region length.

The transfer characteristics for the 600-V SJ-MOSFET structure were also obtained using numerical simulations with a drain bias of 0.1 V at 300 and

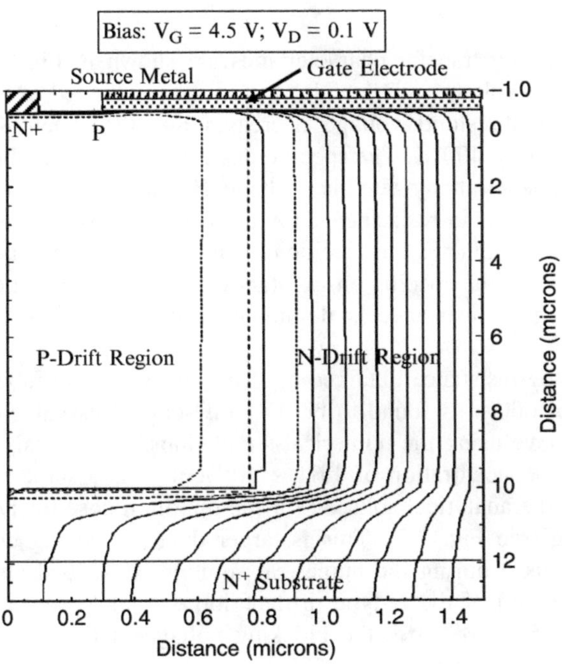


**Fig. 7.33** Current distribution in the 200-V SJ-MOSFET structure

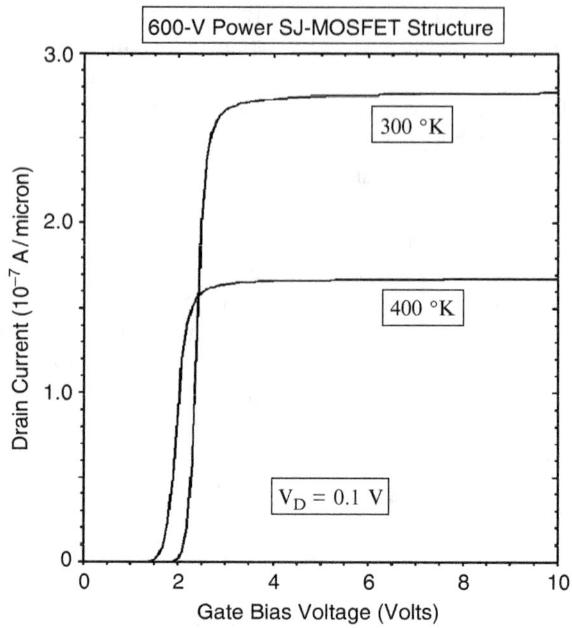


**Fig. 7.34** Transfer characteristics of the 600-V SJ-MOSFET structure

400°K. The resulting transfer characteristics are shown in Fig. 7.34. From this graph, a threshold voltage of 2.1 and 1.6 V can be extracted at 300 and 400°K, respectively. The threshold voltage decreases by 30% when the temperature increases from 300 to 400°K. The specific on-resistance can be obtained from the transfer characteristics at any gate bias voltage. For the case of a gate bias of 4.5 V and 300°K, the specific in-resistance is found to be 5.47 mΩ $cm^2$. For the case of a gate bias of 10 V and 300°K, the specific in-resistance is found to be 5.42 mΩ $cm^2$. The small change in the specific on-resistance with increasing gate bias indicates that the drift region resistance is dominant in the 600-V power SJ-MOSFET structure.

The specific on-resistance obtained by using the analytical model for a breakdown voltage of 600 V is found to be 3.44 mΩ $cm^2$. This value is smaller than that obtained above using the numerical simulations. In the analytical model, an optimum doping concentration of $1.64 \times 10^{16}/cm^3$ was assumed for the N-type drift region per the analytical solution for the optimum dose for two-dimensional charge coupling (see Fig. 7.5). This is larger than the doping concentration of $1.0 \times 10^{16}/cm^3$ used during the numerical solutions described above. When the doping concentration of the N-type drift region was increased in the numerical simulations to $1.5 \times 10^{16}/cm^3$, the blocking voltage of the power SJ-MOSFET structure was found to be reduced to 570 V (see Sect. 7.2).

## 7.4 Output Characteristics

The output characteristics of the power SJ-MOSFET structure are important to the loci for the switching waveforms when it is operating in power circuits. The saturated drain current for the power MOSFET structure is given by:

$$I_{D,sat} = \frac{Z\mu_{ni}C_{OX}}{(L_{CH} - \Delta L_{CH})}(V_G - V_{TH})^2 \quad (7.27)$$

where $\Delta L_{CH}$ is reduction in the channel length due to depletion of the P-base region with increasing drain bias voltage. With sufficiently high doping concentration of the P-base region, the modulation of channel length can be made sufficiently small to ensure a high output resistance. In the power SJ-MOSFET structure, the charge coupling phenomenon suppresses the depletion of the P-base region with increasing drain bias voltage because the voltage is spread into the drift region. This is beneficial for obtaining a reasonable output resistance despite the very short channel length in the device. The short channel length in the power SJ-MOSFET structure produces a high transconductance which is beneficial for reducing switching losses. The saturated drain current in the power SJ-MOSFET structure increases as the square of the gate bias voltage in this model.

### *7.4.1 Simulation Example*

The output characteristics of the 200-V power SJ-MOSFET structure were obtained by using two-dimensional numerical simulations using various gate bias voltages. All the device parameters used for these numerical simulations are the same as those used in the previous sections. The output characteristics of the power SJ-MOSFET obtained using the simulations are shown in Fig. 7.35. The output resistance degrades at larger drain bias voltages [14]. This behavior is related to the pre-breakdown avalanche multiplication current generated by the large uniform electric field along the y-direction in the N-type drift region.

## 7.5 Device Capacitances

For the power MOSFET structures, the switching speed is limited by the device capacitances in practical applications as previously discussed in Chap. 2. The rate at which the power MOSFET structure can be switched between the on- and off-states is determined by the rate at which the input capacitance can be charged or discharged. In addition, the capacitance between the drain and the gate electrodes has been found to play an important role in determining the drain current and voltage transitions during the switching event.

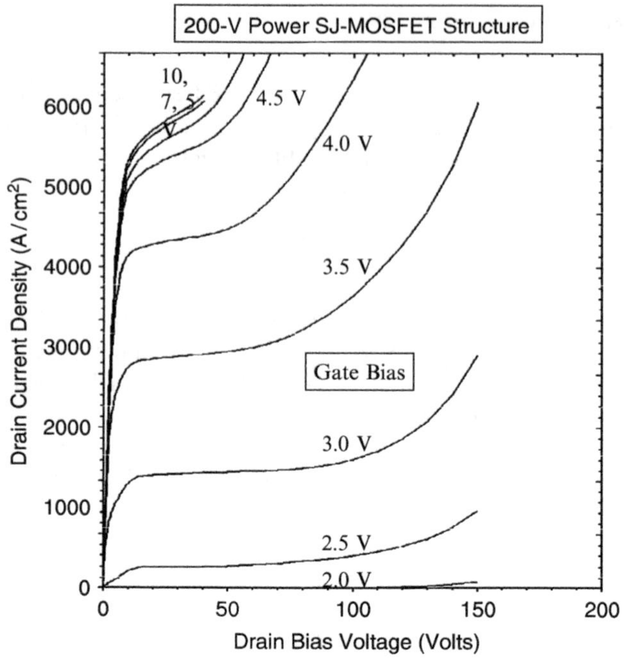


**Fig. 7.35** Output characteristics for the 200-V power SJ-MOSFET structure

The capacitances within the power SJ-MOSFET structure can be analyzed using the same approach used in the textbook [13] for the power D-MOSFET structure. The specific input (or gate) capacitance for the power SJ-MOSFET structure is given by:

$$C_{IN,SP} = C_{N+} + C_P + C_{SM} = \left(\frac{W_P - W_{PW}}{W_{Cell}}\right)\left(\frac{\varepsilon_{OX}}{t_{GOX}}\right) + \frac{W_G}{W_{Cell}}\left(\frac{\varepsilon_{OX}}{t_{IEOX}}\right) \quad (7.28)$$

where $t_{GOX}$ and $t_{IEOX}$ are the thicknesses of the gate and inter-electrode oxides, respectively. When deriving this expression, it has been assumed that the P-base diffusion does not extend beyond the P-drift region. For a 200-V power SJ-MOSFET structure with a cell pitch of 3 μm and polysilicon window width ($W_{PW}$) of 0.6 μm, the specific input capacitance is found to be 25.9 nF/cm$^2$ for a gate oxide thickness of 500 Å and an inter-metal dielectric thickness of 5,000 Å. However, the simulations indicate that the entire gate is coupled to the source region after the formation of the inversion and accumulations layers. In this case, specific input (or gate) capacitance for the power SJ-MOSFET structure is given by:

$$C_{IN,SP} = \left(\frac{W_G}{W_{Cell}}\right)\left(\frac{\varepsilon_{OX}}{t_{GOX}}\right) + \frac{W_G}{W_{Cell}}\left(\frac{\varepsilon_{OX}}{t_{IEOX}}\right) \quad (7.29)$$

The input capacitance obtained by using this equation is 60 nF/cm$^2$ which is close to the value obtained from the numerical simulations with zero drain bias. This value is

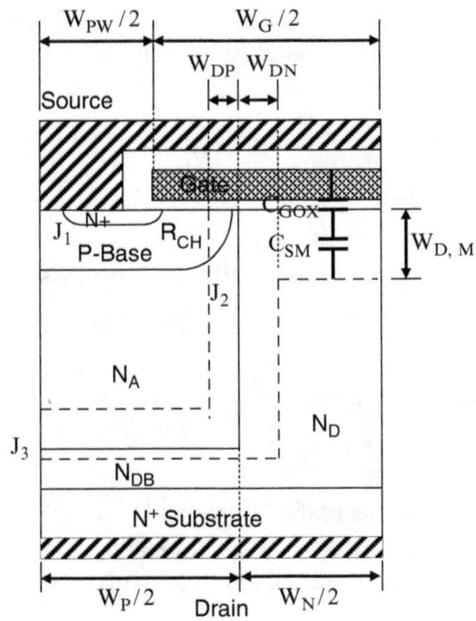


**Fig. 7.36** Depletion boundaries and capacitances within the power SJ-MOSFET structure

larger than 22 nF/cm$^2$ for the power D-MOSFET structure and 47 nF/cm$^2$ for the power U-MOSFET structure.

The capacitance between the gate and drain electrodes (also called the reverse transfer capacitance) for the power SJ-MOSFET structure can be analyzed by examining the extension of the depletion region in the N-drift region with increasing drain bias voltage. The MOS structure over the N-drift region in the power SJ-MOSFET structure operates under deep depletion conditions when a positive voltage is applied to the drain. The depletion region boundaries for this structure are shown in Fig. 7.36 by the dashed lines together with the oxide and semiconductor capacitances needed for the analysis of the gate transfer capacitance. The area under the gate electrode that determines the gate transfer capacitance decreases with increasing drain bias voltage due to the expansion of the depletion region across the vertical junction $J_2$. The gate-drain capacitance for the power SJ-MOSFET structure is given by:

$$C_{GD,SP} = \left(\frac{W_N - 2W_{DN}}{W_{Cell}}\right)\left(\frac{C_{GOX}C_{SM}}{C_{GOX} + C_{SM}}\right) \tag{7.30}$$

where $W_{DN}$ is the depletion width at junction $J_2$ and $C_{SM}$ is the specific capacitance of the semiconductor under the MOS gate electrode.

When computing the semiconductor capacitance, it is not appropriate to use the one-dimensional MOS depletion width because the two-dimensional charge coupling produces a larger extension of the depletion region in the semiconductor.

Since the depletion under the gate electrode occurs with a constant electric field along the y-direction with a value equal to the critical electric field for breakdown for the uniform electric field case:

$$W_{D,M} = \frac{V_D}{E_{CU}} \tag{7.31}$$

The depletion width ($W_{DN}$) in the N-drift region across the vertical junction $J_2$ is given by:

$$W_{DN} = \sqrt{\frac{2\varepsilon_S(V_D+V_{bi})}{q}\left[\frac{N_A}{N_D(N_A+N_D)}\right]} \tag{7.32}$$

When the doping concentrations of the N-type and P-type drift regions are equal the drain voltage is shared equally across the junction and the depletion width ($W_{DN}$) in the N-drift region across the vertical junction $J_2$ is given by:

$$W_{DN} = \sqrt{\frac{\varepsilon_S(V_D+V_{bi})}{qN_D}} \tag{7.33}$$

The depletion region width at the vertical junction $J_2$ becomes equal to the half the width of the N-type drift region at the drain pinch-off voltage ($V_{D,N}$). At this voltage, the effective area for the gate overlap with the drain becomes equal to zero resulting in reduction of the gate transfer capacitance to zero. The drain bias at which the gate transfer capacitance becomes equal to zero can be derived by making the depletion width given by (7.33) equal to half the width of the N-type drift region:

$$V_{D,N} = \frac{qN_D W_N^2}{4\,\varepsilon_S} - V_{bi} \tag{7.34}$$

In the case of the 200-V power SJ-MOSFET structure with an optimum doping concentration of $2 \times 10^{16}/\text{cm}^3$ in the N-type drift region with a width of 1.5 μm, the gate transfer capacitance becomes zero at a drain bias of 16.56 V according to this analytical expression.

The specific gate transfer capacitance obtained by using the above analytical formulae is shown in Fig. 7.37 for the case of the 200-V power SJ-MOSFET structure. A critical electric field of $2.29 \times 10^5$ V/cm for breakdown with uniform electric field distribution for the 200-V case was used here. This structure has a cell pitch of 3 μm, N-type drift region width of 1.5 μm, and an optimum doping concentration of $2 \times 10^{16}/\text{cm}^3$ in the N-type drift region. Starting from a specific

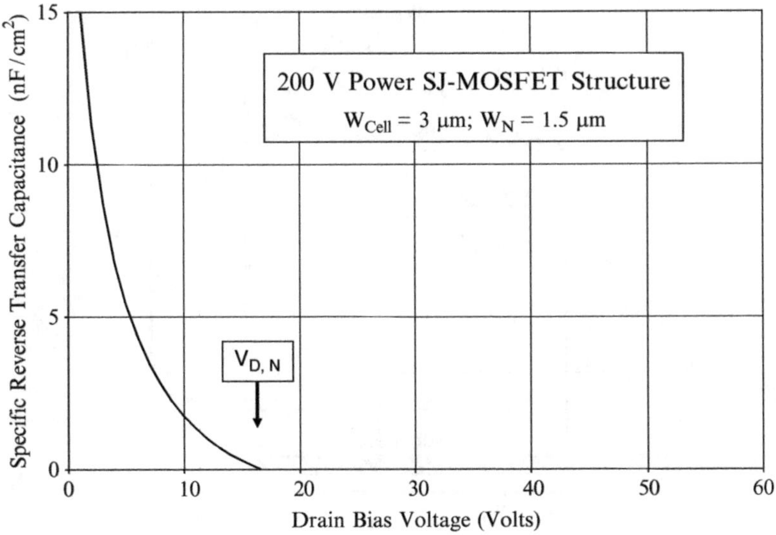


**Fig. 7.37** Gate-drain capacitance for the 200-V power SJ-MOSFET structure

capacitance of 15 nF/cm$^2$ at a gate bias of 1 V, the gate-drain (reverse transfer) capacitance decreases gradually with increasing drain bias voltage until it becomes equal to zero at a drain bias of 16.6 V (the voltage required to completely deplete the N-type drift region). For comparison with the results of numerical simulations, the specific capacitance at a drain bias of 10 V is 1.76 nF/cm$^2$ according to the analytical model.

The output capacitance for the power SJ-MOSFET structure is associated with the capacitance of the vertical junction $J_2$ between the P-type and N-type drift regions and the horizontal junction $J_3$ at the bottom of the P-type drift region. The capacitance at junction $J_2$ decreases with increasing drain bias voltage due to the depletion region extending in the x-direction across the vertical sidewalls of the trenches. At the same time, the area of the vertical junction reduces with increasing drain bias voltage due to the expansion of the depletion layer under the MOS gate structure. In the case of the horizontal junction $J_3$, the capacitance reduces with increasing drain bias voltage due to the upward expansion of the depletion region in the P-type drift region. At the same time, the area of the horizontal junction reduces due to the lateral extension of the depletion region in the P-type drift region.

The depletion layer boundary inside the power SJ-MOSFET structure prior to the complete depletion of the drift regions is shown in Fig. 7.38 by the dashed lines. The specific output capacitance for the power SJ-MOSFET structure is given by:

$$C_{O,SP} = C_{J2,SP} + C_{J3,SP} \tag{7.35}$$

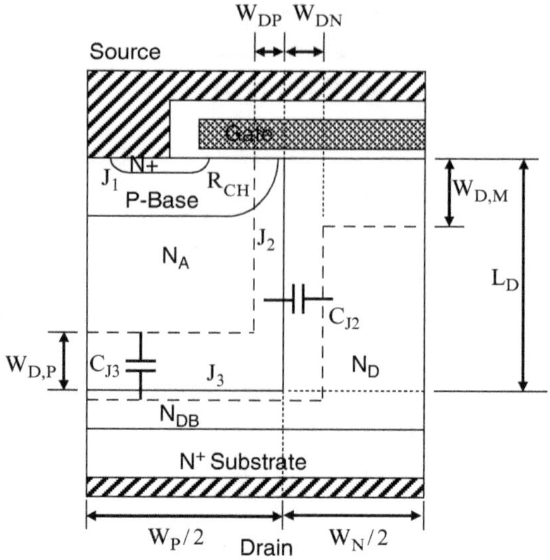


**Fig. 7.38** Depletion region boundaries for determination of the output capacitance for the power SJ-MOSFET structure

The specific output capacitance associated with the vertical junction $J_2$ is given by:

$$C_{J2,SP} = \frac{2\varepsilon_S(L_D - W_{D,M})}{(W_{DP} + W_{DN})W_{Cell}} \tag{7.36}$$

When the doping concentration in the P-type and N-type drift regions are equal, the specific output capacitance associated with the vertical junction $J_2$ is given by:

$$C_{J2,SP} = \frac{\varepsilon_S(L_D - W_{D,M})}{W_{DN}W_{Cell}} \tag{7.37}$$

The specific output capacitance associated with the horizontal junction $J_3$ is given by:

$$C_{J3,SP} = \frac{\varepsilon_S(W_P - 2W_{DP})}{W_{D,P}W_{Cell}} \tag{7.38}$$

In writing this expression, the depletion width in the N-buffer layer was neglected because the depletion width in the P-type drift region is much larger in size.

Since the depletion under the gate electrode occurs with a constant electric field along the y-direction with a value equal to the critical electric field for breakdown for the uniform electric field case:

$$W_{D,M} = \frac{V_D}{E_{CU}} \tag{7.39}$$

Similarly, the upward extension of the depletion region in the P-type drift region also occurs with a constant electric field along the y-direction with a value equal to the critical electric field for breakdown for the uniform electric field case:

$$W_{D,P} = \frac{V_D}{E_{CU}} \tag{7.40}$$

The depletion width ($W_{DN}$) in the N-drift region across the vertical junction $J_2$ is given by:

$$W_{DN} = \sqrt{\frac{2\varepsilon_S(V_D + V_{bi})}{q}\left[\frac{N_A}{N_D(N_A + N_D)}\right]} \tag{7.41}$$

When the doping concentrations of the N-type and P-type drift regions are equal the drain voltage is shared equally across the junction and the depletion width ($W_{DN}$) in the N-drift region across the vertical junction $J_2$ is given by:

$$W_{DN} = \sqrt{\frac{\varepsilon_S(V_D + V_{bi})}{qN_D}} \tag{7.42}$$

Similarly, the depletion width ($W_{DP}$) in the P-drift region across the vertical junction $J_2$ is given by:

$$W_{DP} = \sqrt{\frac{2\varepsilon_S(V_D + V_{bi})}{q}\left[\frac{N_D}{N_A(N_A + N_D)}\right]} \tag{7.43}$$

When the doping concentrations of the N-type and P-type drift regions are equal the drain voltage is shared equally across the junction and the depletion width ($W_{DP}$) in the N-drift region across the vertical junction $J_2$ is given by:

$$W_{DP} = \sqrt{\frac{\varepsilon_S(V_D + V_{bi})}{qN_D}} \tag{7.44}$$

The above analysis is valid until the depletion region at the vertical junction $J_2$ extends through the entire N-type and P-type drift regions at a drain bias given by (7.34). At larger drain bias voltages, the depletion region extends across the entire cross section of the cell structure. Since the depletion region extends for the entire drift region length ($L_D$), the specific output capacitance then becomes:

$$C_{O,SP} = \frac{\varepsilon_S}{L_D} \tag{7.45}$$

According to this analytical model, there is an abrupt reduction in the output capacitance in the power SJ-MOSFET structure when the drain voltage exceeds the voltage required to deplete the drift regions.

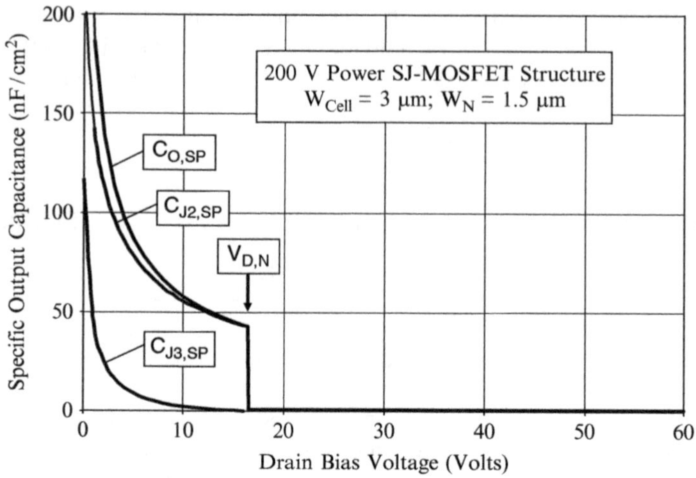


**Fig. 7.39** Output capacitance for the 200-V power SJ-MOSFET structure

The specific output capacitance obtained by using the above analytical model is shown in Fig. 7.39 for the case of the 200-V power SJ-MOSFET structure. This structure has a cell pitch of 3 μm, N-type drift region width of 1.5 μm, and a doping concentration of $2 \times 10^{16}/cm^3$ for the N-type and P-type drift regions. A built-in potential of 0.8 V was assumed for the vertical junction. The specific capacitances associated with the vertical junction $J_2$ and horizontal junction $J_3$ are included in the figure for comparison. It can be observed that the specific output capacitance decreases with increasing drain bias voltage due to expansion of the depletion regions across the junctions. Its value abruptly drops close to zero at a drain bias of 16.5 V. The specific output capacitance at a drain bias of 16.5 V is about 1 nF/cm$^2$ according to (7.45). At a small drain bias of 5 V, the specific output capacitance predicted by the analytical model for this structure is 86 nF/cm$^2$.

## 7.5.1 *Simulation Example*

The capacitances of the 200-V power SJ-MOSFET structure were extracted using two-dimensional numerical simulations with a small AC signal superposed on the DC gate bias voltage. The input capacitances obtained for the 200-V power SJ-MOSFET structure are shown in Fig. 7.40 at a drain bias of 20 V. It is comprised of two components – the first is between the gate electrode and the source electrode ($C_{GS}$) while the second is between the gate electrode and the base electrode ($C_{GB}$). The total input capacitance can be obtained by the addition of

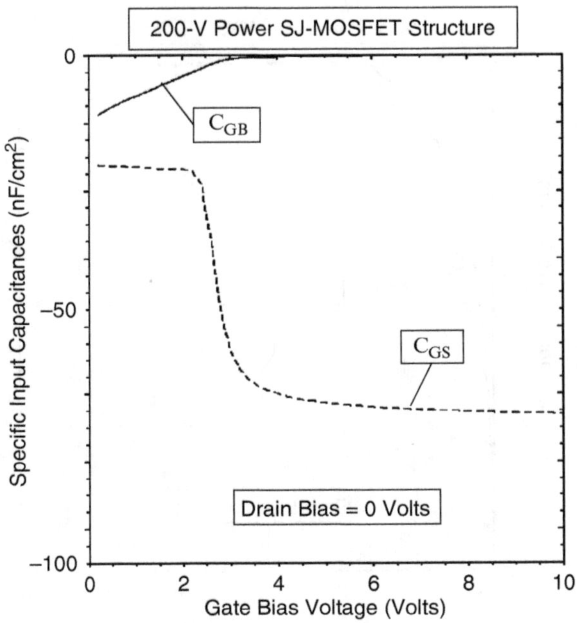


**Fig. 7.40** Input capacitances for the 200-V SJ-MOSFET structure

these capacitances because they are in parallel and share a common contact electrode in the actual power SJ-MOSFET structure. From the figure, a total specific input capacitance of about 65 nF/cm$^2$ is observed for gate bias voltages raging from 2 to 10 V – close to that predicted by the analytical model using the entire width of the gate electrode.

The drain-gate (reverse transfer) capacitance was extracted by performing the numerical simulations with a small AC signal superposed on the DC drain bias voltage. The values obtained for the 200-V power SJ-MOSFET structure are shown in Fig. 7.41. The gate-to-drain and base-to-drain capacitances are shown in the figure for comparison. The gate-to-drain capacitance is much smaller in magnitude than the base-to-drain capacitance. The gate transfer capacitance decreases monotonically until it becomes zero at a drain bias of 16.5 V. This behavior and the numerical values obtained from the simulations are in excellent agreement with the values predicted by the analytical model (see Fig. 7.37) providing validation for the analytical model. The reduction of the gate transfer capacitance to zero implies an abrupt change in drain voltage during the switching of the device as discussed in the next section.

The output capacitance for the 200-V power SJ-MOSFET structure obtained from the numerical simulations is provided in Fig. 7.42. It can be observed that the output capacitance is about 180 nF/cm$^2$ at a drain bias of 1 V and reduces to about 40 nF/cm$^2$ at a drain bias of 16 V. This behavior is in excellent agreement with the

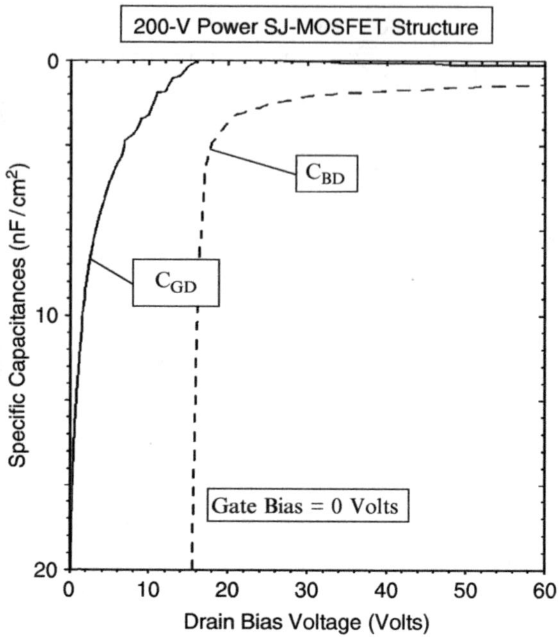


**Fig. 7.41** Reverse transfer and output capacitances for the 200-V SJ-MOSFET structure

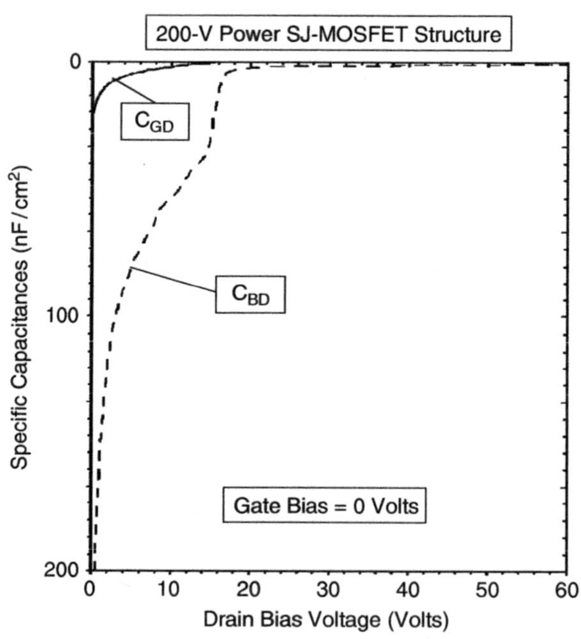


**Fig. 7.42** Reverse transfer and output capacitances for the 200-V SJ-MOSFET structure

values computed by using the analytical model (see Fig. 7.39). The output capacitance then decreases abruptly to almost zero at a drain bias of 16.5 V. This behavior is also predicted very well by the analytical model (see Fig. 7.39). These results demonstrate the importance of taking into account the change in the effective area of the junctions that contribute to the output capacitance in the power SJ-MOSFET structure.

## 7.6 Gate Charge

The most significant gate charge components for assessing the performance of the power MOSFET structures are $Q_{SW}$ (the gate switching charge), $Q_{GD}$ (the gate-drain charge), and $Q_G$ (the total gate charge). In the case of the power SJ-MOSFET structure, these components are given by the same equations derived in the textbook with adjustments made for the unique form for the change in capacitances in this structure with increasing drain bias voltage. The expressions for the gate charge $Q_{GS1}$ and $Q_{GS2}$ are the same as those provided in the textbook. The gate-to-drain capacitance in these equations is zero at the drain supply voltage. For the analysis of the gate transfer charge, the gate transfer capacitance must be modeled by taking into account its zero value at drain bias voltages above the depletion voltage for the N-type drift region. The drain voltage transition in the power SJ-MOSFET structure occurs in two steps – there is initially an abrupt reduction of the drain voltage from the supply voltage to the depletion voltage followed by a gradual reduction of the drain voltage. The gate transfer charge is zero during the abrupt change in the drain voltage.

The gate transfer charge corresponding to the second part of the transition where the drain voltage changes from $V_{D,M}$ to $V_{ON}$ is defined by the solution of the following equation:

$$\frac{dv_D(t)}{dt} = -\frac{J_G}{C_{GD}(v_D)} \tag{7.46}$$

with the variation of the gate transfer capacitance given by (7.30). In (7.30), the semiconductor capacitance $C_{SM}$ and the depletion width $W_{DN}$ are functions of the drain bias voltage. The gate charge is determined largely by the changes in the gate transfer capacitance at low drain bias voltages where its value is large. For simplicity of analysis, it will be assumed that the screening effect can be ignored once the drain voltage becomes less than $V_{D,N}$. Under this assumption, the gate transfer capacitance for the power SJ-MOSFET structure is given by:

$$C_{GD} = \left(\frac{W_N}{W_{Cell}}\right)\left(\frac{\varepsilon_S E_{CU} C_{GOX}}{\varepsilon_S E_{CU} + C_{GOX} v_D(t)}\right) \tag{7.47}$$

Using this expression in (7.46) and performing the integration for the time interval when the drain voltage changes from $V_{D,M}$ to $V_{ON}$ yields an expression for the gate transfer charge:

$$Q_{GD} = \frac{\varepsilon_S E_{CU} W_N}{W_{Cell}} \ln\left(\frac{\varepsilon_S E_{CU} + C_{GOX} V_{D,N}}{\varepsilon_S E_{CU} + C_{GOX} V_{ON}}\right) \tag{7.48}$$

The other gate charge components are given by:

$$Q_{SW} = Q_{GS2} + Q_{GD} \tag{7.49}$$

$$Q_G = C_{GS} V_{GP} + Q_{GD} + [C_{GS} + C_{GD}(V_{ON})](V_G - V_{GP}) \tag{7.50}$$

The gate charge values obtained for the 200-V power SJ-MOSFET structure by using the above analytical equations are: $Q_{GD} = 200$ nC/cm$^2$; $Q_{SW} = 208$ nC/cm$^2$; and $Q_G = 761$ nC/cm$^2$. The gate transfer charge for the power SJ-MOSFET structure is small when compared with the power D-MOSFET and U-MOSFET structures discussed in the previous chapters.

Equations for the gate voltage, drain current, and drain voltage waveforms obtained by using the analytical model are provided in the textbook [13] for the power D-MOSFET structure. The drain voltage waveform for the power SJ-MOSFET structure must be reformulated because of the more complex nature of the gate transfer capacitance. The drain voltage waveform can be derived by integration of (7.46) with the gate transfer capacitance given by (7.47):

$$v_D(t) = \left(\frac{\varepsilon_S E_{CU}}{C_{GOX}} + V_{D,N}\right) e^{-\frac{W_{Cell} J_G}{W_N \varepsilon_S E_{CU}} t} - \frac{\varepsilon_S E_{CU}}{C_{GOX}} \tag{7.51}$$

from $t = t_2$ to $t = t_3$. The drain voltage decreases exponentially from $V_{D,N}$ to $V_{ON}$ according to the analytical solution. This behavior is quite different from that observed for the power D-MOSFET and U-MOSFET structures.

The waveforms obtained for the 200-V power SJ-MOSFET structure using 3-μm cell pitch and an N-type drift region width of 1.5 μm with a gate oxide thickness of 500 Å using the above equations are provided in Fig. 7.43. A gate drive current density of 0.667 A/cm$^2$ was used to turn on the device from a steady-state blocking voltage of 150 V to match the results of two dimensional numerical simulations discussed below.

The gate voltage initially increases linearly with time. After reaching the threshold voltage, the drain current can be observed to increase very quickly because of the large transconductance for this device structure. The drain current density increases until it reaches an on-state current density of 235 A/cm$^2$. The on-state current density determines the gate plateau voltage which has a value of 1.91 V for a threshold voltage of 1.7 V at this drain bias. During the gate voltage plateau phase, the drain voltage decreases abruptly from the drain supply voltage to $V_{D,N}$ followed by an exponentially

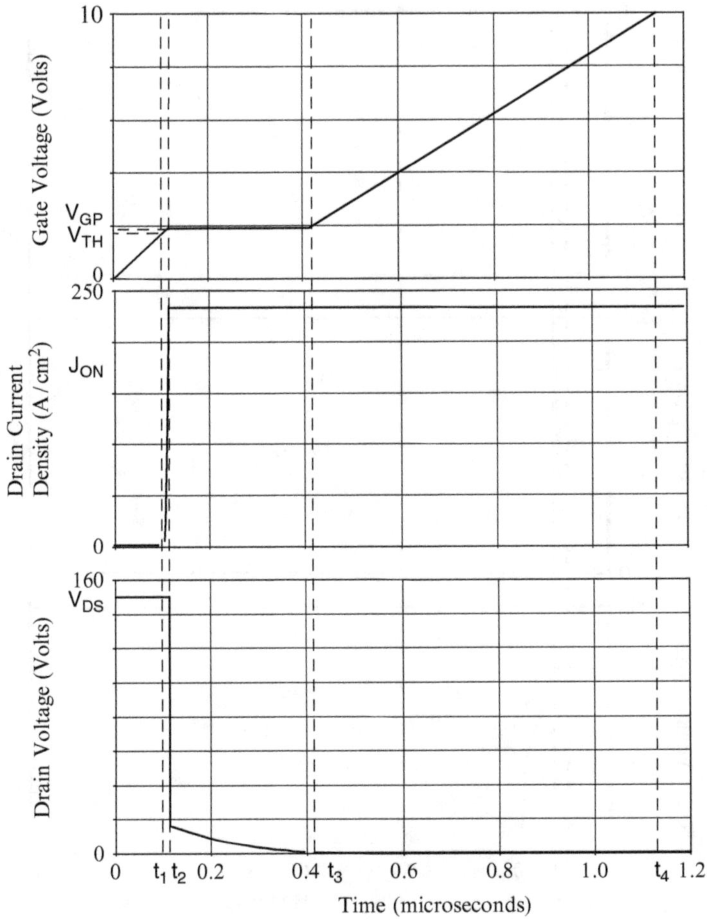


**Fig. 7.43** Analytically computed waveforms for the 200-V power SJ-MOSFET structure

decay to $V_{ON}$. After this time, the gate voltage again increases but at a slower rate than during the initial turn-on phase due to the larger gate transfer capacitance.

## *7.6.1 Simulation Example*

The gate charges for the 200-V power SJ-MOSFET structures were extracted by using the results of two-dimensional numerical simulations of the cell structure with a cell pitch of 3 μm and N-type drift region width of 1.5 μm described in the previous sections. The device was turned-on from blocking state with a drain bias of 150 V by using a gate current of $1 \times 10^{-8}$ A/μm (equivalent to 0.667 A/cm$^2$ for the area of $3.0 \times 10^{-8}$ cm$^2$). Once the drain current density reached the on-state value,

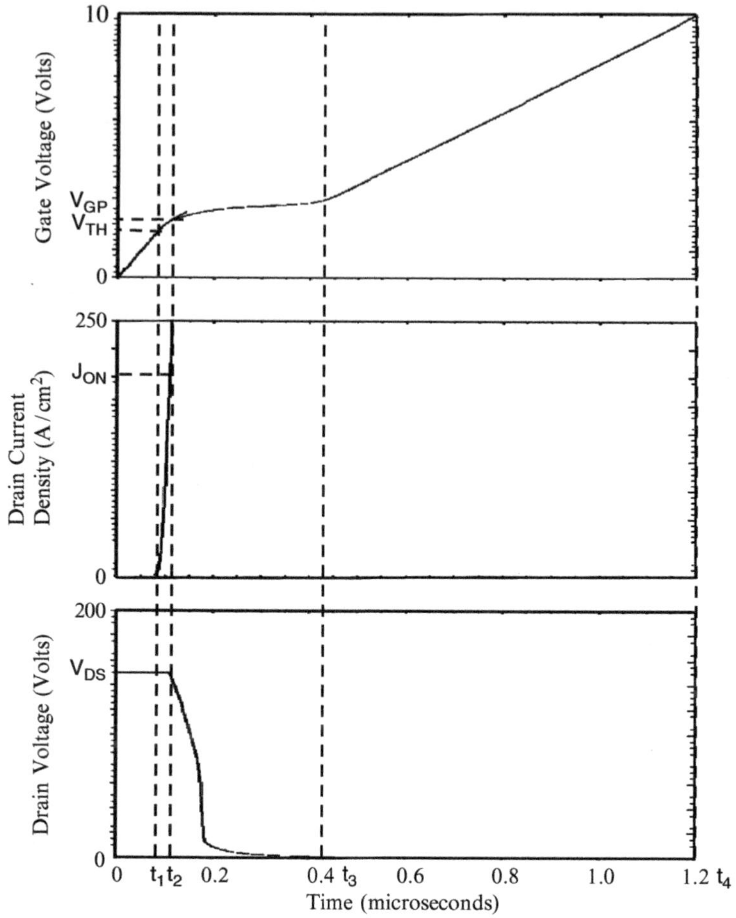


**Fig. 7.44** Turn-on waveforms for the 200-V power SJ-MOSFET structure

the drain current was held constant resulting in a reduction of the drain voltage. Once the drain voltage reached the on-state value, the gate voltage increased to the steady-state value of 10 V.

The gate charge waveforms obtained by using an input gate current density of 0.667 A/cm$^2$ when turning on the power SJ-MOSFET structure from a blocking state with drain bias of 150 V are shown in Fig. 7.44. The on-state current density in this case is 235 A/cm$^2$ at a DC gate bias of 10 V at the end of the turn-on transient. The gate voltage increases at a constant rate at the beginning of the turn-on process as predicted by the analytical model. When the gate voltage reaches the threshold voltage, the drain current begins to increase very rapidly until it reaches the on-state current density of 235 A/cm$^2$.

Once the drain current reaches the on-state value, the gate voltage remains approximately constant at the plateau voltage ($V_{GP}$). The plateau voltage for this

| Specific Gate Charge | Numerical Simulation ($nC/cm^2$) | Analytical Model ($nC/cm^2$) |
|---|---|---|
| $Q_{GS1}$ | 53 | 68 |
| $Q_{GS2}$ | 20 | 8.2 |
| $Q_{GS}$ | 73 | 76.2 |
| $Q_{GD}$ | 207 | 200 |
| $Q_{SW}$ | 227 | 208 |
| $Q_G$ | 800 | 761 |

**Fig. 7.45** Gate charge extracted from numerical simulations for the power SJ-MOSFET structure

structure is 1.9 V for the drain current density of 235 $A/cm^2$ as governed by the transconductance of the device. Unlike the analytical model, the drain voltage does not decrease abruptly at the beginning of the gate plateau phase. This is due to the poor output resistance (see Fig. 7.35) of the power SJ-MOSFET structure. The low output resistance produces an increase in the gate voltage to sustain the same drain current level as can be observed in Fig. 7.44. The drain voltage (7.46) was derived under the assumption that the gate voltage is constant during the plateau phase. This assumption does not hold true for the power SJ-MOSFET structure resulting in a slower reduction of the drain voltage from $V_{DS}$ to $V_{D,N}$. After the drain voltage reaches $V_{D,N}$, the drain voltage decreases exponentially with time during the rest of the gate plateau phase as predicted by the analytical model. After the end of the plateau phase, the gate voltage again increases until it reaches the gate supply voltage.

The values for the various components of the gate charge extracted from the numerical simulations for the 200-V power SJ-MOSFET structure are compared with those calculated by using the analytical model in Fig. 7.45. There is good agreement between these values indicating that the analytical model can be used as a good representation of the physics of turn-on for this device structure. Due to the small values for the gate charge, it can be concluded that the power SJ-MOSFET structure is suitable for high frequency applications.

## 7.7 Device Figures of Merit

Significant power switching losses can arise from the charging and discharging of the large input capacitance in power MOSFET devices at high frequencies. The input capacitance ($C_{IN}$) of the power MOSFET structure must be charged to the gate supply voltage ($V_{GS}$) when turning on the device and then discharged to zero volts when turning off the device during each period of the operating cycle. The total power loss can be obtained by summing the on-state power dissipation for a duty cycle $\delta = t_{ON}/T$ and the switching power losses:

$$P_T = P_{ON} + P_{SW} = \delta R_{ON} I_{ON}^2 + C_{IN} V_{GS}^2 f \tag{7.52}$$

where $R_{ON}$ is the on-resistance of the power MOSFET structure, $I_{ON}$ is the on-state current, and f is the operating frequency. In writing this equation, the switching power losses due to the drain current and voltage transitions has been neglected. A minimum total power loss occurs for each power MOSFET structure at an optimum active area as shown in the textbook [13]. The on-state and switching power losses are equal at the optimum active area. The optimum active area at which the power dissipation is minimized is given by:

$$A_{OPT} = \sqrt{\frac{R_{ON,sp}}{C_{IN,sp}}}\left(\frac{I_{ON}}{V_{GS}}\right)\left(\sqrt{\frac{\delta}{f}}\right) \tag{7.53}$$

From the first term in this expression, a useful technology figure-of-merit can be defined:

$$FOM(A) = \frac{R_{ON,sp}}{C_{IN,sp}} \tag{7.54}$$

In the power electronics community, there is trend towards increasing the operating frequency for switch mode power supplies in order to reduce the size and weight of the magnetic components. The ability to migrate to higher operating frequencies in power conversion circuits is dependent on making enhancements to the power MOSFET technology. From the above equations, an expression for the minimum total power dissipation can be obtained [13]:

$$P_T(\min) = 2 I_{ON} V_{GS} \sqrt{\delta R_{ON,sp} C_{IN,sp} f} \tag{7.55}$$

A second technology figure of merit related to the minimum power dissipation can be defined as:

$$FOM(B) = R_{ON,sp} C_{IN,sp} \tag{7.56}$$

In most applications for power MOSFET structures with high operating frequency, the switching losses associated with the drain current and voltage transitions become a dominant portion of the total power loss. The time period associated with the increase of the drain current and decrease of the drain voltage is determined by the charging of the device capacitances. It is therefore common practice

| Figures of Merit | $V_G$ = 4.5 V | $V_G$ = 10 V |
|---|---|---|
| FOM(A) ($\Omega^2cm^4s^{-1}$) | 35147 | 32361 |
| FOM(B)(ps) | 23.6 | 21.7 |
| FOM(C) (mΩ*nC) | 182 | 168 |
| FOM(D) (mΩ*nC) | 190 | 174 |

**Fig. 7.46** Figures of merit for the 200-V power SJ-MOSFET structure

in the industry to use the following figures-of-merit to compare the performance of power MOSFET products [13]:

$$FOM(C) = R_{ON,sp}Q_{GD,sp} \tag{7.57}$$

and

$$FOM(D) = R_{ON,sp}Q_{SW,sp} \tag{7.58}$$

Although FOM(D) encompasses both the drain current and voltage transitions, it is customary to use FOM(C) because the gate-drain charge tends to dominate in the switching gate charge. One advantage of using these expressions is that the figure-of-merit becomes independent of the active area of the power MOSFET device.

The figures of merit computed for the 200-V power SJ-MOSFET structure are provided in Fig. 7.46. The figure of merit usually used for comparison of device technologies in the literature is FOM(C). Most often, the value for this figure of merit at a gate bias of 4.5 V is utilized for selection of devices in the voltage regulator module application.

In comparison with the power D-MOSFET structure (see Chap. 2), the power SJ-MOSFET structure has a FOM(C) that is ten-times smaller. In comparison with the conventional power U-MOSFET structure (see Chap. 3), the power SJ-MOSFET structure has a FOM(C) that is five times smaller. In comparison with the power SC-MOSFET structure (see Chap. 4), the power SJ-MOSFET structure has a FOM (C) that is three times smaller. Consequently, the power SJ-MOSFET structure offers significant improvement in circuit performance when compared with these structures. However, in comparison with the power GD-MOSFET structure (see Chap. 6), the power SJ-MOSFET structure has a FOM(C) that is 2.5-times larger.

## 7.8 Edge Termination

As in the case of the power GD-MOSFET structure which operates with a breakdown voltage larger than the parallel-plane breakdown voltage of its drift region, a critical challenge for the development of the power SJ-MOSFET structure is the

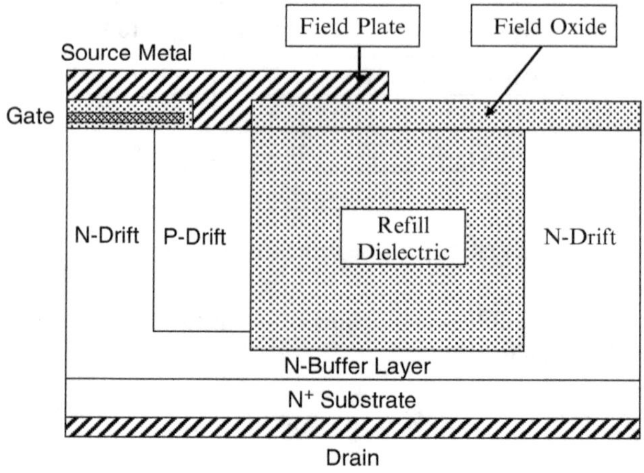


**Fig. 7.47** Edge termination for the SJ-MOSFET structure

formulation of an edge termination that can also support the blocking voltage. Fortunately, an elegant edge termination has been proposed and demonstrated [15, 16] that can be implemented in a manner compatible with the process for the fabrication of the power SJ-MOSFET structure. With this termination structure, the breakdown voltage occurs at the device cell structure rather than at the periphery resulting in performance described above for the cells.

The edge termination for the power SJ-MOSFET structure is illustrated in Fig. 7.47 including a portion of the device with the cell region containing the N-type and P-type drift regions. The $N^+$ source region is not formed on this mesa region. A wide trench is formed at the edge of the device which is slightly deeper than the P-type trench refill region. The wide trench is then refilled with a dielectric. Experimental results for this termination have been demonstrated using Benzo-CycloButene (BCB) polyimide as a spin-on dielectric. A source connected field plate is formed at the periphery of the device extending over the dielectric. This edge termination has the same breakdown voltage capability as the power SJ-MOSFET cell structure because the applied drain bias is supported across the very thick dielectric at the edge without creating a high electric field inside the semiconductor.

## 7.8.1 Simulation Example

The viability of the above edge terminations for power SJ-MOSFET structure can be demonstrated by using the results of two-dimensional numerical simulations for the device with blocking voltage of 200 V. The structure used for the simulations had the same doping ($2 \times 10^{16}$/cm$^3$) and widths (1.5 μm) for the N-type and P-type

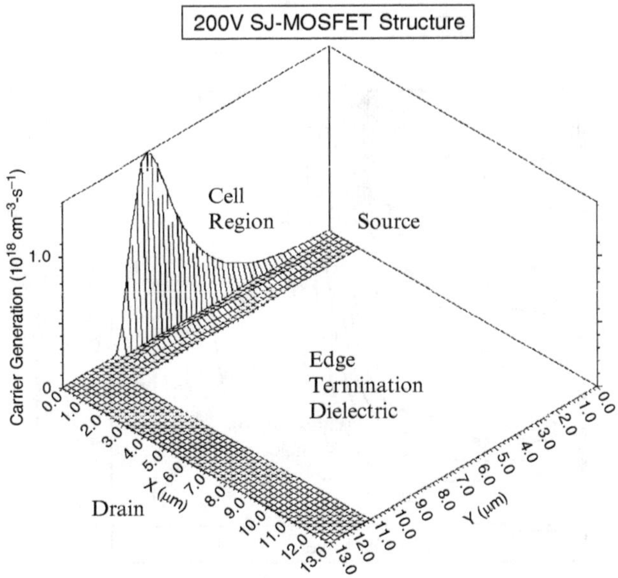


**Fig. 7.48** Carrier generation due to impact ionization at the edge termination of the power SJ-MOSFET structure at a drain bias of 200 V

drift regions as the baseline power SJ-MOSFET structure. The edge termination trench region had a width of 100 μm and a depth of 11 μm, which is slightly deeper than the depth (10 μm) of the P-type drift region. The source connected field plate extended by 50 μm beyond the edge of the trench adjacent to the cell structure.

The breakdown voltage for the edge termination was found to be 230 V which is identical to that for the power SJ-MOSFET cell structure. This demonstrates that the breakdown is occurring in the cell region and not limited by the edge termination. This can be confirmed by examination of the carrier generation due to impact ionization at the edge termination. It can be observed in Fig. 7.48 that the carrier generation due to impact ionization is occurring in the middle of the N-type drift region within the device cell structure at the edge termination as also observed in Fig. 7.17 for the cell structure in the middle of the active area.

The potential distribution obtained from the numerical simulations at the edge termination for the power SJ-MOSFET structure is provided in Figs. 7.49 and 7.50. From the figures, it can be confirmed that the potential distribution within the cell structure at the edge is identical to that for the power SJ-MOSFET cell structure indicating that the charge-coupling phenomenon is occurring effectively at the edge of the device. It can be observed from Fig. 7.49 that the drain bias voltage is supported across the dielectric inside the wide trench region with potential crowding in the vicinity of the edge of the field plate.

A three-dimensional view of the electric field at the edge termination for the power SJ-MOSFET structure is provided in Fig. 7.51. From this figure, it is apparent that there is a high electric field at the edge of the field plate. However, the

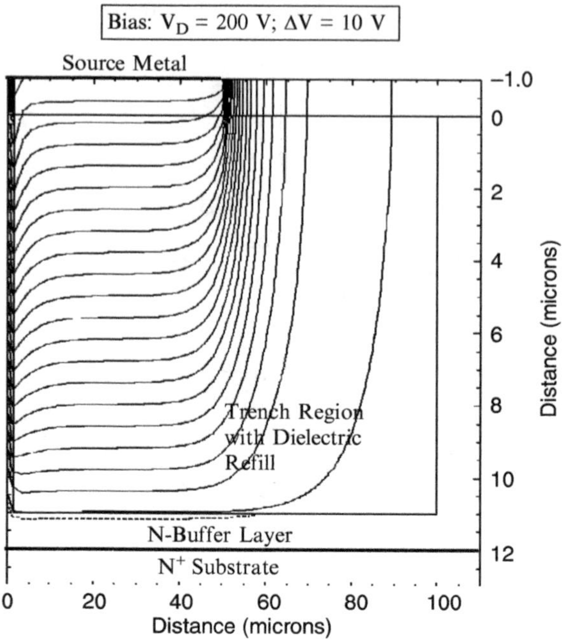


**Fig. 7.49** Potential distribution for the edge termination of the power SJ-MOSFET structure at a drain bias of 200 V

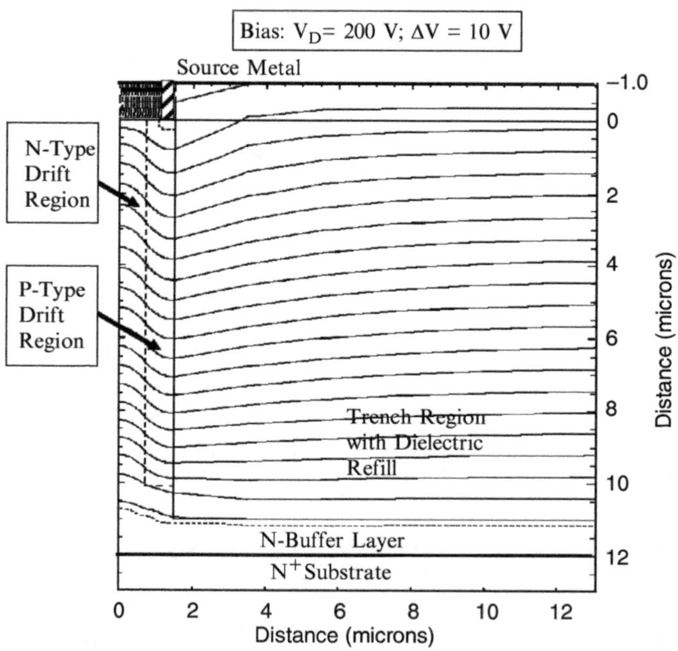


**Fig. 7.50** Potential distribution for the edge termination of the power SJ-MOSFET structure at a drain bias of 200 V

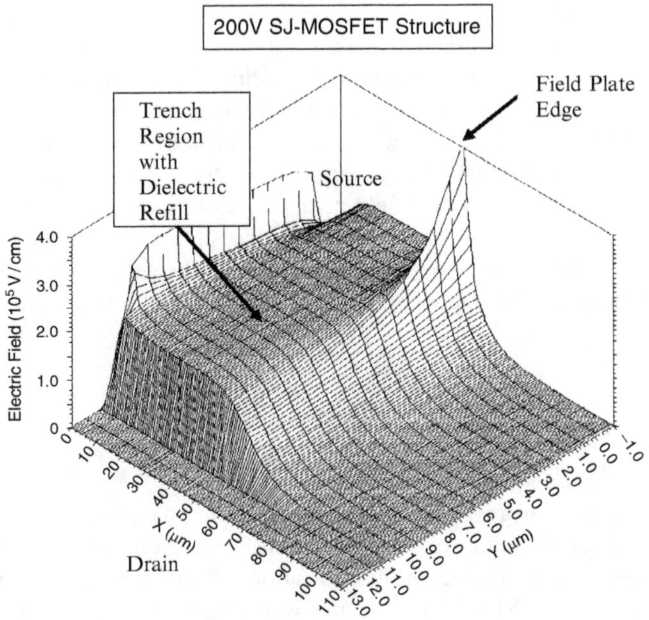


**Fig. 7.51** Electric field distribution at the edge termination of the power SJ-MOSFET structure at a drain bias of 200 V

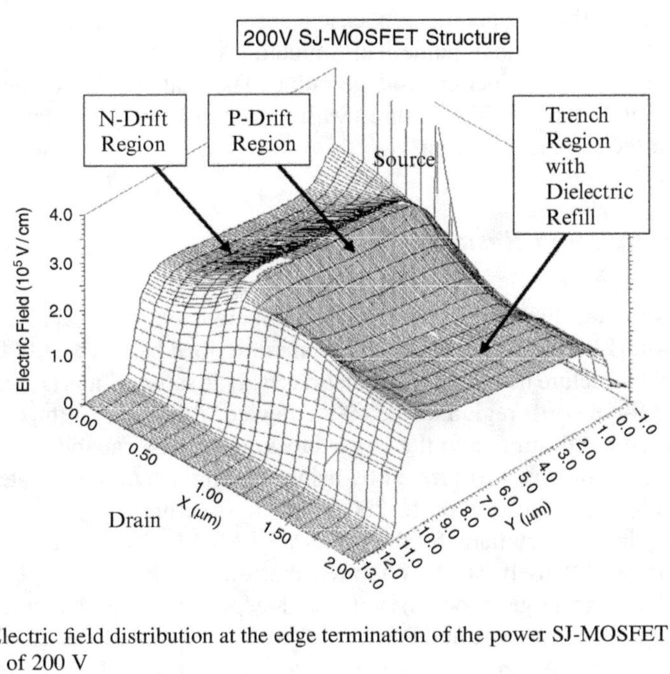


**Fig. 7.52** Electric field distribution at the edge termination of the power SJ-MOSFET structure at a drain bias of 200 V

magnitude of the electric field in the dielectric at the edge of the field plate is only $4 \times 10^5$ V/cm which is sufficiently low to prevent reliability problems. A three-dimensional view of the electric field in the vicinity of the device cell structure is shown in Fig. 7.52. From this figure, it can be observed that there is a smooth transition between the silicon cell region and the dielectric layer in the wide trench.

These results confirm that the edge termination structure with a wide deep trench together with a source connected field plate can achieve the breakdown voltage of the cell structure in super-junction devices. Although the performance of a 200-V device structure was examined in this section of the chapter, this methodology has been demonstrated to be applicable to devices with breakdown voltages of up to 1,200 V.

## 7.9 High Voltage Devices

As demonstrated in the previous sections of this chapter, the charge-coupling concept allows increasing the doping concentration of the drift region well above that dictated by ideal parallel-plane breakdown considerations. This approach has utility in the development of power MOSFET structures with larger breakdown voltages. In fact, the super-junction concept was first proposed for enhancement of the performance of power MOSFET structures with blocking voltages above 500 V [1].

The analytical formulations presented in the previous sections are applicable for any breakdown voltage design when appropriate values for the device parameters are used during the computations. Consequently, this section will focus on the results of two-dimensional numerical simulations for the power SJ-MOSFET structure with 1,000-V blocking voltage rating. The analytically computed values for the power SJ-MOSFET structures with various blocking voltage ratings are provided in the summary section.

### *7.9.1 Simulation Results*

The results of two-dimensional numerical simulations on the power SJ-MOSFET structure with 1,000-V blocking capability are described here. The 1,000-V power SJ-MOSFET structure used for the simulations had the same widths (1.5 μm) for the N-type and P-type drift region as the 600-V device. The depth of the drift regions was increased to 60 μm due to the larger blocking voltage capability. The P-base region for this structure was formed by using ion-implantation to create the same channel profile as in the case of the 200-V device structure.

The blocking characteristic for the 1,000-V SJ-MOSFET cell structure is shown in Fig. 7.53 at 300°K. It can be observed that the cell is capable of supporting 1,020 V. Although larger blocking voltages are possible by reducing the doping concentration for the N-type and P-type drift regions, this produces an undesirable increase in the specific on-resistance or the power SJ-MOSFET structure.

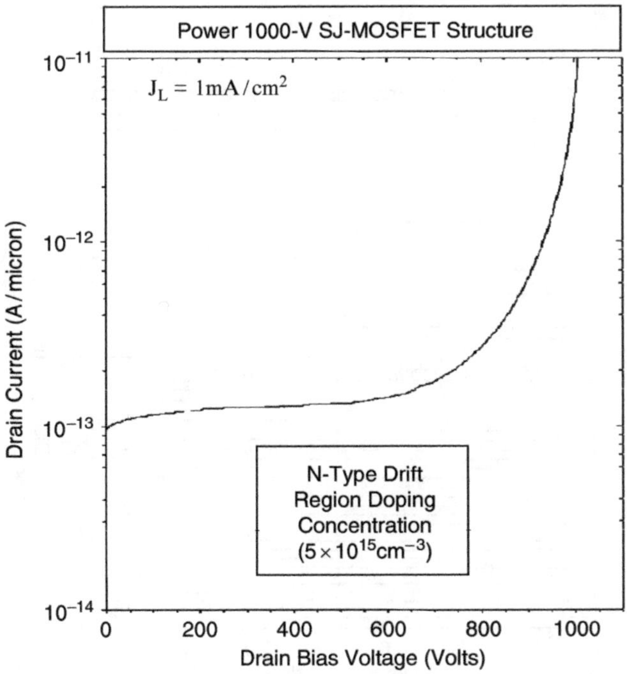


**Fig. 7.53** Blocking characteristics for the 1,000-V SJ-MOSFET structures

In order to further demonstrate the two-dimensional charge coupling phenomenon, it is instructive to examine the potential contours inside the power 1,000-V SJ-MOSFET structure when it is operating in the blocking mode. The potential contours for the 1,000-V SJ-MOSFET structure obtained using the numerical simulations with zero gate bias and a drain bias voltage of 1,000 V are shown in Fig. 7.54. It can be observed that the drain bias voltage is supported in the drift regions with a parallel-plane configuration. The potential contours in Fig. 7.54 are uniformly spaced indicating that the electric field is almost constant in the y-direction at a drain bias of 1,000 V.

The electric field profiles along the vertical direction through the center of the mesa region are shown in Fig. 7.55 for the 1,000-V power SJ-MOSFET structure with a drift region doping concentration of $5 \times 10^{15}/cm^3$. At a drain bias of 20 V, the electric field has a triangular shape representative of a one-dimensional junction with its peak at the bottom of the drift regions. This shape prevails until a drain bias of 300 V, after which the electric field takes a trapezoidal shape resembling that for a punch-through structure.

The transfer characteristics for the 1,000-V SJ-MOSFET structure were obtained using numerical simulations with a drain bias of 0.1 V at 300 and 400°K. The resulting transfer characteristics are shown in Fig. 7.56. The specific on-resistance can be obtained from the transfer characteristics at any gate bias voltage. For the case of a gate bias of 4.5 V and 300°K, the specific in-resistance is found to

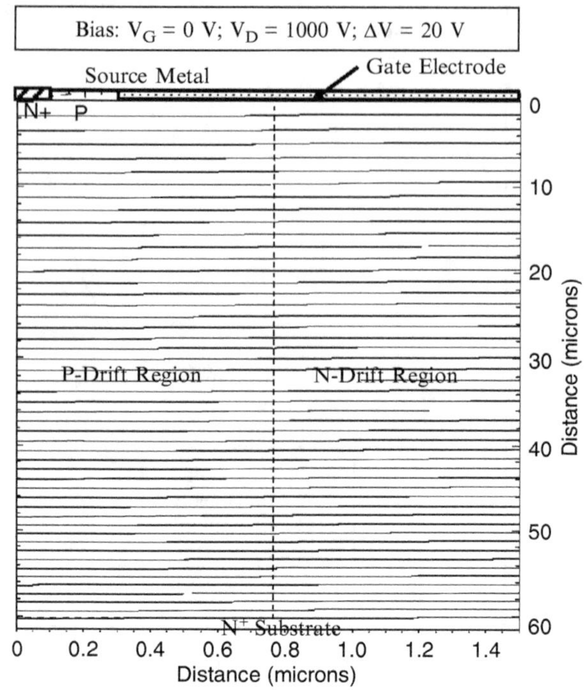


**Fig. 7.54** Potential contours in the 1,000-V SJ-MOSFET structure

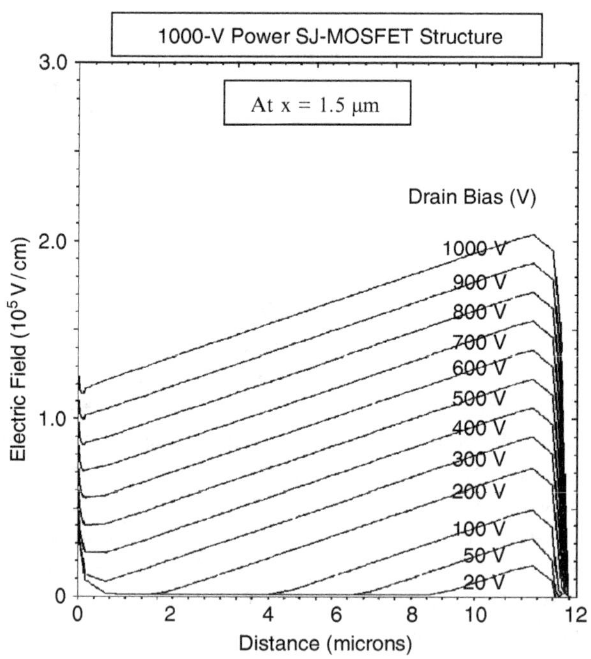


**Fig. 7.55** Electric field profiles in the 1,000-V SJ-MOSFET structure

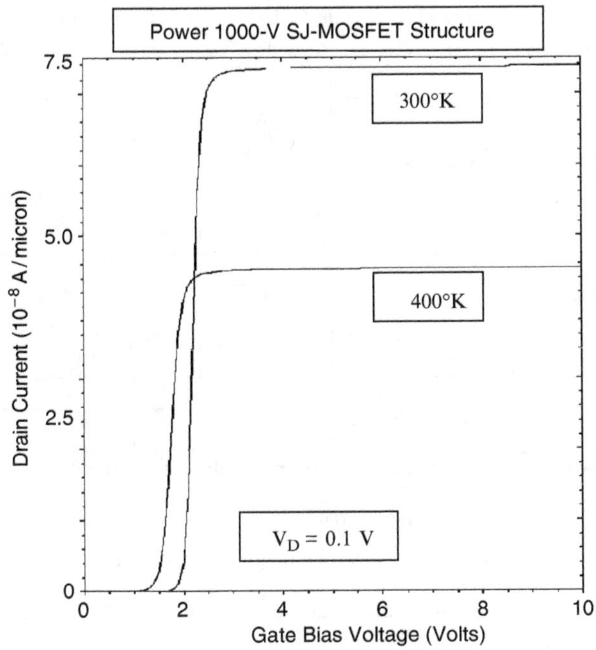


**Fig. 7.56** Transfer characteristics of the 1,000-V SJ-MOSFET structure

be 20.4 mΩ cm$^2$. For the case of a gate bias of 10 V and 300°K, the specific in-resistance is found to be 20.3 mΩ cm$^2$. These values are very close to each other indicating that the resistance of the N-type drift region is dominant in the 1,000-V power SJ-MOSFET structure.

The specific on-resistance obtained by using the analytical model for a breakdown voltage of 1,000 V is found to be 6.57 mΩ cm$^2$. This value is much smaller than that obtained above using the numerical simulations. In the analytical model, an optimum doping concentration of $1.47 \times 10^{16}$/cm$^3$ was assumed for the N-type drift region per the analytical solution for the optimum dose for two-dimensional charge coupling (see Fig. 7.5). This is much larger than the doping concentration of $5 \times 10^{15}$/cm$^3$ used during the numerical solutions described above. When the doping concentration of the N-type drift region was increased in the numerical simulations to $1.5 \times 10^{16}$/cm$^3$, the blocking voltage of the power SJ-MOSFET structure was found to be reduced to only 635 V.

## 7.10 Process Sensitivity Analysis

As described in the previous sections, the doping concentration in the drift region of the power SJ-MOSFET structure must be optimized in order to achieve a desired breakdown voltage. During that analysis, the doping concentrations for the N-type

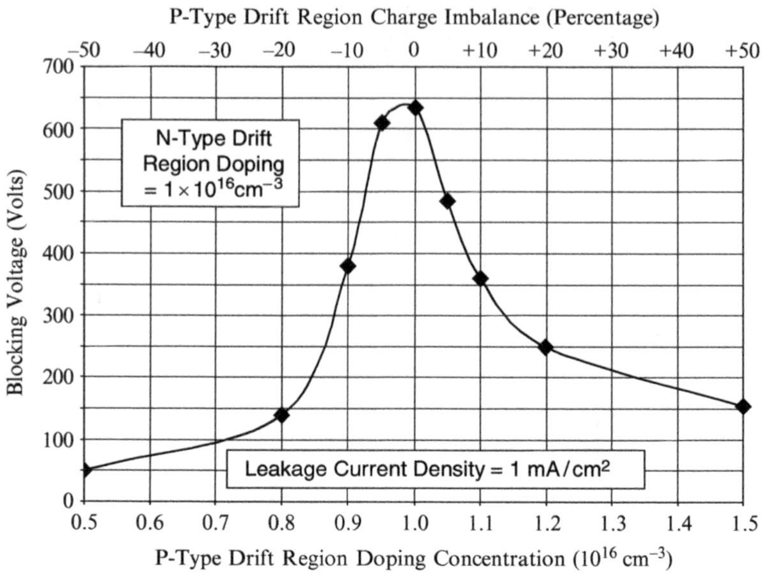


**Fig. 7.57** Sensitivity of breakdown voltage of the power SJ-MOSFET structure due to charge imbalance in the P-type drift region

and P-type drift regions were assumed to be exactly equal. In practice, such an exact match in the doping concentrations of the N-type and P-type drift regions is impossible to achieve during manufacturing of the power SJ-MOSFET structures [17]. In this section, the impact of an imbalance between the doping concentrations between the N-type and P-type drift regions is described based up on the results obtained using the two-dimensional numerical simulations.

The blocking voltage for the 600-V power SJ-MOSFET structures is plotted in Fig. 7.57 as a function of the doping concentration in the P-type drift region for the case of a width of 1.5 μm for the drift regions. The doping concentration of the N-type drift region was assumed to be constant at $1 \times 10^{16}/cm^3$ during this analysis. When making this plot, the blocking voltage was defined at a leakage current density of 1 mA/cm$^2$ because the 'soft' reverse blocking characteristics for the power SJ-MOSFET structure, as observed in Fig. 7.18. With this definition, the blocking voltage is observed to reduce rapidly when there is a relatively small imbalance in the doping concentration of the P-type drift region in either direction. If a process margin of 10% is utilized for the P-type drift region, the blocking voltage for the power SJ-MOSFET structure will range from 360 to 635 V. This power SJ-MOSFET device structure would therefore be considered to have a blocking voltage rating of 350 V rather than 600 V.

A similar reduction of the blocking voltage is observed when the doping concentration for the P-type drift region is held constant and the doping concentration of the N-type drift region is varied as shown in Fig. 7.58. In this case, if a

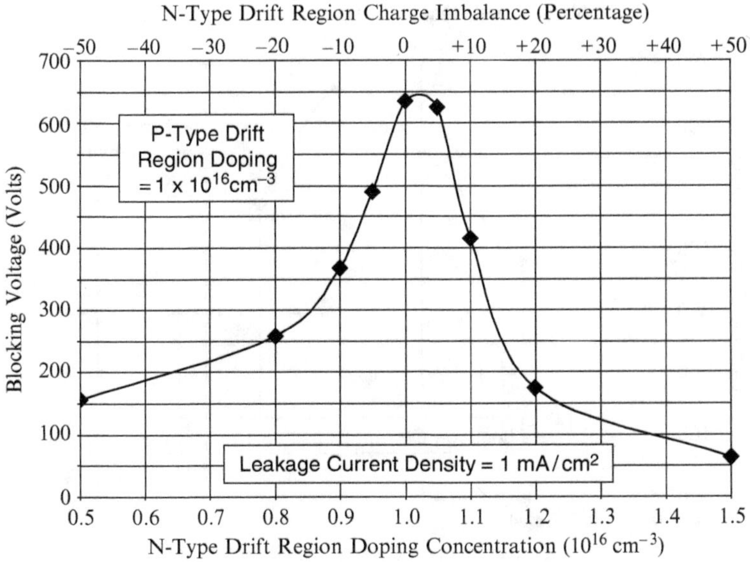


**Fig. 7.58** Sensitivity of breakdown voltage of the power SJ-MOSFET structure due to charge imbalance in the N-type drift region

process margin of 10% is utilized for the N-type drift region, the blocking voltage for the power SJ-MOSFET structure will range from 365 to 635 V. This power SJ-MOSFET device structure would therefore be considered to have a blocking voltage rating of 350 V rather than 600 V.

The change in the blocking voltage of the 600-V power SJ-MOSFET structure due to charge imbalance can be reduced by decreasing the doping concentration of the N-type drift region. The change in the blocking voltage for the power SJ-MOSFET structure is shown in Fig. 7.59 when the doping concentration of the P-type drift region is held at $5 \times 10^{15}/cm^3$. In this case, if a process margin of 10% is utilized for the N-type drift region, the blocking voltage for the power SJ-MOSFET structure will range from 475 to 665 V. This power SJ-MOSFET device structure would therefore be considered to have a blocking voltage rating of 450 V rather than 600 V.

During the charge imbalance study, the specific on-resistance for the power SJ-MOSFET structure does not change when the doping concentration of the P-type drift region is varied if the small impact of the change in depletion width is neglected. However, the specific on-resistance for the power SJ-MOSFET structure changes significantly when the doping concentration of the N-type drift region is varied while keeping the doping concentration of the P-type drift region constant. The variation of the specific on-resistance for the 600-V power SJ-MOSFET structure with doping concentration of the N-type drift region obtained from the numerical simulations is plotted in Fig. 7.60. It can be observed from this plot that

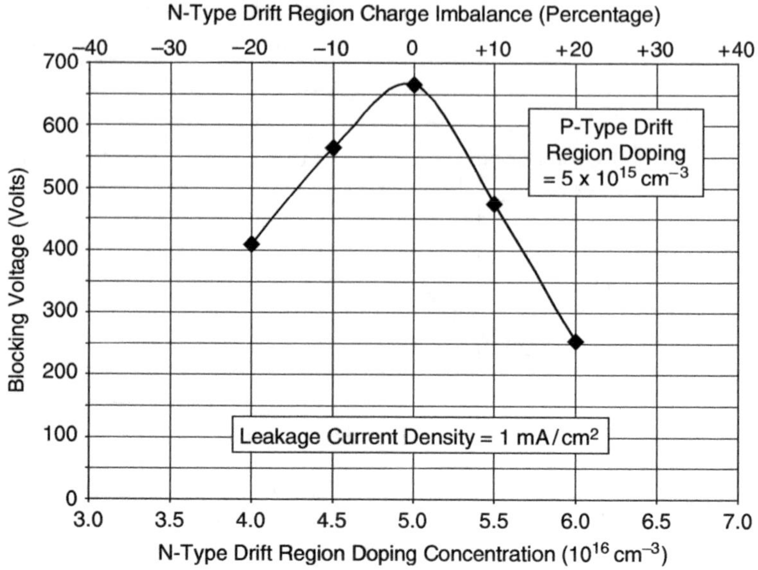


**Fig. 7.59** Sensitivity of breakdown voltage of the power SJ-MOSFET structure charge imbalance in the N-type drift region

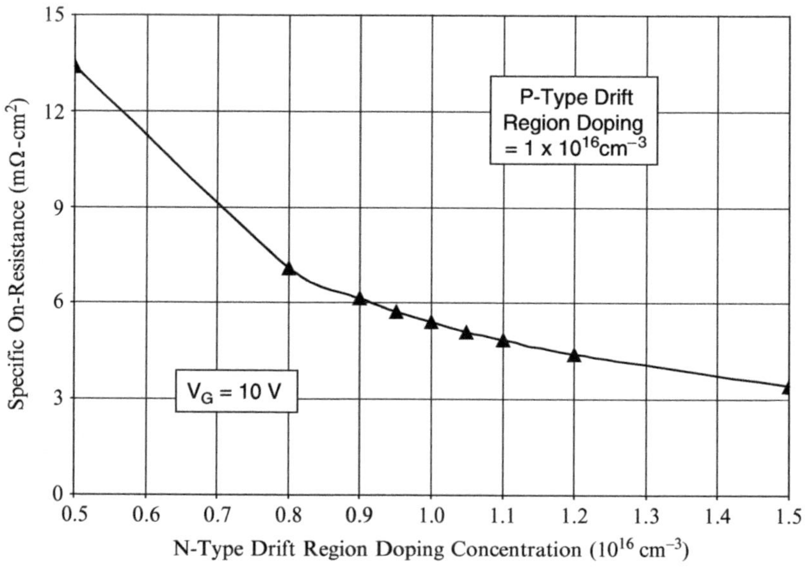


**Fig. 7.60** Sensitivity of specific on-resistance of the 600-V power SJ-MOSFET structure due to charge imbalance in the N-type drift region

the specific on-resistance is varying approximately inversely with the doping concentration of the N-type drift region.

Based up on the above process sensitivity considerations, it becomes apparent that the specific on-resistance for practical power SJ-MOSFET structures is significantly worse than the specific on-resistance that can be achieved with idealized charge balance considerations. The specific on-resistance for a practical power SJ-MOSFET structure with a blocking voltage capability of 500 V is found to be 15 mΩ $cm^2$. This value is a factor of about five times larger than the idealized value for the specific on-resistance for the power SJ-MOSFET structure. A proportionate increase in the device figure-of-merits will also be incurred due to the larger specific on-resistance.

## 7.11 Inductive Load Turn-Off Characteristics

High voltage power MOSFET devices are often used in adjustable speed motor drives which behave as inductive loads. The basic half-bridge circuit utilized in these applications is shown in Fig. 10.1 of the textbook [13]. The waveforms for the current and voltage in the power switch is shown in Fig. 10.2 of the textbook. A high power dissipation occurs during each turn-off event due to the simultaneous high current and voltage during the transient. This power loss is usually characterized as an energy loss per cycle. The operation of a power MOSFET device in an inductive load circuit was illustrated in Fig. 3.42 together with a description of its operation.

Prior to the turn-off transient, the device is operating in its on-state because switch $S_1$ is closed and switch $S_2$ is open. These initial conditions are defined by: $v_G = V_{GS}$; $i_D = I_L$; and $v_D = V_{ON}(V_{GS})$. In order to initiate the turn-off process, switch $S_1$ is opened and switch $S_2$ is subsequently closed by the control circuit. The gate electrode of the power MOSFET device is then connected to the source via the gate resistance to discharge its capacitances. However, no changes in the drain current or voltage can occur until the gate voltage reaches the magnitude required to operate the power MOSFET device at a saturated drain current equal to the load current. (The small increase in the drain voltage, due to the increase in on-resistance resulting from the reduction of the gate bias voltage, has been neglected here). This gate plateau voltage is given by:

$$V_{GP} = V_{TH} + \sqrt{\frac{J_{D,ON} W_{Cell} L_{CH}}{\mu_{ni} C_{GOX}}} \tag{7.59}$$

where $C_{GOX}$ is the gate oxide capacitance. During this time interval, the gate-drain capacitance $C_{GD}(V_{ON})$ remains constant because the drain voltage is constant. Consequently, the time constant for discharging the gate of the power MOSFET device is $R_{G,SP}*[C_{GS} + C_{GD}(V_{ON})]$ and the gate voltage decreases exponentially with time as given by:

$$v_G(t) = V_{GS}e^{-t/R_{G,SP}[C_{GS}+C_{GD}(V_{ON})]} \tag{7.60}$$

The time $t_4$ (using the notation from the textbook) for reaching the gate plateau voltage can be obtained by using this equation with (7.59) for the plateau voltage:

$$t_4 = R_{G,SP}[C_{GS} + C_{GD}(V_{ON})] \ \ln\left[\frac{V_{GS}}{V_{GP}}\right] \tag{7.61}$$

This time can be considered to a *turn-off delay time* before the drain voltage begins to increase after the turn-off is initiated by the control circuit.

The drain voltage begins to increase at time $t_4$ but the drain current remains constant at the load current $I_L$ because the current cannot be transferred to the diode until the voltage at the drain of the MOSFET device exceeds the supply voltage $V_{DS}$ by one diode drop to forward bias the diode. Since the drain current density is constant, the gate voltage also remains constant at the gate plateau voltage. Consequently:

$$J_{GP} = \frac{V_{GP}}{R_{G,SP}} \tag{7.62}$$

where $R_{G,SP}$ is the specific gate resistance. Since all the gate current is used to discharge the gate-drain capacitance during the plateau phase because there is no change in the voltage across the gate-source capacitance:

$$J_{GP} = C_{GD,SP}\frac{dv_D}{dt} \tag{7.63}$$

where $C_{GD,SP}$ is the specific gate transfer capacitance of the power MOSFET structure which is a function of the drain voltage. This voltage dependence of the gate transfer capacitance was not taken into account in the derivation provided in the textbook but is important to include here to allow comparison of the behavior of various power MOSFET structures.

For simplicity of analysis, it will be assumed that the screening effect can be ignored once the drain voltage becomes less than $V_{D,N}$. Under this assumption, the gate transfer capacitance for the power SJ-MOSFET structure is given by:

$$C_{GD} = \left(\frac{W_N}{W_{Cell}}\right)\left(\frac{\varepsilon_S E_{CU} C_{GOX}}{\varepsilon_S E_{CU} + C_{GOX} v_D(t)}\right) \tag{7.64}$$

Using this expression in (7.63) yields the following differential equation for the voltage increase phase of the turn-off transient:

$$dt = \left(\frac{W_N}{W_{Cell}}\right)\frac{1}{J_{GP}}\left[\frac{\varepsilon_S E_{CU} C_{GOX}}{\varepsilon_S E_{CU} + C_{GOX} v_D(t)}\right]dv_D \tag{7.65}$$

Integration of this equation yields:

$$(t-t_4)=\left(\frac{W_N}{W_{Cell}}\right)\frac{\varepsilon_S E_{CU}}{J_{GP}}\ \ln\left[\frac{\varepsilon_S E_{CU}+C_{GOX}v_D(t)}{\varepsilon_S E_{CU}+C_{GOX}V_{ON}}\right] \tag{7.66}$$

In the case of the power SJ-MOSFET structure, the drain voltage increases relatively slowly from the on-state voltage drop ($V_{ON}$) until it reaches the pinch-off voltage ($V_{D,N}$) for the N-type drift region. Once the N-type drift region is pinched-off, the gate-drain transfer capacitance becomes equal to zero according to the analytical model resulting in an abrupt increase in the drain voltage to the drain supply voltage ($V_{DS}$). The voltage rise-time, i.e. the time taken for the voltage to increase from the on-state voltage drop ($V_{ON}$) to the pinch-off voltage ($V_{D,N}$) for the N-type drift region can be derived from the above expression:

$$(t_5-t_4)=t_{V,OFF}=\left(\frac{W_N}{W_{Cell}}\right)\frac{\varepsilon_S E_{CU}}{J_{GP}}\ \ln\left[\frac{\varepsilon_S E_{CU}+C_{GOX}V_{D,N}}{\varepsilon_S E_{CU}+C_{GOX}V_{ON}}\right] \tag{7.67}$$

A closed form solution for the rise in the drain voltage can be obtained from (7.66):

$$v_D(t)=\left(\frac{\varepsilon_S E_{CU}}{C_{GOX}}+V_{ON}\right)e^{\left(\frac{W_{Cell}}{W_N}\right)\left(\frac{J_{GP}}{\varepsilon_S E_{CU}}\right)t}-\left(\frac{\varepsilon_S E_{CU}}{C_{GOX}}\right) \tag{7.68}$$

This equation describes the increase in the drain voltage from the on-state voltage drop until it reaches the drain pinch-off voltage. The drain voltage then abruptly increases to the drain supply voltage according to this analytical model.

At the end of the plateau phase (at time $t_5$), the load current begins to transfer from the power MOSFET device to the free wheeling diode. Since the drain voltage remains constant, the gate-drain capacitance can also be assumed to remain constant during this phase (with a zero value for the power SJ-MOSFET structure). The current flowing through the gate resistance ($R_G$) discharges both the gate-drain and gate-source capacitances leading to an exponential fall in gate voltage from the plateau voltage:

$$v_G(t)=V_{GP}e^{-(t-t_5)/R_{G,SP}C_{GS}} \tag{7.69}$$

The drain current follows the gate voltage as given by:

$$J_D(t)=g_m[v_G(t)-V_{TH}]=\frac{\mu_{ni}C_{OX}}{L_{CH}W_{Cell}}[v_G(t)-V_{TH}]^2 \tag{7.70}$$

The drain current decreases rapidly with time due to the exponential reduction of the gate voltage, as given by (7.68), during the current fall phase. The drain current becomes equal to zero when the gate voltage reaches the threshold voltage. The current fall time can therefore be obtained from (7.69):

$$t_{I,OFF} = R_{G,SP} C_{GS} \ln\left(\frac{V_{GP}}{V_{TH}}\right) \tag{7.71}$$

Specific capacitances should be used in this expression for computation of the current fall time. Beyond this point in time, the gate voltage decreases exponentially until it reaches zero. The time constant for this exponential decay is different from the initial phase due to the smaller (zero) gate-drain capacitance.

The turn-off energy loss per cycle can be obtained using:

$$E_{OFF} = \frac{1}{2} J_{ON} \left(V_{D,N} t_{V,OFF} + V_{DS} t_{I,OFF}\right) \tag{7.72}$$

under the assumption that the drain current and voltage excursions are approximately linear with time. Due to the relatively small value for the drain pinch-off voltage in the power SJ-MOSFET structure, the energy loss during the voltage rise-time interval becomes comparable to the energy loss during the current fall-time interval despite the much larger value for the voltage rise-time versus the current fall-time.

In the case of the 600-V power SJ-MOSFET structure, the typical drain supply voltage is 400 V. Using the specific on-resistance for this device of 5.4 mΩ cm$^2$ with an on-state power dissipation of 100 W/cm$^2$, the on-state current density is found to be 136 A/cm$^2$. The on-state voltage drop at this current density is 0.73 V. The device structure has a cell pitch of 3.0 μm with N-type and P-type drift region widths of 1.5 μm, drift region length of 35 μm, and a gate oxide thickness of 500 Å. The N-type and P-type drift regions have a doping concentration of $1 \times 10^{16}$/cm$^3$. The specific gate resistance used in the turn-off circuit was assumed to have a value of 1.5 Ω cm$^2$.

The specific input capacitance for the 600-V power SJ-MOSFET structure computed is 70 nF/cm$^2$. The specific gate reverse transfer capacitance computed by using (7.30) at the on-state voltage drop is 12.9 nF/cm$^2$. The pinch-off voltage for the N-type drift region ($V_{D,N}$) for the 600-V power SJ-MOSFET structure computed using (7.34) is found to be only 7.88 V. Consequently, the specific gate reverse transfer capacitance is zero at the drain supply voltage. Using these values in (7.61), the time ($t_4$) to reach the gate plateau voltage is found to be 0.205 μs. Using these parameters in (7.67), the voltage rise-time is computed as 0.084 μs. Using these parameters in (7.71), the current fall-time is computed as 0.013 μs. It can be observed that the current fall time is much smaller than the voltage rise-time. The energy loss per cycle obtained by using these values for the voltage rise-time and current fall-time in (7.72) is 0.40 mJ/cm$^2$. Consequently, the turn-off energy loss per cycle for the power SJ-MOSFET structure is very small. However, there is a very abrupt rise in the drain voltage during the turn-off transient. The resulting

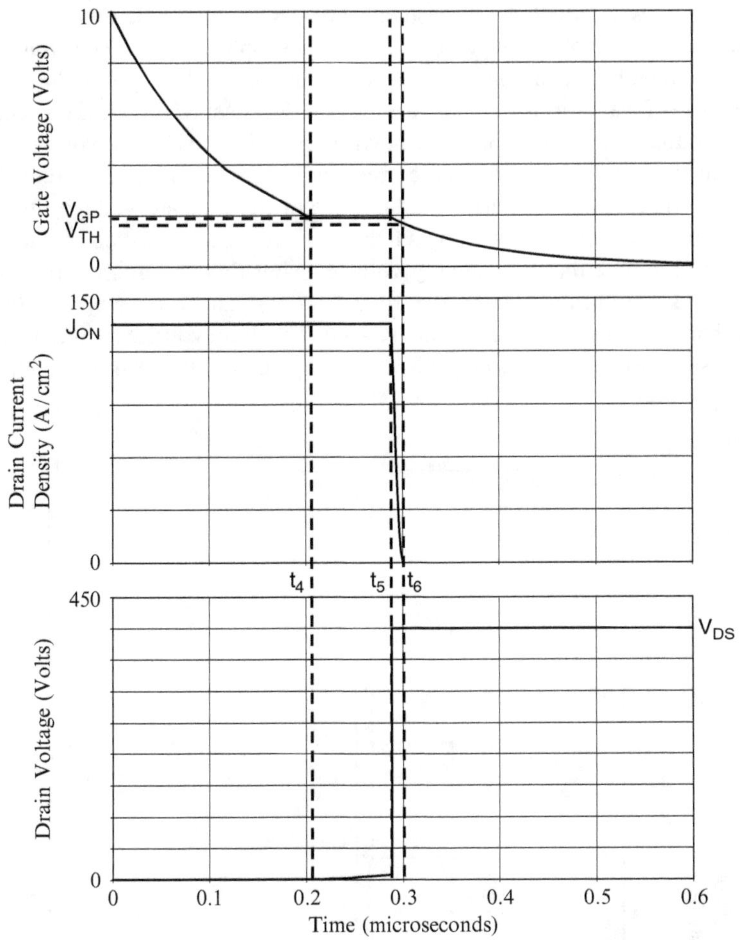


**Fig. 7.61** Analytically computed turn-off waveforms for the 600-V power SJ-MOSFET structure

large turn-off [dV/dt] can be a problem in power circuits. The waveforms that are generated by using the above equations are shown in Fig. 7.61.

### *7.11.1 Simulation Results*

The results of two-dimensional numerical simulations on the turn-off of the 600-V power SJ-MOSFET structure are described below. The drain supply voltage was chosen as 400 V for the turn-off analysis. During the turn-off simulations, the gate voltage was reduced to zero with a gate resistance of $1 \times 10^8\ \Omega\ \mu m$ for the 1.5 μm half-cell structure, which is equivalent to a specific gate resistance of $1.5\ \Omega\ cm^2$. The current density was initially held constant at an on-state current density of 136 A/cm$^2$ allowing the drain voltage to rise to the drain supply voltage. The drain supply voltage was then held constant allowing the drain current density to reduce to zero.

The turn-off waveforms obtained for the 600-V power SJ-MOSFET structure by using the numerical simulations are shown in Fig. 7.62. The gate voltage initially reduces to the gate plateau voltage corresponding to the on-state current density. The drain voltage then increases gradually from the on-state voltage drop to the drain pinch-off voltage followed by an abrupt increase as predicted by the analytical model to almost the drain supply voltage. After this, the drain current rapidly falls to zero. The drain voltage rise-time ($t_5 - t_4$) can be observed to be much greater than the drain current fall time ($t_6 - t_5$). The drain voltage rise-time obtained from the simulations of the power SJ-MOSFET structure is 0.084 ms and the drain

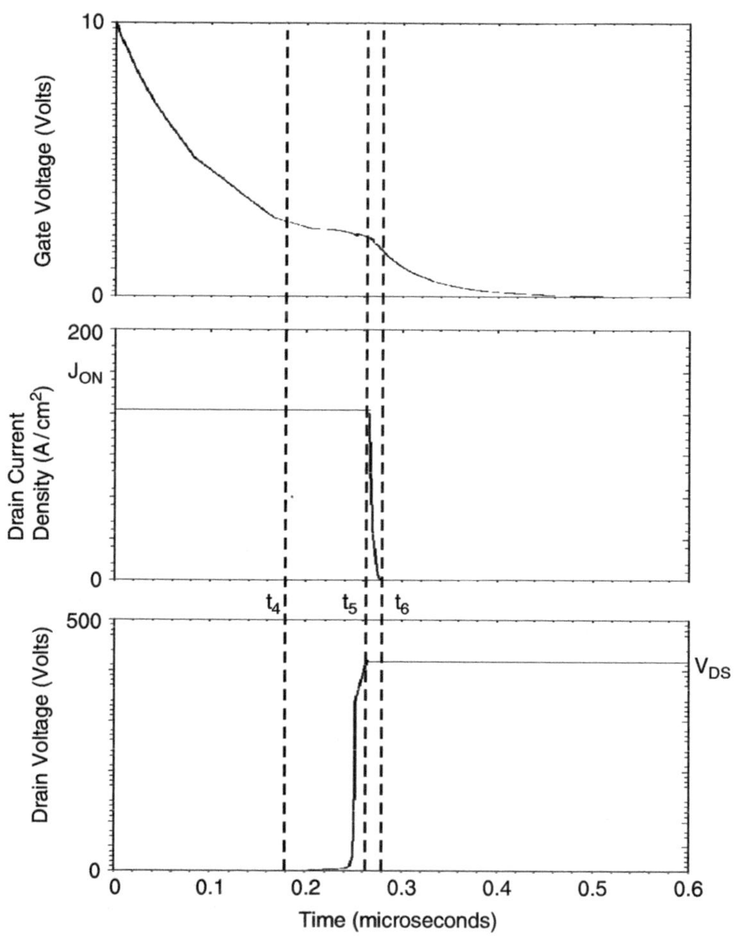


**Fig. 7.62** Turn-off waveforms for the 600-V power SJ-MOSFET structure

current fall-time obtained from the simulations is 0.016 ms. The shape of the waveforms predicted by the analytical model (see Fig. 7.61) for the gate voltage, drain voltage and drain current are in good agreement with the simulation results. The numerical values predicted for the analytical model for the voltage rise-time and the drain fall-time are also in good agreement with those observed in the numerical simulations allowing an accurate computation of the energy loss per cycle for the power SJ-MOSFET structure.

## 7.12 Discussion

The physics of operation and resulting electrical characteristics of the power SJ-MOSFET structure have been described in this chapter. The two-dimensional charge coupling phenomenon in these structures allows supporting blocking voltages well above the one-dimensional parallel-plane breakdown voltage. The specific on-resistance for this device structure is significantly reduced when compared with the power D-MOSFET and U-MOSFET structures due to its very short channel length, small cell pitch, and high doping concentration in the drift region. With the power SJ-MOSFET structure, it is possible to achieve a specific on-resistance that is smaller than the ideal specific on-resistance for silicon. In addition, small reverse transfer capacitance and gate charge can be achieved in the power SJ-MOSFET structure.

The first commercially available power SJ-MOSFET structures were developed with blocking voltage capability of 600 V for motor control applications [1]. Due to the significant reduction in the specific on-resistance achievable with the power SJ-MOSFET structure, there has recently been considerable research activity around the world to develop alternate processes for their fabrication [18,19]. However, the blocking voltage ratings for commercially available devices remain below 1,000 V at this time.

Due to the relatively high doping levels in the drift region, the specific on-resistance for the power SJ-MOSFET structure is reduced to below that for the ideal parallel-plane junction case. With this device structure, it is possible to reduce the specific on-resistance to 5.7-times smaller than the ideal specific on-resistance at a breakdown voltage of 200 V. An even greater improvement over the ideal specific on-resistance is obtained for devices with larger breakdown voltages: a factor of 21-times for a 600-V device; and factor of 42-times for a 1,000-V device. The commercially developed 600-V power SJ-MOSFET devices have been found to exhibit specific on-resistances considerably larger than these values because of the previously discussed process tolerance issues due to charge imbalance problems.

For purposes of comparison with the power MOSFET structures discussed in subsequent chapters, the analysis of the power SJ-MOSFET structure is provided here for blocking voltages ranging from 30 to 1,000 V. In this analysis, the power SJ-MOSFET structure was assumed to have the following parameters: (a) $N^+$

source junction depth of 0.1 μm; (b) P-base junction depth of 0.38 μm; (c) gate oxide thickness of 500 Å; (d) N-type and P-type drift region widths of 1.5 μm; (e) threshold voltage of 2 V; (f) gate drive voltage of 10 V; (g) inversion mobility of 450 $cm^2$/V s; (h) accumulation mobility of 1,000 $cm^2$/V s; (i) cell width of 3 μm. The contributions from the contacts and the $N^+$ substrate were neglected during the analysis. The doping concentration and thickness of the drift region were determined under the assumption that the edge termination limits the breakdown voltage to the breakdown voltage of the cells, which exceeds the parallel-plane breakdown voltage. The device parameters pertinent to each blocking voltage are provided in Fig. 7.63. It can be observed that the doping concentration of the drift region is much larger than that for the conventional power MOSFET structures. The analytical model described in Sect. 7.3 for computing the specific on-resistance was used for all of the power SJ-MOSFET structures. These values are provided in Fig. 7.64.

The specific on-resistance for the power SJ-MOSFET structure can be compared with the ideal specific on-resistance obtained by using Baliga's power law for the

| Blocking Voltage (V) | Drift Doping Concentration ($cm^{-3}$) | Drift Region Width (microns) | Drift Region Thickness (microns) |
|---|---|---|---|
| 30 | $2.71 \times 10^{16}$ | 1.5 | 1.0 |
| 60 | $2.41 \times 10^{16}$ | 1.5 | 2.0 |
| 120 | $2.15 \times 10^{16}$ | 1.5 | 4.5 |
| 200 | $1.98 \times 10^{16}$ | 1.5 | 9.5 |
| 300 | $1.85 \times 10^{16}$ | 1.5 | 15 |
| 600 | $1.64 \times 10^{16}$ | 1.5 | 35 |
| 1000 | $1.51 \times 10^{16}$ | 1.5 | 60 |

**Fig. 7.63** Device parameters for the power SJ-MOSFET structures

| Blocking Voltage (V) | Cell Pitch (microns) | Specific On-Resistance (mΩ-$cm^2$) |
|---|---|---|
| 30 | 3.0 | 0.143 |
| 60 | 3.0 | 0.219 |
| 120 | 3.0 | 0.419 |
| 200 | 3.0 | 0.812 |
| 300 | 3.0 | 1.36 |
| 600 | 3.0 | 3.43 |
| 1000 | 3.0 | 6.35 |

**Fig. 7.64** Specific on-resistances for the power SJ-MOSFET structures

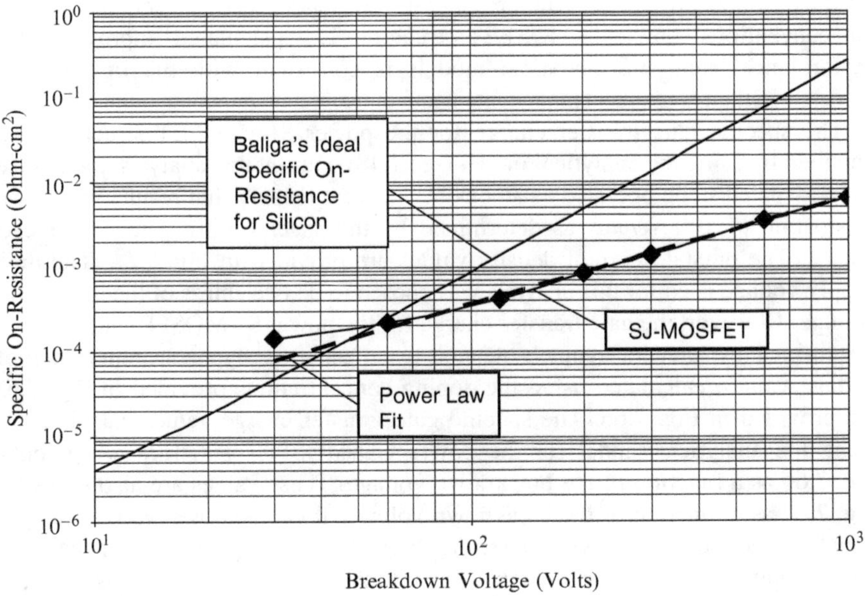


**Fig. 7.65** Specific on-resistances for the power SJ-MOSFET structures

impact ionization coefficients in Fig. 7.65. From this figure, it can be concluded that the specific on-resistance for the SJ-MOSFET structure is less than the ideal specific on-resistance for blocking voltages above 60 V. The reduction of the specific on-resistance for the power SJ-MOSFET structure below the ideal specific on-resistance becomes larger with increasing breakdown voltage. The figure clearly demonstrates that extremely low specific on-resistance, well below the ideal specific on-resistance of one-dimensional device structures, can be achieved by using the power SJ-MOSFET structure at high blocking voltages. The specific on-resistance increases in a linear fashion on this log–log graph for breakdown voltages ranging from 120 to 1,000 V indicating a power law relationship. The power law relationship that fits the data is shown in the figure by the dashed line. The equation for this line is:

$$R_{on,sp} = 1.05 \times 10^{-6}\ BV^{1.26} \tag{7.73}$$

where the specific on-resistance has units of Ω cm$^2$.

As discussed in Sect. 7.11, the specific on-resistance that can be achieved in practical power SJ-MOSFET structures is constrained by process sensitivity issues. For a blocking voltage of 500 V, the specific on-resistance of a practical power SJ-MOSFET structure increases to 18.3 mΩ cm$^2$. In comparison, the specific on-resistance of a practical power U-MOSFET structure is 63.5 mΩ cm$^2$ (see Chap. 3). Consequently, at a blocking voltage of 500 V, the specific on-resistance for the power SJ-MOSFET structure is 3.5-times smaller than that achievable using the conventional power U-MOSFET structure. This calculated improvement in performance using the analytical models is consistent with the improvement reported in the literature [1].

The specific gate transfer charge for the power SJ-MOSFET structures was obtained by using the analytical model (see 7.48). During this analysis, the devices were assumed to be operated at an on-state current density that results in a power dissipation of 100 W/cm$^2$ as determined by the specific on-resistance for each device. The on-state current density values are provided in Fig. 7.66. The drain supply voltage used for this analysis was chosen to be two-thirds of the breakdown voltage. However, the gate transfer charge for the power SJ-MOSFET structures is determined by the drain pinch-off voltage. Consequently, the drain pinch-off voltage values, calculated using the doping concentrations provided in Fig. 7.63, are provided in Fig. 7.66. The specific gate transfer charge values calculated by using the analytical model for the power GD-MOSFET structure are given in Fig. 7.66 as a function of the breakdown voltage. These values are also plotted in Fig. 7.67 as a function of the breakdown voltage. It can be observed that the gate transfer charge decreases with increasing breakdown voltage in spite of the larger drain supply voltage during the transient for higher voltage devices. This reduction of specific gate transfer charge is related to the decrease in the drain pinch-off voltage with increasing breakdown voltage due to the smaller N-type drift region doping concentration. The specific gate transfer charge reduces in a linear fashion on this log–log graph for breakdown voltages ranging from 120 to 1,000 V indicating a power law relationship. The power law relationship that fits the data is shown in the figure by the dashed line. The equation for this line is:

$$Q_{GD} = 699\ BV^{-0.243} \tag{7.74}$$

where the specific gate transfer charge has units of nC/cm$^2$.

| **Blocking Voltage (V)** | **Drain Pinch-Off Voltage (Volts)** | **On-State Current Density (A / cm$^2$)** | **Specific Gate Transfer Charge (nC / cm$^2$)** |
|---|---|---|---|
| 30 | 22.7 | 836 | 281 |
| 60 | 20.1 | 674 | 248 |
| 120 | 17.9 | 489 | 218 |
| 200 | 16.4 | 345 | 197 |
| 300 | 15.3 | 271 | 181 |
| 600 | 13.4 | 170 | 153 |
| 1000 | 12.0 | 123 | 130 |

**Fig. 7.66** Parameters used for gate transfer charge analysis for the power SJ-MOSFET structures

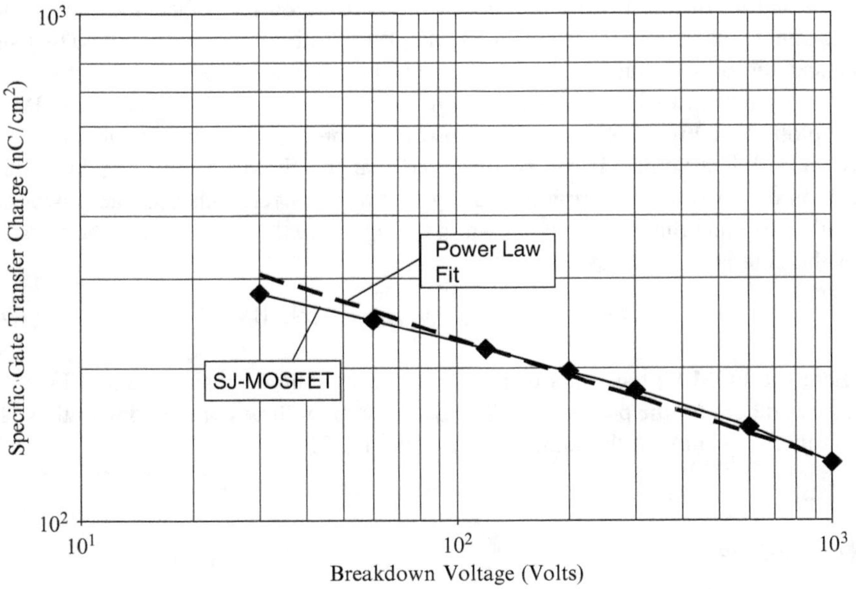


**Fig. 7.67** Specific gate transfer charge for the power SJ-MOSFET structures

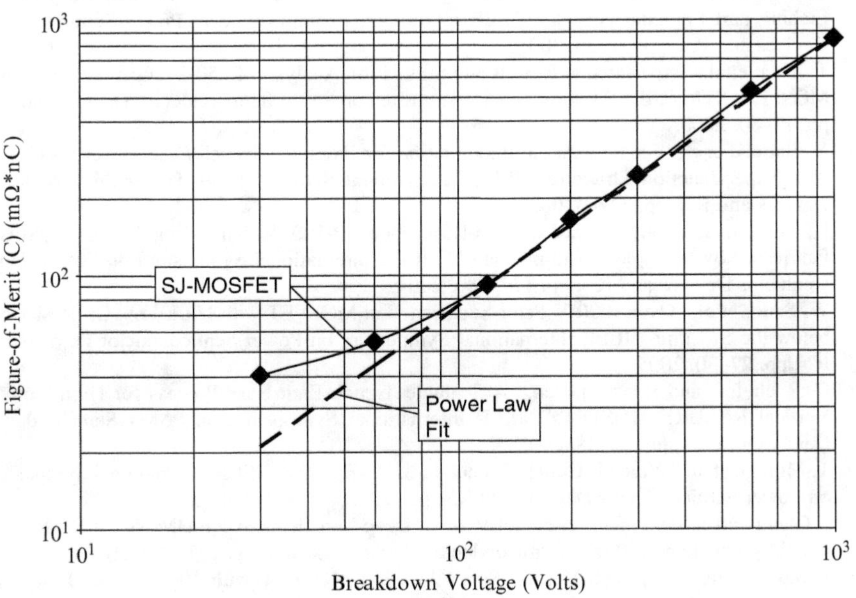


**Fig. 7.68** Figure-of-merit (C) for the power SJ-MOSFET structures

The figure-of merit (C) – product of the specific on-resistance and the specific gate transfer charge – for the power SJ-MODSFET structures was obtained by using the specific on-resistance and specific gate transfer charge values calculated using the analytical model as described above. The resulting values for the FOM(C) are plotted in Fig. 7.68 as a function of the breakdown voltage of the power SJ-MOSFET structure. It can be observed that the FOM(C) increases in a linear fashion on this log–log graph indicating a power law relationship. The power law relationship that fits the data is shown in the figure by the dashed line. The equation for this line is:

$$\mathrm{FOM(C)} = R_{ON,sp}Q_{GD,sp} = 0.593\ BV^{1.05} \tag{7.75}$$

where the FOM(C) has units of mΩ nC. The specific on-resistance and FOM(C) discussed here for the power SJ-MOSFET structure will be compared with those for the other structures in the final chapter of the book.

## References

1. L. Lorenz, et al, "COOLMOS – A New Milestone in High Voltage Power MOS", IEEE International Symposium on Power Semiconductor Devices and ICs, pp. 3–10, 1999.
2. T. Fujihira and Y. Miyasaka, "Simulated Superior Performances of Semiconductor Superjunction Devices", IEEE International Symposium on Power Semiconductor Devices and ICs, pp. 423–426, 1998.
3. S. Iwamoto, et al, "Above 500-V Class Super-Junction MOSFETs Fabricated by Deep Trench Etching and Epitaxial Growth", IEEE International Symposium on Power Semiconductor Devices and ICs, pp. 31–34, 2005.
4. T. Nitta, et al, "Experimental Results and Simulation Analysis of 250-V Super-Trench Power MOSFET (STM)", IEEE International Symposium on Power Semiconductor Devices and ICs, pp. 77–80, 2000.
5. Y. Hattori, et al, "Shallow Angle Implantation for Extended Trench Gate Power MOSFETs with Super-Junction Structure", IEEE International Symposium on Power Semiconductor Devices and ICs, pp. 427–430, 2001.
6. H. Ninomiya, et al, "Ultra-Low On-Resistance 60-100 V Super-Junction U-MOSFETs Fabricated by Multiple Ion-Implantation", IEEE International Symposium on Power Semiconductor Devices and ICs, pp. 177–180, 2004.
7. W. Saito, et al, "Over 1000-V Semi-Superjunction MOSFET with Ultra-Low On-Resistance below the Si-Limit", IEEE International Symposium on Power Semiconductor Devices and ICs, pp. 27–30, 2005.
8. C. Rochefort and R. van Dalen, "A Scalable Trench Etch Base Process for High Voltage Vertical RESURF MOSFETs", IEEE International Symposium on Power Semiconductor Devices and ICs, pp. 35–38, 2005.
9. T. Minato, et al, "Which is Cooler, Trench or Multi-Epitaxy", IEEE International Symposium on Power Semiconductor Devices and ICs, pp. 73–76, 2000.
10. J. Glenn and J. Siekkinen, "A Novel Vertical Deep Trench RESURF DMOS", IEEE International Symposium on Power Semiconductor Devices and ICs, pp. 197–200, 2000.
11. J. Sakakibara, et al, "600-V class Super Junction MOSFET with High Aspect Ratio P/N Columns Structure", IEEE International Symposium on Power Semiconductor Devices and ICs, pp. 299–302, 2008.

12. D. Disney and G. Dolny, "JFET Depletion in Super-Junction Devices", IEEE International Symposium on Power Semiconductor Devices and ICs, pp. 157–160, 2008.
13. B.J. Baliga, "Fundamentals of Power Semiconductor Devices", Springer-Science, New York, 2008.
14. B. Zhang, Z. Xu, and A.Q. Huang, "Analysis of the Forward Biased Safe Operating Area of the Super Junction MOSFET", IEEE International Symposium on Power Semiconductor Devices and ICs, pp. 61–64, 2000.
15. H. Mahfoz-Kotb, et al, "Feasibility study of a Junction Termination using Deep Trench Isolation Technique for the Realization of DT-SJMOSFETs", IEEE International Symposium on Power Semiconductor Devices and ICs, pp. 303–306, 2008.
16. L. Theolier, et al, "A New Junction Termination Technique: Deep Trench Termination ($DT^2$)", IEEE International Symposium on Power Semiconductor Devices and ICs, pp. 176–179, 2009.
17. P.M. Shenoy, A. Bhalla, and G.M. Dolny, "Analysis of the Effect of Charge Imbalance on the Static and Dynamic Characteristics of the Super-Junction MOSFET", IEEE International Symposium on Power Semiconductor Devices and ICs, pp. 99–102, 1999.
18. M. Rub, et al, "A Novel Trench Concept for the Fabrication of Compensation Devices", IEEE International Symposium on Power Semiconductor Devices and ICs, pp. 203–206, 2003.
19. M. Buzzo, et al, "Characterization of Unconventional Engineering Solutions for Super-Junction Devices", IEEE International Symposium on Power Semiconductor Devices and ICs, pp. 149–152, 2006.

# Chapter 8
# Integral Diode

The power MOSFET device is often used in circuits which produce current flow through the structure in the third quadrant of its $i - v$ characteristics. Two prominent examples of such circuits are the voltage regulator module (VRM) used to deliver power to microprocessors in computers and the H-bridge motor control circuits used to achieve adjustable speed drives. One of the advantages of the power MOSFET structure is an inherent reverse conducting diode within the structure which allows carrying current in the third quadrant of operation. Unfortunately, the switching speed of this diode is very slow in as-fabricated devices producing excessive power losses that limit the circuit operating frequency. This problem was first overcome by the use of electron irradiation to control the minority carrier lifetime to achieve improved reverse recovery characteristics for the body diode in the power D-MOSFET structure [1].

A circuit approach that has been utilized in VRMs to addressing the poor reverse recovery behavior of the integral diode in power MOSFET devices is the parallel connection of an external Schottky diode. This approach has been found to be ineffective due to the inductance between the separately packaged power MOSFET and Schottky diode which results in some of the current flowing through the body diode of the power MOSFET structure. A preferred approach to improving the characteristics of the reverse conducting diode is by incorporation of a Schottky diode within the power MOSFET chip [2]. The reverse leakage current from the integral Schottky diode can be minimized by utilizing the junction barrier control concept [3] for the planar gate power MOSFET structure and the MOS barrier control concept [4], [5] for the trench gate power MOSFET structure.

In the case of high voltage power MOSFET devices, it is possible to utilize an integrated Schottky structure to create a Merged-PiN-Schottky rectifier [6] which has superior reverse recovery characteristics when compared with the P-N body diode. With advent of the super-junction devices discussed in Chap. 7, the performance of the body diode has to be re-examined due to very large junction area introduced in these structures [7]. The reverse recovery characteristics of the body diode in the super-junction power MOSFET structures can also be controlled by using electron irradiation [8].

B.J. Baliga, *Advanced Power MOSFET Concepts*,
DOI 10.1007/978-1-4419-5917-1_8, © Springer Science+Business Media, LLC 2010

## 8.1 Power MOSFET Body Diode

A cross-section of the basic cell structure for the power U-MOSFET structure is illustrated in Fig. 8.1 with the body diode highlighted. An equivalent circuit for the power MOSFET with its body-diode is included in the figure. The P-base region of the power MOSFET structure serves as the P-anode region of the integral diode. When a negative drain bias is applied to the power MOSFET structure, the integral diode becomes forward biased leading to the injection of holes from the P-base region into the N-drift region. A large concentration of holes is stored within the N-drift region due to this minority injection phenomenon. When the drain voltage of the power MOSFET switches from negative to a positive, the stored charge in the drift region must be removed before the P-N junction $J_1$ is able to support voltage. The removal of the stored charge produces a large reverse recovery current which results in power dissipation and loss of circuit operating efficiency.

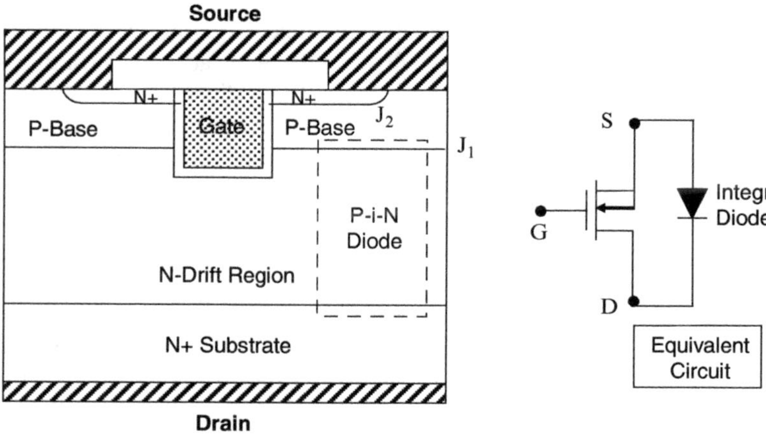


**Fig. 8.1** The body diode within the power U-MOSFET structure

## 8.2 Computer Power Supplies

One of the major applications for silicon power MOSFET devices is in computer power supplies that provide power to microprocessors. The power delivery in a computer consists of conversion of the high voltage AC input power to a DC back-plane power source with a typical voltage of 17 V. The power delivery to the microprocessor is performed using a voltage regulator module (VRM) that converts the 17-V DC power to the 1-V level DC power required by modern microprocessors. Typical current levels required to operate the microprocessor can range above 100 A. This large current delivery is accomplished by using VRMs operated in parallel with each VRM delivering approximately 20 A.

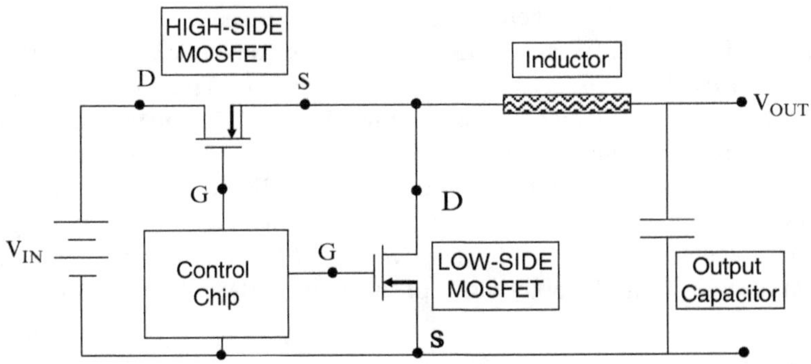


**Fig. 8.2** Sync-buck DC–DC converter circuit used for the VRM

The commonly used sync-buck topology for the DC-to-DC voltage conversion in the VRM is shown in Fig. 8.2. Due to the relatively low operating voltage in this circuit, power MOSFET devices are typically used as the switch in the high-side and low-side locations as illustrated in the figure. When the high-side MOSFET is turned on by the control circuit, current flows from the DC input source through the inductor to the load connected at the output terminals. When the high-side transistor is switched off by the control circuit, the load current circulates through the MOSFET connected on the low-side and the inductor. Since the current in the low-side power MOSFET flows from the source to the drain, it is operated in the third quadrant (i.e. with a negative drain bias) under these circuit conditions.

The regulation of the DC output voltage can be achieved by adjusting the on-time of the high-side transistor [9]. The power MOSFET on the low-side can be replaced by a Schottky diode [10] but the power MOSFET is preferred in order to reduce on-state losses. However, in this case, the gate signal must be synchronized with the voltage across the low-side power MOSFET so that it is turned-on only when its drain bias is negative. The low-side MOSFET is therefore also referred to as a synchronous rectifier.

The current flow in the low-side MOSFET is in the opposite direction to the normal operation of power MOSFET devices. If the low-side power MOSFET is turned on by the control circuit, the current flow occurs with a voltage drop determined by the on-resistance of the MOSFET structure. This voltage drop is usually much smaller than the built-in potential of the junction between the P-base region and the N-drift region (the body-diode of the MOSFET). The high-side and low-side MOSFET devices cannot be simultaneously turned-on during circuit operation to prevent short-circuiting of the input power source. This requires delaying the turn-on of the low-side transistor after the gate bias to the high-side transistor has been turned-off. During this delay-time interval, the current in the low-side MOSFET flows through its body-diode. The maximum operating frequency of the converter can be limited by the slow reverse recovery of the body diode.

Three power MOSFET structural options are discussed in this section. The first is the power U-MOSFET structure as a bench-mark conventional power MOSFET structure that has been widely utilized in VRM circuits. The second is the power CC-MOSFET structure which has been shown in Chap. 5 to have a much smaller specific on-resistance when compared with the power U-MOSFET structure. The third option is a power MOSFET structure, called the JBSFET structure, with a Schottky diode integrated within the cell structure. The monolithic integration of the Schottky diode within a planar power MOSFET structure suppresses the injection of holes from the body diode of the power MOSFET structure.

### *8.2.1 Power U-MOSFET Structure*

The power U-MOSFET structure was illustrated in Fig. 8.1 with its body diode highlighted. The body diode in the power U-MOSFET structure can be analyzed using the physics described for the P-i-N rectifier in the textbook [11]. The on-state current density ($J_T$) for the P-i-N rectifier can be related to the total on-state voltage drop ($V_{ON}$) by:

$$J_T = \frac{2qD_a n_i}{d} F\left(\frac{d}{L_a}\right) e^{\frac{qV_{ON}}{2kT}} \tag{8.1}$$

where d is half the thickness of the drift region in the power MOSFET structure, $L_a$ is the ambipolar diffusion length, $D_a$ is the ambipolar diffusion co-efficient, $n_i$ is the intrinsic carrier concentration, k is Boltzmann's constant, and T is the absolute temperature.

Analytical expressions for the reverse recovery waveforms were also derived in the textbook allowing determination of the peak reverse recovery current and the stored charge. These equations can also be applied to the analysis of the reverse recovery of the body diode in the power U-MOSFET structure. The analysis demonstrates that the peak reverse recovery current and the stored charge can be reduced by reduction of the minority carrier lifetime. This has been experimentally demonstrated by using electron irradiation [1]. The operation of the 30-V power U-MOSFET structure is described in this section using the results of numerical simulations.

The current density flowing through the body diode in its on-state is the same value as the current density flowing through the power MOSFET structure when the gate bias has been applied because they are both determined by the load current in the VRM. The on-state current density will therefore be assumed to be limited by the on-state power dissipation to 100 W/cm$^2$ as determined by the specific on-resistance of the power MOSFET structure:

$$J_{ON} = \sqrt{\frac{P_{D,ON}}{R_{ON,SP}}} \tag{8.2}$$

### 8.2.1.1 Simulation Results

The results of two-dimensional numerical simulations on the body diode within the 30 V power U-MOSFET structure are described here to provide a detailed understanding of the underlying device physics and operation. The structure used for the numerical simulations had the same cell parameters as the 30-V power U-MOSFET structure discussed in Chap. 3. Consequently, the body diode had a drift region thickness of 3 μm below the P-base region with a doping concentration of $1.6 \times 10^{16}/cm^3$. For the numerical simulations, half the cell (with a width of 1.5 μm) was utilized as a unit cell that is representative of the structure.

The forward conduction characteristics of the body diode within the power U-MOSFET structure were obtained by application of a negative bias to the drain terminal while using zero gate bias. This is representative of the condition in the VRM circuit when both the high-side and low-side power MOSFET switches are turned-off. The load current then flows through the body diode within the power MOSFET structure. The results of the numerical simulations are shown in Fig. 8.3 for the case of 5 values for the minority carrier lifetime. During the simulations, the lifetimes for holes ($\tau_{p0}$) in heavily doped N-type material and electrons ($\tau_{n0}$) in heavily doped P-type material were assumed to be equal to values provided in the figure.

The $i - v$ characteristics follow the diode law under high level injection conditions (see (8.1)) over a broad range of current levels for the larger lifetime values. The on-state voltage drop, at the on-state current density determined by power

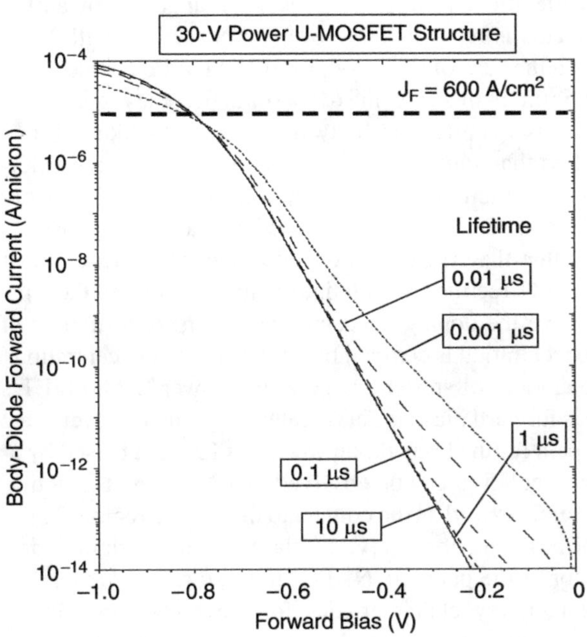


**Fig. 8.3** On-state characteristics of the body-diode within the power U-MOSFET structure

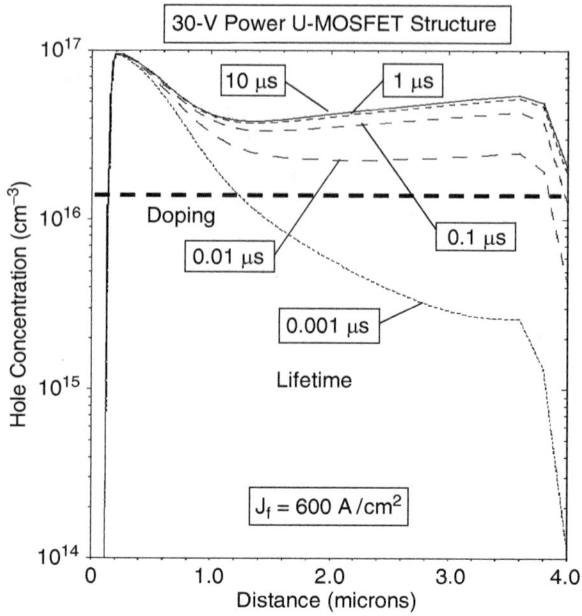


**Fig. 8.4** Hole distribution in the 30-V power U-MOSFET structure

dissipation considerations (see (8.2)), does not change significantly with reduction of the lifetime. This is representative of a P-i-N rectifier with (d/La) ratio of less than unity [11]. In the case of the 30-V power U-MOSFET structure, the parameter d has a value of 1.5 μm. In this case, the (d/La) ratio becomes equal to unity at a lifetime of 0.0015 μs. Consequently, the body diode behaves like a P-i-N rectifier with (d/La) ratio of less than unity over the entire range of lifetime values in the figure.

The hole concentration profiles in the power U-MOSFET structure are shown in Fig. 8.4 when the body diode is forward biased with a current density of 600 A/cm$^2$. It can be observed that the drift region operates under high-level injection conditions like that for a P-i-N rectifier for lifetime values as low as 0.01 μs with the hole concentration above the doping concentration. Even for a lifetime of 0.001 μs, the injected hole concentration is comparable to the doping concentration. Consequently, there is a large amount of stored charge in the power U-MOSFET structure when the body diode is forward biased. This creates a significant reverse recovery current.

The current path for the body diode when it is forward biased by a negative drain bias is shown in Fig. 8.5. It can be observed that the current flow lines pass through the N$^+$ source region as well as the contact to the P-base region. This occurs because the N$^+$ source/P-base junction is reverse biased when the drain bias is negative and acts as a collector of the parasitic N$^+$-P-N transistor.

The reverse recovery characteristics for the body diode in the 30-V power U-MOSFET structure were obtained by initially forward biasing the body diode with an on-state current density of 600 A/cm$^2$ corresponding with a on-state power

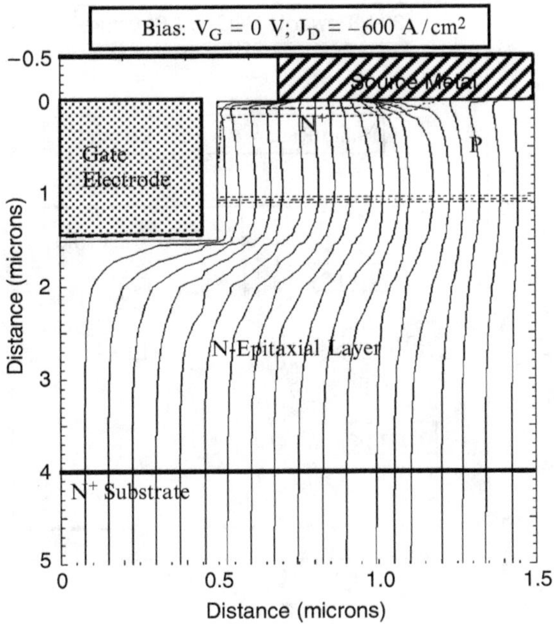


**Fig. 8.5** Body diode current path within the 30-V power U-MOSFET structure

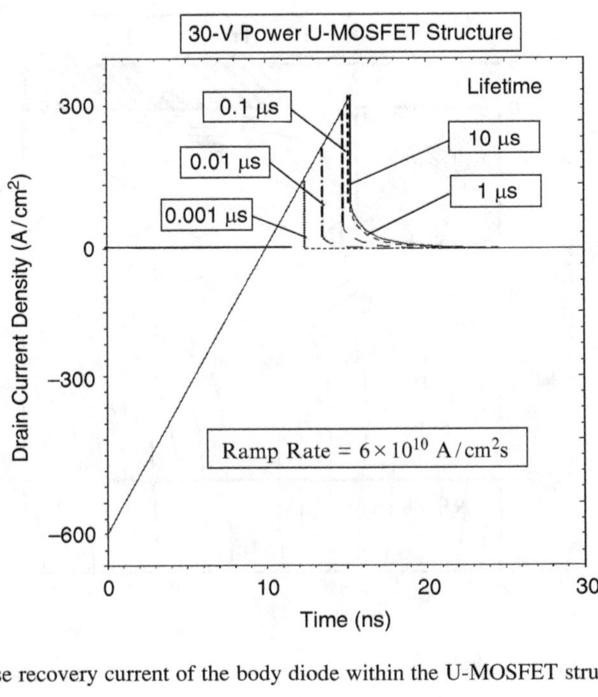


**Fig. 8.6** Reverse recovery current of the body diode within the U-MOSFET structure

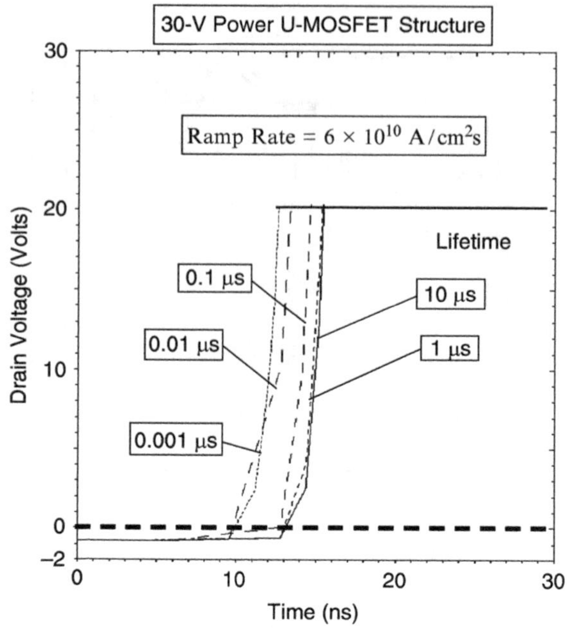


**Fig. 8.7** Reverse recovery voltage of the body diode within the U-MOSFET structure

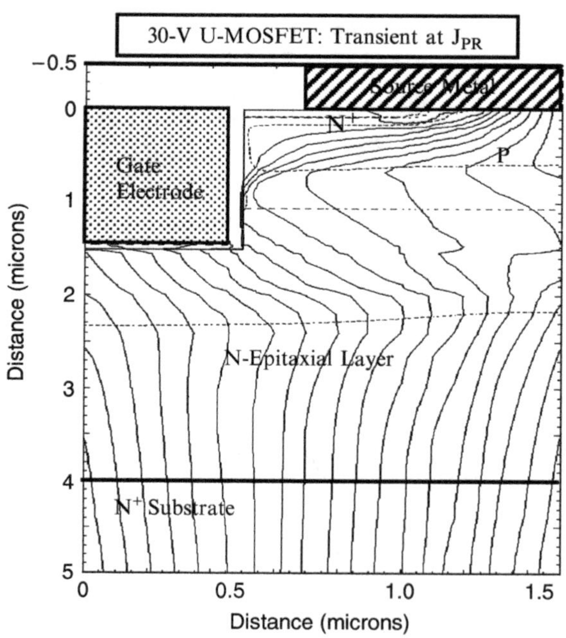


**Fig. 8.8** Current path during reverse recovery of the 30-V power U-MOSFET structure

dissipation of 100 W/cm$^2$ based up on the specific on-resistance of this structure. The drain current was ramped from negative to positive at a ramp rate of 60,000 A/cm$^2$ μs. The resulting waveforms for the drain current and drain voltage are shown in Figs. 8.6 and 8.7 for various lifetime values in the drift region. It can be observed that the peak reverse recovery current density ($J_{PR}$) is 312 A/cm$^2$ for the case of a lifetime of 10 μs. It is reduced to 140 A/cm$^2$ when the lifetime is reduced to 0.001 μs. Such small lifetime values would require extremely large electron irradiation doses which would damage the gate oxide. Consequently, it is difficult to significantly reduce the peak reverse recovery current in the body diode for the 30-V power U-MOSFET structure by using lifetime control processes.

The current path within the 30-V power U-MOSFET structure during the reverse recovery transient when the drain current reaches the peak reverse recovery current is shown in Fig. 8.8. It can be observed that the current flow lines do not pass through the N$^+$ source region under these conditions because the N$^+$ source/P-base junction is forward biased when the drain bias is positive. In addition, a significant current flow occurs via the gate electrode because of the large rate of change of the drain voltage [$dV_D/dt$] as can be seen in Fig. 8.7.

## *8.2.2 Power CC-MOSFET Structure*

The power CC-MOSFET structure with source electrode in the trenches was illustrated in Fig. 5.2. Its body diode is formed between the P-base region and the N-drift region as in the case of the power U-MOSFET structure discussed above. Consequently, the body diode in the power CC-MOSFET structure can also be analyzed using the physics described for the P-i-N rectifier in the textbook [11]. However, the doping concentration of the drift region in the power CC-MOSFET structure is much larger than in the power U-MOSFET structure. Consequently, the body diode operates under low-level injection conditions over a broader range of on-state current densities including the on-state current density of the body diode. This produces superior reverse recovery behavior for the body diode in the power CC-MOSFET structure.

Due to low-level injection conditions for the body diode in the power CC-MOSFET structure, the on-state current density ($J_T$) can be related to the on-state voltage drop ($V_{ON}$) by:

$$J_T = \frac{qD_p p_{ON}}{L_p}\left(e^{qV_F/kT} - 1\right) \tag{8.3}$$

where $p_{ON}$ is the equilibrium hole concentration in the N-drift region and $L_p$ is the diffusion length for holes in the N-drift region.

Due to low-level injection conditions in the drift region, the reverse recovery for the body diode in the power CC-MOSFET structure is determined by the establishment of a depletion region in the N-drift region without stored charge. The charge in the depletion region is equal to the ionized donor concentration ($N_D$) in the drift region. At any point in time after the drain current traverses zero (in the case of the body diode reverse recovery, the drain current changes from negative to positive at zero crossing), the charge removed by the drain current flow in a small time interval (dt) must be equal to the charge removed from a small incremental change in the depletion layer width ($dW_D$):

$$J_D(t)dt = qN_D dW_D \tag{8.4}$$

Since the drain current density is changing at a fixed ramp rate (a):

$$(at)dt = qN_D dW_D \tag{8.5}$$

Performing the integration on both sides yields:

$$W_D(t) = \frac{at^2}{2qN_D} \tag{8.6}$$

The drain voltage is related to the depletion region width:

$$V_D(t) = \frac{qN_D}{2\,\varepsilon_S} W_D^2(t) = \frac{a^2t^4}{8q\,\varepsilon_S N_D} \tag{8.7}$$

by using (8.6). The peak reverse recovery current occurs at time ($t_2$) when the drain voltage becomes equal to the drain supply voltage ($V_{DS}$). Using this criterion in (8.7) yields an equation for the time after zero crossing when the drain current reaches its peak reverse recovery value:

$$t_{PR} = \left(\frac{8q\varepsilon_S N_D V_{DS}}{a^2}\right)^{1/4} \tag{8.8}$$

The peak reverse recovery current density can be obtained by multiplying this time with the ramp rate:

$$J_{PR} = at_{PR} = \left(8qa^2\varepsilon_S N_D V_{DS}\right)^{1/4} \tag{8.9}$$

Based up on this equation, it can be concluded that the peak reverse recovery current for the body diode in the power CC-MOSFET structure will increase with increasing ramp rate and drain supply voltage. Note that the reverse recovery current is independent of the lifetime according to this analysis. In the case of a doping concentration in the drift region of $1 \times 10^{17}$/cm$^3$, a drain supply voltage of 20 V, and a ramp rate of $6 \times 10^{10}$ A/cm$^2$ s, the peak reverse recovery current density predicted by this expression is 313 A/cm$^2$.

#### 8.2.2.1 Simulation Results

The results of two-dimensional numerical simulations on the body diode within the 30 V power CC-MOSFET structure are described here to provide a detailed understanding of the underlying device physics and operation. The structure used for the numerical simulations had the same cell parameters as the 30-V power CC-MOSFET structure discussed in Chap. 5. Consequently, the body diode had a drift region thickness of 2 μm below the P-base region with a doping concentration of $1 \times 10^{17}/\text{cm}^3$. For the numerical simulations, half the cell (with a width of 0.5 μm) was utilized as a unit cell that is representative of the structure.

The forward conduction characteristics of the body diode within the power CC-MOSFET structure were obtained by application of a negative bias to the drain terminal while using zero gate bias. This is representative of the condition in the VRM circuit when both the high-side and low-side power MOSFET switches are turned-off. The load current then flows through the body diode within the power MOSFET structure. The results of the numerical simulations are shown in Fig. 8.9 for the case of 5 values for the minority carrier lifetime. During the simulations, the lifetimes for holes ($\tau_{p0}$) in heavily doped N-type material and electrons ($\tau_{n0}$) in heavily doped P-type material were assumed to be equal to values provided in the figure. The $i - v$ characteristics follow the diode law under low-level injection conditions (see (8.3)) over a broad range of current levels for the larger lifetime values. The on-state voltage drop, at the on-state current density determined by

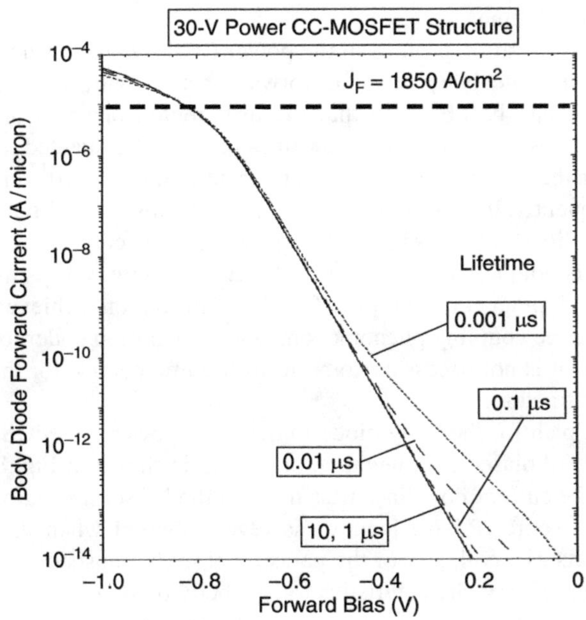


**Fig. 8.9** On-state characteristics of the body-diode within the CC-MOSFET structure

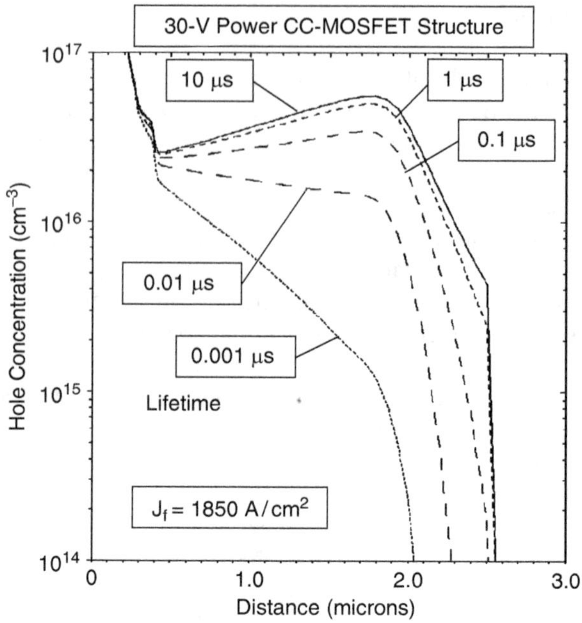


**Fig. 8.10** Hole distribution in the 30-V power CC-MOSFET structure

power dissipation considerations (see (8.2)), does not change significantly with reduction of the lifetime.

The hole concentration profiles in the power CC-MOSFET structure are shown in Fig. 8.10 when the body diode is forward biased with a current density of 1,850 A/cm$^2$. It can be observed that the drift region operates under low-level injection conditions for all the lifetime values because the injected hole concentration is less than the relatively high doping concentration ($1 \times 10^{17}$/cm$^3$) of the drift region. Consequently, there is no stored charge in the power CC-MOSFET structure when the body diode is forward biased in spite of the injection of minority carriers. This behavior is unique among power MOSFET structures and is associated with the very high doping concentration for the drift region achieved by the two dimensional charge coupling phenomenon. Due to operation under low-level injection conditions, it is not necessary to perform lifetime control for the 30-V power CC-MOSFET structure.

The current path for the body diode in the 30-V power CC-MOSFET structure when it is forward biased by a negative drain bias is shown in Fig. 8.11. It can be observed that the current flow lines pass through the $N^+$ source region. This occurs because the $N^+$ source/P-base junction is reverse biased when the drain bias is negative and acts as a collector of the parasitic $N^+$-P-N transistor.

The reverse recovery characteristics for the body diode in the 30-V power CC-MOSFET structure were obtained by initially forward biasing the body diode with an on-state current density of 1,850 A/cm$^2$ corresponding with a on-state power

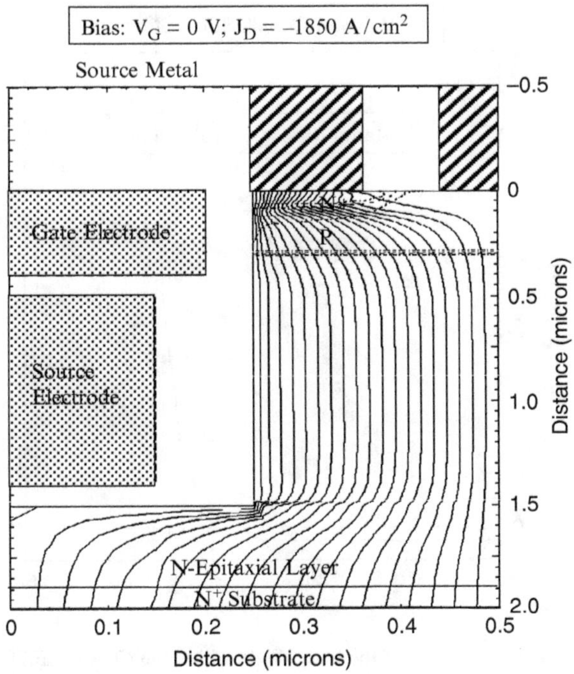


**Fig. 8.11** Body diode current path within the 30-V power CC-MOSFET structure

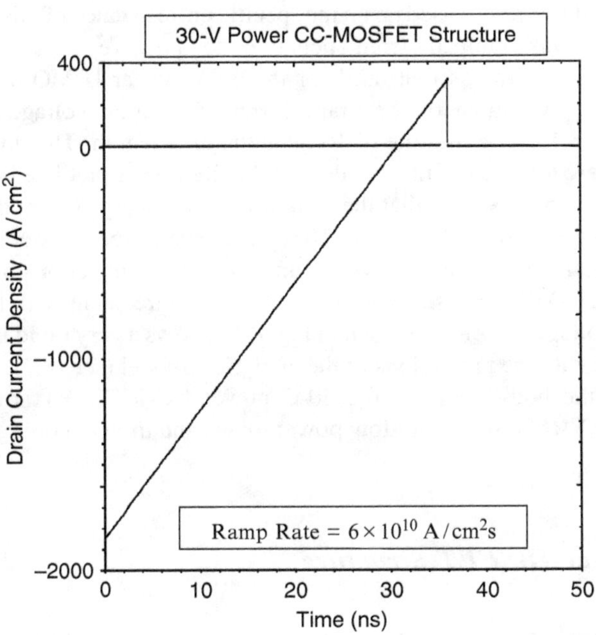


**Fig. 8.12** Reverse recovery current of the body diode within the CC-MOSFET structure

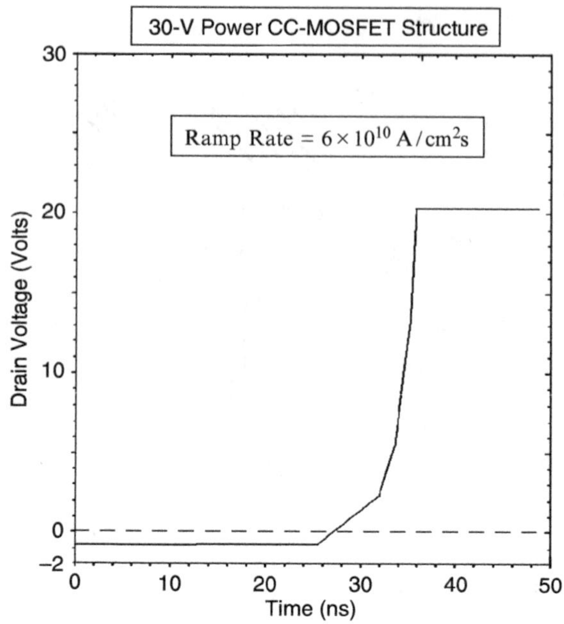


**Fig. 8.13** Reverse recovery voltage of the body diode within the CC-MOSFET structure

dissipation of 100 W/cm$^2$ based up on the specific on-resistance of this structure. The drain current was ramped from negative to positive at a ramp rate of 60,000 A/cm$^2$ μs (same as that used for the 30-V power U-MOSFET structure). The resulting waveforms for the drain current and drain voltage are shown in Figs. 8.12 and 8.13 for a lifetime of 10 μs in the drift region. The reverse recovery waveforms were found to be independent of the lifetime as predicted by the analytical model. It can be observed that the peak reverse recovery current density ($J_{PR}$) is 310 A/cm$^2$ for the case of a lifetime of 10 μs. The reverse recovery current predicted by the analytical model (see (8.9)) based up on low-level injection conditions in the 30-V power CC-MOSFET structure is in excellent agreement with the simulation results. The voltage waveform shown in Fig. 8.13 shows a very rapid increase which is also consist with the predictions of the analytical model (see (8.7)). These results indicate that the body diode of the 30-C power CC-MOSFET structure can be utilized in the VRM circuit with low power losses and high operating speed.

### *8.2.3 Power JBSFET Structure*

As described earlier in this chapter, the use of an externally connected anti-parallel Schottky barrier diode across the synch-FET in the VRM circuit has been found

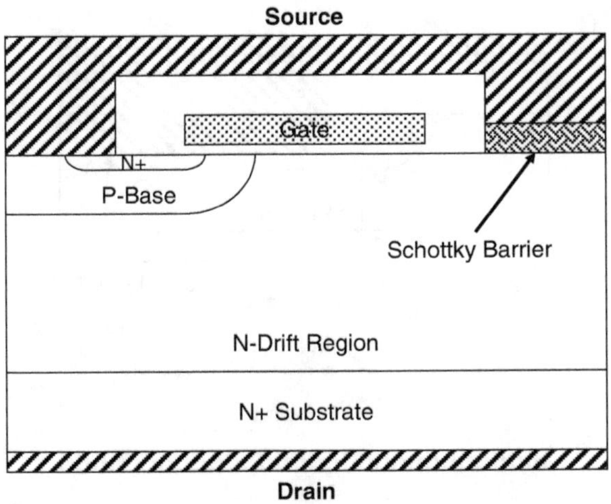


**Fig. 8.14** Schottky diode integrated within the power D-MOSFET structure

to be in-effective due to the parasitic inductance in the package and connections. A superior approach is by the integration of the Schottky rectifier inside the power MOSFET structure.

One approach to the integration of a Schottky rectifier with the power MOSFET structure is illustrated in Fig. 8.14. In this case, a part of chip area is assigned to the Schottky rectifier [2]. Due to large area required for the Schottky contact in order to carry the same current as the power MOSFET structure when it is turned-on, a separate Schottky metal can be utilized to optimize the on-state voltage drop and leakage current of the Schottky rectifier. The relatively large leakage current of the Schottky rectifier due to the barrier lowering phenomenon [11] is a significant disadvantage of this approach. In addition, this approach increases the processing complexity and cost.

A superior approach to the integration of the Schottky rectifier with the power MOSFET structure is by merging the Schottky rectifier with the cell structure for the power MOSFET structure. This approach was first successfully implemented with the power SSCFET structure previously described in Chap. 4 to create the power JBSFET structure [3]. The power JBSFET structure is illustrated in Fig. 8.15. It be seen that the Schottky contact is located in the middle of the P-base region within the power MOSFET cell structure. The area of the Schottky contact can be adjusted to optimize the on-state voltage drop and the leakage current. One of the challenges with this approach is that it is only practical to use the source contact metal to form the Schottky contact. The source contact process is usually optimized to create a good ohmic contact to the $N^+$ source region. In contrast, the same metal and its annealing process must create a sufficiently low leakage current at the Schottky contact in the JBSFET structure. Fortunately, this is possible because the P-body region of the power MOSFET structure is utilized to shield the Schottky

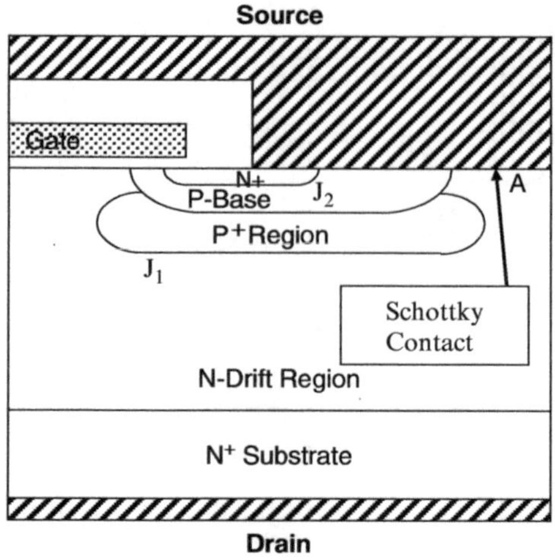


**Fig. 8.15** The power JBSFET structure

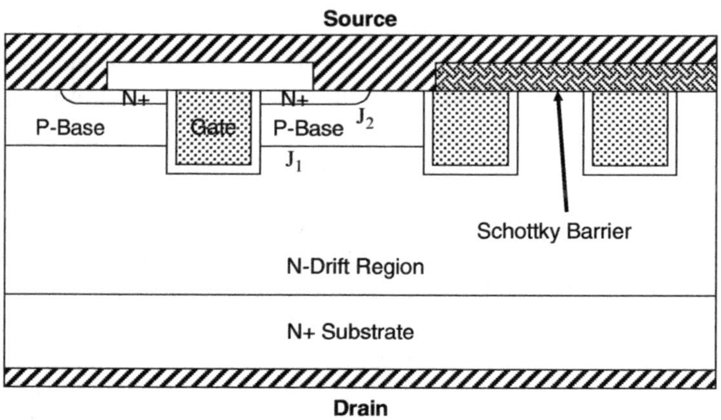


**Fig. 8.16** The power U-MOSFET structure with integrated TMBS rectifier

contact from high electric fields developed in the N-drift region by the drain bias when the device is supporting large voltages in the blocking mode. Schottky contacts that are shielded by using P-N junctions are referred to as Junction-Barrier-controlled Schottky (JBS) rectifiers [10] from which the name of the power JBSFET structure is derived.

Yet another approach to the integration of a Schottky rectifier which is suitable for the power U-MOSFET structure [4, 5] is illustrated in Fig. 8.16. This approach utilized a separate area for the Schottky contact like the structure previously shown in Fig. 8.14. However, the leakage current due to Schottky barrier lowering is mitigated by using the MOS-barrier concept originally proposed for discrete trench-MOS Barrier Schottky (TMBS) rectifiers [10].

In all the power MOSFET structures with integrated Schottky rectifiers, it is possible to completely suppress the injection of minority carriers into the drift region by providing sufficient area for the Schottky contact for unipolar current flow. The physics of operation of these types of rectifiers is treated in detail in another companion book [10]. In this section, the advantages of using this approach will be illustrated using the power JBSFET structure which is available as a commercially viable product.

### 8.2.3.1 Simulation Results

The results of two-dimensional numerical simulations on the body diode within the 30 V power JBSFET structure are described here to provide a detailed understanding of the underlying device physics and operation. The structure used for the numerical simulations had similar cell parameters and doping profiles as the 30-V power SC-MOSFET structure discussed in Chap. 4. Consequently, the body diode had a drift region thickness of 3 μm below the P-base region with a doping concentration of $1.6 \times 10^{16}/cm^3$. For the numerical simulations, half the cell (with a width of 1.5 μm) was utilized as a unit cell that is representative of the power JBSFET structure. A three-dimensional view of the doping profile for the power JBSFET structure is provided in Fig. 8.17. The MOSFET region is located on

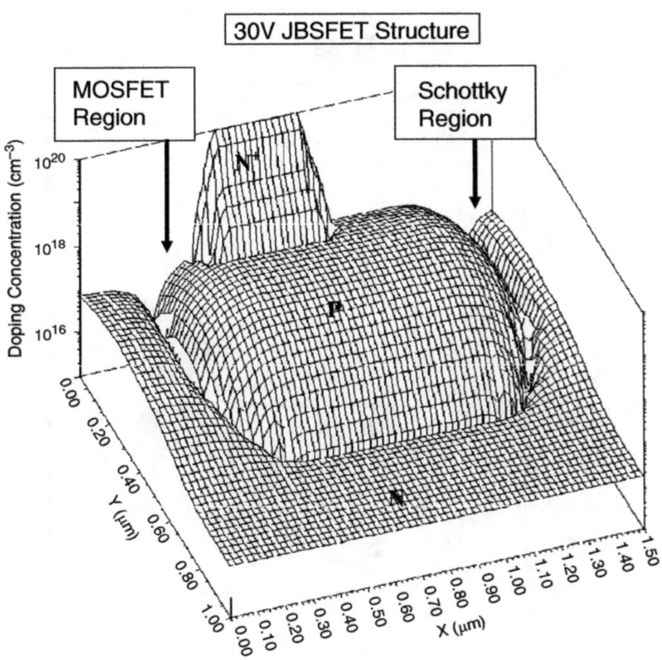


**Fig. 8.17** Doping profile within the power JBSFET structure

the upper left-hand-side while the Schottky contact is located on the upper right-hand-side. The P-base region of the power MOSFET portion is also used to form the $P^+$ shielding region that protects the Schottky contact from high electric fields in the N-drift region. It is worth pointing out that the same retrograde N-type doping profile used in the JFET region of the MOSFET portion has been incorporated under the Schottky contact for process simplicity. This can be observed in the doping profile shown in Fig. 8.18 under the Schottky contact within the 30-V power JBSFET structure. In spite of this larger doping concentration in the vicinity of the Schottky contact, the electric field is still suppressed under the Schottky contact as shown by numerical simulations below.

The transfer characteristics for the 30-V power JBSFET structure were obtained using numerical simulations with a drain bias of 0.1 V at 300 and 400°K. The resulting transfer characteristics are shown in Fig. 8.19. From this graph, a threshold voltage of 2.5 and 2.0 V can be extracted at 300 and 400°K, respectively. The threshold voltage is the same as that observed for the power SC-MOSFET structure. The specific on-resistance can be obtained from the transfer characteristics at any gate bias voltage. For the case of a gate bias of 4.5 V and 300°K, the specific in-resistance is found to be 0.295 mΩ cm$^2$. For the case of a gate bias of 10 V and 300°K, the specific in-resistance is found to be 0.272 mΩ cm$^2$. These values are larger than those for the power SC-MOSFET structure because the cell pitch has been increased in the power JBSFET to accommodate the Schottky contact.

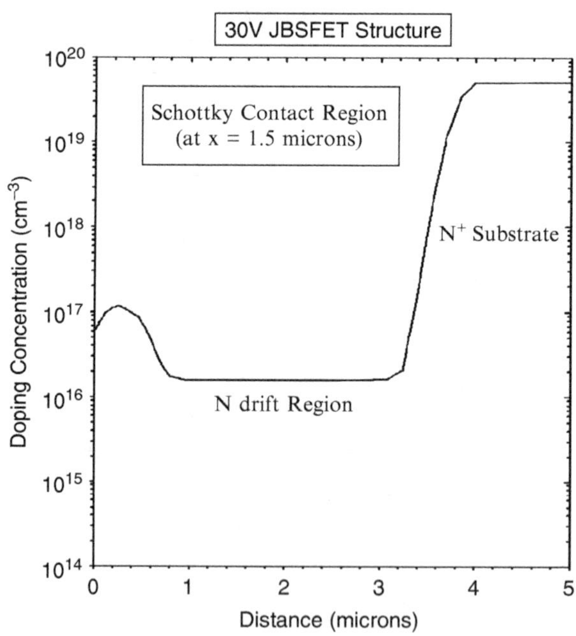


**Fig. 8.18** Doping profile under the Schottky contact within the power JBSFET structure

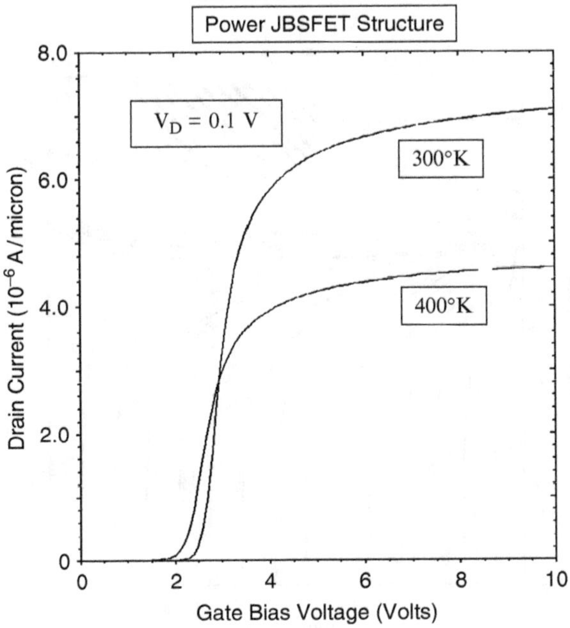


**Fig. 8.19** Transfer characteristics of the JBSFET structure

The on-state current flow pattern within the power JBSFET structure at a small positive drain bias of 0.1 V and a gate bias of 4.5 V is shown in Fig. 8.20. In the figure, the depletion layer boundary is shown by the dotted lines and the junction boundary is delineated by the dashed line. It can be observed that the current flows through the MOSFET region in the same manner as observed within the power SC-MOSFET structure (see Fig. 4.9). No current flows through the Schottky contact as expected because the metal-semiconductor junction is reverse biased under these bias conditions.

The blocking characteristic for the power JBSFET cell structure is shown in Fig. 8.21 at 300°K for a Schottky metal work function of 4.7 eV. It can be observed that the cell is capable of supporting 40 V similar to that observed for the power SC-MOSFET cell structure. This provides enough margin to achieve a device blocking voltage capability of slightly over 30 V after accounting for the reduction due to the edge termination. However, the leakage current in the power JBSFET structure is much larger due to the presence of the Schottky contact. The leakage current density in the power JBSFET structure is still at an acceptable level (3 mA/cm$^2$) at 300°K.

It is instructive to examine the potential contours inside the power JBSFET structure when it is operating in the blocking mode. This allows determination of the voltage distribution within the structure. The potential contours for the JBSFET structure obtained using the numerical simulations with zero gate bias and a drain

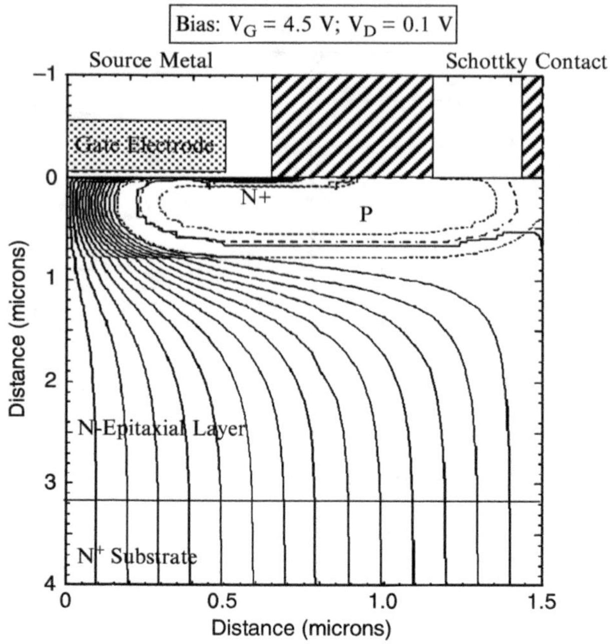


**Fig. 8.20** Current distribution in the power JBSFET structure in the first quadrant

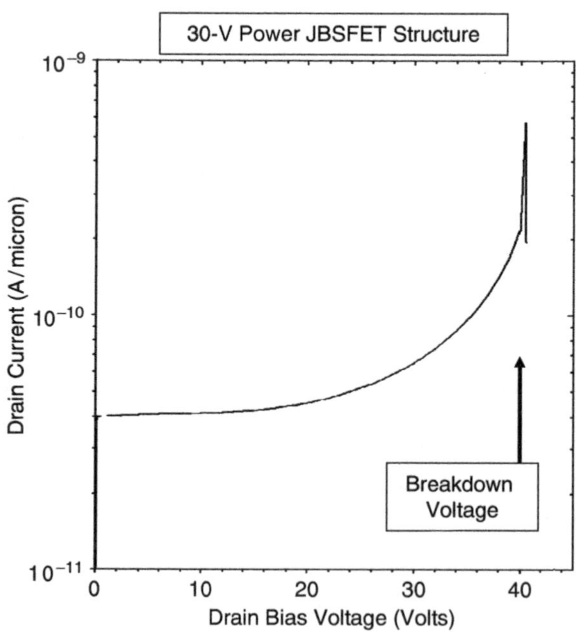


**Fig. 8.21** Blocking characteristic for the 30-V power JBSFET structure

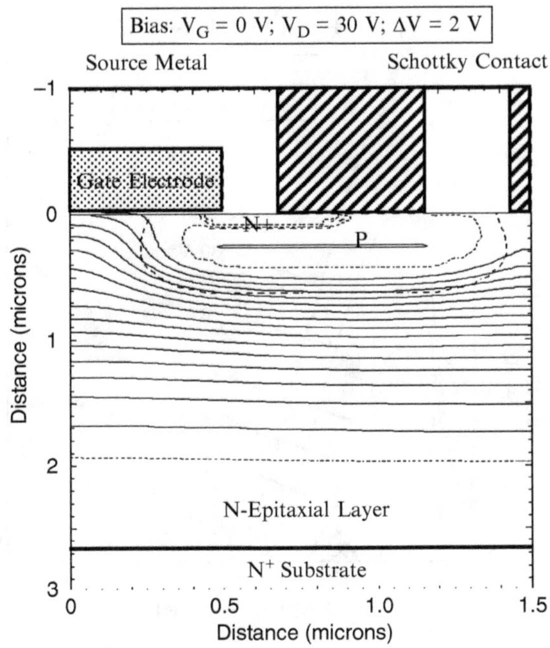


**Fig. 8.22** Potential contours in the power JBSFET structure

bias voltage of 30 V are shown in Fig. 8.22. From this figure, it can be observed that the surface region under the Schottky contact is screened from the drain bias by the $P^+$ shielding region. This suppresses the Schottky barrier lowering phenomenon and reduces the leakage current.

It is insightful to also examine the electric field profile inside the power JBSFET structure when it is operating in the blocking mode. The electric field profile obtained through the middle of the Schottky contact is shown in Fig. 8.23. It can be observed that the maximum electric field occurs at the junction at a depth of 0.6 μm from the surface. The electric field at the Schottky contact is greatly reduced ($7 \times 10^4$ V/cm) when compared the peak electric field in the bulk ($2.7 \times 10^5$ V/cm) at a drain bias of 30 V. The suppression of the electric field at the Schottky contact allows operation of the power JBSFET structure with a low leakage current despite using the source ohmic contact metal for making the Schottky contact.

The forward conduction characteristics of the body diode within the power JBSFET structure were obtained by application of a negative bias to the drain terminal while using zero gate bias. This is representative of the condition in the VRM circuit when both the high-side and low-side power MOSFET switches are turned-off. The load current then flows through the body diode within the power MOSFET structure. The results of the numerical simulations are shown in Fig. 8.24 for the case of a minority carrier lifetime of 10 μs. When compared with the P-N

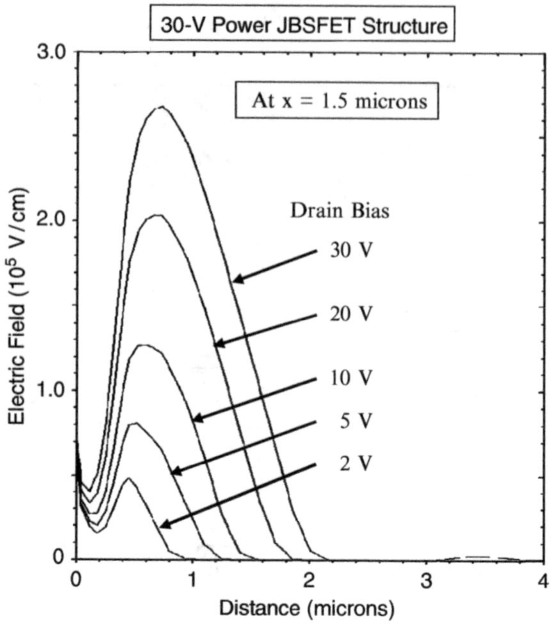


**Fig. 8.23** Electric field distribution in the power JBSFET structure

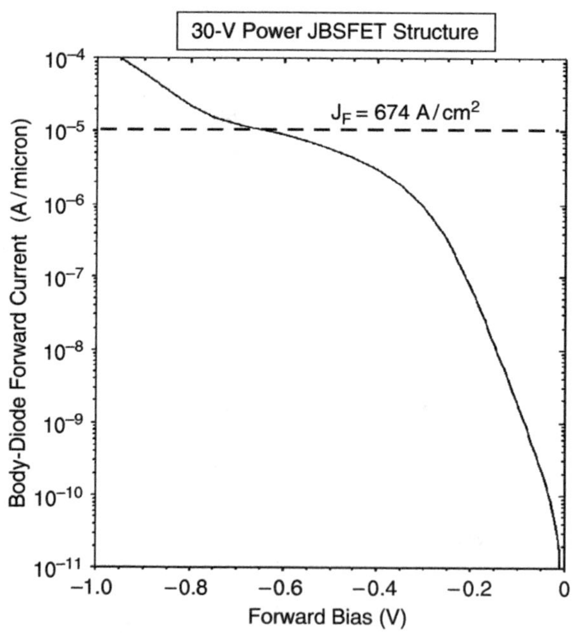


**Fig. 8.24** On-state characteristics of the body-diode within the power JBSFET structure

body-diodes of the previously described power MOSFET structures in this chapter, the $i - v$ characteristic for the power JBSFET structure is shifted towards lower voltages due to current flow via the Schottky contact. An inflection in the $i - v$ characteristics is observed at a drain bias of about 0.8 V indicating the on-set of minority carrier injection from the P-N junction. However, the area of the Schottky contact in the power JBSFET structure used for the numerical simulations is sufficiently large that the on-state voltage drop is only 0.625 V at even the relatively high on-state current density of 674 A/cm$^2$ for the device in the MOSFET mode of operation.

The hole concentration profile in the power JBSFET structure is shown in Fig. 8.25 when the body diode is forward biased with a current density of 674 A/cm$^2$. It can be observed that the drift region operates under low-level injection conditions for even for a lifetime of 10 μs. Consequently, there is very little stored charge in the power JBSFET structure when the body diode is forward biased. This creates good reverse recovery characteristics without using lifetime control processes. The reason for suppression of the injection of holes can be analyzed by observing the current path for the body diode in the power JBSFET structure, when it is forward biased by a negative drain bias, as shown in Fig. 8.26. It can be seen that the current flow is confined to the Schottky contact despite its small area within the device cell resulting in insufficient forward bias across the P-N junction to inject holes.

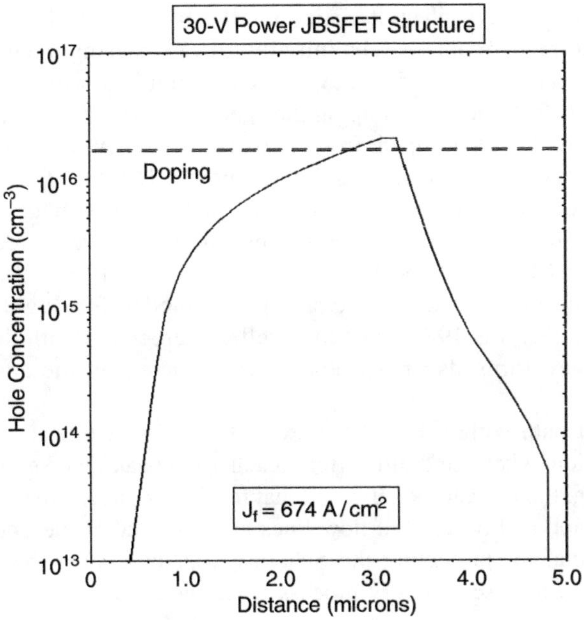


**Fig. 8.25** Hole distribution in the 30-V power JBSFET structure

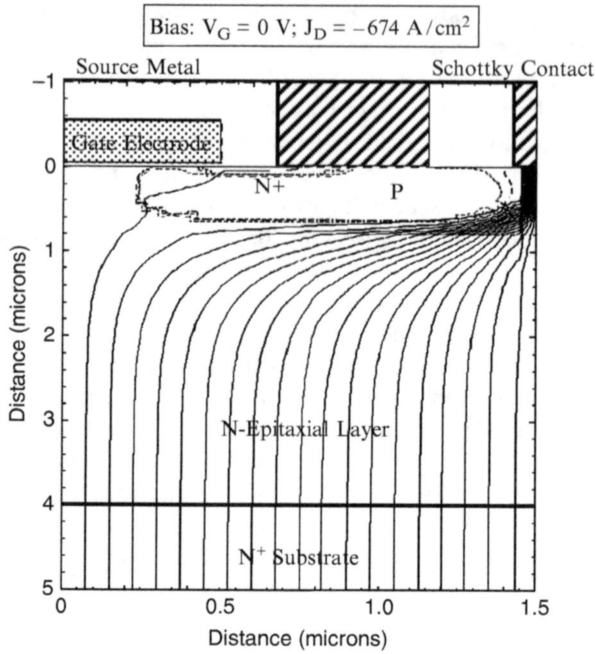


**Fig. 8.26** Body diode current path within the 30-V power JBSFET structure

The reverse recovery characteristics for the body diode in the 30-V power JBSFET structure were obtained by initially forward biasing the body diode with an on-state current density of 674 A/cm$^2$ corresponding with a on-state power dissipation of 100 W/cm$^2$ based up on the specific on-resistance of this structure. The drain current was ramped from negative to positive at a ramp rate of 60,000 A/cm$^2$ μs (the same rate as for the power U-MOSFET structure). The resulting waveforms for the drain current and drain voltage are shown in Figs. 8.27 and 8.28. It can be observed that the peak reverse recovery current density ($J_{PR}$) is 182 A/cm$^2$ for the case of a lifetime of 10 μs. The peak reverse recovery current predicted by the analytical model for low-level injection conditions (see (8.9)) is 198 A/cm$^2$ in excellent agreement with the simulations. The voltage waveform also rises rapidly as predicted by the analytical model (see (8.7)).

The current path within the 30-V power JBSFET structure during the reverse recovery transient when the drain current reaches the peak reverse recovery current is shown in Fig. 8.29. It can be observed that the current flow lines all pass through the P-base region and no current flow lines are observed at the Schottky contact. This is beneficial for protecting the Schottky contact against the high electrical stress during the reverse recovery process when the drain current and voltage are simultaneously large.

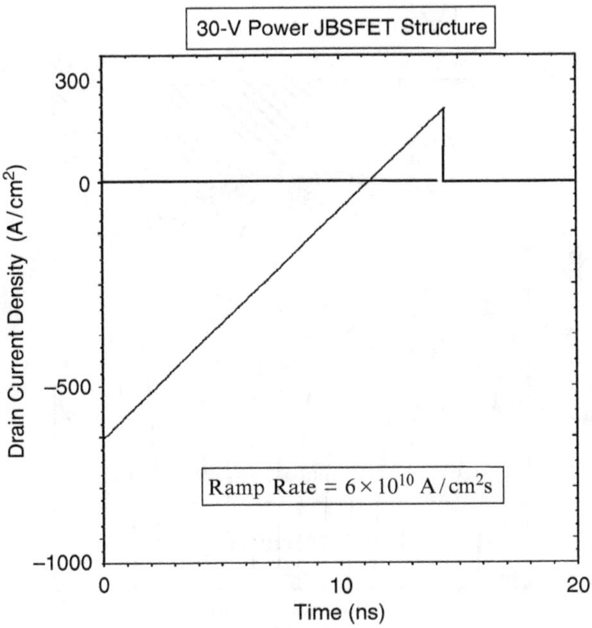


**Fig. 8.27** Reverse recovery current of the body diode within the power JBSFET structure

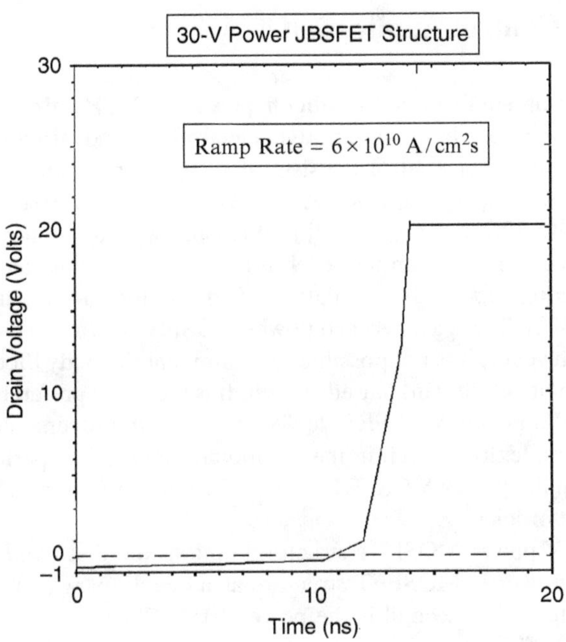


**Fig. 8.28** Reverse recovery voltage of the body diode within the JBSFET structure

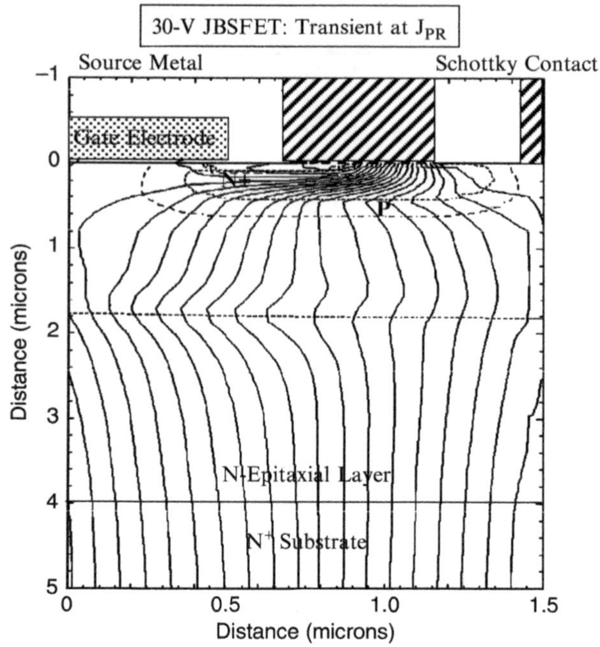


**Fig. 8.29** Current path during reverse recovery of the 30-V power JBSFET structure

## 8.3 Motor Control Application

The second major application for silicon power MOSFET devices is in motor control for disc-drives, fans, automotive electronics, and air-conditioning. The power delivery in motor control consists of conversion of the high voltage DC input power to a variable frequency AC voltage by the inverter. The commonly used H-bridge topology used for adjustable speed motor control is shown in Fig. 8.30. The current flow through each branch in the circuit can occur in either direction depending up on the regulation of the motor current. It is possible to utilize anti-parallel diodes across each power MOSFET device to carry this reverse current flow. However, it is not possible to ensure that the body diode in the power MOSFET does not get forward biased as well. It is therefore preferable to utilize the body diode in the power MOSFET devices to carry the reverse current to avoid the cost and complexity of adding the additional diodes. The performance of the integral diode in the power MOSFET devices then becomes important for determination of circuit efficiency.

Several 600 V power MOSFET structural options are discussed in this section. The first is the power U-MOSFET structure as a bench-mark conventional power MOSFET structure. The second is the power JBSFET structure which is a planar gate power MOSFET structure with a Schottky contact integrated into the cell structure. The third option is a power GD-MOSFET structure, which was shown in

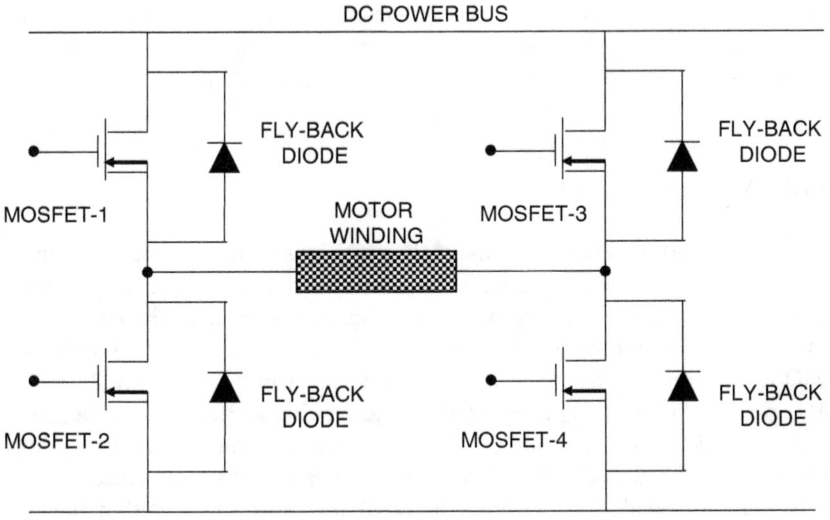


**Fig. 8.30** H-bridge circuits used for motor control

Chap. 6 to have a very low specific on-resistance. The fourth option is the power GD-MOSFET structure with an integrated Schottky contact to reduce the stored charge. The fifth option is the power SJ-MOSFET structure, which was shown in Chap. 7 to have a very low specific on-resistance. The sixth option is the power SJ-MOSFET structure with an integrated Schottky contact to reduce the stored charge.

## *8.3.1 Power U-MOSFET Structure*

The power U-MOSFET structure was illustrated in Fig. 8.1 with its body diode highlighted. As mentioned in Sect. 8.2.1, the body diode in the power U-MOSFET structure can be analyzed using the physics described for the P-i-N rectifier in the textbook [11]. Analytical expressions for the reverse recovery waveforms were also derived in the textbook allowing determination of the peak reverse recovery current and the stored charge. These equations can also be applied to the analysis of the reverse recovery of the body diode in the power U-MOSFET structure. The analysis demonstrates that the peak reverse recovery current and the stored charge can be reduced by reduction of the minority carrier lifetime. This has been experimentally demonstrated by using electron irradiation [1]. The operation of the 600-V power U-MOSFET structure is described in this section using the results of numerical simulations.

The current density flowing through the body diode in its on-state is the same value as the current density flowing through the power MOSFET structure when the gate bias has been applied because they are both determined by the current flowing through the motor windings. The on-state current density will therefore be assumed to be limited by the on-state power dissipation to 100 W/cm$^2$ as determined by the specific on-resistance of the power MOSFET structure (see (8.2)).

#### 8.3.1.1 Simulation Results

The results of two-dimensional numerical simulations on the body diode within the 600 V power U-MOSFET structure are described here to provide a detailed understanding of the underlying device physics and operation. The structure used for the numerical simulations had the same cell parameters as the 600-V power U-MOSFET structure discussed in Chap. 3. Consequently, the body diode had a drift region thickness of 60 μm below the P-base region with a doping concentration of $2.4 \times 10^{14}/\text{cm}^3$. For the numerical simulations, half the cell (with a width of 1.5 μm) was utilized as a unit cell that is representative of the structure.

The forward conduction characteristics of the body diode within the power U-MOSFET structure were obtained by application of a negative bias to the drain terminal while using zero gate bias. This is representative of the condition in the motor control H-bridge circuit when both the high-side and low-side power MOSFET switches are turned-off. The motor winding current then flows through the body diode within one of the power MOSFET structures. The results of the numerical simulations are shown in Fig. 8.31 for the case of 5 values for the

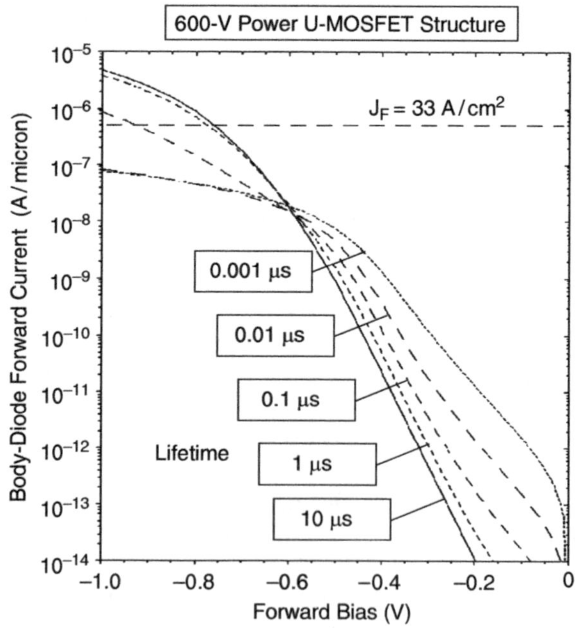


**Fig. 8.31** On-state characteristics of the body-diode within the power U-MOSFET structure

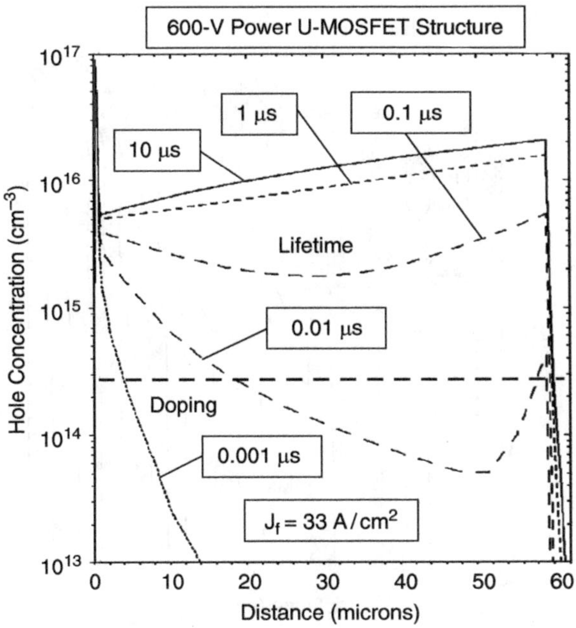


**Fig. 8.32** Hole distribution in the 600-V power U-MOSFET structure

minority carrier lifetime. During the simulations, the lifetimes for holes ($\tau_{p0}$) in heavily doped N-type material and electrons ($\tau_{n0}$) in heavily doped P-type material were assumed to be equal to values provided in the figure.

The $i - v$ characteristics follow the diode law under high level injection conditions (see (8.1)) over a broad range of current levels for the larger lifetime values. The on-state voltage drop, at the on-state current density determined by power dissipation considerations (see (8.2)), increases significantly when the lifetime is reduced below 1 microsecond. This is representative of a P-i-N rectifier with (d/La) ratio of greater than unity [11]. In the case of the 600-V power U-MOSFET structure, the parameter d has a value of 30 μm. In this case, the (d/La) ratio becomes equal to unity at a lifetime of 0.6 μs. Consequently, the body diode behaves like a P-i-N rectifier with (d/La) ratio of more than unity when the lifetime is reduced to 0.1 μs and lower.

The hole concentration profiles in the 600-V power U-MOSFET structure are shown in Fig. 8.32 when the body diode is forward biased with a current density of 33 A/cm$^2$. It can be observed that the drift region operates under high-level injection conditions like that for a P-i-N rectifier for lifetime values as low as 0.1 μs with the hole concentration above the doping concentration. At smaller lifetime values, a significant portion of the drift region remains un-modulated by the injected carriers. Consequently, the on-state voltage drop then becomes insensitive to the lifetime albeit at a large value.

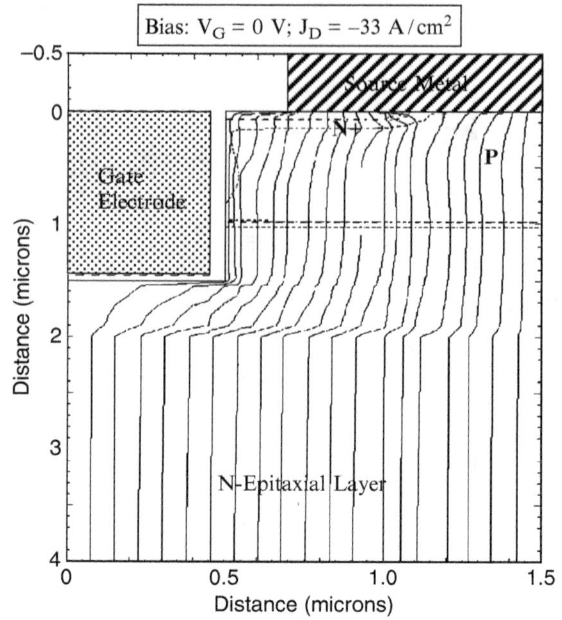


**Fig. 8.33** Body diode current path within the 600-V power U-MOSFET structure

The current path for the body diode in the 600-V power U-MOSFET structure, when it is forward biased by a negative drain bias, is shown in Fig. 8.33. It can be observed that the current flow lines pass through the $N^+$ source region as well as the contact to the P-base region. This occurs because the $N^+$ source/P-base junction is reverse biased when the drain bias is negative and acts as a collector of the parasitic $N^+$-P-N transistor.

The reverse recovery characteristics for the body diode in the 600-V power U-MOSFET structure were obtained by initially forward biasing the body diode with an on-state current density of 33 A/cm$^2$ corresponding with a on-state power dissipation of 100 W/cm$^2$ based up on the specific on-resistance of this structure. The drain current was ramped from negative to positive at a ramp rate of 3,300 A/cm$^2$ μs. The resulting waveforms for the drain current and drain voltage are shown in Figs. 8.34 and 8.35 for various lifetime values in the drift region. It can be observed that the peak reverse recovery current density ($J_{PR}$) is 263 A/cm$^2$ for the case of a lifetime of 10 μs. This value is eight-times larger than the on-state current density. Such a large reverse recovery current density creates significant power dissipation in the power MOSFET structure making it un-acceptable. The reverse recovery current density is reduced to 33 A/cm$^2$ when the lifetime is reduced to 0.001 μs which is acceptable from a power dissipation point of view. However, such small lifetime values would require extremely large electron irradiation doses which would damage the gate oxide.

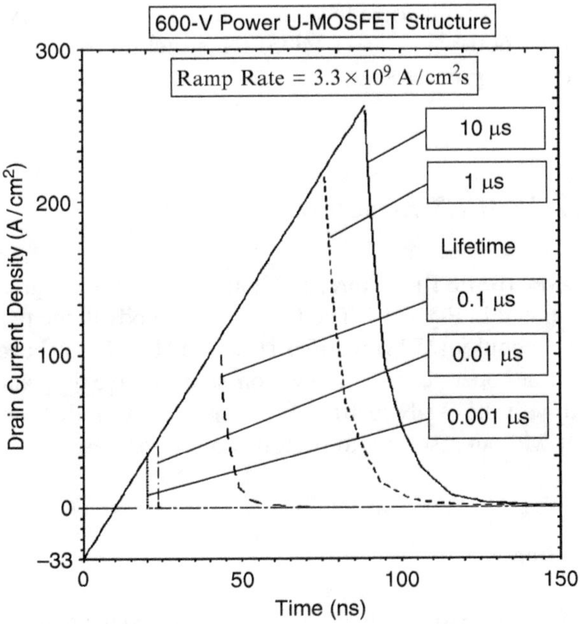


**Fig. 8.34** Reverse recovery current of the body diode within the U-MOSFET structure

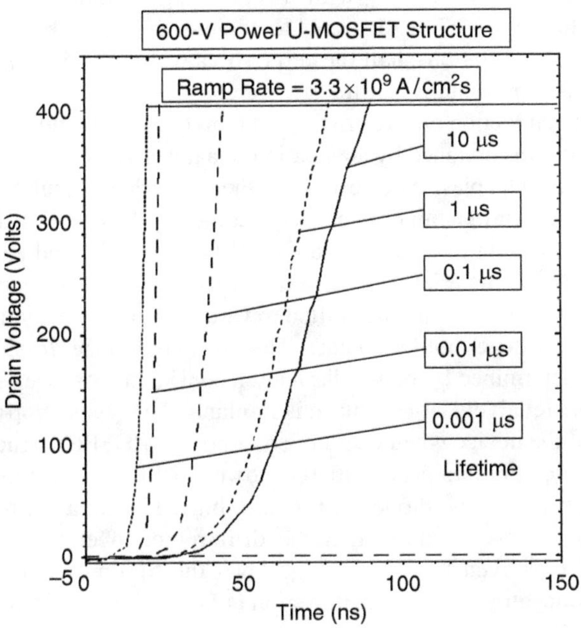


**Fig. 8.35** Reverse recovery voltage of the body diode within the U-MOSFET structure

Consequently, it is difficult to reduce the peak reverse recovery current in the body diode for the 600-V power U-MOSFET structure to acceptable levels by using lifetime control processes.

## 8.3.2 Power JBSFET Structure

In the 600-V power JBSFET structure, a Schottky contact is integrated into the cell structure as illustrated in Fig. 8.15. The P-N junction body diode then operates like the Merged-PiN-Schottky (MPS) rectifier structure [10]. It has been demonstrated that the MPS rectifier operates with a lower on-state voltage drop and stored charge when compared with the P-i-N rectifier [12]. This allows reduction of the reverse recovery current without resorting to lifetime control processes.

### 8.3.2.1 Simulation Results

The results of two-dimensional numerical simulations on the body diode within the 600 V power JBSFET structure are described here to provide a detailed understanding of the underlying device physics and operation. The structure used for the numerical simulations had the same cell parameters as the 30-V power JBSFET structure discussed in Sect. 8.2.3 with the drift region thickness of 60 μm and doping concentration of $2.7 \times 10^{14}/\text{cm}^3$ to achieve the larger breakdown voltage. For the numerical simulations, half the cell (with a width of 1.5 μm) was utilized as a unit cell that is representative of the structure.

The forward conduction characteristic of the body diode within the 600-V power JBSFET structure was obtained by application of a negative bias to the drain terminal while using zero gate bias. The results of the numerical simulations are shown in Fig. 8.36 for the case of a minority carrier lifetime of 10 μs. During the simulations, the lifetimes for holes ($\tau_{p0}$) in heavily doped N-type material and electrons ($\tau_{n0}$) in heavily doped P-type material were assumed to be equal to values provided in the figure. The $i - v$ characteristics are shifted towards lower on-state voltage drops due to current flow via the Schottky contact. The on-state voltage drop, at the on-state current density determined by power dissipation considerations (see (8.2)), is 3.33 V. Although this is a relatively large value, it is similar to the voltage drop (3.09 V) across the resistance of the device when operating as a power MOSFET structure.

The hole concentration profile in the power JBSFET structure is shown in Fig. 8.37 when the body diode is forward biased with a current density of 32.4 A/cm$^2$. It can be observed that the drift region operates under low-level injection conditions even for a large high-level the lifetime value of 10 μs. The injected hole concentration in the drift region is far less than the doping concentration ($2.7 \times 10^{14}/\text{cm}^3$) of the drift region. Consequently, there is no stored charge in the power JBSFET structure when the body diode is forward biased. Due to

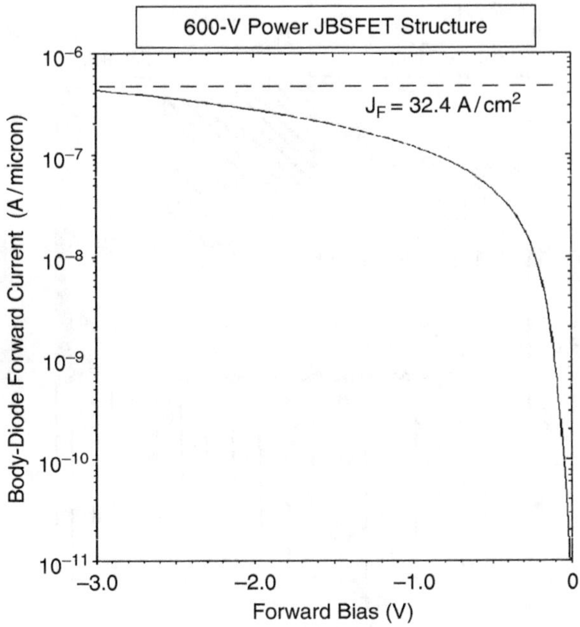


**Fig. 8.36** On-state characteristics of the body-diode within the JBSFET structure

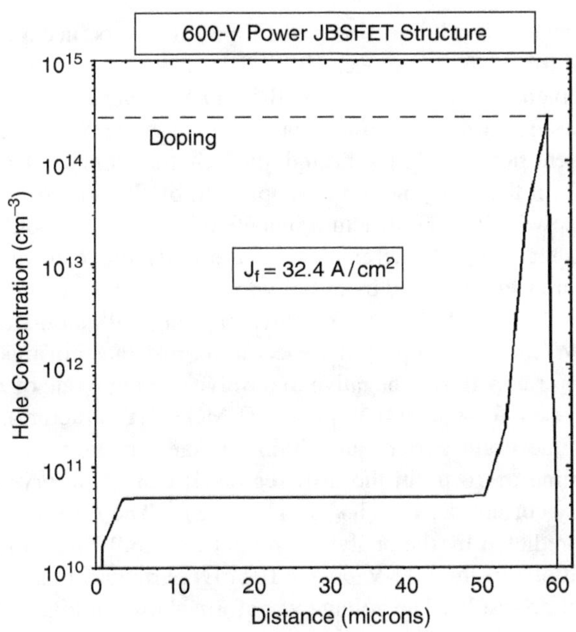


**Fig. 8.37** Hole distribution in the 600-V power JBSFET structure

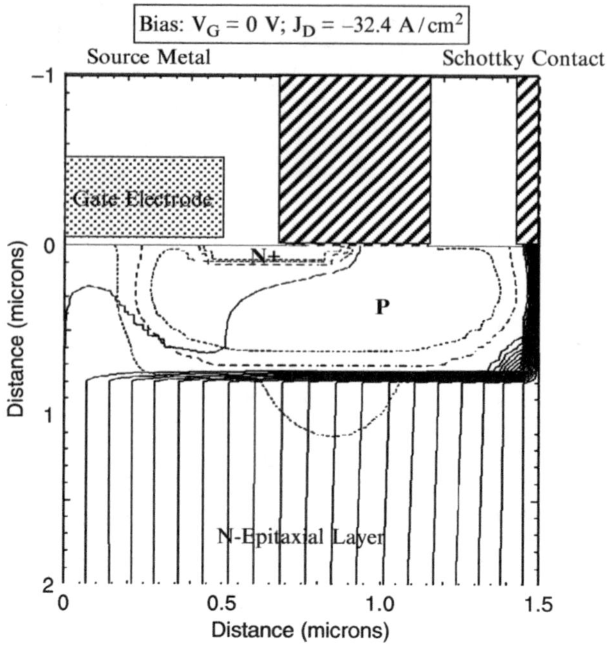


**Fig. 8.38** Body diode current path within the 600-V power JBSFET structure

operation under low-level injection conditions, it is not necessary to perform lifetime control for the 600-V power JBSFET structure.

The current path for the body diode in the 600-V power JBSFET structure when it is forward biased by a negative drain bias is shown in Fig. 8.38. It can be observed that all the current flow lines pass through the Schottky contact. This demonstrates that injection from the P-N junction is suppressed by the presence of the Schottky contact in the power JBSFET structure in spite of the small Schottky contact area.

The reverse recovery characteristics for the body diode in the 600-V power JBSFET structure were obtained by initially forward biasing the body diode with an on-state current density of 32.4 A/cm$^2$ corresponding with a on-state power dissipation of 100 W/cm$^2$ based up on the specific on-resistance of this structure. The drain current was ramped from negative to positive at a ramp rate of 3,300 A/cm$^2$ μs (same as that used for the 600-V power U-MOSFET structure). The resulting waveforms for the drain current and drain voltage are shown in Figs. 8.39 and 8.40 for a lifetime of 10 μs in the drift region. It can be observed that the peak reverse recovery current density ($J_{PR}$) is 41.7 A/cm$^2$. The reverse recovery current (35.3 A/cm$^2$) predicted by the analytical model (see (8.9)) based up on low-level injection conditions in the 600-V power JBSFET structure is in good agreement with the simulation results. The voltage waveform shown in Fig. 8.40 shows a very rapid increase which is also consistent with the predictions of the analytical model (see (8.7)). These results indicate that the body diode of the 600-V power JBSFET

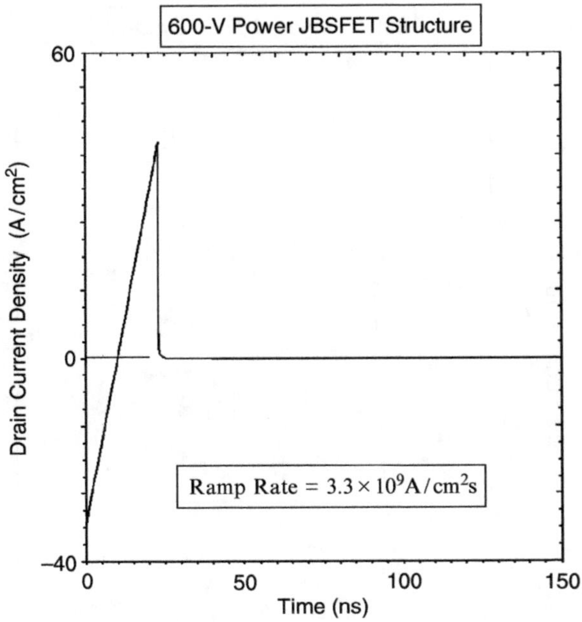


**Fig. 8.39** Reverse recovery current of the body diode within the JBSFET structure

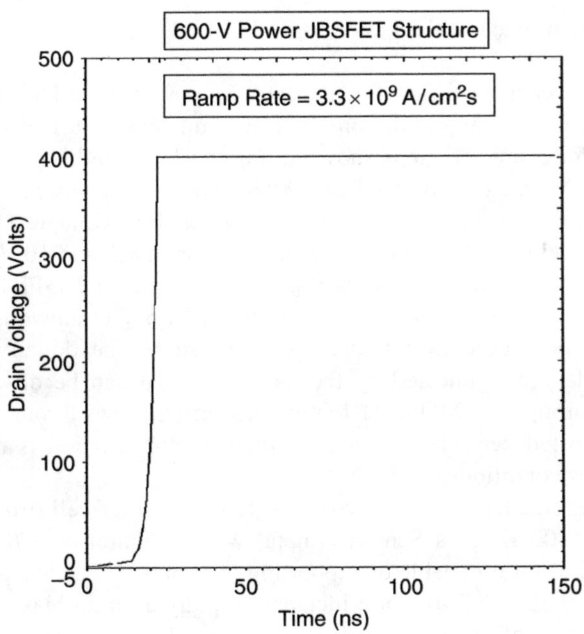


**Fig. 8.40** Reverse recovery voltage of the body diode within the JBSFET structure

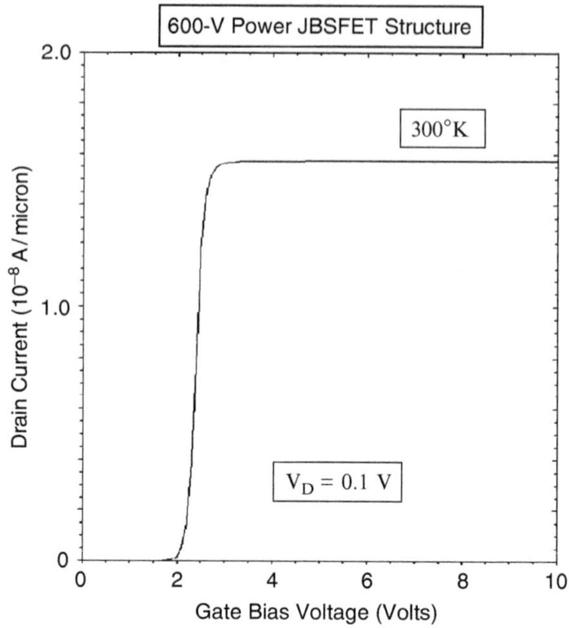


**Fig. 8.41** Transfer characteristics of the 600-V JBSFET structure

structure can be utilized in the motor control H-bridge circuit without performing any lifetime control processes.

The transfer characteristics for the 600-V power JBSFET structure were obtained using numerical simulations with a drain bias of 0.1 V at 300°K. The resulting transfer characteristic is shown in Fig. 8.41. From this graph, a threshold voltage of 2.1 V can be extracted at 300°K. The specific on-resistance can be obtained from the transfer characteristics at any gate bias voltage. The specific in-resistance is found to be 95.3 mΩ $cm^2$ at a gate bias of 4.5 and 10 V.

The on-state current flow pattern within the 600-V power JBSFET structure at a small positive drain bias of 0.1 V and a gate bias of 4.5 V is shown in Fig. 8.42. In the figure, the depletion layer boundary is shown by the dotted lines and the junction boundary is delineated by the dashed line. It can be observed that the current flows through the MOSFET region. No current flows through the Schottky contact as expected because the metal-semiconductor junction is reverse biased under these bias conditions.

The blocking characteristic for the 600-V power JBSFET cell structure is shown in Fig. 8.43 at 300°K for a Schottky metal work function of 4.7 eV. It can be observed that the cell is capable of supporting over 600 V. The leakage current for the 600-V power JBSFET structure increases rapidly at drain bias voltages above 300 V leading to a soft breakdown characteristics. The components of the leakage current are provided in the figure. It can be observed that the leakage current

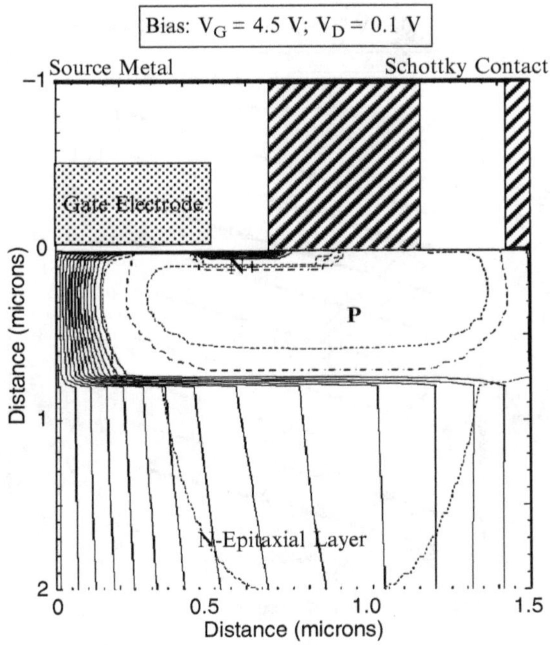


**Fig. 8.42** Current distribution in the power JBSFET structure in the first quadrant

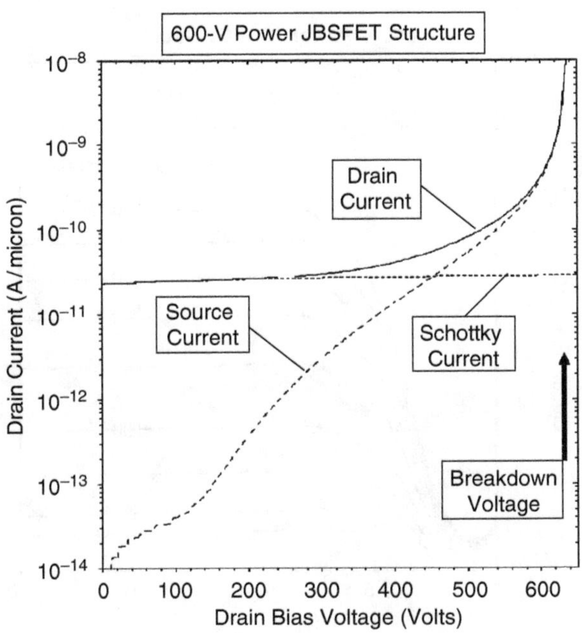


**Fig. 8.43** Blocking characteristic for the 600-V power JBSFET structure

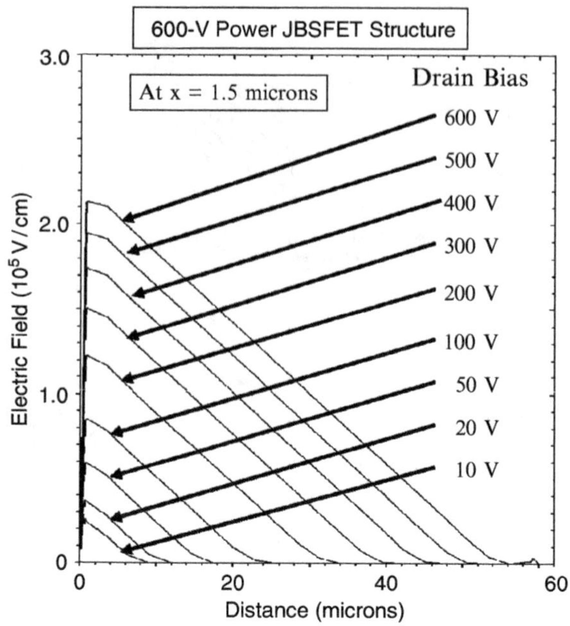


Fig. 8.44 Electric field distribution in the 600-V power JBSFET structure

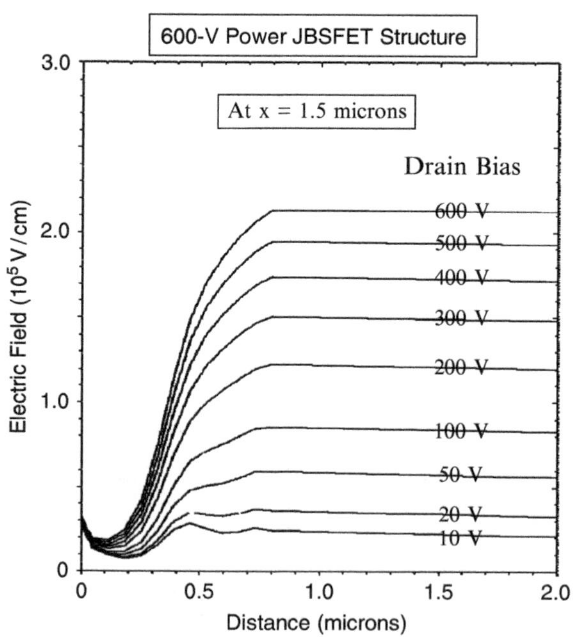


Fig. 8.45 Electric field distribution in the 600-V power JBSFET structure

flowing through the Schottky contact (dotted line) is not increasing with increasing drain bias voltage above 300 V. This is consistent with the low electric field at the Schottky contact due to the shielding by the P-N junction. The leakage current increase is found to be associated with the source current due to the short channel length in the structure.

It is insightful to examine the electric field profile inside the 600-V power JBSFET structure when it is operating in the blocking mode. The electric field profile obtained through the middle of the Schottky contact is shown in Figs. 8.44 and 8.45. It can be observed that the electric field has a triangular distribution with a maximum value at a depth of about 1 micron from the surface. The electric field at the Schottky contact is greatly reduced ($3 \times 10^4$ V/cm) when compared the peak electric field in the bulk ($2.2 \times 10^5$ V/cm) at a drain bias of 600 V. Moreover, the electric field at the Schottky contact does not increase with increasing drain bias voltage. The suppression of the electric field at the Schottky contact allows operation of the power JBSFET structure with a low leakage current despite using the source ohmic contact metal for making the Schottky contact.

### 8.3.3 Power GD-MOSFET Structure

The power GD-MOSFET structure was illustrated in Fig. 6.1 with its linearly graded doping profile. This device has a body diode formed between the P-base region and the N-drift region like in the case of the power U-MOSFET structure. However, the doping concentration in the 600-V power GD-MOSFET increases from $2 \times 10^{15}$/cm$^3$ at the P-N junction to $2 \times 10^{17}$/cm$^3$ at the N$^+$ substrate. When the body diode is forward biased, minority carriers ae injected into the drift region. Their concentration exceeds the doping concentration in the vicinity of the junction resulting in high level injection conditions. However, low-level injection conditions will prevail when proceeding away from the junction due to the higher doping concentrations. This results in reduced stored charge and superior reverse recovery behavior for the 600-V power GD-MOSFET structure.

The current density flowing through the body diode in its on-state is the same value as the current density flowing through the power MOSFET structure when the gate bias has been applied because they are both determined by the current flowing through the motor windings. The on-state current density will therefore be assumed to be limited by the on-state power dissipation to 100 W/cm$^2$ as determined by the specific on-resistance of the power MOSFET structure (see (8.2)).

Due to low-level injection conditions in the drift region, the reverse recovery for the body diode in the power GD-MOSFET structure is determined by the establishment of a depletion region in the N-drift region without stored charge. The presence of the source electrode within the trenches introduces a significant capacitance between the drain and source which must be charged during the reverse recovery process. In this section, the reverse recovery waveforms will

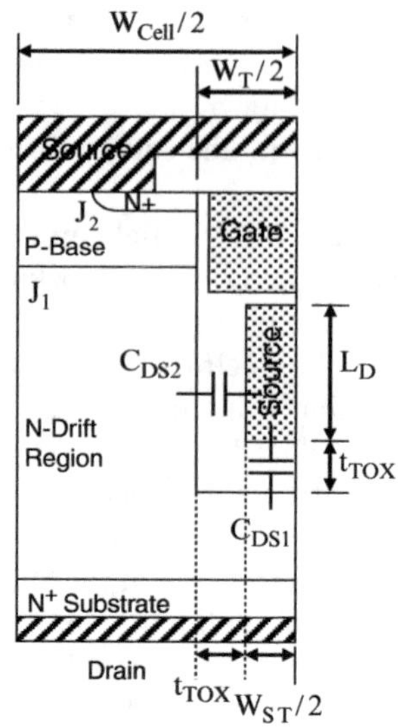


**Fig. 8.46** Output capacitance in the 600-V power GD-MOSFET structure

be analyzed under the assumption that the charging of the drain-source capacitance is dominant.

The specific drain-source capacitance for the power GD-MOSFET structure can be derived by using the device structural parameters shown in Fig. 8.46. The drain-source capacitance consists of two components: the first component ($C_{DS1}$) is associated with the bottom of the source electrode in the trenches while the second component ($C_{DS2}$) is associated with the sidewall of the source electrode in the trenches:

$$C_{DS,SP} = C_{DS1} + C_{DS2} = \frac{\varepsilon_{OX}}{t_{TOX}}\left(\frac{W_{ST}}{W_{Cell}}\right) + \frac{\varepsilon_{OX}}{t_{TOX}}\left(\frac{2L_D}{W_{Cell}}\right) \tag{8.10}$$

where $W_{ST}$ is the width of source electrode in the trenches. This capacitance is independent of the drain voltage. The drain current flow during the reverse-recovery process charges the drain-source capacitance during the voltage rise-time:

$$J_D(t) = C_{DS,SP}\frac{dV_D}{dt} \tag{8.11}$$

Since the drain current density is changing at a fixed ramp rate (a):

$$a\,t = C_{DS,SP}\frac{dV_D}{dt} \tag{8.12}$$

Performing the integration on both sides yields an expression for the drain voltage transient:

$$V_D(t) = \frac{a\,t^2}{2C_{DS,SP}} \tag{8.13}$$

According to the analytical model, the drain voltage increases quadratically with time.

The peak reverse recovery current occurs at time ($t_2$) when the drain voltage becomes equal to the drain supply voltage ($V_{DS}$). Using this criterion in (8.13) yields an equation for the time after zero crossing when the drain current reaches its peak reverse recovery value:

$$t_{PR} = \sqrt{\frac{2C_{DS,SP}V_{DS}}{a}} \tag{8.14}$$

The peak reverse recovery current density can be obtained by multiplying this time with the ramp rate:

$$J_{PR} = a t_{PR} = \sqrt{2aC_{DS,SP}V_{DS}} \tag{8.15}$$

Based up on this equation, it can be concluded that the peak reverse recovery current for the body diode in the power GD-MOSFET structure will increase with increasing ramp rate and drain supply voltage. Note that the reverse recovery current is independent of the lifetime according to this analysis. For the 600-V power GD-MOSFET structure with a cell pitch of 6.8 μm and drift region length of 35 μm with a trench oxide thickness of 3 μm, the specific drain-source capacitance obtained by using (8.10) is 11.75 nF/cm$^2$. Using this drain-source specific capacitance for the case of a drain supply voltage of 400 V, and a ramp rate of $3.3 \times 10^9$ A/cm$^2$ s, the peak reverse recovery current density predicted by this expression for the power GD-MOSFET structure is found to be 176 A/cm$^2$ which is comparable to the on-state current density of 125 A/cm$^2$. From this result, it can be concluded that the body diode in the 600-V power GD-MOSFET structure has acceptable reverse recovery characteristics without the need for lifetime control.

#### 8.3.3.1 Simulation Results

The results of two-dimensional numerical simulations on the body diode within the 600 V power GD-MOSFET structure are described here to provide a detailed understanding of the underlying device physics and operation. The structure used for the numerical simulations had the same cell parameters as the 600-V power GD-MOSFET structure discussed in Chap. 6. Consequently, the body diode had a drift region thickness of 40 μm below the P-base region with a linearly graded doping profile with a gradient of $5 \times 10^{19}$/cm$^4$ (see Fig. 6.51). For the numerical

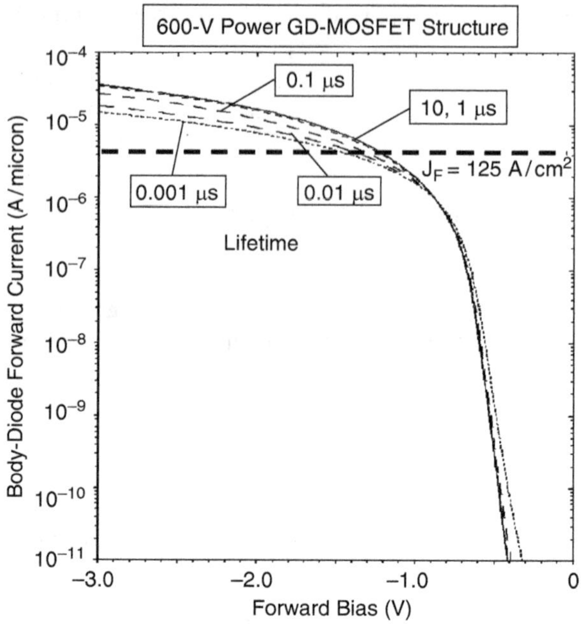


**Fig. 8.47** On-state characteristics of the body-diode within the power GD-MOSFET structure

simulations, half the cell (with a width of 3.4 μm) was utilized as a unit cell that is representative of the structure.

The forward conduction characteristics of the body diode within the 600-V power GD-MOSFET structure were obtained by application of a negative bias to the drain terminal while using zero gate bias. This is representative of the condition in the motor control H-bridge circuit when both the high-side and low-side power MOSFET switches are turned-off. The motor winding current then flows through the body diode within one of the power MOSFET structures. The results of the numerical simulations are shown in Fig. 8.47 for the case of 5 values for the minority carrier lifetime. During the simulations, the lifetimes for holes ($\tau_{p0}$) in heavily doped N-type material and electrons ($\tau_{n0}$) in heavily doped P-type material were assumed to be equal to values provided in the figure. The $i - v$ characteristics follow the diode law under low-level injection conditions over a broad range of current levels for all lifetime values. The on-state voltage drop, at the on-state current density determined by power dissipation considerations (see (8.2)), increases only slightly when the lifetime is reduced.

The hole concentration profiles in the 600-V power GD-MOSFET structure are shown in Fig. 8.48 when the body diode is forward biased with a current density of 125 A/cm$^2$. By comparison with the doping profile shown in Fig. 6.51, it can be concluded that the drift region operates under low-level injection conditions with the hole concentration below the doping concentration for all lifetime values at

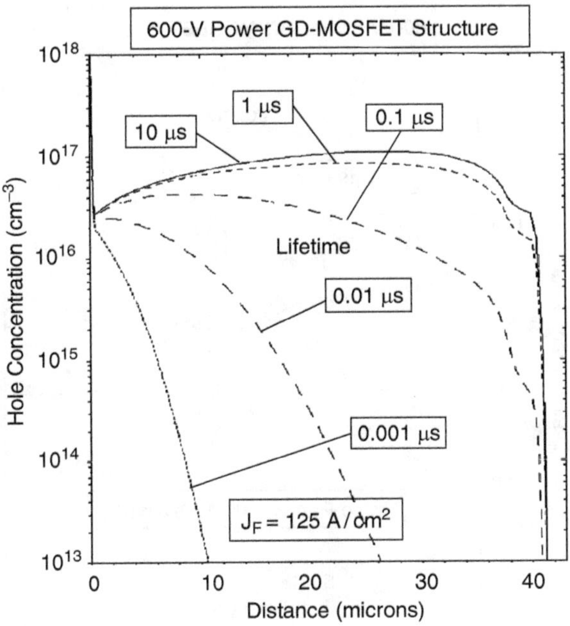


**Fig. 8.48** Hole distribution in the 600-V power GD-MOSFET structure

depths below 20 μm from the top surface. At smaller lifetime values, a greater portion of the drift region remains un-modulated by the injected carriers. Consequently, the on-state voltage drop becomes insensitive to the lifetime and has a low value due to high doping concentration in the drift region.

The reverse recovery characteristics for the body diode in the 600-V power GD-MOSFET structure were obtained by initially forward biasing the body diode with an on-state current density of 125 A/cm$^2$ corresponding with a on-state power dissipation of 100 W/cm$^2$ based up on the specific on-resistance of this structure. The drain current was ramped from negative to positive at a ramp rate of 3,300 A/cm$^2$ μs. The resulting waveforms for the drain current and drain voltage are shown in Figs. 8.49 and 8.50 for various lifetime values in the drift region. It can be observed that the peak reverse recovery current density ($J_{PR}$) is 200 A/cm$^2$ for the case of a lifetime of 10 μs. This value is only 1.6-times larger than the on-state current density making it acceptable from the point of view of power dissipation in the motor control circuit. The reverse recovery current density is reduced to 145 A/cm$^2$ when the lifetime is reduced to 0.001 μs. The analytical model based up on charging the drain-source capacitance provides a good prediction of the peak reverse current density.

The current path within the 600-V power GD-MOSFET structure during the reverse recovery process is shown in Fig. 8.51 when the current reaches its maximum peak reverse value. It can be observed from this figure that the current

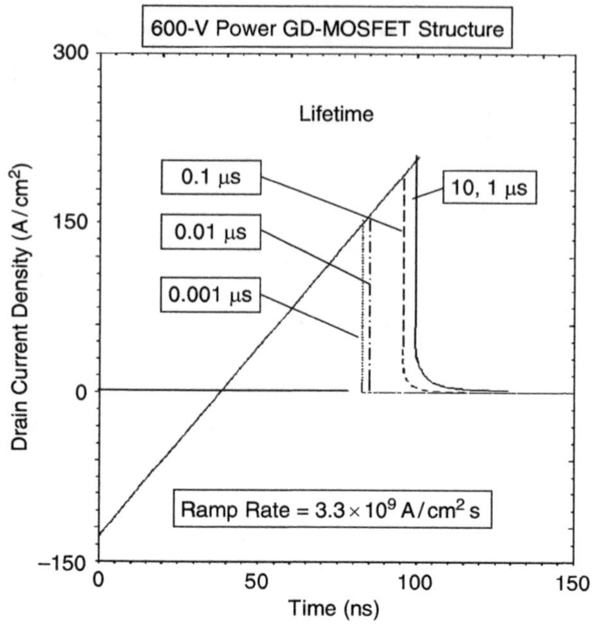


Fig. 8.49 Reverse recovery current of the body diode within the GD-MOSFET structure

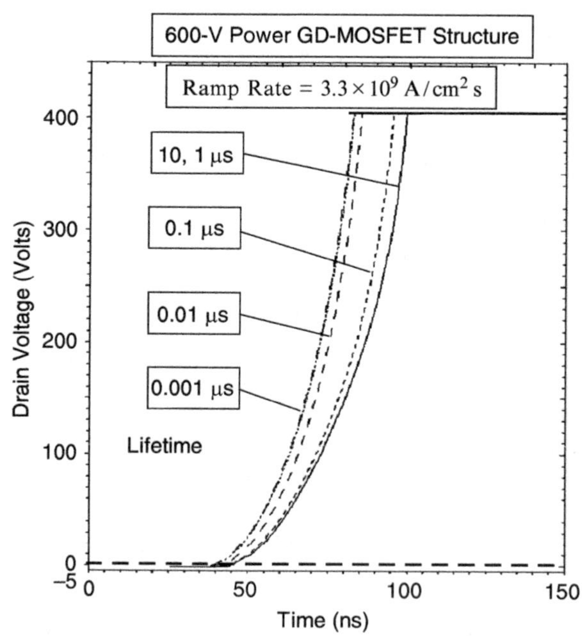


Fig. 8.50 Reverse recovery voltage of the body diode within the GD-MOSFET structure

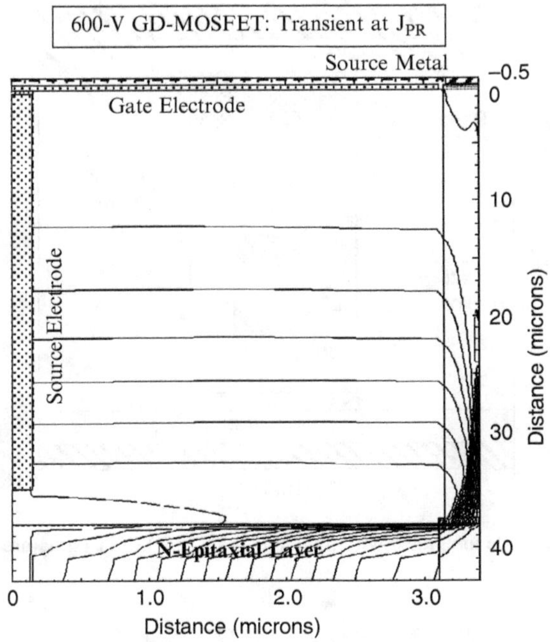


**Fig. 8.51** Current path during reverse recovery of the body diode within the 600-V power GD-MOSFET structure

flow lines extend from the drain to the source electrode located inside the trench. This validates the assumption that the drain current is charging the output capacitance of the 600-V power GD-MOSFET structure as the basis for deriving (8.15) for the peak reverse recovery current.

### *8.3.4 Power GD-JBSFET Structure*

The reverse recovery characteristics of the integral diode in the high voltage power GD-MOSFET structure can be enhanced by the incorporation of a Schottky contact into the cell structure. One approach to the integration of a Schottky rectifier with the power GD-MOSFET structure is illustrated in Fig. 8.52. In this case, the Schottky contact is formed in a separate mesa region from the MOSFET structure. The same mesa region parameters, i.e. the doping gradient and thickness for the drift region, can then be chosen under the Schottky contact and the MOSFET diffusions. A superior approach to the integration of the Schottky rectifier with the power MOSFET structure is by merging the Schottky contact with the base region of the MOSFET to create a JBS rectifier. This power GD-JBSFET structure is illustrated in Fig. 8.53. It be seen that the Schottky contact is located in the middle

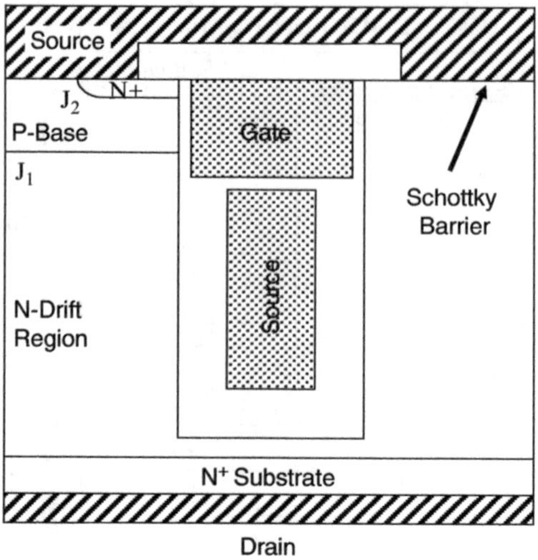


**Fig. 8.52** Schottky diode integrated within the power GD-MOSFET (structure 1)

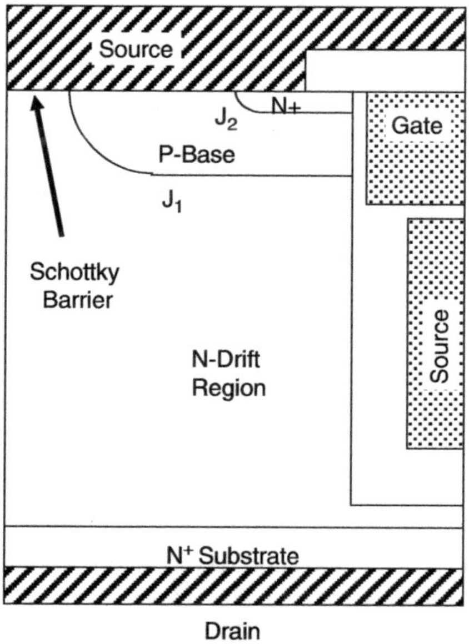


**Fig. 8.53** Schottky diode integrated within the power GD-MOSFET (structure 2)

of the P-base region within the power GD-MOSFET cell structure. A common mesa region is therefore used for both the MOSFET portion and the Schottky rectifier portion. The area of the Schottky contact can be adjusted to optimize the on-state voltage drop and the leakage current. One of the challenges with this approach is that it is only practical to use the source contact metal to form the Schottky contact. The source contact process is usually optimized to create a good ohmic contact to the $N^+$ source region. In contract, the same metal and its annealing process must create a sufficiently low leakage current at the Schottky contact in the JBSFET structure. Fortunately, this is possible because the P-body region of the power MOSFET structure is utilized to shield the Schottky contact from high electric fields developed in the N-drift region by the drain bias when the device is supporting large voltages in the blocking mode.

#### 8.3.4.1 Simulation Results

The results of two-dimensional numerical simulations on the body diode within the 600 V power GD-JBSFET structures are described here to provide a detailed understanding of the underlying device physics and operation. The first structure used for the numerical simulations has the cell structure illustrated in Fig. 8.52 with a separate mesa region for the Schottky contact. The need for two mesa region enlarges the cell size from 3.4 μm for the basic 600-V power GD-MOSFET structure (see Sect. 8.3.3) to 6.8 μm for the first 600-V power GD-JBSFET structure.

The doping profile under the Schottky contact in the first 600-V power GD-JBSFET structure shown in Fig. 8.54 by the solid line. The doping

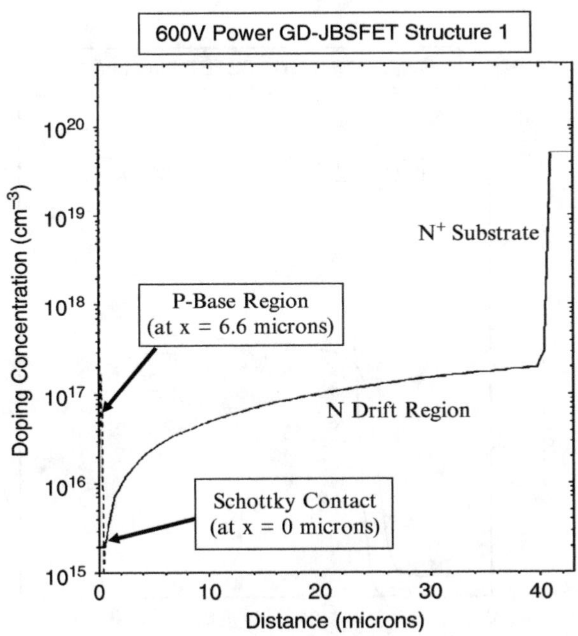


**Fig. 8.54** Doping profile under the Schottky contact and the MOSFET portion within the power GD-JBSFET structure 1

concentration at the Schottky contact is low ($2 \times 10^{15}/cm^3$) which favors making a good metal-semiconductor rectifying junction. At the same time, the doping concentration increases linearly to $2 \times 10^{17}/cm^3$ near the $N^+$ substrate making the resistance in series with Schottky contact small. The doping profile at the mesa with the P-base region is included in this figure, as shown by the dashed lines, for comparison.

The transfer characteristics for the first 600-V power GD-JBSFET structure were obtained using numerical simulations with a drain bias of 0.1 V at 300 and 400°K. The resulting transfer characteristics are shown in Fig. 8.55. From this graph, a threshold voltage of 2.5 and 2.0 V can be extracted at 300 and 400°K, respectively. The specific on-resistance can be obtained from the transfer characteristics at any gate bias voltage. For the case of a gate bias of 4.5 V and 300°K, the specific in-resistance is found to be 13.8 mΩ cm$^2$. For the case of a gate bias of 10 V and 300°K, the specific in-resistance is found to be 12.8 mΩ cm$^2$. These values are twice as large as those for the 600-V power GD-MOSFET structure because the cell pitch has been increased in the 600-V power GD-JBSFET to accommodate the separate mesa region with the Schottky contact. The drain current observed at 400°K below the threshold voltage is due to the leakage current contributed by the Schottky contact.

The on-state current flow pattern within the 600-V power GD-JBSFET structure at a small positive drain bias of 0.1 V and a gate bias of 4.5 V is shown in Fig. 8.56. In the figure, the depletion layer boundary is shown by the dotted lines and the

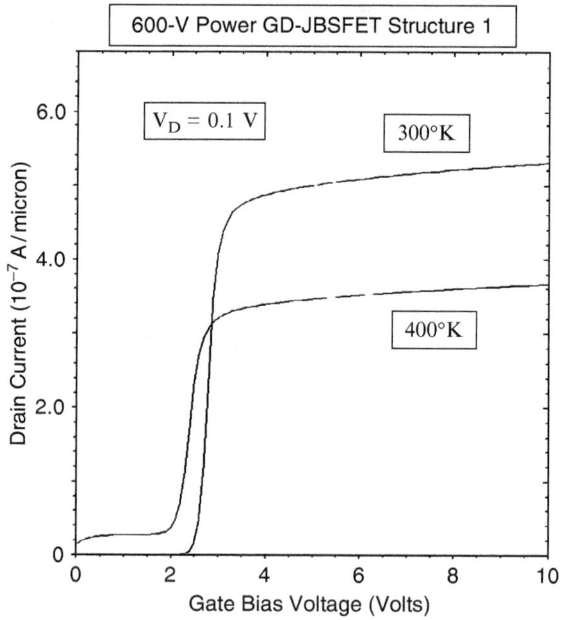


**Fig. 8.55** Transfer characteristics of the 600-V power GD-JBSFET structure 1

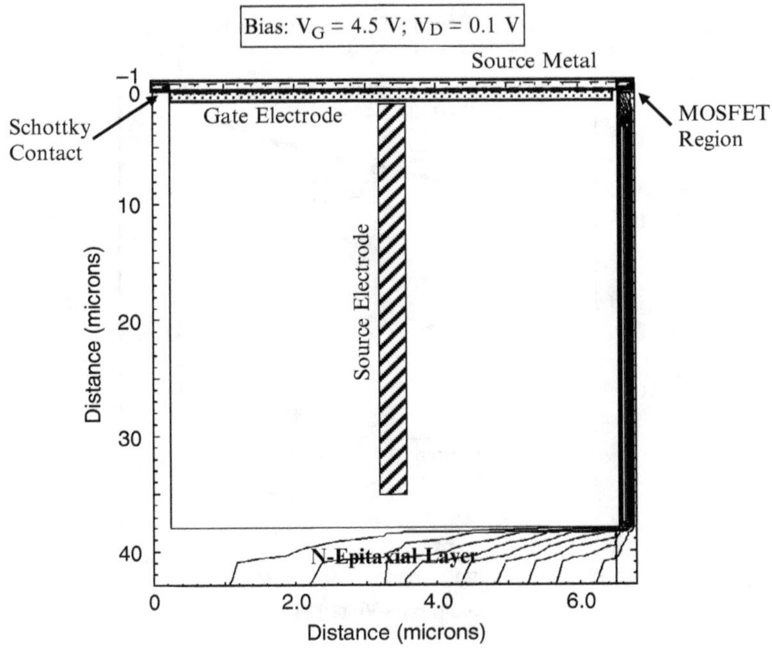


**Fig. 8.56** Current distribution in the 600-V power GD-JBSFET structure 1 in the first quadrant

junction boundary is delineated by the dashed line. It can be observed that the current flows through the MOSFET region in the same manner as observed within the 600-V power GD-MOSFET structure. No current flows through the Schottky contact as expected because the metal-semiconductor junction is reverse biased under these bias conditions.

The blocking characteristic for the first 600-V power GD-JBSFET cell structure is shown in Fig. 8.57 at 300°K for a Schottky metal work function of 4.7 eV. It can be observed that the cell is capable of supporting 600 V similar to that observed for the 600-V power GD-MOSFET cell structure. However, the leakage current in the power JBSFET structure is much larger due to the presence of the Schottky contact. The figure also shows the current through the source contact and the Schottky contact. It can be concluded that the current flow through the Schottky contact is determining the leakage current in the first 600-V power GD-JBSFET structure.

It is instructive to examine the potential contours inside the first 600-V power GD-JBSFET structure when it is operating in the blocking mode. This allows determination of the voltage distribution within the structure. The potential contours for the device, obtained using the numerical simulations with zero gate bias and a drain bias voltage of 600 V, are shown in Fig. 8.58. From this figure, it can be observed that the surface region under the Schottky contact has a smaller electric field. This suppresses the Schottky barrier lowering phenomenon and reduces the leakage current.

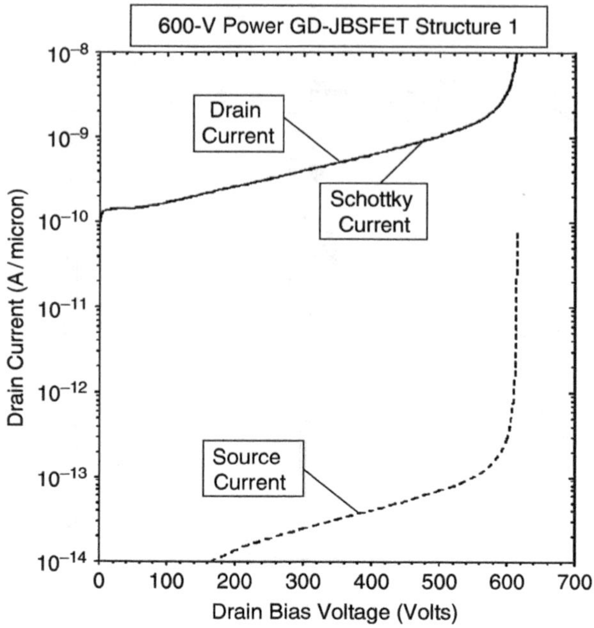


**Fig. 8.57** Blocking characteristic for the 600-V power GD-JBSFET structure 1

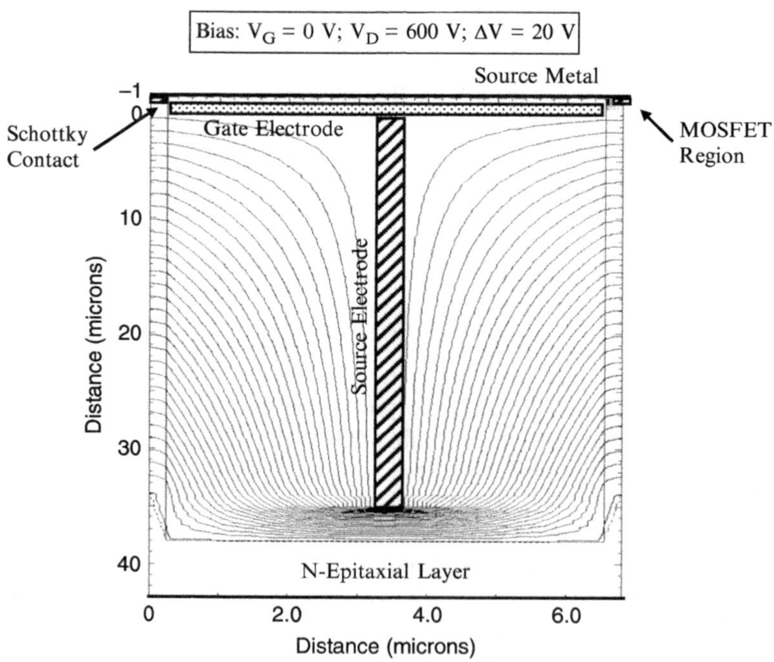


**Fig. 8.58** Potential contours in the 600-V power GD-JBSFET structure 1

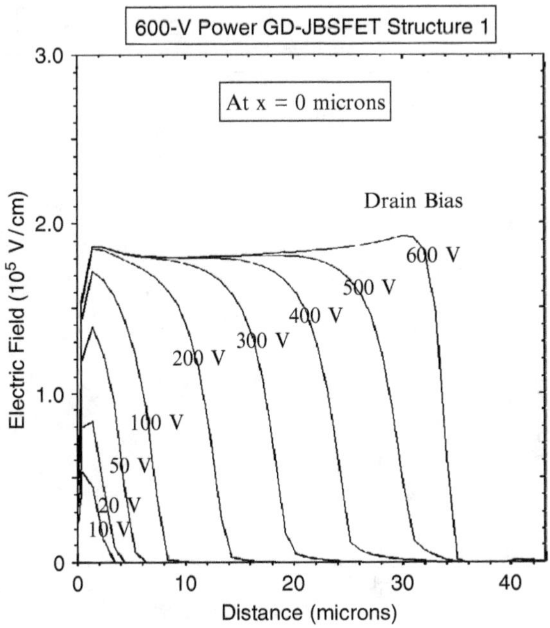


**Fig. 8.59** Electric field distribution in the 600-V power GD-JBSFET structure 1 at the Schottky contact

It is insightful to also examine the electric field profile inside the first 600-V power GD-JBSFET structure when it is operating in the blocking mode. The electric field profile obtained through the middle of the Schottky contact is shown in Fig. 8.59. It can be observed that the maximum electric field occurs at a depth of 2 μm from the surface. The electric field at the Schottky contact is greatly reduced ($7 \times 10^4$ V/cm) when compared the peak electric field in the bulk ($1.9 \times 10^5$ V/cm) at a drain bias of 600 V. The suppression of the electric field at the Schottky contact allows operation of the 600-V power GD-JBSFET structure with a low leakage current despite using the source ohmic contact metal for making the Schottky contact.

The forward conduction characteristics of the body diode within the first 600-V power GD-JBSFET structure were obtained by application of a negative bias to the drain terminal while using zero gate bias. This is representative of the condition in the motor control circuit when both the high-side and low-side power MOSFET switches are turned-off. The motor current then flows through the body diode within the power MOSFET structure. The results of the numerical simulations are shown in Fig. 8.60 for the case of a minority carrier lifetime of 10 μs. When compared with the P-N body-diodes of the previously described power MOSFET structures in this chapter, the $i - v$ characteristics for the power GD-JBSFET structure is shifted towards lower voltages due to current flow via the Schottky contact. The area of the Schottky contact in the 600-V power GD-JBSFET structure used for the numerical

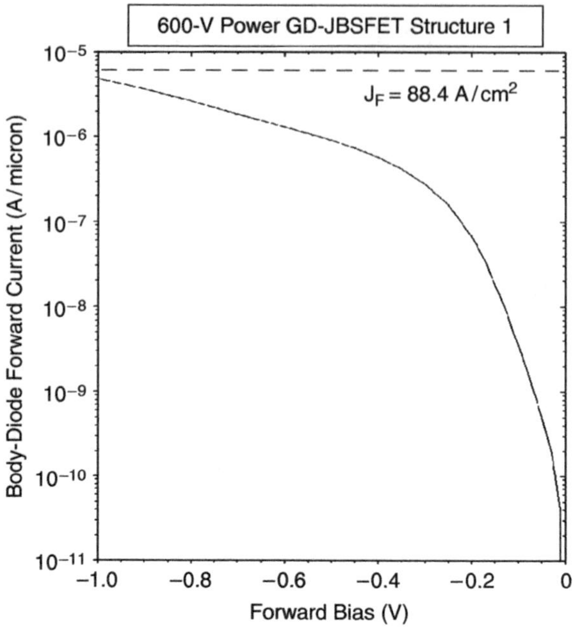


**Fig. 8.60** On-state characteristics of the body-diode within the 600-V power GD-JBSFET structure 1

simulations is sufficiently large that the on-state voltage drop is only 1.09 V at even the relatively high on-state current density of 88.4 A/cm$^2$ for the device in the MOSFET mode of operation.

The hole concentration profiles in the mesa with the Schottky contact and the mesa with the MOSFET portion in the 600-V power GD-JBSFET structure are shown in Fig. 8.61 when the body diode is forward biased with a current density of 88.4 A/cm$^2$. The doping profile is also shown in the plot for comparison. It can be observed that the drift region operates under low-level injection conditions over most of the drift region. The hole concentration in the mesa with the Schottky contact is very small as expected. However, the injected hole concentration in the mesa with the MOSFET portion is not substantially less than in the 600-V power GD-MOSFET structure (see Fig. 8.48). This is because the Schottky contact in the first 600-V power GD-JBSFET structure is not located in the vicinity of the P-N body diode of the MOSFET portion and the voltage drop at the Schottky contact exceeds 0.8 V.

The current flow path inside the 600-V power GD-JBSFET structure is shown in Fig. 8.62 when the device is operated in the third quadrant with a current density of 88.4 A/cm$^2$. It can be observed that the drain current splits equally into two components: one via the Schottky contact and a second one via the P-N body diode of the MOSFET portion. Consequently, the injection of holes from the P-N body diode is not substantially reduced by the addition of the mesa with the Schottky contact.

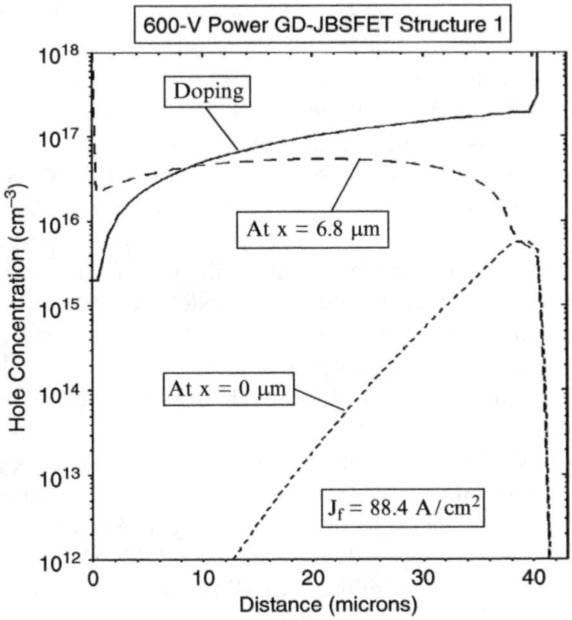


**Fig. 8.61** Hole distribution in the 600-V power GD-JBSFET structure 1

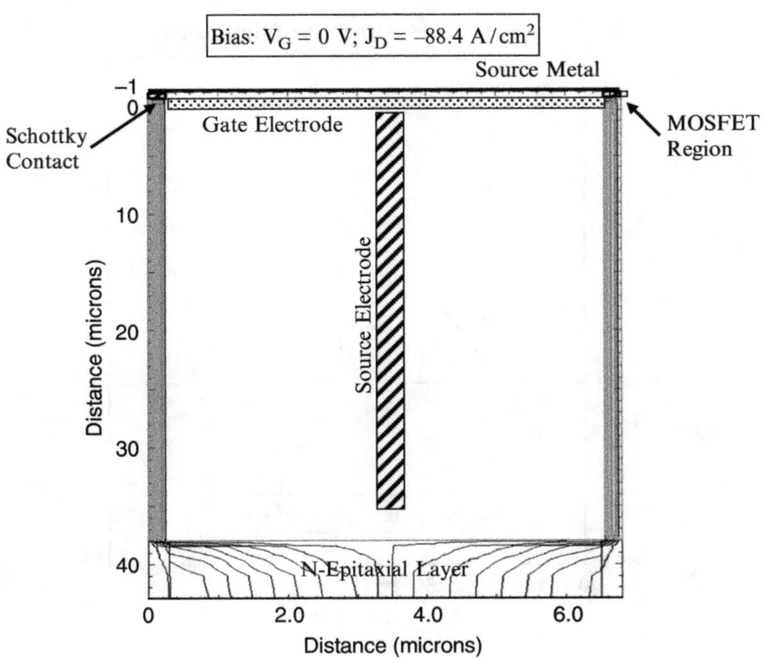


**Fig. 8.62** Body diode current path within the 600-V power GD-JBSFET structure 1

The reverse recovery characteristics for the body diode in the first 600-V power GD-JBSFET structure were obtained by initially forward biasing the body diode with an on-state current density of 88.4 A/cm$^2$ corresponding with a on-state power dissipation of 100 W/cm$^2$ based up on the specific on-resistance of this structure. The drain current was ramped from negative to positive at a ramp rate of 3,300 A/cm$^2$ μs (the same rate as for the 600-V power U-MOSFET structure). The resulting waveforms for the drain current and drain voltage are shown in Figs. 8.63 and 8.64. It can be observed that the peak reverse recovery current density ($J_{PR}$) is 168 A/cm$^2$ for the case of a lifetime of 10 μs. The peak reverse recovery current predicted by the analytical model for low-level injection conditions (see (8.15)) for the first 600-V power GD-JBSFET structure is 176 A/cm$^2$ in excellent agreement with the simulations. The voltage waveform also rises rapidly as predicted by the analytical model (see (8.13)) with a square-law behavior as a function of time.

The results of two-dimensional numerical simulations on the body diode within the first 600 V power GD-JBSFET structure (see Fig. 8.52) demonstrate that the reverse recovery performance is not significantly enhanced. A much superior outcome can be achieved by utilizing the second 600 V power GD-JBSFET structure (see Fig. 8.53). The results of two-dimensional numerical simulations of this structure are next described to provide a detailed understanding of the underlying device physics and operation. The incorporation of the Schottky contact

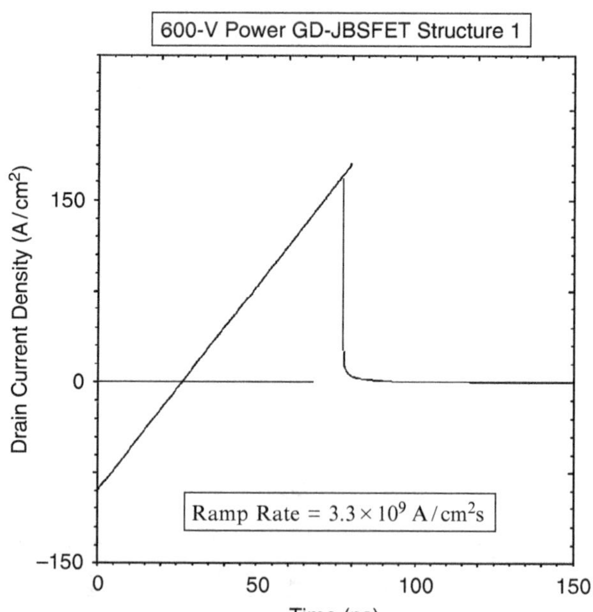


**Fig. 8.63** Reverse recovery current of the body diode within the 600-V power GD-JBSFET structure 1

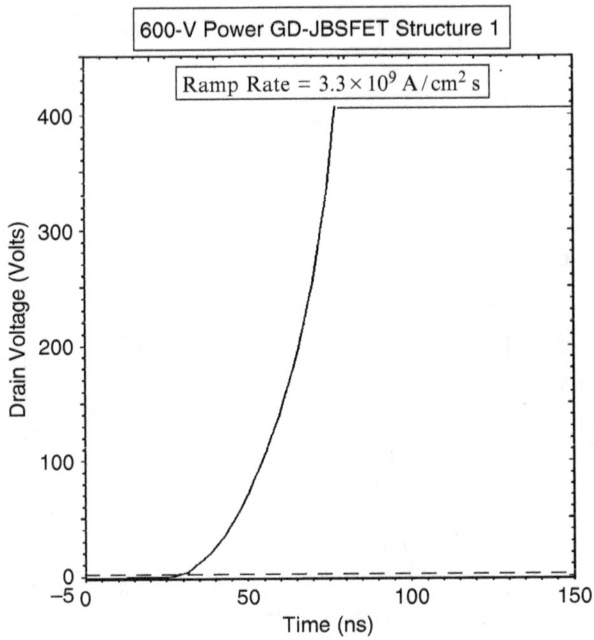


**Fig. 8.64** Reverse recovery voltage of the body diode within the 600-V power GD-JBSFET structure 1

enlarges the cell size for this structure to 7.4 μm in comparison with 6.8 μm for the 600-V power GD-MOSFET structure.

The doping profile under the Schottky contact in the second 600-V power GD-JBSFET structure is the same as that shown in Fig. 8.54. The doping concentration at the Schottky contact is low ($2 \times 10^{15}/cm^3$) which favors making a good metal-semiconductor rectifying junction. At the same time, the doping concentration increases linearly to $2 \times 10^{17}/cm^3$ near the $N^+$ substrate making the resistance in series with Schottky contact small. The doping profile at the P-base region was included in this figure, as shown by the dashed lines, for comparison.

The transfer characteristics for the second 600-V power GD-JBSFET structure were obtained using numerical simulations with a drain bias of 0.1 V at 300 and 400°K. The resulting transfer characteristics are shown in Fig. 8.65. From this graph, a threshold voltage of 2.5 and 2.0 V can be extracted at 300 and 400°K, respectively. The specific on-resistance can be obtained from the transfer characteristics at any gate bias voltage. For the case of a gate bias of 4.5 V and 300°K, the specific in-resistance is found to be 7.5 mΩ cm$^2$. For the case of a gate bias of 10 V and 300°K, the specific in-resistance is found to be 6.8 mΩ cm$^2$. These values are slightly larger than those for the 600-V power GD-MOSFET structure because the cell pitch has been increased in the 600-V power GD-JBSFET to accommodate the Schottky contact.

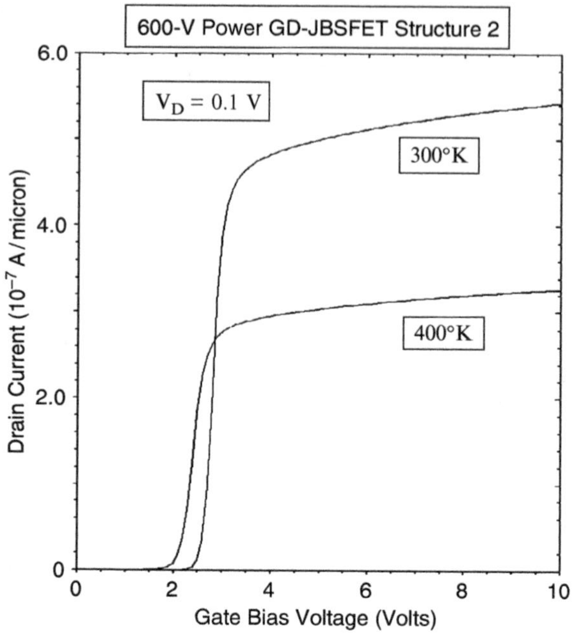


**Fig. 8.65** Transfer characteristics of the 600-V power GD-JBSFET structure 2

The on-state current flow pattern within the second 600-V power GD-JBSFET structure at a small positive drain bias of 0.1 V and a gate bias of 4.5 V is shown in Fig. 8.66. In the figure, the depletion layer boundary is shown by the dotted lines and the junction boundary is delineated by the dashed line. Only the upper portion of the structure is shown for clarity. It can be observed that the current flows through the MOSFET region in the same manner as observed within the 600-V power GD-MOSFET structure. No current flows through the Schottky contact as expected because the metal-semiconductor junction is reverse biased under these bias conditions.

The blocking characteristic for the second 600-V power GD-JBSFET cell structure is shown in Fig. 8.67 at 300°K for a Schottky metal work function of 4.7 eV. It can be observed that the cell is capable of supporting 600 V similar to that observed for the 600-V power GD-MOSFET cell structure. However, the leakage current in the power GD-JBSFET structure is much larger due to the presence of the Schottky contact. The figure also shows the current through the source contact and the Schottky contact. It can be concluded that the current flow through the Schottky contact is determining the leakage current in the second 600-V power GD-JBSFET structure at lower drain bias voltages. However, the leakage current through the MOSFET region determines the increase in the leakage current with increasing drain bias voltage. This increase can be suppressed by increasing the P-base thickness to reduce the short-channel effects. The impact of an increase in the

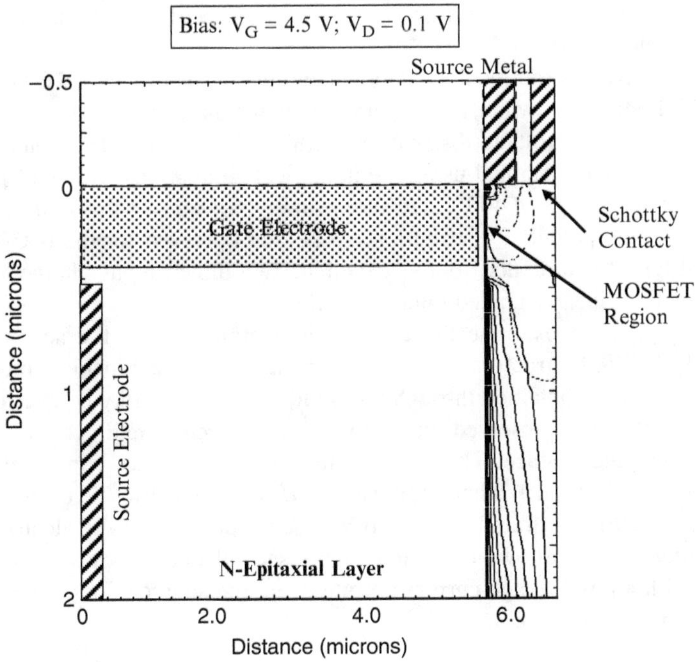


**Fig. 8.66** Current distribution in the 600-V power GD-JBSFET structure 2 in the first quadrant

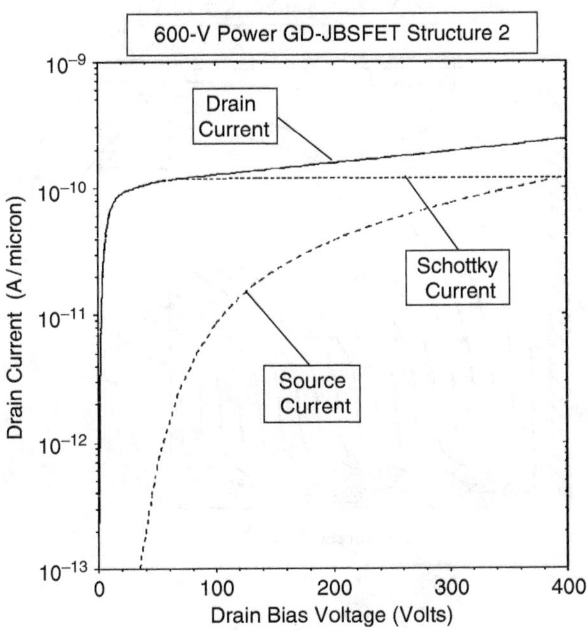


**Fig. 8.67** Blocking characteristic for the 600-V power GD-JBSFET structure 2

channel length is small for the high-voltage device structure where the drift region resistance is dominant.

It is instructive to examine the potential contours inside the second 600-V power GD-JBSFET structure when it is operating in the blocking mode. This allows determination of the voltage distribution within the structure. The potential contours for the device, obtained using the numerical simulations with zero gate bias and a drain bias voltage of 400 V, are shown in Fig. 8.68. From this figure, it can be observed that the potential distribution is like that observed for the 600-V power GD-MOSFET structure due to the excellent two-dimensional charge coupling achieved with a linearly graded doping profile.

It is insightful to examine the electric field profile inside the second 600-V power GD-JBSFET structure when it is operating in the blocking mode. The electric field profile obtained through the middle of the Schottky contact is shown in Fig. 8.69. It can be observed that the maximum electric field occurs at a depth of 2 μm from the surface. The electric field at the Schottky contact is greatly reduced ($6 \times 10^4$ V/cm) when compared the peak electric field in the bulk ($1.7 \times 10^5$ V/cm) at a drain bias of 400 V. The suppression of the electric field at the Schottky contact allows operation of the second 600-V power GD-JBSFET structure with a low leakage current despite using the source ohmic contact metal for making the Schottky contact.

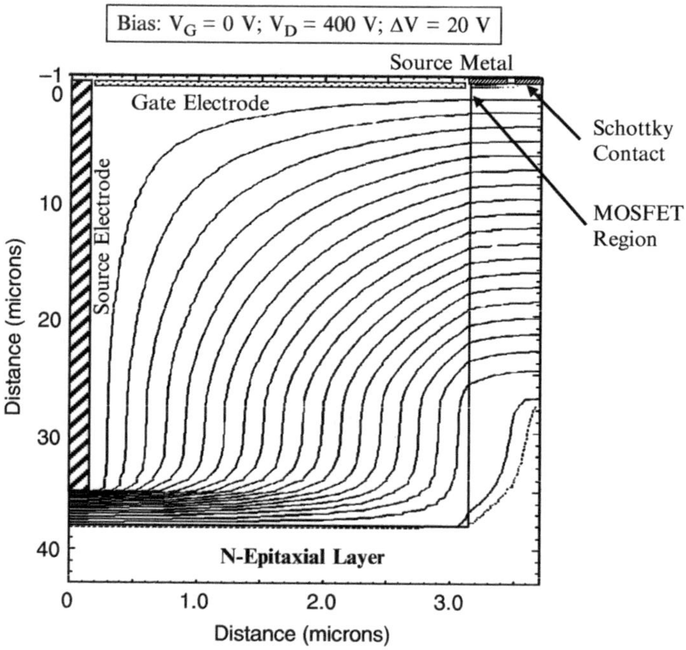


**Fig. 8.68** Potential contours in the 600-V power GD-JBSFET structure 2

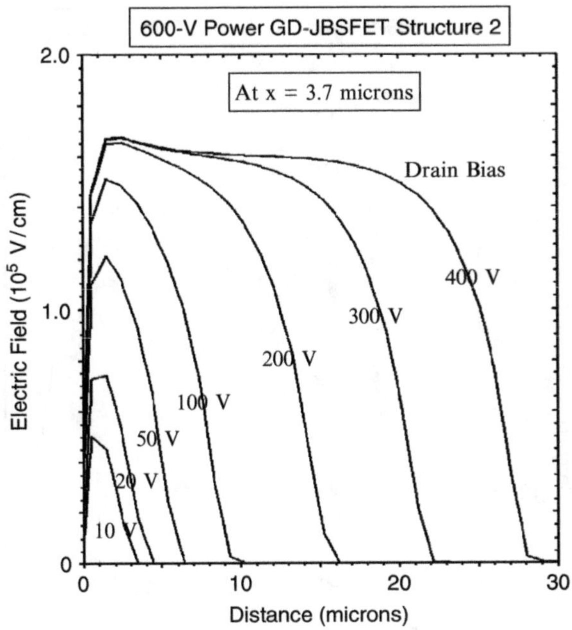


**Fig. 8.69** Electric field distribution in the 600-V power GD-JBSFET structure 2 at the Schottky contact

The forward conduction characteristics of the integral diode within the second 600-V power GD-JBSFET structure were obtained by application of a negative bias to the drain terminal while using zero gate bias. This is representative of the condition in the motor control circuit when both the high-side and low-side power MOSFET switches are turned-off. The motor current then flows through the body diode within the power MOSFET structure. The results of the numerical simulations are shown in Fig. 8.70 for the case of a minority carrier lifetime of 10 μs. When compared with the P-N body-diodes of the previously described high voltage power MOSFET structures in this chapter, the $i - v$ characteristics for the power GD-JBSFET structure is shifted towards lower voltages due to current flow via the Schottky contact. The area of the Schottky contact in the 600-V power GD-JBSFET structure used for the numerical simulations is sufficiently large that the on-state voltage drop is only 1.22 V at even the relatively high on-state current density of 121 A/cm$^2$ for the device in the MOSFET mode of operation.

The hole concentration profile in the mesa at the P-N junction in the first 600-V power GD-JBSFET structure is shown in Fig. 8.71 when the body diode is forward biased with a current density of 121 A/cm$^2$. The doping profile is also shown in the plot for comparison. It can be observed that the drift region operates under low-level injection conditions over the entire drift region. The presence of the Schottky contact suppresses injection from the P-N body diode as expected based up on the physics of operation of MPS rectifiers [10]. The injected hole concentration in the

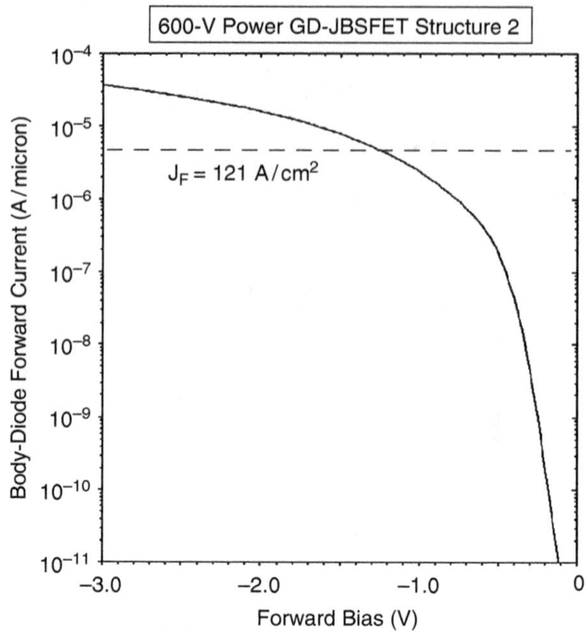


**Fig. 8.70** On-state characteristics of the integral-diode within the 600-V power GD-JBSFET structure 2

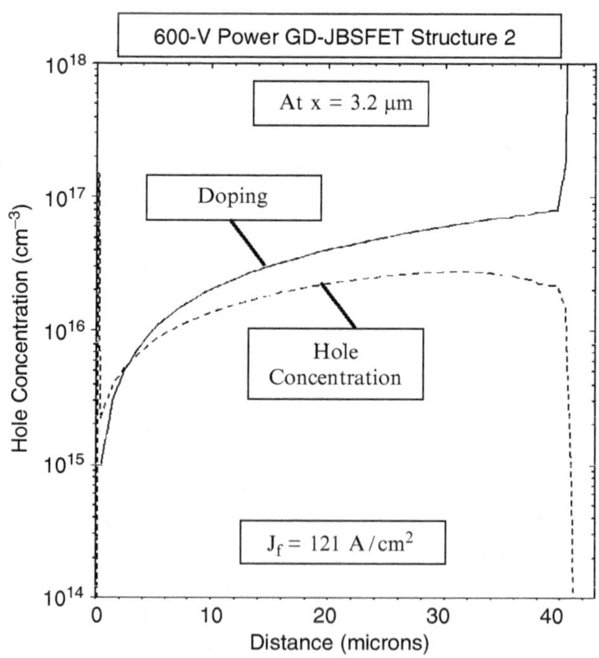


**Fig. 8.71** Hole distribution in the 600-V power GD-JBSFET structure 2

mesa is substantially less than in the 600-V power GD-MOSFET structure (see Fig. 8.48). These results demonstrate that the Schottky contact must be located in the vicinity of the P-N body diode of the MOSFET portion in order to suppress the injection of holes into the drift region during operation of the device in the third quadrant.

The current flow path inside the second 600-V power GD-JBSFET structure is shown in Fig. 8.72 when the device is operated in the third quadrant with a current density of 121 A/cm$^2$. Only the upper portion of the structure is shown in the figure to allow discerning the current path at the contacts. It can be observed that the drain current flows mainly via the Schottky contact with a small part flowing through the P-N junction. Consequently, the injection of holes from the P-N body diode is substantially reduced by the presence of the Schottky contact.

The reverse recovery characteristics for the body diode in the second 600-V power GD-JBSFET structure were obtained by initially forward biasing the body diode with an on-state current density of 121 A/cm$^2$ corresponding with a on-state power dissipation of 100 W/cm$^2$ based up on the specific on-resistance of this structure. The drain current was ramped from negative to positive at a ramp rate of 3300 A/cm$^2$ μs (the same rate as for the 600-V power U-MOSFET structure). The resulting waveforms for the drain current and drain voltage are shown in Figs. 8.73 and 8.74. It can be observed that the peak reverse recovery current density ($J_{PR}$)

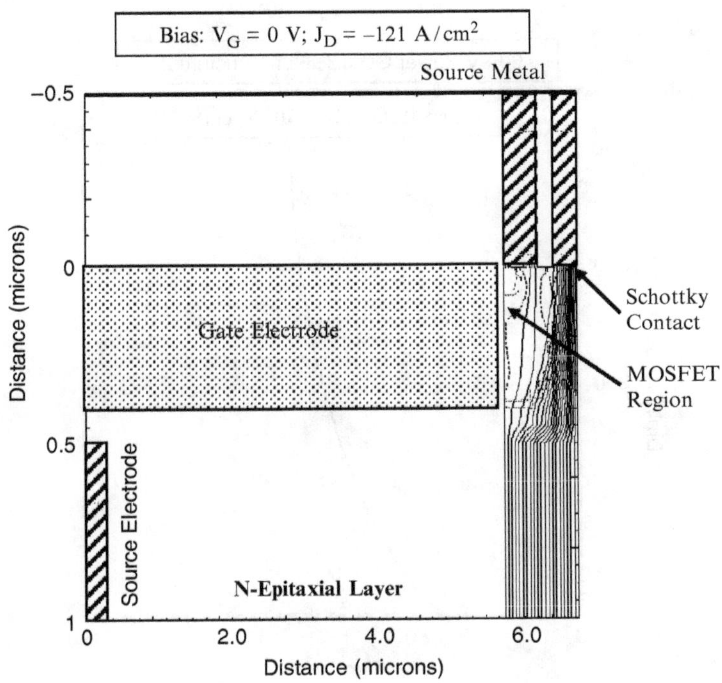


**Fig. 8.72** Body diode current path within the 600-V power GD-JBSFET structure 2

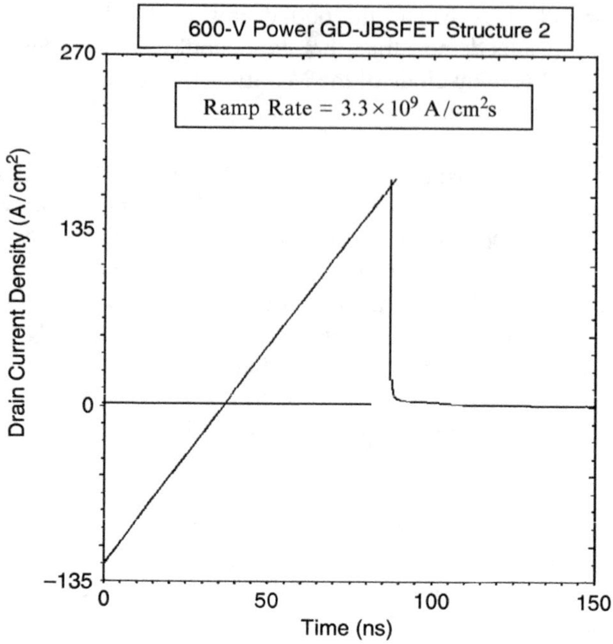


Fig. 8.73 Reverse recovery current of the body diode within the 600-V power GD-JBSFET structure 2

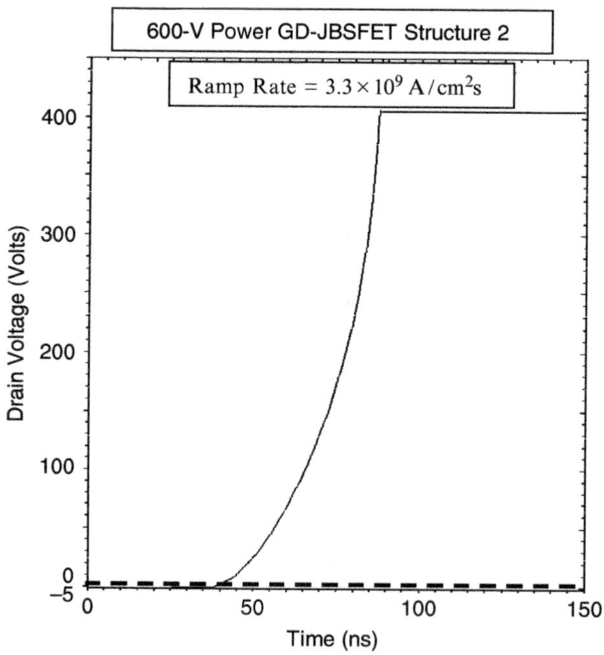


Fig. 8.74 Reverse recovery voltage of the body diode within the 600-V power GD-JBSFET structure 2

is 172 A/cm$^2$ for the case of a lifetime of 10 μs. The peak reverse recovery current predicted by the analytical model for low-level injection conditions (see (8.15)) for the second 600-V power GD-JBSFET structure is 169 A/cm$^2$ in excellent agreement with the simulations. The voltage waveform also rises rapidly as predicted by the analytical model (see (8.13)) with a square-law behavior as a function of time.

The peak reverse recovery current densities for the first and second 600-V power GD-JBSFET structures are nearly equal according to the analytical solution and the numerical simulations. However, it is worth pointing out that the second structure operates at twice the on-state current density of the first structure. Consequently, the reverse recovery characteristics of the second structure are much superior to those for the first structure. In addition, the specific on-resistance for the second structure is half that for the first structure. It can therefore be concluded that the second 600-V GD-JBSFET structure is the preferred device for motor control applications.

### 8.3.5 Power SJ-MOSFET Structure

The power SJ-MOSFET structure was illustrated in Fig. 7.1. This device has a body diode formed between the P-type drift region and the N-type drift region. The doping concentrations on both sides of the junction are equal for the optimum charge-coupled design. Moreover, the doping concentrations in the P-type drift and the N-type drift regions are low. Consequently, when the body diode is forward biased, minority carriers are injected into both the P-type drift and the N-type drift regions. The injected carrier concentration exceeds the doping concentration resulting in high level injection conditions. This results in significant stored charge that adversely impacts the reverse recovery behavior for the 600-V power SJ-MOSFET structure.

The current density flowing through the body diode in its on-state is the same value as the current density flowing through the power MOSFET structure when the gate bias has been applied because they are both determined by the current flowing through the motor windings. The on-state current density will therefore be assumed to be limited by the on-state power dissipation to 100 W/cm$^2$ as determined by the specific on-resistance of the power MOSFET structure (see (8.2)).

With low-level injection conditions in the drift region when the lifetime is small in the drift region of the 600-V power SJ-MOSFET structure, the reverse recovery for the body diode is determined by the establishment of a depletion region in the P-type and N-type drift regions without stored charge. The vertical P-N junction introduces a significant capacitance between the drain and source which must be charged during the reverse recovery process. In this section, the reverse recovery waveforms will be analyzed under the assumption that the charging of the drain-source capacitance is dominant.

The specific drain-source capacitance for the power SJ-MOSFET structure can be derived by using the device structural parameters shown in Fig. 8.75. The drain-

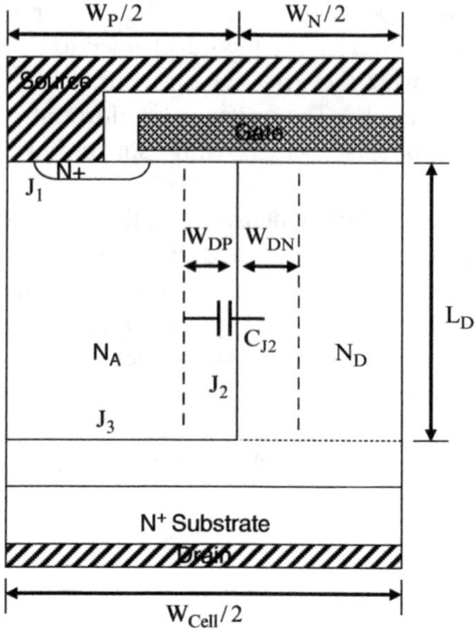


**Fig. 8.75** Output capacitance in the 600-V power SJ-MOSFET structure

source capacitance consists of two components: a component associated with the bottom junction ($J_3$) and the vertical junction ($J_2$). Due to the narrow drift region widths in comparison to the large drift region length in the high voltage power SJ-MOSFET structures, it will be assumed that the capacitance of the vertical junction is dominant. The specific drain-source capacitance is then given by:

$$C_{DS,SP} = C_{J2,SP} = \frac{\varepsilon_S}{W_D}\left(\frac{2L_D}{W_{Cell}}\right) \tag{8.16}$$

where $W_D$ is the total depletion width at the vertical junction. The depletion width at the junction is given by:

$$W_D = W_{DN} + W_{DP} = 2\sqrt{\frac{\varepsilon_S V_D}{qN_D}} \tag{8.17}$$

if the doping concentration in the P-type and N-type drift regions are equal. Using this equation in (8.16) yields:

$$C_{DS,SP} = \left(\frac{L_D}{W_{Cell}}\right)\sqrt{\frac{q\varepsilon_S N_D}{V_D}} \tag{8.18}$$

The drain current flow during the reverse-recovery process charges the drain-source capacitance during the voltage rise-time:

$$J_D(t) = C_{DS,SP}\frac{dV_D}{dt} \tag{8.19}$$

Since the drain current density is changing at a fixed ramp rate (a):

$$a\,t = C_{DS,SP}\frac{dV_D}{dt} \tag{8.20}$$

Substituting (8.18) for the specific drain-source capacitance into this equation:

$$a\,t\,dt = \left(\frac{L_D}{W_{Cell}}\right)\sqrt{\frac{q\varepsilon_S N_D}{V_D}}dV_D \tag{8.21}$$

Performing the integration on both sides yields an expression for the drain voltage transient:

$$V_D(t) = \frac{a^2}{16q\ \varepsilon_S N_D}\left(\frac{W_{Cell}}{L_D}\right)^2 t^4 \tag{8.22}$$

According to the analytical model, the drain voltage increases as the fourth power of time after the current crosses zero.

The peak reverse recovery current occurs at time ($t_{PR}$) when the drain voltage becomes equal to the drain pinch-off voltage ($V_{D,N}$) because the voltage then abruptly increases to the drain supply voltage ($V_{DS}$). Using this criterion in (8.22) yields an equation for the time after zero crossing when the drain current reaches its peak reverse recovery value:

$$t_{PR} = \left[\frac{16q\varepsilon_S N_D V_{D,N}}{a^2}\left(\frac{L_D}{W_{Cell}}\right)^2\right]^{1/4} \tag{8.23}$$

The peak reverse recovery current density can be obtained by multiplying this time with the ramp rate:

$$J_{PR} = at_{PR} = \left[16q\ \varepsilon_S N_D V_{D,N} a^2\left(\frac{L_D}{W_{Cell}}\right)^2\right]^{1/4} \tag{8.24}$$

Based up on this equation, it can be concluded that the peak reverse recovery current for the body diode in the power SJ-MOSFET structure will increase with increasing ramp rate. Note that the reverse recovery current is independent of the lifetime according to this analysis. For the 600-V power SJ-MOSFET structure with a cell pitch of 3 μm and drift region length of 35 μm, a pinch-off voltage of

16.5 V, and a ramp rate of $3.3 \times 10^9$ A/cm$^2$ s, the peak reverse recovery current density predicted by this expression for the power SJ-MOSFET structure is found to be 160 A/cm$^2$ which is comparable to the on-state current density of 136 A/cm$^2$. From this result, it can be concluded that the body diode in the 600-V power SJ-MOSFET structure has acceptable reverse recovery characteristics after lifetime control [8].

#### 8.3.5.1 Simulation Results

The results of two-dimensional numerical simulations on the body diode within the 600 V power SJ-MOSFET structure are described here to provide a detailed understanding of the underlying device physics and operation. The structure used for the numerical simulations had the same cell parameters as the 600-V power SJ-MOSFET structure discussed in Chap. 7. Consequently, the P-type and N-type drift region had a vertical length of 35 μm, a width of 1.5 μm and equal doping concentration of $1 \times 10^{16}$/cm$^3$. For the numerical simulations, half the cell (with a width of 1.5 μm) was utilized as a unit cell that is representative of the structure.

The forward conduction characteristics of the body diode within the 600-V power SJ-MOSFET structure were obtained by application of a negative bias to the drain terminal while using zero gate bias. This is representative of the condition in the motor control H-bridge circuit when both the high-side and low-side power MOSFET switches are turned-off. The motor winding current then flows through the body diode within one of the power MOSFET structures. The results of the numerical simulations are shown in Fig. 8.76 for the case of 5 values for the minority carrier lifetime. During the simulations, the lifetimes for holes ($\tau_{p0}$) in heavily doped N-type material and electrons ($\tau_{n0}$) in heavily doped P-type material were assumed to be equal to values provided in the figure. The $i - v$ characteristics follow the diode law under low-level injection conditions over a broad range of current levels for all lifetime values. The on-state voltage drop, at the on-state current density determined by power dissipation considerations (see (8.2)), increases only slightly when the lifetime is reduced.

The electron concentration profiles in the P-type drift region within the 600-V power SJ-MOSFET structure are shown in Fig. 8.77 when the body diode is forward biased with a current density of 136 A/cm$^2$. By comparison with the doping profile shown in the figure, it can be concluded that the drift region operates under high-level injection conditions with the hole concentration above the doping concentration for the larger lifetime values. When the lifetime is reduced to 0.1 μs and below, the body diode operates with low-level injection conditions. The analytical model for reverse recovery described in this section is valid for these lower lifetime cases.

The reverse recovery characteristics for the body diode in the 600-V power SJ-MOSFET structure were obtained by initially forward biasing the body diode with an on-state current density of 136 A/cm$^2$ corresponding with a on-state

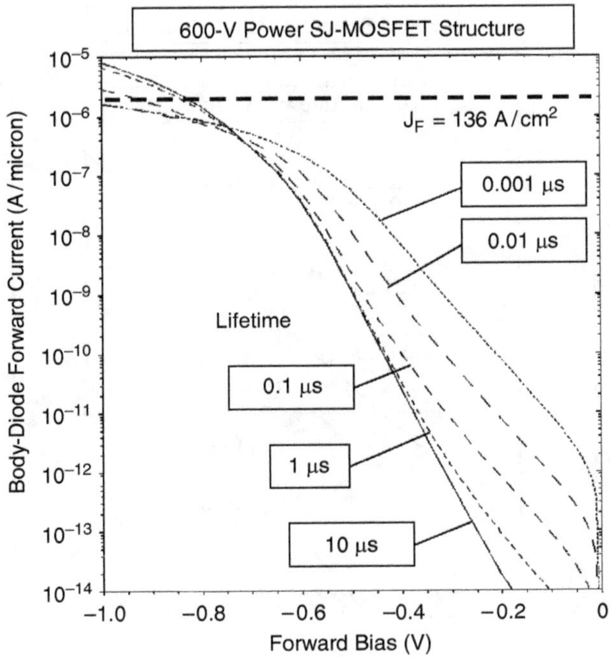


**Fig. 8.76** On-state characteristics of the body-diode within the 600-V power SJ-MOSFET structure

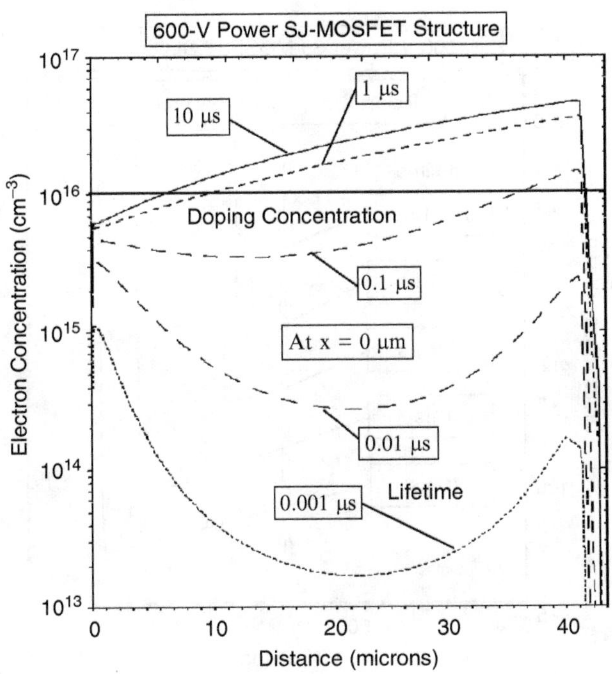


**Fig. 8.77** Electron distribution in the P-type drift region of the 600-V power SJ-MOSFET structure

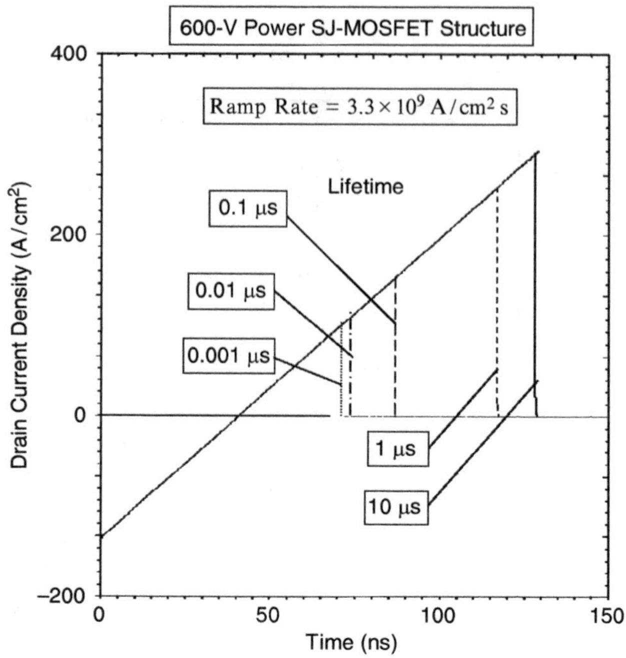


**Fig. 8.78** Reverse recovery current of the body diode within the SJ-MOSFET structure

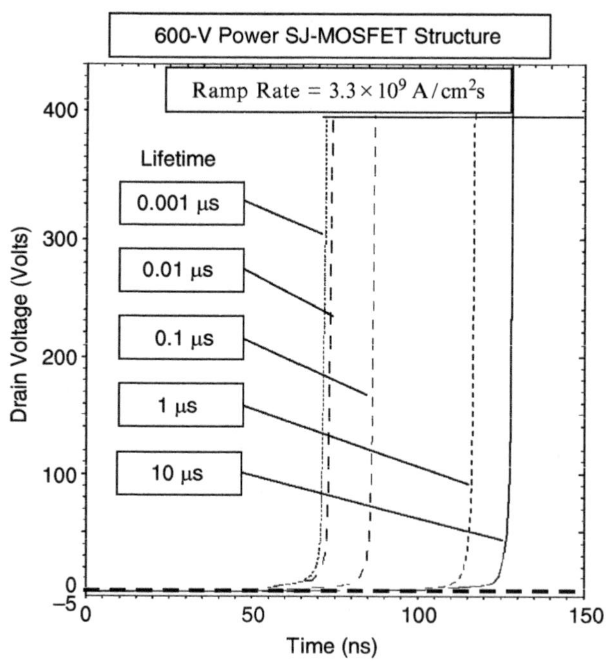


**Fig. 8.79** Reverse recovery voltage of the body diode within the SJ-MOSFET structure

power dissipation of 100 W/cm$^2$ based up on the specific on-resistance of this structure. The drain current was ramped from negative to positive at a ramp rate of 3,300 A/cm$^2$ μs. The resulting waveforms for the drain current and drain voltage are shown in Figs. 8.78 and 8.79 for various lifetime values in the drift region. It can be observed that the peak reverse recovery current density ($J_{PR}$) is 290 A/cm$^2$ for the case of a lifetime of 10 μs. This value is only 2.1-times larger than the on-state current density making it acceptable from the point of view of power dissipation in the motor control circuit. The reverse recovery current density is reduced to 104 A/cm$^2$ when the lifetime is reduced to 0.001 μs. The analytical model based up on charging the drain-source capacitance provides a good prediction of the peak reverse current density for the low lifetime cases.

The reverse recovery voltage waveforms show a gradual increase in the voltage until it reaches about 15 V followed by an abrupt increase to the drain supply voltage. This occurs because the stored charge in the P-type and N-type drift region is almost completely removed when the drain bias reaches the drain pinch-off voltage ($V_{D,N}$). The drain pinch-off voltage for a drift region doping concentration of $1 \times 10^{16}$/cm$^3$ and width of 1.5 μm is 16.5 V. Once the stored charge is removed, the drain voltage can increase rapidly to the drain supply voltage.

### 8.3.6 Power SJ-JBSFET Structure

The power SJ-JBSFET structure is illustrated in Fig. 8.80. A Schottky contact is integrated into the power SJ-MOSFET structure by bringing the source metal into

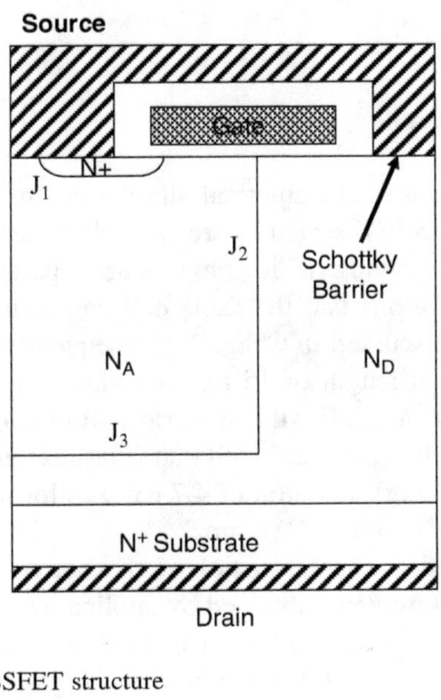


**Fig. 8.80** The power SJ-JBSFET structure

contact with the N-type drift region. The doping concentrations and widths of the N-type and P-type drift regions can be kept at the same values as for the power SJ-MOSFET structure to achieve the desired two-dimensional charge coupling phenomenon. Due to the relatively high doping concentration in the N-drift region, the current flow in the third quadrant occurs mostly through the Schottky contact suppressing the injection of minority carriers at the vertical junction ($J_2$).

The current density flowing through the body diode in its on-state is the same value as the current density flowing through the power MOSFET structure when the gate bias has been applied because they are both determined by the current flowing through the motor windings. The on-state current density will therefore be assumed to be limited by the on-state power dissipation to 100 W/cm$^2$ as determined by the specific on-resistance of the power MOSFET structure (see (8.2)).

Low-level injection conditions can be assumed in the drift regions of the 600-V power SJ-JBSFET structure under the assumption that the Schottky contact can sufficiently suppress the injection of minority carriers to below the relatively high doping concentrations of the P-type and N-type drift regions. The reverse recovery for the body diode is then determined by the establishment of a depletion region in the P-type and N-type drift regions without stored charge. As in the case of the power SJ-MOSFET structure with small lifetime values, the reverse recovery process is then determined by the charging of the output capacitance determined by junction ($J_2$). The reverse recovery waveforms will be then be defined by the equations derived for the power SJ-MOSFET structure in the previous section. For the 600-V power SJ-JBSFET structure with a cell pitch of 3 μm and drift region length of 35 μm, a pinch-off voltage of 16.5 V, and a ramp rate of $3.3 \times 10^9$ A/cm$^2$ s, the peak reverse recovery current density predicted by the analytical model is found to be 160 A/cm$^2$ which is comparable to the on-state current density of 136 A/cm$^2$.

#### 8.3.6.1 Simulation Results

The results of two-dimensional numerical simulations on the body diode within the 600 V power SJ-JBSFET structure are described here to provide a detailed understanding of the underlying device physics and operation. The structure used for the numerical simulations had the same cell parameters as the 600-V power SJ-MOSFET structure discussed in Chap. 7. Consequently, the P-type and N-type drift region had a vertical length of 35 μm, a width of 1.5 μm and equal doping concentration of $1 \times 10^{16}$/cm$^3$. For the numerical simulations, half the cell (with a width of 1.5 μm) was utilized as a unit cell that is representative of the structure. A Schottky contact with a work function of 4.7 eV was located at the middle of the N-type drift region with a width of 0.6 μm.

The forward conduction characteristics of the body diode within the 600-V power SJ-JBSFET structure were obtained by application of a negative bias to the drain terminal while using zero gate bias. This is representative of the condition in the motor control H-bridge circuit when both the high-side and low-side power

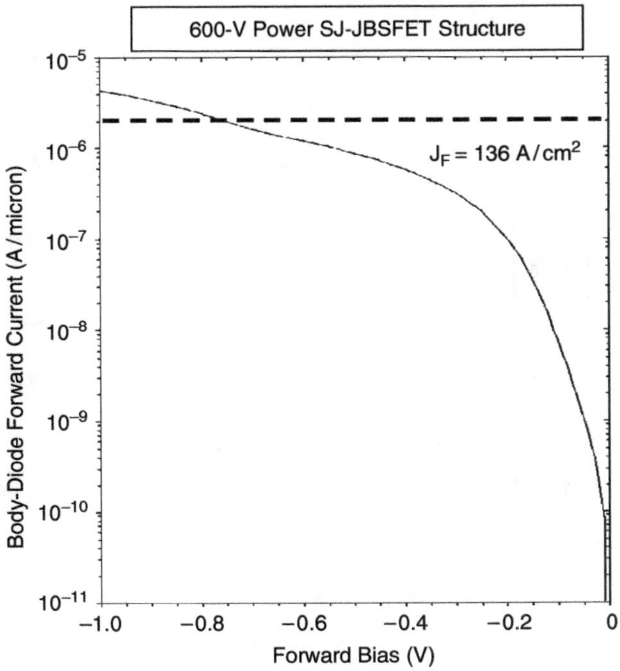


**Fig. 8.81** On-state characteristics of the body-diode within the 600-V power SJ-JBSFET structure

MOSFET switches are turned-off. The motor winding current then flows through the body diode within one of the power MOSFET structures. The results of the numerical simulations are shown in Fig. 8.81 for the case of a lifetime of 10 μs. During the simulations, the lifetimes for holes ($\tau_{p0}$) in heavily doped N-type material and electrons ($\tau_{n0}$) in heavily doped P-type material were assumed to be equal. The $i - v$ characteristics are shifted to lower on-state voltage drops due to the presence of the Schottky contact. The on-state voltage drop is only 0.755 V at the on-state current density of 136 A/cm$^2$. Although this is lower than the built-in potential of 0.8 V for typical P-N junctions, the built-in potential for the vertical junction ($J_2$) in the power SJ-JBSFET structure is only 0.7 V due to the low doping concentrations of the N-type and P-type drift regions on both sides of the junction. Consequently, an inflection can be observed in the forward $i - v$ characteristics of the integral diode in the power SJ-JBSFET structure at 0.7 V indicated some injection from the P-N junction.

The hole concentration profile in the N-type drift region within the 600-V power SJ-JBSFET structure is shown in Fig. 8.82 when the integral diode is forward biased with a current density of 136 A/cm$^2$. By comparison with the doping profile shown in the figure, it can be concluded that the drift region operates under low-level injection conditions with the hole concentration below the doping concentration despite the large lifetime value of 10 μs in the drift regions. The analytical

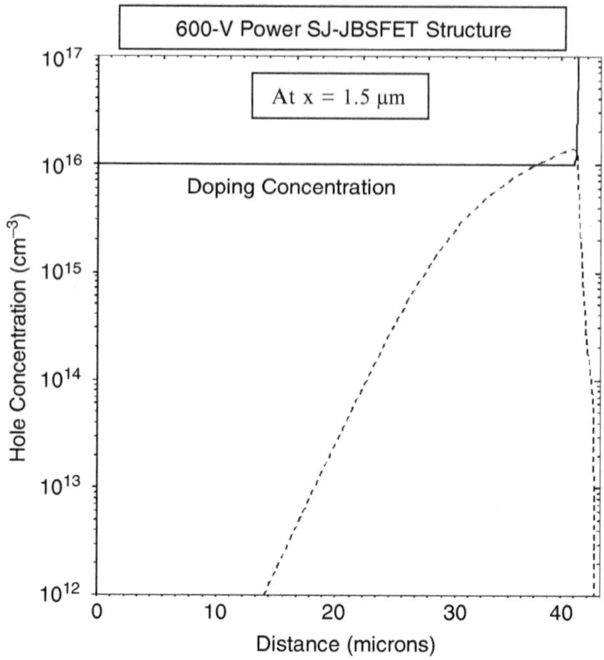


**Fig. 8.82** Hole distribution in the N-type drift region of the 600-V power SJ-JBSFET structure

model for reverse recovery described in the previous section for the power SJ-MOSFET structure is valid for this case.

The reverse recovery characteristics for the body diode in the 600-V power SJ-JBSFET structure were obtained by initially forward biasing the body diode with an on-state current density of 136 A/cm$^2$ corresponding with a on-state power dissipation of 100 W/cm$^2$ based up on the specific on-resistance of this structure. The drain current was ramped from negative to positive at a ramp rate of 3,300 A/cm$^2$ μs. The resulting waveforms for the drain current and drain voltage are shown in Figs. 8.83 and 8.84 for a lifetime of 10 μs in the drift regions. It can be observed that the peak reverse recovery current density ($J_{PR}$) is 127 A/cm$^2$ for the case of a lifetime of 10 μs. This value is equal to the on-state current density without lifetime control processes making it acceptable from the point of view of power dissipation in the motor control circuit. The analytical model based up on charging the drain-source capacitance provides a good prediction of the peak reverse current density.

The suppression of the injection of minority carriers across the vertical P-N junction due to the presence of the Schottky contact in the power SJ-JBSFET structure can also be confirmed by examination of the current flow-lines within the structure when the device is operated in the third quadrant. The current flow-lines for the power SJ-JBSFET structure are shown in Fig. 8.85 for a forward

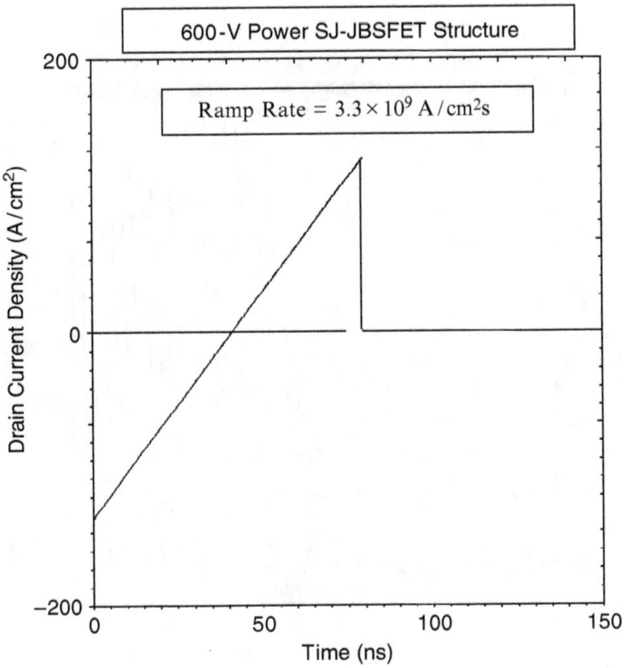


**Fig. 8.83** Reverse recovery current of the body diode within the SJ-JBSFET structure

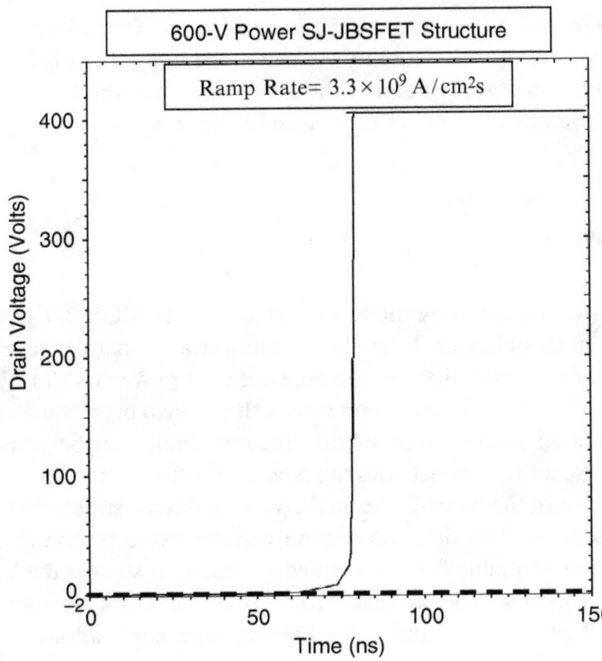


**Fig. 8.84** Reverse recovery voltage of the body diode within the SJ-JBSFET structure

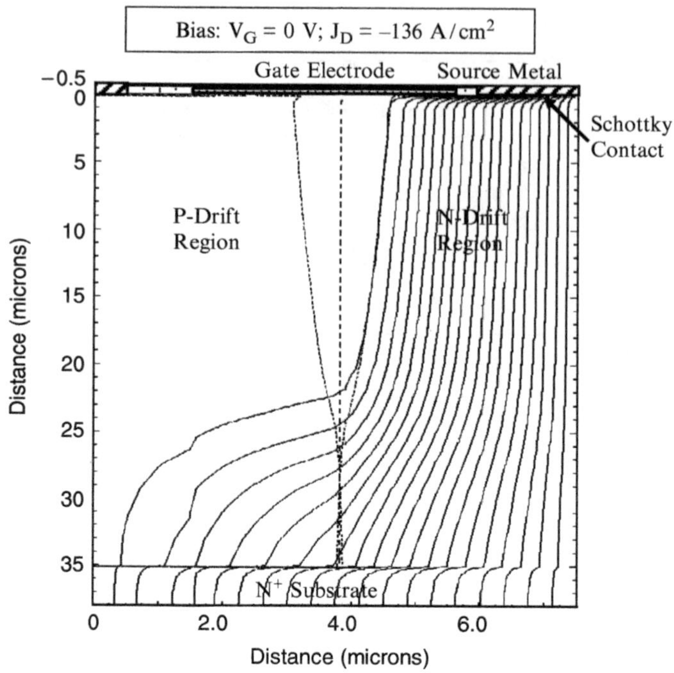


**Fig. 8.85** Current flow path for the body diode within the SJ-JBSFET structure

current density of 136 A/cm$^2$ through the integral diode. It can be observed that all the current flows through the Schottky contact at the upper portion of the structure. Some current flow is observed across the P-N junction at the bottom leading to a small concentration of injected holes as seen in Fig. 8.82.

## 8.4 Discussion

The characteristics of the body diode in various power MOSFET structures have been reviewed in this chapter. It has been demonstrated that the reverse recovery characteristics of the body diode in the conventional power U-MOSFET structure can be improved by lifetime control processes. It has also been shown that substantial improvement in performance of the integral diode can be achieved by the integration of a Schottky contact with the power MOSFET structure. A comparison of the performance of the body diodes in the various device structures is provided in the summary section. The first sub-section considers the performance of power MOSFET structures suitable for low-voltage applications such as the VRMs used to power microprocessors. The second sub-section considers the performance of power MOSFET structures suitable for high-voltage applications such as motor control.

In order to compare the performance of the various devices, it is important to take into account the difference in the operating current density of the integral diode in the on-state. This can be done by computation of the ratio of the peak reverse recovery current to the on-state current density when comparing the devices. In addition, it is useful to define two new figures-of-merit related to the reverse recovery performance of the integral diodes.

The first new figure-of-merit is obtained by taking the product of the specific on-resistance and the peak reverse recovery current density:

$$\mathrm{FOM(E)} = R_{ON,sp} J_{PR} \tag{8.25}$$

This figure-of-merit is independent of the device area and has units of volts. However, the peak reverse recovery current is dependent up on the reverse ramp rate. It is therefore important to extract the peak reverse recovery current using the same ramp rate for all the device structure (as done in Sects. 8.2 and 8.3).

The second new figure-of-merit is obtained by taking the product of the specific on-resistance and the specific reverse recovery charge density:

$$\mathrm{FOM(F)} = R_{ON,sp} Q_{RR,sp} \tag{8.26}$$

The specific reverse recovery charge can be extracted from the area under the drain current versus time plot for the reverse recovery waveform. This figure-of-merit is also independent of the device area and has units of volt-seconds. An advantage of this figure-of-merit is that the specific reverse recovery charge is not dependent on the reverse recovery ramp rate.

## *8.4.1 Low-Voltage Devices*

In this section, the performance of the integral diode in low-voltage power MOSFET structures suitable for the VRM application will be compared. The performance of the integral diodes within these devices was described in Sect. 8.2. The device structures considered here are: the 30-V power U-MOSFET structure; the 30-V power CC-MOSFET structure; and the 30-V power JBSFET structure. The performance of the integral diode within these structures is summarized in Figs. 8.86 and 8.87 for the case of a lifetime of 10 μs in the drift region. This is representative of devices without lifetime control. The on-state voltage drop provided in the Fig. 8.86 is for the integral diode and in Fig. 8.87 for the MOSFET mode.

From the information in the above figures, it can be concluded that the best performance is observed with the 30-V power CC-MOSFET structure. It has an exceptionally small specific on-resistance which allows operation at a very high on-state current density. The chip area required to serve any application is

| Device Structure | $R_{DSON,SP}$ (mΩ-cm²) | $J_{ON}$ (A/cm²) | $V_{ON}$ (V) | $J_{PR}$ (A/cm²) | $J_{PR}/J_{ON}$ | FOM(E) (V) |
|---|---|---|---|---|---|---|
| U-MOSFET | 0.280 | 600 | 0.797 | 312 | 0.52 | 0.0874 |
| CC-MOSFET | 0.029 | 1850 | 0.819 | 310 | 0.17 | 0.0090 |
| JBSFET | 0.220 | 674 | 0.635 | 182 | 0.27 | 0.0400 |

**Fig. 8.86** Integral diode parameters for the 30-V power MOSFET structures

| Device Structure | $R_{DSON,SP}$ (mΩ-cm²) | $J_{ON}$ (A/cm²) | $V_{ON}$ (V) | $Q_{RR}$ (nC/cm²) | $Q_{RR}/J_{ON}$ (ns) | FOM(F) (pV-s) |
|---|---|---|---|---|---|---|
| U-MOSFET | 0.280 | 600 | 0.15 | 811 | 1.35 | 227 |
| CC-MOSFET | 0.029 | 1850 | 0.05 | 806 | 0.436 | 23.4 |
| JBSFET | 0.220 | 674 | 0.15 | 282 | 0.418 | 62.0 |

**Fig. 8.87** Integral diode parameters for the 30-V power MOSFET structures

proportionately reduced. The peak reverse recovery current for the 30-V power CC-MOSFET structure is only 17% of the on-state current density. The figures-of-merit for the 30-V power CC-MOSFET structure are also an order of magnitude superior to those of the conventional 30-V power U-MOSFET structure.

As a planar technology, the 30-V power JBSFET structure also offers significant advantages when compared with the conventional 30-V power U-MOSFET structure. The peak reverse recovery current for the 30-V power JBSFET structure is only 27% of the on-state current density which is half that for the conventional 30-V power U-MOSFET structure. The figures-of-merit for the 30-V power JBSFET structure are also superior to those of the conventional 30-V power U-MOSFET structure by a factor of between two and four times.

### *8.4.2 High-Voltage Devices*

In this section, the performance of the integral diode in 600-V power MOSFET structures suitable for the motor-control application will be compared. The performance of the integral diodes within these devices was described in Sect. 8.3. The device structures considered here are: the 600-V power U-MOSFET structure; the 600-V power JBSFET structure; the 600-V power GD-MOSFET structure; the 600-V power GD-JBSFET structure; the 600-V power SJ-MOSFET structure; and the 600-V power SJ-JBSFET structure. The performance of the integral diode within these structures is summarized in Figs. 8.88 and 8.89 for the case of a lifetime of 10 μs in the drift region. This is representative of devices without lifetime control.

| Device Structure | $R_{DSON,SP}$ (mΩ-cm²) | $J_{ON}$ (A/cm²) | $V_{ON}$ (V) | $J_{PR}$ (A/cm²) | $J_{PR}/J_{ON}$ | FOM(E) (V) |
|---|---|---|---|---|---|---|
| U-MOSFET | 94.0 | 33 | 0.758 | 263 | 7.97 | 24.7 |
| JBSFET | 95.3 | 32.4 | 3.33 | 42 | 1.30 | 4.00 |
| GD-MOSFET | 6.42 | 125 | 1.21 | 200 | 1.60 | 1.28 |
| GD-JBSFET | 6.81 | 121 | 1.22 | 172 | 1.42 | 1.17 |
| SJ-MOSFET | 5.42 | 136 | 0.827 | 291 | 2.14 | 1.58 |
| SJ-JBSFET | 5.42 | 136 | 0.827 | 127 | 0.93 | 0.69 |

**Fig. 8.88** Integral diode parameters for the 600-V power MOSFET structures

| Device Structure | $R_{DSON,SP}$ (mΩ-cm²) | $J_{ON}$ (A/cm²) | $V_{ON}$ (V) | $Q_{RR}$ (μC/cm²) | $Q_{RR}/J_{ON}$ (μs) | FOM(F) (nV-s) |
|---|---|---|---|---|---|---|
| U-MOSFET | 94.0 | 33 | 3.10 | 14.5 | 0.44 | 1363 |
| JBSFET | 95.3 | 32.4 | 3.09 | 0.27 | 0.0083 | 25.7 |
| GD-MOSFET | 6.42 | 125 | 0.80 | 6.15 | 0.049 | 39.5 |
| GD-JBSFET | 6.81 | 121 | 0.82 | 4.42 | 0.037 | 30.1 |
| SJ-MOSFET | 5.42 | 136 | 0.74 | 12.63 | 0.093 | 68.5 |
| SJ-JBSFET | 5.42 | 136 | 0.74 | 2.43 | 0.018 | 13.2 |

**Fig. 8.89** Integral diode parameters for the 600-V power MOSFET structures

The on-state voltage drop provided in Fig. 8.88 is for the integral diode and in Fig. 8.89 for the MOSFET mode.

From the information in the above figures, it can be concluded that the best performance is observed with the 600-V power GD-JBSFET structure and the 600-V power SJ-JBSFET structure. They have an exceptionally small specific on-resistance which allows operation at a high on-state current density. The chip area required to serve any application is proportionately reduced. The peak reverse recovery current for the 600-V power GD-JBSFET and 600-V power SJ-JBSFET structures are only 142 and 93% of the on-state current density. The figures-of-merit for the 600-V power GD-JBSFET and 600-V power SJ-JBSFET structures are also an order of magnitude superior to those of the conventional 600-V power U-MOSFET structure.

As a planar technology, the 600-V power JBSFET structure also offers significant advantages when compared with the conventional 600-V power U-MOSFET structure. The peak reverse recovery current for the 600-V power JBSFET structure is only 130% of the on-state current density which is six-times smaller than that for the conventional 600-V power U-MOSFET structure. The figures-of-merit for the 600-V power JBSFET structure are also much superior to those of the conventional 600-V power U-MOSFET structure.

## References


1. B.J. Baliga and J.P Walden, "Improving the Reverse Recovery of Power MOSFET Integral Diodes by Electron Irradiation", Solid State Electronics, Vol. 26, pp. 1133–1141, 1983.
2. K. Shenai and B.J. Baliga, "Monolithically Integrated Power MOSFET and Schottky Diode with Improved Reverse Recovery Characteristics", IEEE Transactions on Electron Devices, Vol. 37, pp. 1167–1169, 1990.
3. B.J. Baliga and D.A. Girdhar, "Paradigm Shift in Planar Power MOSFET Technology", Power Electronics Technology, pp. 24–32, November 2003.
4. D. Calafut, "Trench Power MOSFET Low-side Switch with Optimized Integrated Schottky Diode", IEEE International Symposium on Power Semiconductor Devices, Paper 7.2, pp. 397–400, 2004.
5. S. Ono et al. "High Density MOSBD (UMOS with Built-in Trench Schottky Barrier Diode) for Synchronous Buck Converters", IEEE International Symposium on Power Semiconductor Devices, Paper 6.1, pp. 77–80, 2006.
6. B.J. Baliga, "Analysis of the High Voltage Merged P-i-N/Schottky (MPS) Rectifier", IEEE Electron Device Letters, Vol. EDL-8, pp. 407–409, 1987.
7. M. Saggio, D. Fagone, and S. Musumeci, "MDMesh: Innovative Technology for High Voltage Power MOSFETs", IEEE International Symposium on Power Semiconductor Devices, pp. 65–68, 2000.
8. M. Schmitt et al. "A Comparison of Electron, Proton and Helium Ion Irradiation for the Optimization of the CoolMOS Body Diode", IEEE International Symposium on Power Semiconductor Devices, pp. 229–232, 2002.
9. N. Mohan, T.M. Undeland, and W.P. Robbins, "Power Electronics", pp. 164–172, Wiley, New York, 1995.
10. B.J. Baliga, "Advanced Power Rectifier Concepts", Springer-Science, New York, 2009.
11. B.J. Baliga, "Fundamentals of Power Semiconductor Devices", Springer-Science, New York, 2008.
12. B.J. Baliga, "Analysis of the High-Voltage Merger P-i-N/Schottky (MPS) Rectifier", IEEE Electron Device Letters, Vol. EDL-8, pp. 407–409, 1987.

# Chapter 9
# SiC Planar MOSFET Structures

In Chap. 1, it was demonstrated that the specific on-resistance of power MOSFET devices can be greatly reduced by replacing silicon with wide band gap semiconductors. Among wide band gap semiconductors, the most progress with creating power MOSFET structures has been achieved using silicon carbide. Silicon carbide power device structures have been discussed in detail in a previous book [1]. In that book, it was shown that the conventional planar power D-MOSFET structure, developed and widely utilized for silicon, is not suitable for the development of silicon carbide devices. Two problems are encountered when utilizing the conventional power D-MOSFET structure for silicon carbide. The first problem is the much larger threshold voltage required to create an inversion layer in silicon carbide due to its much greater band gap. The doping concentration required in the P-base region to achieve a typical threshold voltage of 2 V is so low that the device cannot sustain a high blocking voltage due to reach-through of the depletion layer in the base region. The second problem is the very high electric field generated in the gate oxide because the electric field in the silicon carbide drift region under the gate is an order to magnitude larger than for silicon devices. This leads to rupture of the gate oxide at large blocking voltages.

For a planar power MOSFET structure, these issues can be addressed in a satisfactory manner by shielding the channel from the high electric field developed in the drift region. The concept of shielding of the channel region in a planar power MOSFET structure was first proposed [2] at PSRC in the early 1990s with a U.S. patent issued in 1996. The shielding was accomplished by formation of either a $P^+$ region under the channel or by creating a high resistivity conduction barrier region under the channel. The shielding approach also allowed the creation of a new power MOSFET structure called the ACCUFET where an accumulation layer is utilized to create the channel. The accumulation mode of operation allows achieving the desired typical threshold voltage and also provides a much larger channel mobility to reduce the channel resistance contribution.

In this chapter, the basic principles of operation of the shielded planar inversion-mode and accumulation-mode silicon carbide power MOSFET structures are described. The impact of shielding on ameliorating the reach-through breakdown

B.J. Baliga, *Advanced Power MOSFET Concepts*,
DOI 10.1007/978-1-4419-5917-1_9, © Springer Science+Business Media, LLC 2010

in silicon carbide is described. The difference between the threshold voltage for inversion and accumulation mode silicon carbide structures is then analyzed based up on fundamental considerations. The results of the analysis of the shielded planar silicon carbide MOSFET structures by using two-dimensional numerical simulations are described in this chapter as in the case of all the silicon devices. It is shown that the shielding concept also enables reduction of the electric field developed in the gate oxide leading to the possibility of fully utilizing the breakdown field strength of the underlying semiconductor drift region. The shielded accumulation-mode MOSFET structures (named the ACCUFET) discussed in this chapter have very promising characteristics for high voltage motor control applications.

## 9.1 Shielded Planar Inversion-Mode MOSFET Structure

The basic structure of the shielded planar inversion-mode power MOSFET structure is shown in Fig. 9.1. The structure contains a sub-surface $P^+$ shielding region which extends under both the $N^+$ source region and the P-base region. The $P^+$ shielding region is shown to extend beyond the edge of the P-base region in the figure. However, the P-base and the $P^+$ shielding region can also be formed by using a self-aligned ion-implantation process. The space between the $P^+$ shielding regions, indicated in the figure as the JFET region, is optimized to obtain a low specific on-resistance while simultaneously shielding the gate oxide interface and

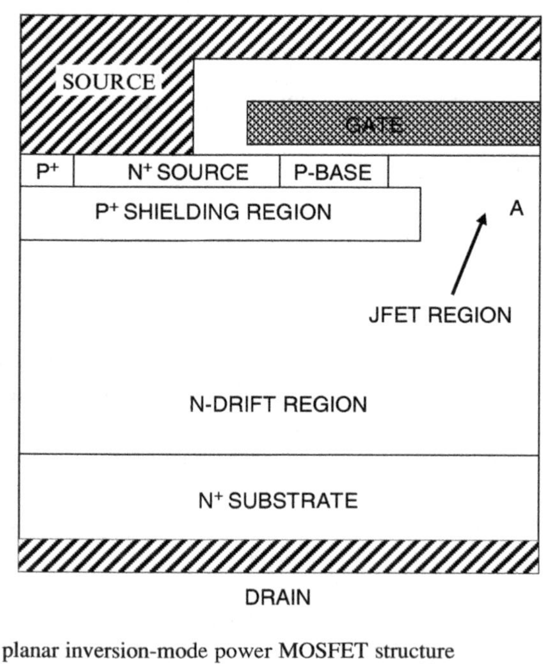


**Fig. 9.1** Shielded planar inversion-mode power MOSFET structure

the P-base region from the high electric field in the drift region. A potential barrier is formed at location A after the JFET region becomes depleted by the applied drain bias in the blocking mode. This barrier prevents the electric field from becoming large at the gate oxide interface and at the P-base/N-drift junction. When a positive bias is applied to the gate electrode, an inversion layer channel is formed at the surface of the P-base region in the structure enabling the conduction of drain current with a low specific on-resistance. The specific on-resistance of the silicon carbide inversion-mode power MOSFET structure is limited by the channel resistance as described below.

### *9.1.1 Blocking Mode*

In the forward blocking mode of the silicon carbide shielded planar inversion-mode power MOSFET structure, the voltage is supported by a depletion region formed on both sides of the $P^+$ region/N-drift junction. The maximum blocking voltage is determined by the electric field at this junction becoming equal to the critical electric field for breakdown if the parasitic $N^+$/P/N bipolar transistor is completely suppressed. This suppression is accomplished by short-circuiting the $N^+$ source and $P^+$ regions using the source metal as shown on the upper left hand side of the cross-section. If the doping concentration of the $P^+$ region is large, the reach-through breakdown problem is completely eliminated. In addition, the high doping concentration in the $P^+$ region promotes the depletion of the JFET region at lower drain voltages providing enhanced shielding of the channel and gate oxide.

In the conventional power D-MOSFET structure, the minimum P-base thickness and doping concentration are constrained by the reach-through limitation. This does not occur in the silicon carbide shielded planar inversion-mode power MOSFET structure due to shielding of the P-base region from the drain potential by the $P^+$ shielding region. This allows reducing the channel length to less than 1 μm. In addition, the doping concentration of the P-base region can be reduced to achieve a desired threshold voltage without reach-through induced breakdown. The smaller channel length and threshold voltage reduce the channel resistance contribution.

As in the case of the planar silicon power D-MOSFET structure, the maximum blocking voltage capability of the silicon carbide shielded inversion-mode planar MOSFET structure is determined by the drift region doping concentration and thickness. However, to fully utilize the high breakdown electric field strength available in silicon carbide, it is necessary to screen the gate oxide from the high field within the semiconductor. In the shielded planar MOSFET structure, this is achieved by the formation of a potential barrier at location A by the depletion of the JFET region at a low drain bias voltage. The maximum electric field in the gate oxide can be made not only below its rupture strength but lower than values required for reliable operation over long time durations.

#### 9.1.1.1 Simulation Results

The results of two-dimensional numerical simulations on the 600 V shielded 4H-SiC planar power MOSFET structure are described here to provide a more detailed understanding of the underlying device physics and operation during the blocking mode. The structure used for the numerical simulations had a drift region thickness of 4 μm below the $P^+$ shielding region with a doping concentration of $5 \times 10^{16}$ cm$^{-3}$. The $P^+$ region extended from a depth of 0.2 to 1.0 μm with a doping concentration of $1 \times 10^{19}$ cm$^{-3}$. The P-base and $N^+$ source regions were formed within the 0.2 μm of the N-drift region located above the $P^+$ region. The doping concentration of the P-base region was $5 \times 10^{16}$ cm$^{-3}$. For the numerical simulations, the cell structure (with a width of 4 μm) illustrated in Fig. 9.1 was utilized as a unit cell that is representative of the structure. Due to high doping concentration in the drift region for the 600-V 4H-SiC devices, the doping concentration in the JFET region was not enhanced as is usually required for silicon devices.

A three dimensional view of the doping distribution in the 600 V shielded 4H-SiC planar power MOSFET structure is shown in Fig. 9.2 with the upper surface of the structure located on the right hand side in order to display the doping concentration in the vicinity of the channel. The highly doped $P^+$ shielding region is prominently located just below the surface. The P-base region can be observed to have a much lower doping concentration. The junction between the P-base region and N-JFET region is also visible.

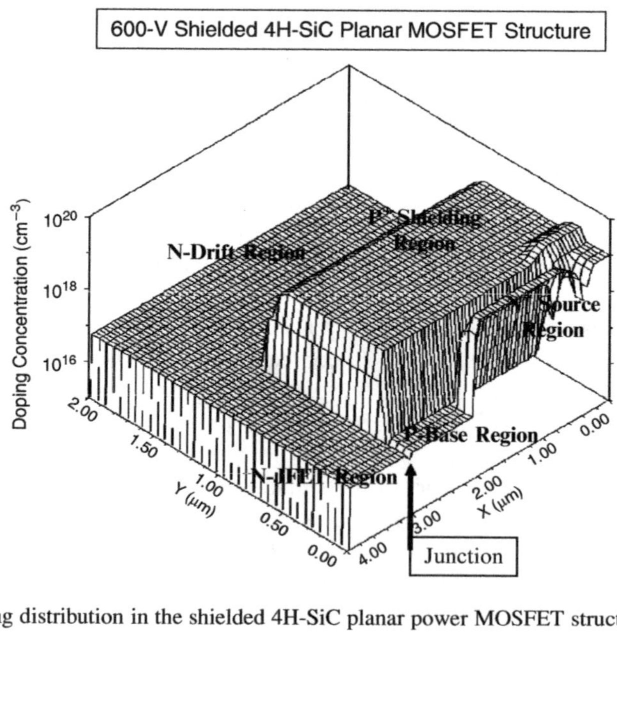


**Fig. 9.2** Doping distribution in the shielded 4H-SiC planar power MOSFET structure

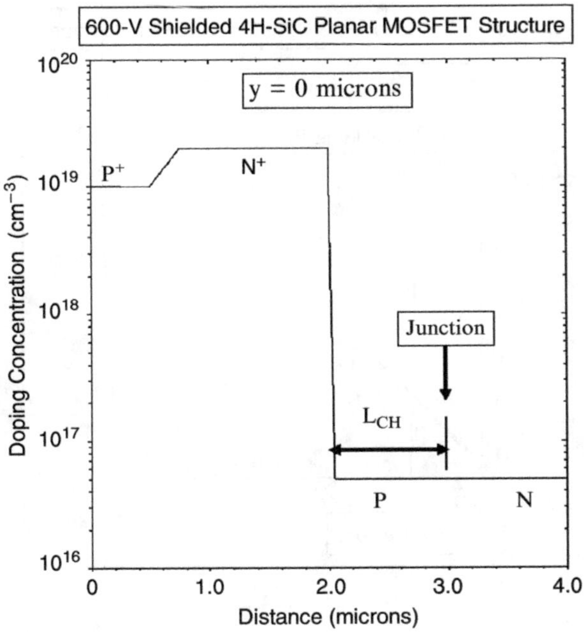


**Fig. 9.3** Channel doping profile for the shielded 4H-SiC planar power MOSFET structure

The lateral doping profile taken along the surface of the 600 V shielded 4H-SiC planar power MOSFET structure is shown in Fig. 9.3. From the profile, it can be observed that the channel extends from 2 to 3 μm creating a channel length of 1 μm in the P-base region. The doping concentration of the JFET region is the same as that of the N-drift region. The $N^+$ source region and the $P^+$ contact region for shorting the source to the base region are visible on the left-hand-side. All the regions were defined with uniform doping with abrupt interfaces between them due to the low diffusion rates for dopants in 4H-SiC material.

The vertical doping profile taken at two positions within the 600 V shielded 4H-SiC planar power MOSFET structure are provided in Fig. 9.4. From the profile taken at x = 1 μm through the $N^+$ source region (solid line), it can be observed that the doping concentration of the $P^+$ shielding region has a maximum value of $1 \times 10^{19}$ cm$^{-3}$ at a depth ranging from 0.2 to 1.0 μm. The P-base region is located between the $N^+$ source region and the $P^+$ shielding region with a doping concentration of $5 \times 10^{16}$ cm$^{-3}$. From the profile taken at x = 4 μm (dashed line), it can be observed that the JFET region has doping concentration of $5 \times 10^{16}$ cm$^{-3}$.

The blocking characteristics for the 600 V shielded 4H-SiC planar power MOSFET structure were obtained by using zero gate bias. Due to the very small intrinsic concentration in 4H-SiC, no substantial leakage current is observed at room temperature. This also confirms that the reach-through of the P-base region has been suppressed by the $P^+$ shielding region. The potential contours within the

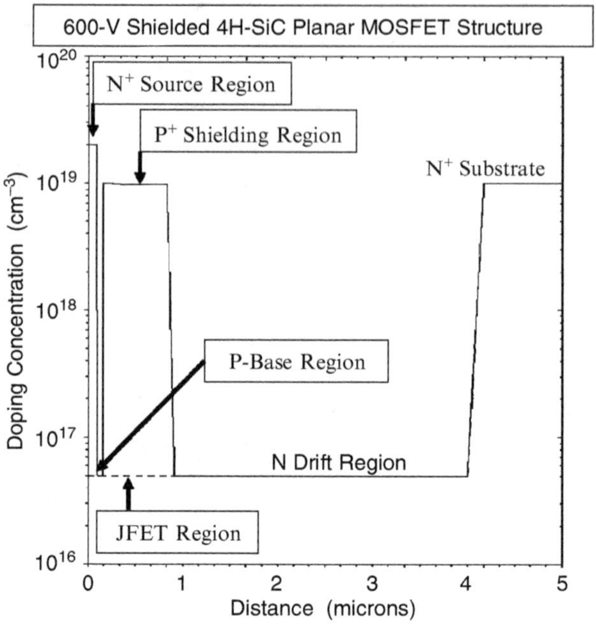


**Fig. 9.4** Vertical doping profiles in the shielded 4H-SiC planar power MOSFET structure

shielded 4H-SiC planar power MOSFET structure at a drain bias of 600 V are provided in Fig. 9.5. It can be observed that the drain voltage is supported below the $P^+$ shielding region. The potential contours do not extend into the P-base region indicating that it is shielded from the drain potential by the $P^+$ shielding region. The potential contours are crowding at the edge of the $P^+$ shielding region indicating an enhanced electric field. This can be clearly observed in Fig. 9.6 which provides a three-dimensional view of the electric field distribution. In this figure, it can also be observed that the electric field in the JFET region, and most importantly at the surface under the gate oxide, has been greatly reduced by the presence of the $P^+$ shielding region.

It is insightful to examine the electric field profile in the JFET region within the 600 V shielded 4H-SiC planar power MOSFET structure when it is operating in the blocking mode. The electric field profiles obtained through the middle of the JFET region are shown in Fig. 9.7 at various drain bias voltages. It can be observed that the maximum electric field in the JFET region occurs at a depth of 1.5 μm from the surface. This reduces the electric field at the surface under the gate oxide to about one-third of the electric field in the bulk below the $P^+$ shielding region. Consequently, the electric field in the gate oxide is reduced to about $3 \times 10^6$ V/cm. The low electric field in the gate oxide for the 600 V shielded 4H-SiC planar power MOSFET structure prevents gate oxide rupture and allows stable device performance over long periods of time. An even further reduction of electric field in the

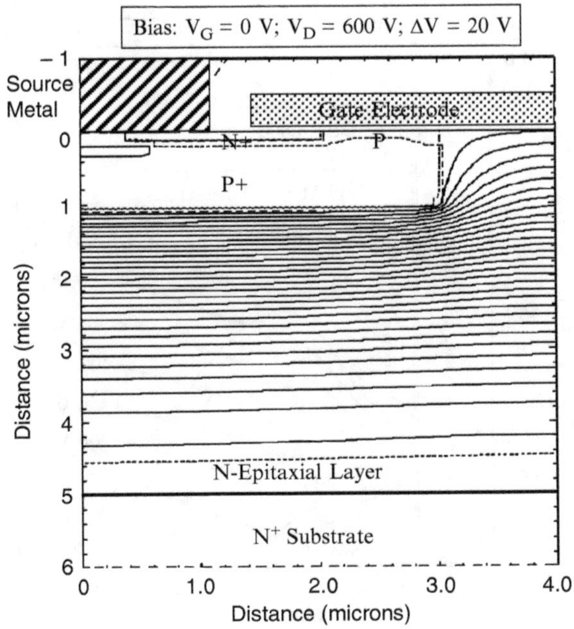


**Fig. 9.5** Potential contours in the shielded 4H-SiC planar power MOSFET structure

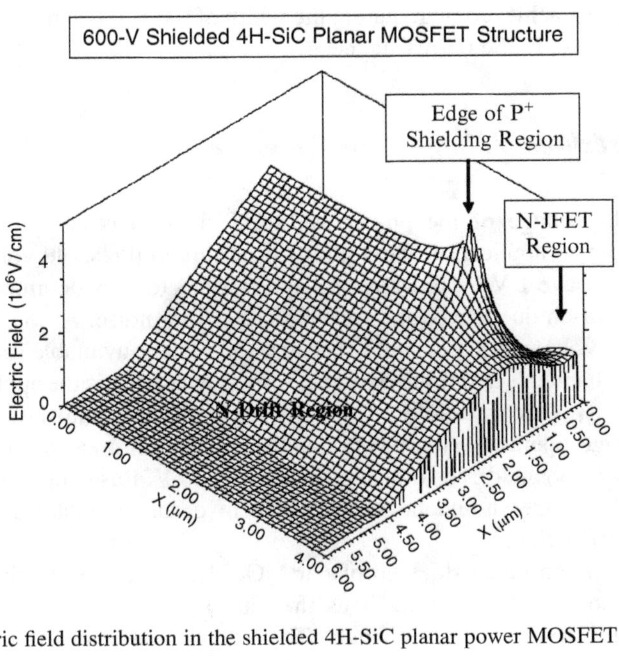


**Fig. 9.6** Electric field distribution in the shielded 4H-SiC planar power MOSFET structure

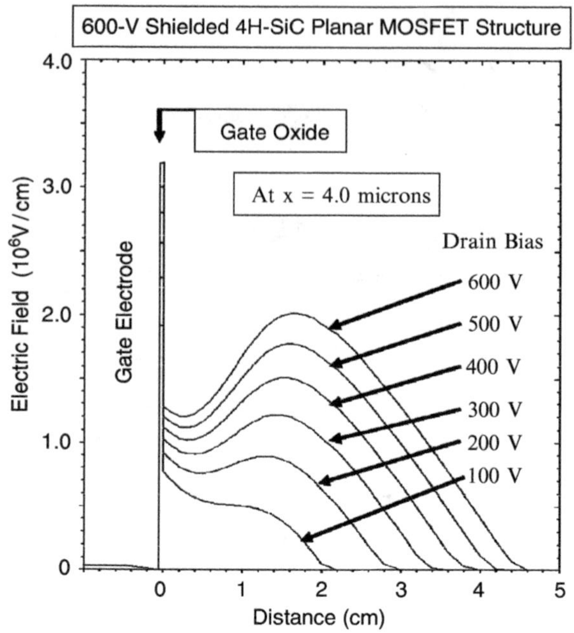


**Fig. 9.7** Electric field distribution under the gate region in the shielded 4H-SiC planar power MOSFET structure

gate oxide can be achieved by reducing the width of the JFET region at the expense of a small increase in the on-resistance.

## 9.1.2 Threshold Voltage

The threshold voltage of the power MOSFET devices is an important design parameter from an application stand-point. A minimum threshold voltage must be maintained at above 1 V for most system applications to provide immunity against inadvertent turn-on due to voltage spikes arising from noise. At the same time, a high threshold voltage is not desirable because the voltage available for creating the charge in the channel is determined by ($V_G - V_{TH}$) where $V_G$ is the applied gate bias voltage and $V_{TH}$ is the threshold voltage. Most power electronic systems designed for high voltage operation (the most suitable application area for silicon carbide devices) provide a gate drive voltage of only up to 10 V. Based upon this criterion, the threshold voltage should be kept at 2 V in order to obtain a low channel resistance contribution.

For the inversion-mode shielded planar MOSFET structure, the threshold voltage can be modeled by defining it as the gate bias at which on-set of *strong inversion* begins to occur in the channel. This voltage can be determined using [3]

$$V_{TH} = \frac{\sqrt{4\varepsilon_S kTN_A \ln(N_A/n_i)}}{C_{ox}} + \frac{2kT}{q} \ln\left(\frac{N_A}{n_i}\right) \tag{9.1}$$

where $N_A$ is the doping concentration of the P-base region, k is Boltzmann's constant, and $T$ is the absolute temperature. The presence of positive fixed oxide charge shifts the threshold voltage in the negative direction by

$$\Delta V_{TH} = \frac{Q_F}{C_{ox}} \tag{9.2}$$

A further shift of the threshold voltage in the negative direction by 1 V can be achieved by using heavily doped N-type polysilicon as the gate electrode as routinely done for silicon power MOSFET structures.

The analytically calculated threshold voltage for 4H-SiC inversion-mode MOSFET structures are provided in Fig. 9.8 for the case of a gate oxide thickness of 0.05 μm as a function of the P-base doping concentration with the inclusion of a metal-semiconductor work-function difference of 1 V. It can be observed that the model predicts a threshold voltage of about 6.5 V even for a relatively low P-base doping concentration of $5 \times 10^{16}$ cm$^{-3}$. It is difficult to reduce the threshold voltage below 3 V for the inversion-mode structure. This is one reason for the development of accumulation-mode power MOSFET structures for silicon carbide.

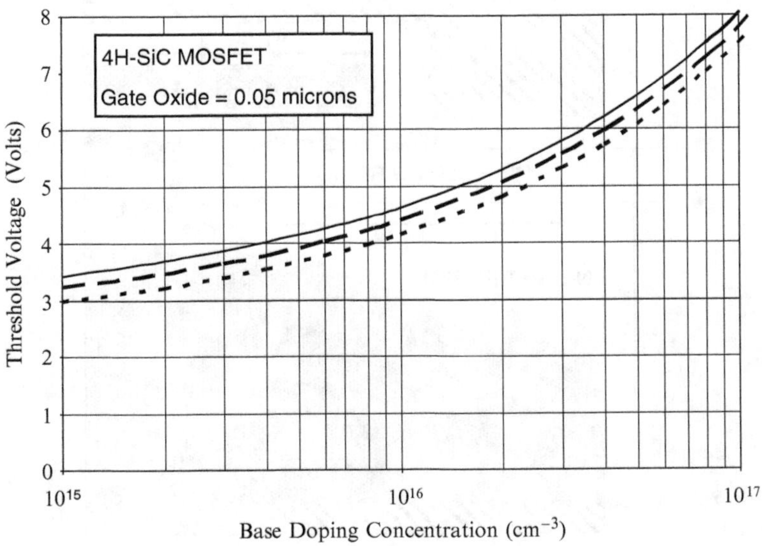


**Fig. 9.8** Threshold voltage of 4H-SiC inversion-mode MOSFET structures (*solid line*: 300°K; *dash line*: 400°K; *dotted line*: 500°K)

### 9.1.3 *On-State Resistance*

In the silicon carbide shielded inversion-mode planar MOSFET structure, current flow between the drain and source can be induced by creating an inversion layer channel on the surface of the P-base region. The current flows through the channel formed due to the applied gate bias into the JFET region via the accumulation layer formed above it under the gate oxide. It then spreads into the N-drift region at a 45° angle and becomes uniform through the rest of the structure. The total on-resistance for the silicon carbide shielded inversion-mode planar SiC MOSFET structure is determined by the resistance of the components in the current path

$$R_{on,sp} = R_{CH} + R_A + R_{JFET} + R_D \tag{9.3}$$

where $R_{CH}$ is the channel resistance, $R_A$ is the accumulation region resistance, $R_{JFET}$ is the resistance of the JFET region, $R_D$ is the resistance of the drift region after taking into account current spreading from the JFET region. For consistency with previous chapters, the resistance of the $N^+$ substrate has been omitted in the above analysis even though the substrate contribution for 4H-SiC can be very large unless its thickness is reduced to below 50 μm. The resistances can be analytically modeled by using the current flow pattern indicated by the shaded regions in Fig. 9.9.

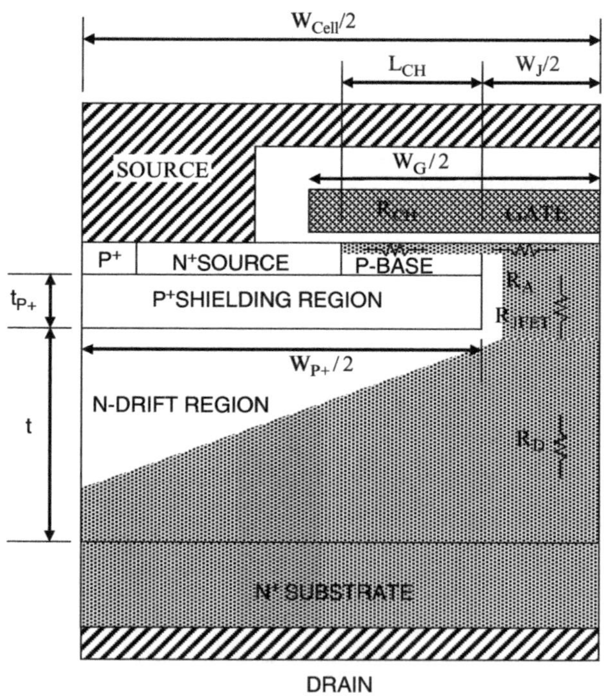


**Fig. 9.9** Current path and resistances in the shielded planar SiC inversion-mode power MOSFET structure

#### 9.1.3.1 Channel-Resistance

For the shielded planar SiC MOSFET structure with the P-base region, the specific channel resistance is given by

$$R_{CH} = \frac{(L_{CH} W_{Cell})}{2\ \mu_{inv} C_{ox} (V_G - V_{TH})} \tag{9.4}$$

where $L_{CH}$ is the channel length as shown in Fig. 9.9, $\mu_{inv}$ is the mobility for electrons in the inversion layer channel, $C_{ox}$ is the specific capacitance of the gate oxide, $V_G$ is the applied gate bias, and $V_{TH}$ is the threshold voltage. Although an inversion layer mobility of 165 $cm^2$/V-s has been observed in lateral MOSFET structures [4], the inversion layer mobility in high voltage 4H-SiC power MOSFET structure [5] is usually only 10–20 $cm^2$/V-s. The relatively low inversion layer mobility makes the channel resistance component dominant in the shielded planar SiC MOSFET structure.

In the case of the 600 V shielded planar 4H-SiC MOSFET structure, the cell width will be assumed to be 8 μm with a channel length of 1 μm. The threshold voltage for the inversion mode structure is 4.5 V despite the low doping concentration of $5 \times 10^{16}$ $cm^{-3}$ for the P-base region. Using a gate oxide thickness is 500 Å, and an inversion layer mobility of 12 $cm^2$/V-s in the above equation, the specific resistance contributed by the channel at a gate bias of 10 V is found to be 8.88 mΩ $cm^2$. This value is much larger than the ideal specific on-resistance of 0.0262 mΩ $cm^2$ for the drift region of a 600-V 4H-SiC device.

#### 9.1.3.2 Accumulation-Resistance

In the shielded planar SiC MOSFET structure, the current flowing through the inversion channel enters the JFET region at the edge of the P-base junction. The current spreads downwards from the edge of the P-base junction into the JFET region. The current spreading phenomenon is aided by the formation of an accumulation layer in the semiconductor below the gate oxide due to the positive gate bias applied to turn-on the device. The specific on-resistance contributed by the accumulation layer in the shielded planar SiC MOSFET structure is given by

$$R_{A,SP} = K_A \frac{W_J W_{Cell}}{4 \mu_{nA} C_{OX} (V_G - V_{TH})} \tag{9.5}$$

In writing this expression, a coefficient $K_A$ has been introduced to account for the current spreading from the accumulation layer into the JFET region. A typical value for this coefficient is 0.6 based upon the current flow observed from numerical simulations of shielded planar SiC MOSFET structures. The threshold voltage in the expression is for the on-set of formation of the accumulation layer. A zero threshold voltage will be assumed here when performing the analytical computations. Note that the width of the JFET region defines the length of the accumulation region.

In the case of n-channel 4H-SiC MOSFET structures, accumulation layer mobility values of 100–200 $cm^2$/V-s have been experimentally observed [6]. Using an accumulation layer mobility of 180 $cm^2$/V-s for the 600-V shielded planar SiC MOSFET structure with a cell width of 8 μm and JFET width of 2 μm, the specific resistance contributed by the accumulation layer at a gate bias of 10 V is found to be 0.196 mΩ $cm^2$ if the gate oxide thickness is 500 Å.

#### 9.1.3.3 JFET-Resistance

The electrons entering from the channel into the drift region are distributed into the JFET region via the accumulation layer formed under the gate electrode. The spreading of current in this region was accounted for by using a constant $K_A$ of 0.6 for the accumulation layer resistance. Consequently, the current flow through the JFET region can be treated with a uniform current density. In the shielded planar SiC MOSFET structure, the cross-sectional area for the JFET region is uniform with the width given by

$$a = (W_J - 2W_0) \tag{9.6}$$

where $W_0$ is the zero-bias depletion width for the JFET region. The zero-bias depletion width ($W_0$) in the JFET region can be computed by using its doping concentrations on both sides of the junction

$$W_0 = \sqrt{\frac{2\varepsilon_S V_{bi}}{qN_{DJ}}} \tag{9.7}$$

where $N_{DJ}$ is the doping concentration in the JFET region. The built-in potential is also related to the doping concentrations on both sides of the junction

$$V_{bi} = \frac{kT}{q} \ln\left(\frac{N_A N_{DJ}}{n_i^2}\right) \tag{9.8}$$

where $N_A$ is the doping concentration in the $P^+$ shielding region. Compared with silicon devices, the built-in potential for 4H-SiC is about 3-times larger.

The specific on-resistance contributed by the JFET region in the shielded planar SiC MOSFET structure can be obtained by using

$$R_{JFET,SP} = \frac{\rho_{JFET} t_{P+} W_{Cell}}{(W_J - 2W_0)} \tag{9.9}$$

where $\rho_{JFET}$ is the resistivity of the JFET region given by

$$\rho_{JFET} = \frac{1}{q\mu_n N_{DJ}} \tag{9.10}$$

where $\mu_n$ is the bulk mobility appropriate to the doping level of the JFET region. In the case of the 600-V shielded planar 4H-SiC MOSFET structure with a doping concentration of $5 \times 10^{16}$ cm$^{-3}$ in the N-drift and JFET regions, the resistivity for the JFET region is found to be 0.156 Ω cm. The zero-bias depletion width in the JFET region for this JFET doping concentration is 0.254μm based up on a built-in potential of 3 V. For the 600-V shielded planar SiC MOSFET structure with a cell width of 8 μm and JFET width of 2 μm, the specific resistance contributed by the JFET region is found to be 0.067 mΩ cm$^2$ based up on using the above parameters with a P$^+$ region thickness ($t_{P+}$) of 0.8 μm.

#### 9.1.3.4 Drift-Resistance

The resistance contributed by the drift region in the shielded planar SiC MOSFET structure is enhanced above that for the ideal drift region due to current spreading from the JFET region. The cross-sectional area for the current flow in the drift region increases from the width 'a' of the JFET region at a 45° angle as illustrated in Fig. 9.9 by the shaded area. For the 600-V shielded planar SiC MOSFET structure, the current paths in the drift region overlap at a depth of $W_{P+}/2$ from the bottom of the P$^+$ shielding region. The specific on-resistance contributed by the drift region with this model is given by

$$R_{D,SP} = \frac{\rho_D W_{Cell}}{2} \ln\left[\frac{a + W_{P+}}{a}\right] + \rho_D[t - (W_{P+}/2)] \tag{9.11}$$

For the parameters given above for this structure, the dimension 'a' in the equation is found to be 1.49 μm. For the 600-V shielded planar 4H-SiC MOSFET structure with a cell width of 8 μm and JFET width of 2 μm, the specific resistance contributed by the drift region is then found to be 0.117 mΩ cm$^2$ by using a resistivity of the drift region of 0.156 Ω cm (based upon a doping concentration of $5 \times 10^{16}$ cm$^{-3}$) and a drift region thickness of 4 μm below the P$^+$ shielding region.

#### 9.1.3.5 Total On-Resistance

The total specific on-resistance for the 600-V shielded planar 4H-SiC MOSFET structure with a cell width of 8 μm and JFET width of 2 μm is obtained by adding the above components of the resistances within the device structure. For a gate bias of 10 V, the total specific on-resistance is found to be 9.26 mΩ cm$^2$. The channel

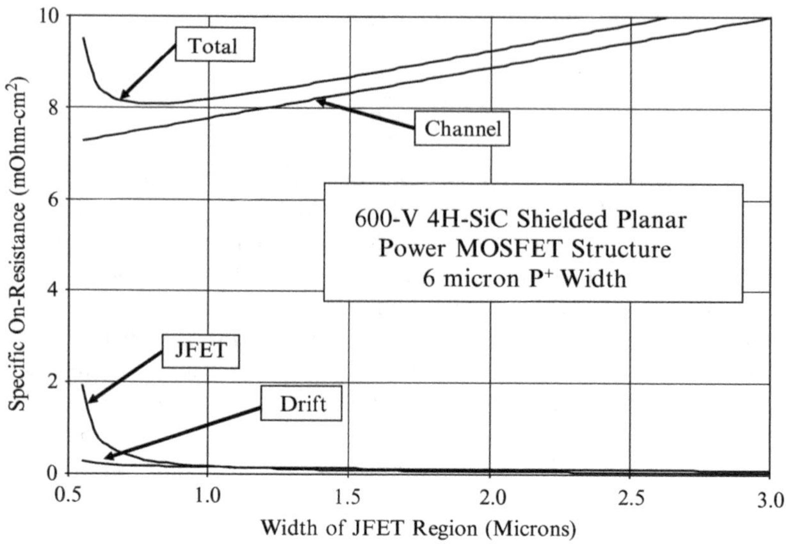


**Fig. 9.10** On-resistance for the 600-V 4H-SiC shielded planar MOSFET structures

resistance constitutes 96% of the total specific on-resistance due to the poor mobility for the electrons in the inversion layer. The specific on-resistance for the 600-V shielded planar 4H-SiC MOSFET structure is an order of magnitude smaller than that for the silicon 600-V power D-MOSFET and U-MOSFET structures. However, specific on-resistances lower than those of the 600-V shielded planar 4H-SiC MOSFET structure can be obtained by using the silicon 600-V GD-MOSFET and SJ-MOSFET structures. Consequently, the 600-V shielded 4H-SiC power MOSFET structure is not competitive with the best silicon technology unless the channel mobility can be improved.

The impact of changing the width of the JFET region on the specific on-resistance of the shielded planar 4H-SiC MOSFET structure can be determined by using the above analytical model. The results obtained for the case of a 600-V device structure with a $P^+$ shielding region width of 6 μm are provided in Fig. 9.10. For this analysis, an inversion layer mobility of 12 $cm^2/V$-s was used. It can be observed that the specific on-resistance goes through a minimum as the width of the JFET region is increased. At very small widths for the JFET region, the resistance from the JFET region and the drift region become comparable to the channel contribution producing an increase in the total specific on-resistance. When the width of the JFET region is increased beyond 1 μm, the channel resistance becomes dominant producing a monotonic increase in the specific on-resistance. The increase in the specific on-resistance between a JFET width of 1 and 2 μm is relatively small. Consequently, this width can be optimized from the point of view of shielding the gate oxide and P-base region from the high electric fields in the drift region.

#### 9.1.3.6 Impact of Breakdown Voltage

The specific on-resistance for the shielded planar 4H-SiC MOSFET structure is plotted in Fig. 9.11 as a function of the breakdown voltage by using the analytical model. In performing the modeling, it is important to recognize that the thickness of the drift region (parameter 't' in Fig. 9.9) can become smaller than half the width of the $P^+$ shielding region at lower breakdown voltages. Under these conditions, the current does not distribute across the entire drift region under the $P^+$ shielding region. The device parameters used for the plot are: channel inversion mobility of 12 $cm^2/V\text{-}s$; a fixed JFET doping concentration of $5 \times 10^{16}$ $cm^{-3}$; a width of 6 μm for the $P^+$ shielding region; a width of 2 μm for the JFET region; a cell width of 8 μm; gate bias of 10 V; threshold voltage of 4.5 V; and a gate oxide thickness of 500 Å. From Fig. 9.11, it can be seen that specific on-resistance of 4H-SiC shielded planar MOSFET structure is limited by the channel resistance for breakdown voltages below 3,000 V due to the poor inversion layer mobility. The drift region resistance can be observed to be close to the ideal specific on-resistance over most the range of breakdown voltages. The total specific on-resistance of the 4H-SiC shielded planar MOSFET structure approaches that of the drift region only when the breakdown voltage exceeds 10,000 V.

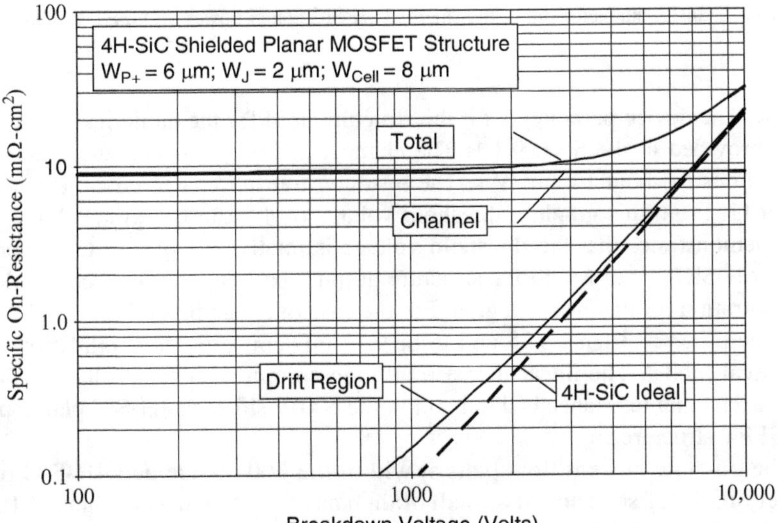


**Fig. 9.11** Specific on-resistance for the 4H-SiC shielded planar MOSFET structure

#### 9.1.3.7 Simulation Results

The transfer characteristic for the 600-V shielded 4H-SiC planar power MOSFET structure was obtained using numerical simulations with a drain bias of 0.1 V at

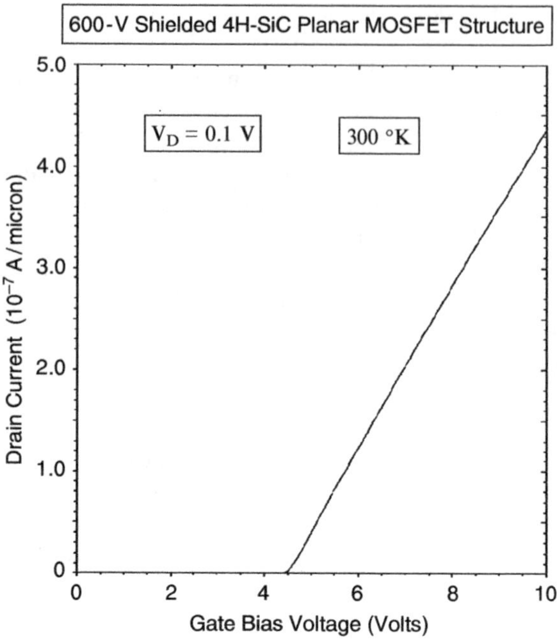


**Fig. 9.12** Transfer characteristic of the shielded 4H-SiC planar power MOSFET structure

300°K. The device parameters for the structure used for the numerical simulations were provided in the Sect. 9.1.1. The channel mobility was degraded during the simulations to about 12 cm$^2$/V-s. The resulting transfer characteristic is shown in Fig. 9.12. From this graph, a threshold voltage of 4.5 can be extracted at 300°K. This demonstrates that the threshold voltage is relatively large for the inversion-mode 4H-SiC power MOSFET structure despite the low ($5 \times 10^{16}$ cm$^{-3}$) doping concentration for the P-base region. For the case of a gate bias of 10 V and 300°K, the specific on-resistance is found to be 9.17 mΩ cm$^2$ providing validation of the analytical model. The linear increase in drain current with gate voltage indicates that the channel resistance is dominant in the 600-V shielded 4H-SiC planar power MOSFET structure.

The on-state current flow pattern within the 600-V shielded 4H-SiC planar power MOSFET structure at a small drain bias of 0.1 V and a gate bias of 10 V is shown in Fig. 9.13. In the figure, the depletion layer boundary is shown by the dotted lines and the junction boundary is delineated by the dashed line. The depletion layer width ($W_0$) in the JFET region is about 0.25 μm in good agreement with the value computed using the analytical model. It can be observed that the current flows from the channel and distributes into the JFET region via the accumulation layer. Within the JFET region, the cross-sectional area is approximately constant with a width (a/2) of 0.75 μm. From the figure, it can be seen that the current spreads from the JFET region to the drift region at a

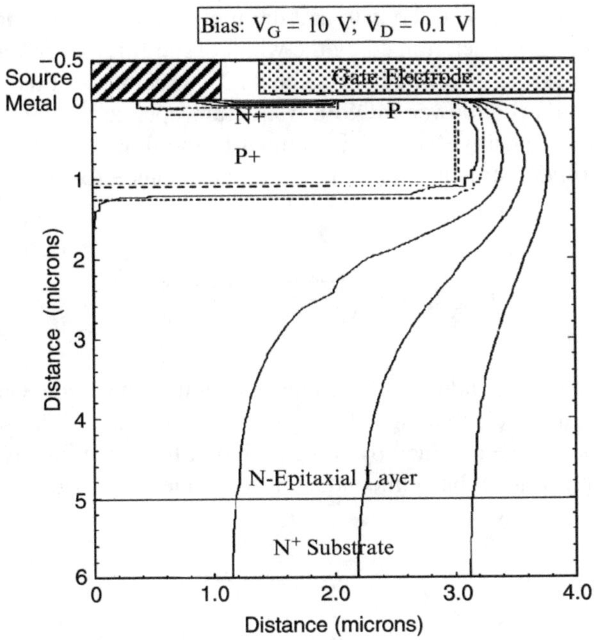


**Fig. 9.13** Current distribution in the shielded 4H-SiC planar power MOSFET structure

45° angle as assumed in the model and becomes uniform for the last 1 μm of the drift region.

### *9.1.4 Capacitances*

The capacitances within the shielded planar 4H-SiC MOSFET structure can be analytically modeled using the same approach as used for the power SC-MOSFET structure in Chap. 4. The specific input (or gate) capacitance for the shielded planar 4H-SiC MOSFET structure is given by

$$C_{IN,SP} = C_{N+} + C_P + C_{SM} = \frac{(W_G - W_J)}{W_{Cell}}\left(\frac{\varepsilon_{OX}}{t_{OX}}\right) + \frac{W_G}{W_{Cell}}\left(\frac{\varepsilon_{OX}}{t_{IEOX}}\right) \quad (9.12)$$

where $t_{OX}$ and $t_{IEOX}$ are the thicknesses of the gate and inter-electrode oxides, respectively. For a 600-V shielded planar 4H-SiC power MOSFET structure with a cell width ($W_{CELL}$ in Fig. 9.9) of 8 μm, JFET region width of 2 μm, and gate electrode width of 5 μm, the specific input capacitance is found to be 30 nF cm$^{-2}$ for a gate oxide thickness of 500 Å and an inter-metal dielectric thickness of 5,000 Å.

The capacitance between the gate and drain electrodes (also called the reverse transfer capacitance) is determined by the width of the JFET region where the gate electrode overlaps the N-drift region. The MOS structure in this portion of the shielded planar 4H-SiC power MOSFET structure operates under deep depletion conditions when a positive voltage is applied to the drain. As in the case of the power SC-MOSFET structure, the gate-drain capacitance for the shielded planar 4H-SiC MOSFET power structure is given by

$$C_{GD,SP} = \frac{(W_G - W_J)}{W_{Cell}} \left( \frac{C_{OX} C_{S,M}}{C_{OX} + C_{S,M}} \right) \tag{9.13}$$

where $C_{S,M}$ is the semiconductor capacitance under the gate oxide, which decreases with increasing drain bias voltage. The specific capacitance of the semiconductor depletion region can be obtained by computation of the depletion layer width. The depletion layer width in the semiconductor under the gate oxide can be obtained using

$$W_{D,MOS} = \frac{\varepsilon_S}{C_{OX}} \left\{ \sqrt{1 + \frac{2 V_D C_{OX}^2}{q\ \varepsilon_S N_{DJ}}} - 1 \right\} \tag{9.14}$$

where $N_{DJ}$ is the doping concentration of the JFET region. The specific capacitance for the semiconductor is then obtained using

$$C_{S,M} = \frac{\varepsilon_S}{W_{D,MOS}} \tag{9.15}$$

The gate-drain (or reverse transfer) capacitance can be computed by using (9.13) with the above equations to determine the semiconductor capacitance as a function of the drain bias voltage.

However, as in the case of the power SC-MOSFET structure, the above equation is only valid until the depletion region from the $P^+$ shielding regions pinches-off the JFET region in the shielded planar 4H-SiC power MOSFET structure. The gate-drain capacitance then decreases at a different rate because the gate is screened from the drain. The drain voltage at which the JFET region is pinched-off is given by

$$V_{P,JFET} = \frac{q N_{DJ}}{8\ \varepsilon_S} W_J^2 \tag{9.16}$$

For the 600-V shielded planar 4H-SiC power MOSFET structure with JFET region doping concentration ($N_{DJ}$) of $5 \times 10^{16}$ cm$^{-3}$ and JFET width ($W_J$) of 2 μm, the JFET region pinch-off voltage is 46.5 V. After the JFET region is pinched-off, the gate-drain capacitance is determined by the edge of the depletion region located below the $P^+$ shielding region. This distance below the gate oxide is given by

$$W_S = t_{P+} + \sqrt{\frac{2\varepsilon_S V_D}{qN_D}} \tag{9.17}$$

The specific capacitance for the semiconductor below the gate oxide can then be obtained using (9.15) with the width $W_S$.

The output capacitance for the shielded planar 4H-SiC power MOSFET structure is associated with the capacitance of the junction between the $P^+$ shielding region and the N-drift region. Due to pinch-off of the JFET region with increasing drain bias voltage, it is necessary to examine the change in the depletion region boundary with applied voltage. The depletion layer boundary inside the shielded planar 4H-SiC power MOSFET structure prior to the pinch-off of the JFET region is similar to that previously shown by the dashed lines in Fig. 4.19 for the power SC-MOSFET structure. It can be observed that the depletion region has a vertical boundary inside the JFET region and a horizontal boundary below the $P^+$ shielding region. The capacitances associated with each of these regions are indicated in the figure as $C_{S1}$ and $C_{S2}$. The specific junction capacitance associated with the JFET region is given by

$$C_{S1,SP} = \frac{\varepsilon_S}{W_{DJ}}\left(\frac{2L_{P+}}{W_{Cell}}\right) \tag{9.18}$$

where the depletion region thickness ($W_{DJ}$) in the JFET region is related to the drain bias voltage

$$W_{DJ} = \sqrt{\frac{2\varepsilon_S(V_D + V_{bi})}{qN_{DJ}}} \tag{9.19}$$

The specific junction capacitance associated with the bottom of the $P^+$ shielding region is given by [1]

$$C_{S2,SP} = \frac{\varepsilon_S}{W_{DD}}\left(\frac{W_{P+}}{W_{Cell}}\right) \tag{9.20}$$

where the depletion region thickness ($W_{DD}$) in the drift region is related to the drain bias voltage

$$W_{DD} = \sqrt{\frac{2\varepsilon_S(V_D + V_{bi})}{qN_D}} \tag{9.21}$$

where $N_D$ is the doping concentration of the drift region. The specific output capacitance for the shielded planar 4H-SiC power MOSFET structure can then be obtained by combining the above values

$$C_{O,SP} = C_{S1,SP} + C_{S2,SP} \tag{9.22}$$

### 9.1.5 Gate Charge

The gate charge components for the shielded planar 4H-SiC power MOSFET structure are given by similar equations to those derived in Chap. 4 for the power SC-MOSFET structure. The gate transfer capacitance for this device structure is small during the transition from a drain bias of $V_{DS}$ to the pinch-off voltage ($V_{P,JFET}$) for the JFET region. Due to the small gate transfer capacitance, a more rapid drop in drain voltage occurs from $V_{DS}$ to $V_{P,JFET}$ at the beginning of the gate plateau phase. If the screening effect is neglected, the gate transfer charge corresponding to the transition where the drain voltage changes from $V_{DS}$ to $V_{ON}$ is given by

$$Q_{GD} = \frac{2K_G q \varepsilon_S N_D}{C_{OX}}\left[\sqrt{1+\frac{2V_{DS}C_{OX}^2}{q\ \varepsilon_S N_D}} - \sqrt{1+\frac{2V_{ON}C_{GOX}^2}{q\ \varepsilon_S N_D}}\right] \tag{9.23}$$

where the parameter $K_G$ is given by

$$K_G = \frac{(W_G - W_J)}{W_{Cell}} \tag{9.24}$$

The other components of the gate charge are similar to those already provided in the textbook [3]

$$Q_{SW} = Q_{GS2} + Q_{GD} \tag{9.25}$$

$$Q_G = [C_{GS} + C_{GD}(V_{DS})]V_{GP} + Q_{GD} + [C_{GS} + C_{GD}(V_{ON})](V_G - V_{GP}) \tag{9.26}$$

Equations for the gate voltage, drain current, and drain voltage waveforms for the shielded planar 4H-SiC power MOSFET structure are similar to those derived in Chap. 4 for the power SC-MOSFET structure. If the screening effect is neglected, the drain voltage is determined by the gate transfer capacitance given by (9.13)

$$v_D(t) = \frac{q\varepsilon_S N_D}{2C_{OX}^2}\left\{\left[\sqrt{1+\frac{2V_{DS}C_{Ox}^2}{q\ \varepsilon_S N_D}} - \frac{J_G C_{OX}(t-t_2)}{2K_G q\ \varepsilon_S N_D}\right]^2 - 1\right\} \tag{9.27}$$

from $t = t_2$ to $t = t_3$.

### 9.1.5.1 Simulation Example

The gate charge for the 600-V shielded 4H-SiC planar power MOSFET structure was extracted by using the results of two-dimensional numerical simulations of the cell described in the previous sections. The device was turned-on from blocking state with a drain bias of 400 V by driving it using a gate current of $1 \times 10^{-8}$ A/μm (equivalent to 0.25 A $cm^{-2}$ for the area of $4 \times 10^{-8}$ $cm^2$). Once the drain current density reached 125 A $cm^{-2}$, the drain current was held constant resulting in a reduction of the drain voltage. The gate plateau voltage for this drain current density was found to be 8.2 V due to the poor transconductance of the device as a result of the low inversion layer mobility. Once the drain voltage reached the on-state value corresponding to the gate plateau voltage, the gate voltage increased to the steady-state value of 10 V.

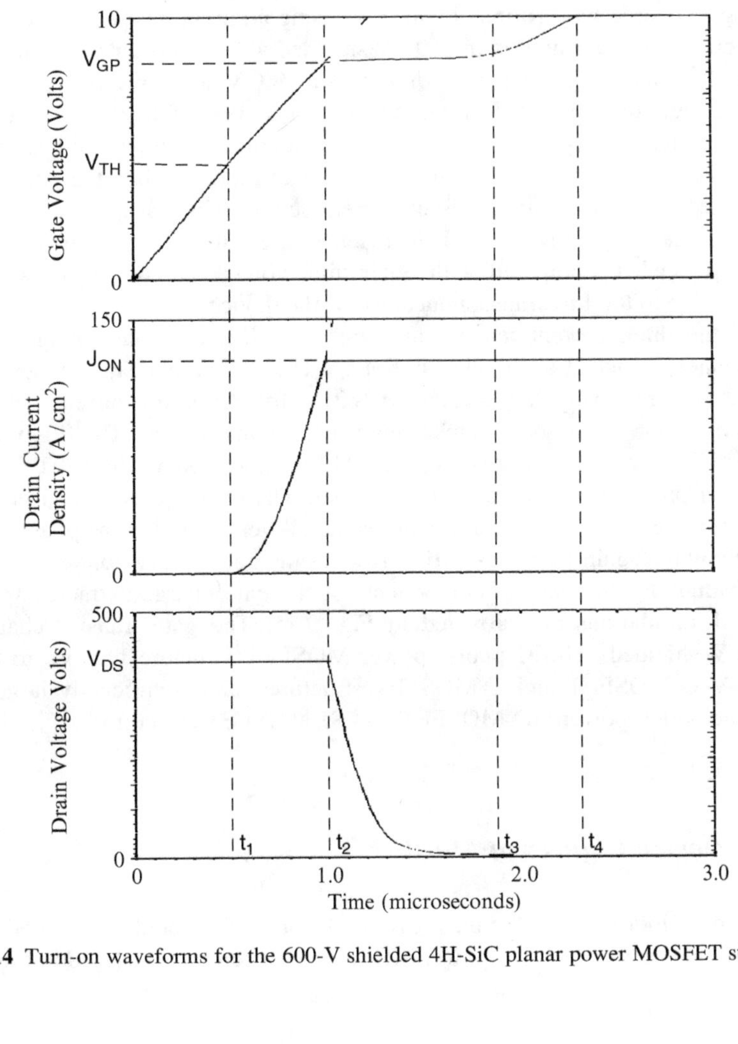


**Fig. 9.14** Turn-on waveforms for the 600-V shielded 4H-SiC planar power MOSFET structure

| Specific Gate Charge | Numerical Simulation ($nC/cm^2$) |
|---|---|
| $Q_{GS1}$ | 125 |
| $Q_{GS2}$ | 125 |
| $Q_{GS}$ | 250 |
| $Q_{GD}$ | 200 |
| $Q_{SW}$ | 325 |
| $Q_G$ | 575 |

**Fig. 9.15** Gate charge extracted from numerical simulations for the 600-V shielded 4H-SiC planar power MOSFET structure

The gate charge waveforms obtained by using an input gate current density of 0.25 A $cm^{-2}$ when turning on the 600-V shielded 4H-SiC planar power MOSFET structure from a blocking state with drain bias of 400 V are shown in Fig. 9.14. The on-state current density is 125 A $cm^{-2}$ at a DC gate bias of 10 V at the end of the turn-on transient. The gate voltage increases at a constant rate at the beginning of the turn-on process as predicted by the analytical model. When the gate voltage reaches the threshold voltage, the drain current begins to increase. The drain current increases relatively slowly until it reaches the on-state current density of 125 A $cm^{-2}$ when compared with other high voltage silicon power MOSFET structures due to the low transconductance of the device.

Once the drain current reaches the on-state value, the gate voltage remains approximately constant at the plateau voltage ($V_{GP}$). The plateau voltage for this structure is a relatively large value of 8.2 V for the drain current density of 125 A $cm^{-2}$ due to the poor transconductance of the device. The drain voltage drops rapidly from the supply voltage of 400 V of about 50 V (the JFET pinch-off voltage) as predicted by the analytical model. The drain voltage then decreases during the plateau phase in a non-linear manner. After the end of the plateau phase, the gate voltage again increases until it reaches the gate supply voltage.

The values for the various components of the gate charge extracted from the numerical simulations are provided in Fig. 9.15. The gate transfer charge for the 600-V shielded 4H-SiC planar power MOSFET structure is close to that of the 600-V D-MOSFET and U-MOSFET structures, and significantly larger than that of the 600-V power GD-MOSFET and SJ-MOSFET structures.

### *9.1.6 Device Figures of Merit*

The figures of merit (defined in the previous chapters of the book) computed for the 600-V shielded planar 4H-SiC power MOSFET structure are provided in Fig. 9.16.

| Figures of Merit | $V_G$ = 10 V |
|---|---|
| FOM(A) ($\Omega^2 cm^4 s^{-1}$) | 306, 667 |
| FOM(B)(ps) | 276 |
| FOM(C) (mΩ*nC) | 1840 |
| FOM(D) (mΩ*nC) | 2990 |

**Fig. 9.16** Figures of merit for the 600-V 4H-SiC shielded planar MOSFET structures

The figure of merit usually used for comparison of device technologies in the literature is FOM(C). In comparison with the 600-V power D-MOSFET and U-MOSFET structures, the 600-V shielded planar 4H-SiC power MOSFET structure has a FOM(C) that is an order of magnitude smaller. However, in comparison with the 600-V power GD-MOSFET and SJ-MOSFET structures, the 600-V shielded planar 4H-SiC power MOSFET structure has a FOM(C) that is five-times larger. Consequently, 600-V shielded planar 4H-SiC power MOSFET structure offers significant improvement in circuit performance compared with the conventional silicon power MOSFET structures but is not competitive with the new silicon power MOSFET technology based up on the charge coupling concept. This conclusion is a consequence of the low inversion layer mobility.

### *9.1.7 Inductive Load Turn-Off Characteristics*

As discussed in preceding chapters, high voltage power MOSFET devices are often used in adjustable speed motor drives which behave as inductive loads. The operation of the shielded planar 4H-SiC power MOSFET structure in an inductive load circuit can be analyzed using the same approach as used for the power SJ-MOSFET structure. The gate plateau voltage for the shielded planar 4H-SiC power MOSFET structure is given by

$$V_{GP} = V_{TH} + \sqrt{\frac{J_{D,ON} W_{Cell} L_{CH}}{\mu_{ni} C_{GOX}}} \tag{9.28}$$

where $C_{GOX}$ is the gate oxide capacitance. The time constant for discharging the gate of the shielded planar 4H-SiC power MOSFET structure is $R_{G,SP}$* $[C_{GS} + C_{GD}(V_{ON})]$ and the gate voltage decreases exponentially with time as given by

$$v_G(t) = V_{GS} e^{-t/R_{G,SP}[C_{GS}+C_{GD}(V_{ON})]} \tag{9.29}$$

The time $t_4$ (using the notation from the textbook) for reaching the gate plateau voltage can be obtained by using this equation with (9.28) for the plateau voltage

$$t_4 = R_{G,SP}[C_{GS} + C_{GD}(V_{ON})]\ln\left[\frac{V_{GS}}{V_{GP}}\right] \tag{9.30}$$

This time can be considered to a *turn-off delay time* before the drain voltage begins to increase after the turn-off is initiated by the control circuit.

The drain voltage begins to increase at time $t_4$ but the drain current remains constant at the load current $I_L$ because the current cannot be transferred to the diode until the voltage at the drain of the MOSFET device exceeds the supply voltage $V_{DS}$ by one diode drop to forward bias the diode. Since the drain current density is constant, the gate voltage also remains constant at the gate plateau voltage. Consequently

$$J_{GP} = \frac{V_{GP}}{R_{G,SP}} \tag{9.31}$$

where $R_{G,SP}$ is the specific gate resistance. Since all the gate current is used to discharge the gate-drain capacitance during the plateau phase because there is no change in the voltage across the gate-source capacitance

$$J_{GP} = C_{GD,SP}\frac{dv_D}{dt} \tag{9.32}$$

where $C_{GD,SP}$ is the specific gate transfer capacitance of the power MOSFET structure which is a function of the drain voltage. This voltage dependence of the gate transfer capacitance was not taken into account in the derivation provided in the textbook but is important to include here to allow comparison of the behavior of various power MOSFET structures.

For simplicity of analysis, it will be assumed that any screening effect can be ignored. The gate transfer capacitance for the shielded planar 4H-SiC power MOSFET structure is then given by

$$C_{GD,SP} = \frac{(W_G - W_J)}{W_{Cell}}\left(\frac{C_{OX}C_{S,M}}{C_{OX} + C_{S,M}}\right) \tag{9.33}$$

Using this expression in (9.32) yields the following differential equation for the voltage increase phase of the turn-off transient

$$dt = \left(\frac{W_G - W_J}{W_{Cell}}\right)\frac{1}{J_{GP}}\left[\frac{C_{GOX}}{\sqrt{1 + \frac{2v_D(t)C_{GOX}^2}{q\,\varepsilon_S N_D}}}\right]dv_D \tag{9.34}$$

Integration of this equation yields

$$(t - t_4) = \left(\frac{W_G - W_J}{W_{Cell}}\right) \frac{4q\varepsilon_S N_D}{J_{GP} C_{GOX}} \left[\sqrt{1 + \frac{2v_D(t) C_{GOX}^2}{q\ \varepsilon_S N_D}} - \sqrt{1 + \frac{2V_{ON} C_{GOX}^2}{q\ \varepsilon_S N_D}}\right] \quad (9.35)$$

In the case of the shielded planar 4H-SiC power MOSFET structure, the drain voltage increases from the on-state voltage drop ($V_{ON}$) until it reaches the drain-supply voltage ($V_{DS}$). The voltage rise-time, i.e. the time taken for the voltage to increase from the on-state voltage drop ($V_{ON}$) to the drain supply voltage ($V_{DS}$)

$$t_{V,OFF} = (t_5 - t_4) = \left(\frac{W_G - W_J}{W_{Cell}}\right) \frac{4q\varepsilon_S N_D}{J_{GP} C_{GOX}} \left[\sqrt{1 + \frac{2V_{DS} C_{GOX}^2}{q\ \varepsilon_S N_D}} - \sqrt{1 + \frac{2V_{ON} C_{GOX}^2}{q\ \varepsilon_S N_D}}\right] \quad (9.36)$$

A closed form solution for the rise in the drain voltage can be obtained from (9.35)

$$v_D(t) = \frac{q\varepsilon_S N_D}{2C_{GOX}^2} \left\{\left[\frac{J_{GP} C_{GOX}}{4q\ \varepsilon_S N_D} \left(\frac{W_{Cell}}{W_G - W_J}\right)(t - t_4) + \sqrt{1 + \frac{2V_{ON} C_{GOX}^2}{q\ \varepsilon_S N_D}}\right]^2 - 1\right\} \quad (9.37)$$

This equation describes the increase in the drain voltage from the on-state voltage drop until it reaches the drain supply voltage. The drain voltage has an approximately quadratic shape as a function of the time after $t_4$.

At the end of the plateau phase (at time $t_5$), the load current begins to transfer from the power MOSFET device to the free wheeling diode. Since the drain voltage remains constant, the gate-drain capacitance can also be assumed to remain constant during this phase. The current flowing through the gate resistance ($R_G$) discharges both the gate-drain and gate-source capacitances leading to an exponential fall in gate voltage from the plateau voltage

$$v_G(t) = V_{GP} e^{-(t - t_5)/R_{G,SP} C_{GS}} \quad (9.38)$$

The drain current follows the gate voltage as given by

$$J_D(t) = g_m[v_G(t) - V_{TH}] = \frac{\mu_{ni} C_{OX}}{L_{CH} W_{Cell}} [v_G(t) - V_{TH}]^2 \quad (9.39)$$

The drain current decreases rapidly with time due to the exponential reduction of the gate voltage, as given by (9.38), during the current fall phase. The drain current becomes equal to zero when the gate voltage reaches the threshold voltage. The current fall time can therefore be obtained from (9.38)

$$t_{I,OFF} = R_{G,SP} C_{GS} \ln\left(\frac{V_{GP}}{V_{TH}}\right) \tag{9.40}$$

Specific capacitances should be used in this expression for computation of the current fall time. Beyond this point in time, the gate voltage decreases exponentially until it reaches zero. The time constant for this exponential decay is different from the initial phase due to the smaller gate-drain capacitance.

The turn-off energy loss per cycle can be obtained using

$$E_{OFF} = \frac{1}{2} J_{ON} V_{DS} \left(t_{V,OFF} + t_{I,OFF}\right) \tag{9.41}$$

under the assumption that the drain current and voltage excursions are approximately linear with time. The energy loss during the voltage rise-time interval is comparable to the energy loss during the current fall-time interval for the shielded planar 4H-SiC power MOSFET structure.

#### 9.1.7.1 Simulation Results

The results of two-dimensional numerical simulations on the turn-off of the 600 V shielded 4H-SiC planar power MOSFET structure are described here. The drain supply voltage was chosen as 400 V for the turn-off analysis. During the turn-off simulations, the gate voltage was reduced to zero with a gate resistance of $1 \times 10^8\ \Omega\ \mu m$ for the 4 μm half-cell structure, which is equivalent to a specific gate resistance of $4\ \Omega\ cm^2$. The current density was initially held constant at an on-state current density of $104\ A\ cm^{-2}$ allowing the drain voltage to rise to the drain supply voltage. The drain voltage was then held constant allowing the drain current density to reduce to zero.

The turn-off waveforms obtained for the 600-V shielded 4H-SiC planar power MOSFET structure by using the numerical simulations are shown in Fig. 9.17. The gate voltage initially reduces to the gate plateau voltage corresponding to the on-state current density. The drain voltage then increases quadratically from the on-state voltage drop to the drain supply voltage as predicted by the analytical model. After this, the drain current reduces exponentially. The drain current fall time is much larger than observed for the silicon devices due to the poor transconductance of the shielded 4H-SiC planar power MOSFET structure. The drain voltage rise-time ($t_5 - t_4$) and the drain current fall time ($t_6 - t_5$) are comparable for the shielded 4H-SiC planar power MOSFET structure. The drain voltage rise-time obtained from the simulations of the 600-V shielded 4H-SiC planar power MOSFET structure is 0.13 μs and the drain current fall-time obtained from the simulations is 0.06 μs. The energy loss per cycle for the shielded 4H-SiC planar

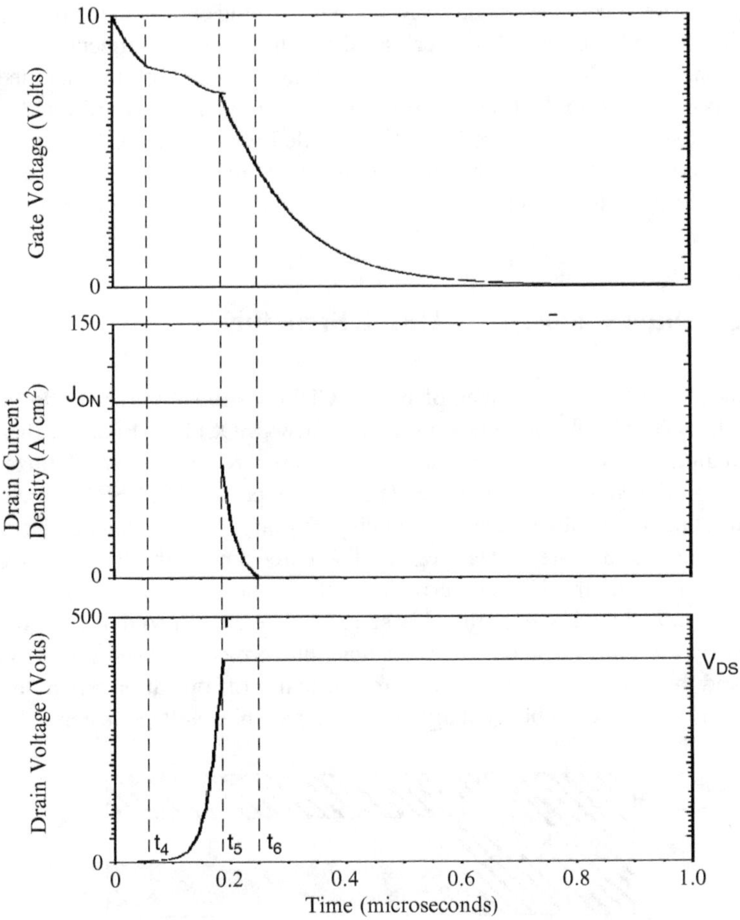


**Fig. 9.17** Turn-off waveforms for the 600-V shielded 4H-SiC planar power MOSFET structure

power MOSFET structure computed by using the values from the simulations is 3.95 mJ cm$^{-2}$. This value is ten-times larger than that observed for the 600-V silicon power SJ-MOSFET structure and twice as large as that observed for the 600-V silicon power GD-MOSFET structure.

### *9.1.8 Body-Diode Characteristics*

The shielded planar 4H-SiC power MOSFET structure has an internal body-diode formed between the P$^+$ shielding region and the N-drift region. The on-state voltage drop for this body diode is about 3 V compared with 1 V for silicon devices.

Large power loss occurs when the body diode is carrying current in the third quadrant for the shielded planar 4H-SiC power MOSFET structure. Consequently, it is necessary to connect a 4H-SiC Schottky rectifier in anti-parallel with the shielded planar 4H-SiC power MOSFET structure to carry the current in the third quadrant. The Schottky diode can be integrated into the shielded planar 4H-SiC power MOSFET structure. In this case, the $P^+$ shielding region can be utilized to protect the Schottky contact by forming a junction-barrier controlled Schottky (JBS) rectifier.

## 9.2 Shielded Planar ACCUFET Structure

The basic structure of the shielded planar ACCUFET structure is shown in Fig. 9.18. Unlike the shielded planar inversion-mode power MOSFET structure, there is no P-base region in this structure. Instead, a lightly doped N-base region is utilized. As in the case of the shielded planar inversion-mode power MOSFET structure, the structure contains a sub-surface $P^+$ shielding region which extends under both the $N^+$ source region and the N-base region. The extension of the $P^+$ shielding region beyond the edge of the $N^+$ source region defines the channel length. The doping concentration and thickness of the N-base region are chosen so that it is completely depleted by the built-in potential of the junction formed between the $P^+$ shielding region and the N-base region. This ensures normally-off operation where the device can support a large drain bias voltage when the gate bias voltage is zero.

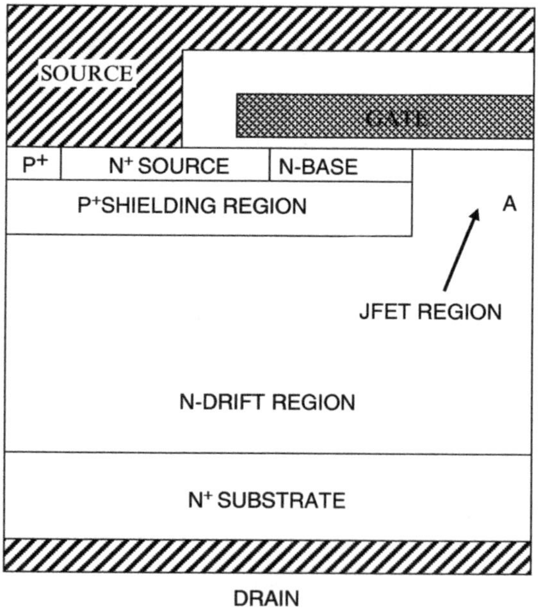


**Fig. 9.18** Shielded planar accumulation-mode power MOSFET structure

The space between the $P^+$ shielding regions, indicated in the figure as the JFET region, is optimized to obtain a low specific on-resistance while simultaneously shielding the gate oxide interface and the N-base region from the high electric field in the drift region. A potential barrier is formed at location A after the JFET region becomes depleted by the applied drain bias in the blocking mode. This barrier prevents the electric field from becoming large at the gate oxide interface.

When a positive bias is applied to the gate electrode, an accumulation layer channel is formed at the surface of the N-base region in the shielded planar accumulation-mode power MOSFET structure enabling the conduction of drain current with a low specific on-resistance. The name for this device derives from the formation of the accumulation layer channel. The specific on-resistance for the shielded planar accumulation-mode power MOSFET structure is less than that for the inversion-mode structure because of a lower threshold voltage and a larger mobility for electrons in an accumulation layer. Since the specific on-resistance for the 4H-SiC planar power MOSFET structure was demonstrated to be limited by the channel resistance, a reduction of the channel resistance achieved with the shielded planar accumulation-mode power MOSFET structure creates devices with superior specific on-resistance.

## 9.2.1 *Blocking Mode*

In the forward blocking mode of the silicon carbide shielded planar accumulation-mode power MOSFET structure, the voltage is supported by a depletion region formed on both sides of the $P^+$ region/N-drift junction. The maximum blocking voltage is determined by the electric field at this junction becoming equal to the critical electric field for breakdown if the parasitic $N^+$/P/N bipolar transistor is completely suppressed. This suppression is accomplished by short-circuiting the $N^+$ source and $P^+$ regions using the source metal as shown on the upper left hand side of the cross-section. If the doping concentration of the $P^+$ region is large, the reach-through breakdown problem is completely eliminated. In addition, the high doping concentration in the $P^+$ region promotes the depletion of the JFET region at lower drain voltages providing enhanced shielding of the channel and gate oxide.

In the silicon carbide planar accumulation-mode power MOSFET structure, shielding of the N-base region from the drain potential by the $P^+$ shielding region allows reducing the channel length to less than 1 μm. In addition, the doping concentration of the N-base region can be adjusted to achieve any desired threshold voltage. The smaller channel length and threshold voltage, as well as the larger mobility for electrons in an accumulation layer, reduce the channel resistance contribution.

As in the case of the planar silicon power D-MOSFET, the maximum blocking voltage capability of the silicon carbide shielded accumulation-mode planar

MOSFET structure is determined by the drift region doping concentration and thickness. However, to fully utilize the high breakdown electric field strength available in silicon carbide, it is necessary to screen the gate oxide from the high field within the semiconductor. In the shielded planar MOSFET structure, this is achieved by the formation of a potential barrier at location A by the depletion of the JFET region at a low drain bias voltage. The maximum electric field in the gate oxide can be made not only below its rupture strength but lower than values required for reliable operation over long time durations.

#### 9.2.1.1 Simulation Results

The results of two-dimensional numerical simulations on the 600 V shielded 4H-SiC planar power ACCUFET structure are described here to provide a more detailed understanding of the underlying device physics and operation during the blocking mode. As in the case of the inversion-mode device structure, the 600-V ACCUFET structure used for the numerical simulations had a drift region thickness of 4 μm below the $P^+$ shielding region with a doping concentration of $5 \times 10^{16}$ cm$^{-3}$. The $P^+$ region extended from a depth of 0.2 to 1.0 μm with a doping concentration of $1 \times 10^{19}$ cm$^{-3}$. The N-base and $N^+$ source regions were formed within the 0.2 μm of the N-drift region located above the $P^+$ region. The doping concentration of the N-base region was kept at the same value as the drift region ($5 \times 10^{16}$ cm$^{-3}$). In general, the doping concentration for the N-base region can be adjusted independently by using the shallow ion-implantation of phosphorus. For the numerical simulations, the cell structure (with a width of 4 μm) illustrated in Fig. 9.18 was utilized as a unit cell that is representative of the structure. Due to high doping concentration in the drift region for the 600-V 4H-SiC devices, the doping concentration in the JFET region was not enhanced as is usually required for silicon devices.

A three dimensional view of the doping distribution in the 600 V shielded 4H-SiC planar power ACCUFET structure is shown in Fig. 9.19 with the upper surface of the structure located on the right hand side in order to display the doping concentration in the vicinity of the channel. The highly doped $P^+$ shielding region is prominently located just below the surface. The N-base region can be observed to have a much lower doping concentration. Unlike the inversion-mode device, there is no junction between the N-base region and N-JFET region.

The lateral doping profile taken along the surface of the 600 V shielded 4H-SiC planar power ACCUFET structure is shown in Fig. 9.20. The channel extends from 2 to 3 μm creating a channel length of 1 μm because the $P^+$ shielding region extends to 3 μm. The doping concentration of the JFET region is the same as that of the N-drift region. The $N^+$ source region and the $P^+$ contact region for shorting the source to the base region are visible on the left-hand-side. All the regions were defined with uniform doping with abrupt interfaces between them due to the low diffusion rates for dopants in 4H-SiC material. The vertical doping profile within the 600 V shielded 4H-SiC planar power ACCUFET structure is similar to that previously

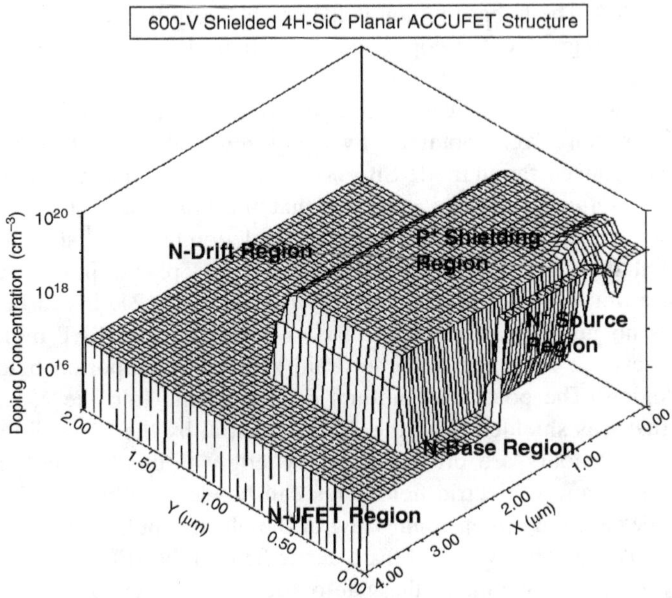


**Fig. 9.19** Doping distribution in the shielded 4H-SiC planar power ACCUFET structure

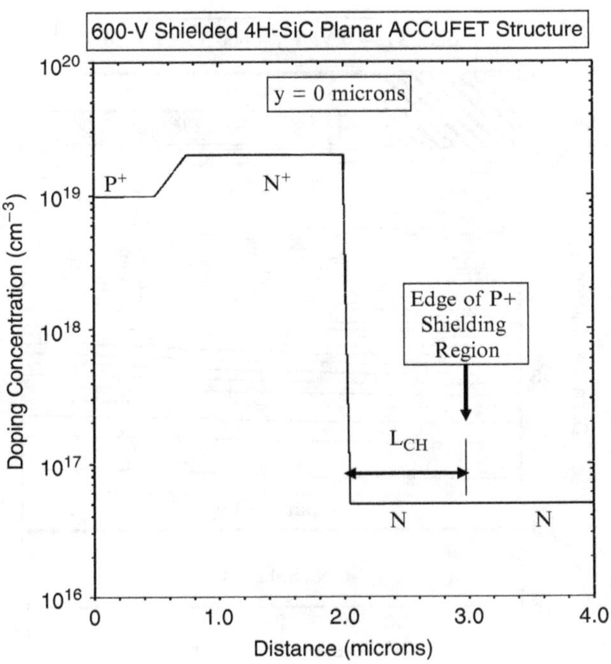


**Fig. 9.20** Channel doping profile for the shielded 4H-SiC planar power ACCUFET structure

shown in Fig. 9.4. The N-base region is located between the N$^+$ source region and the P$^+$ shielding region with a doping concentration of $5 \times 10^{16}$ cm$^{-3}$ and thickness of 0.1 μm.

The blocking characteristics for the 600 V shielded 4H-SiC planar power ACCUFET structure were obtained by using zero gate bias. Due to the very small intrinsic concentration in 4H-SiC, no substantial leakage current is observed at room temperature. This also confirms that the bulk channel through of the N-base region has been suppressed due to its depletion by the P$^+$ shielding region. The potential contours within the shielded 4H-SiC planar power ACCUFET structure at a drain bias of 600 V are provided in Fig. 9.21. It can be observed that there is no junction between the base region and the JFET region in this structure. It can be observed that the drain voltage is supported below the P$^+$ shielding region. The potential contours do not extend into the N-base region indicating that it is shielded from the drain potential by the P$^+$ shielding region. The potential contours are crowding at the edge of the P$^+$ shielding region indicating an enhanced electric field. This can be clearly observed in Fig. 9.22 which provides a three-dimensional view of the electric field distribution. In this figure, it can also be observed that the electric field in the JFET region, and most importantly at the surface under the gate oxide, has been greatly reduced by the presence of the P$^+$ shielding region.

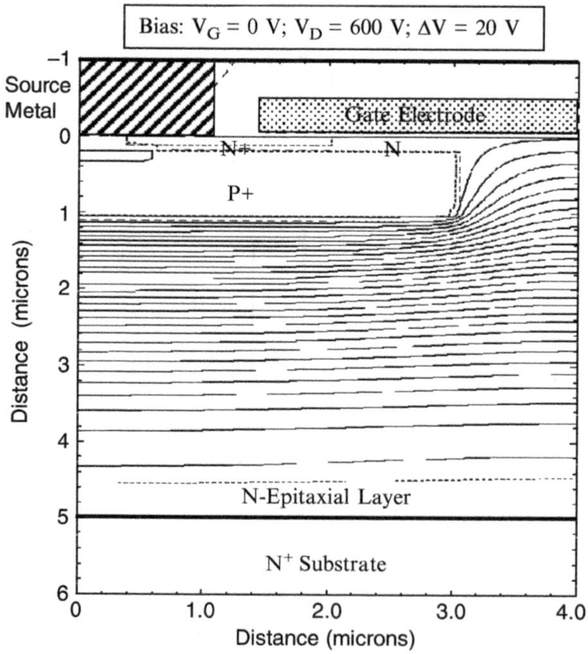


**Fig. 9.21** Potential contours in the shielded 4H-SiC planar power ACCUFET structure

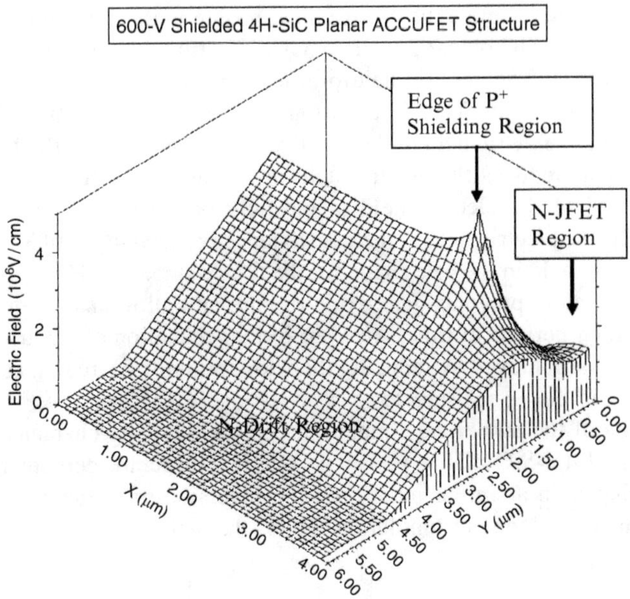


**Fig. 9.22** Electric field distribution in the 600-V shielded 4H-SiC planar power ACCUFET structure

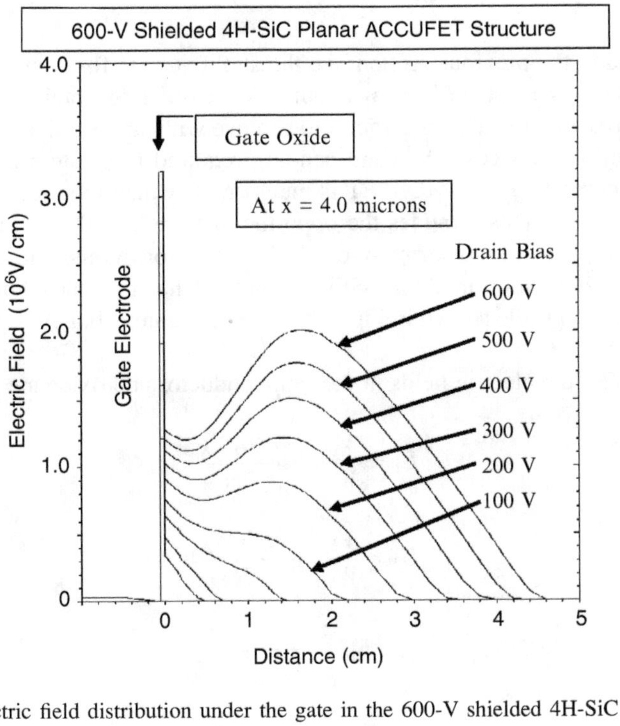


**Fig. 9.23** Electric field distribution under the gate in the 600-V shielded 4H-SiC planar power ACCUFET structure

It is insightful to examine the electric field profile in the JFET region within the 600 V shielded 4H-SiC planar power ACCUFET structure when it is operating in the blocking mode. The electric field profiles obtained through the middle of the JFET region are shown in Fig. 9.23 at various drain bias voltages. It can be observed that the maximum electric field in the JFET region occurs at a depth of 1.5 μm from the surface. This reduces the electric field at the surface under the gate oxide to about one-third of the electric field in the bulk below the $P^+$ shielding region. Consequently, the electric field in the gate oxide is reduced to about $3 \times 10^6$ V/cm. The low electric field in the gate oxide for the 600 V shielded 4H-SiC planar power ACCUFET structure prevents gate oxide rupture and allows stable device performance over long periods of time. An even further reduction of electric field in the gate oxide can be achieved by reducing the width of the JFET region at the expense of an increase in the on-resistance.

The results of the numerical simulations on the blocking characteristics of the 600 V shielded 4H-SiC planar power ACCUFET structure demonstrate that the structure behaves in a similar manner to the inversion-mode device. This allows taking advantage of an accumulation-mode channel for reducing the channel resistance.

### 9.2.2 *Threshold Voltage*

As discussed in the previous section, the threshold voltage for high voltage power MOSFET devices must be kept at about 2 V in order to obtain a low channel resistance contribution. This is difficult to achieve with an inversion-mode channel in silicon carbide devices. The band bending required to create a channel in the accumulation-mode planar MOSFET is much smaller than required for the inversion mode device. This provides the opportunity to reduce the threshold voltage while obtaining the desired normally-off device behavior. A model for the threshold voltage of accumulation-mode MOSFET structures has been developed [7] using the electric field profile shown in Fig. 9.24 when the gate is biased at the threshold voltage.

In Fig. 9.24, the electric fields in the semiconductor and oxide are given by

$$E_1 = \frac{V_{bi}}{W_N} - \frac{qN_D W_N}{2\varepsilon_S} \tag{9.42}$$

$$E_2 = \frac{V_{bi}}{W_N} + \frac{qN_D W_N}{2\varepsilon_S} \tag{9.43}$$

$$E_{ox} = \frac{\varepsilon_S}{\varepsilon_{ox}} E_1 \tag{9.44}$$

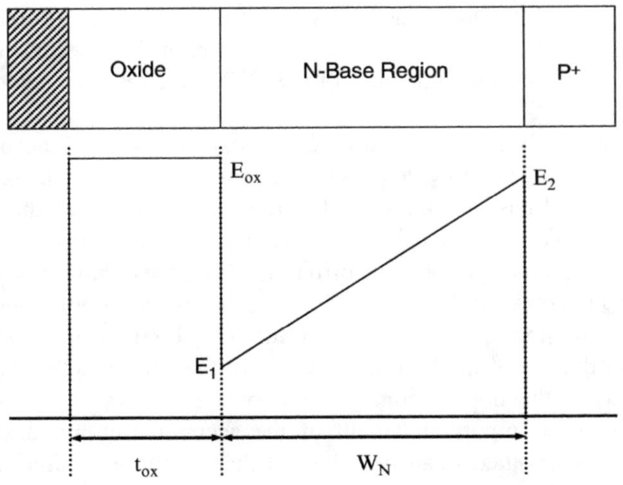


**Fig. 9.24** Electric field profile for the accumulation-mode MOSFET structure

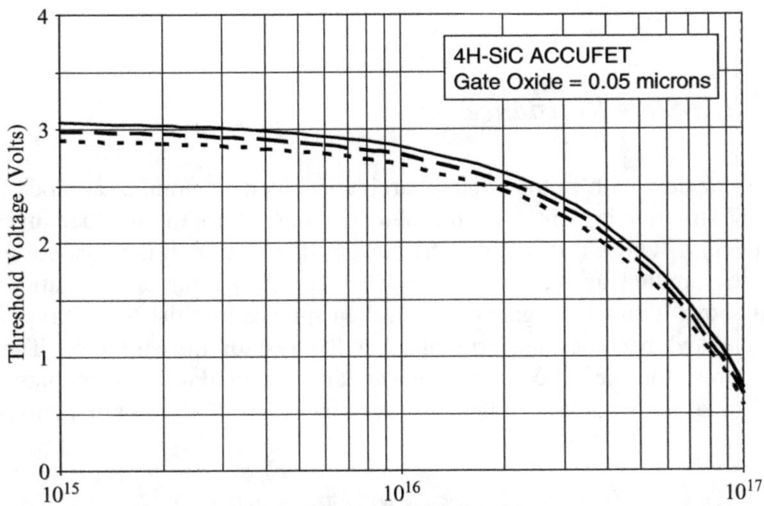


**Fig. 9.25** Threshold voltage of the 4H-SiC ACCUFET structure (*solid line*: 300°K; *dash line*: 400°K; *dotted line*: 500°K)

Note that this model is based upon neglecting any voltage supported within the $P^+$ region under the assumption that it is very heavily doped. Using these electric fields, the threshold voltage is found to be given by

$$V_{TH} = \phi_{MS} + \left( \frac{\varepsilon_S V_{bi}}{\varepsilon_{ox} W_N} - \frac{q N_D W_N}{2\ \varepsilon_{ox}} \right) t_{ox} \tag{9.45}$$

The first term in this equation accounts for the work function difference between the gate material and the lightly doped N-base region. The second term represents the effect of the built-in potential of the $P^+/N$ junction that depletes the N-base region.

The analytically calculated threshold voltage for 4H-SiC accumulation-mode MOSFET structure are provided in Fig. 9.25 for the case of a gate oxide thickness of 0.05 μm, and N-base thickness of 0.2 μm as a function of the N-base doping concentration with the inclusion of a metal-semiconductor work-function difference of 1 V. A strikingly obvious difference from the behavior of inversion-mode devices is a decrease in the threshold voltage for the accumulation-mode structure with increasing doping concentration in the N-base region. This occurs due to the declining influence of the built-in potential of the $P^+/N$ junction at the gate oxide interface when the doping concentration of the N-base region is increased. Of course, the most important benefit of the accumulation-mode structure is that lower threshold voltages can be achieved than in the inversion-mode structures. For the chosen thickness of the N-base region, the desired threshold voltage of 2 V can be obtained for the accumulation-mode structure by using an N-base doping concentration of $5 \times 10^{16}$ cm$^{-3}$.

## 9.2.3 *On-State Resistance*

The current flow path in the silicon carbide shielded accumulation-mode planar MOSFET structure is identical to that shown in Fig. 9.9 for the shielded inversion-mode planar MOSFET structure. The current flows through the channel formed due to the applied gate bias into the JFET region via the accumulation layer formed above it under the gate oxide. It then spreads into the N-drift region at a 45° angle and becomes uniform through the rest of the structure. The total on-resistance for the silicon carbide shielded accumulation-mode planar SiC MOSFET structure is determined by the resistance of the components in the current path

$$R_{on,sp} = R_{CH} + R_A + R_{JFET} + R_D \tag{9.46}$$

where $R_{CH}$ is the channel resistance, $R_A$ is the accumulation region resistance, $R_{JFET}$ is the resistance of the JFET region, $R_D$ is the resistance of the drift region after taking into account current spreading from the JFET region. For consistency with previous chapters, the resistance of the $N^+$ substrate has been omitted in the above analysis even though the substrate contribution for 4H-SiC can be very large unless its thickness is reduced to below 50 μm.

The resistances in the shielded accumulation-mode planar MOSFET structure can be analytically modeled by using the current flow pattern indicated by the shaded regions in Fig. 9.9. With the exception of the channel, all the other resistance components are identical to those already provided for the shielded inversion-mode planar MOSFET structure in the previous section. Consequently, only the channel resistance component will be analyzed in this section for the shielded accumulation-mode planar MOSFET structure. The impact of the improved channel resistance on the total specific on-resistance of the shielded accumulation-mode planar MOSFET structure is also provided here.

### 9.2.3.1 Channel-Resistance

For the shielded accumulation-mode planar MOSFET structure with the N-base region, the specific channel resistance is given by

$$R_{CH} = \frac{(L_{CH} W_{Cell})}{2\ \mu_{acc} C_{ox} (V_G - V_{TH})} \tag{9.47}$$

where $L_{CH}$ is the channel length as shown in Fig. 9.9, $\mu_{acc}$ is the mobility for electrons in the accumulation layer channel, $C_{ox}$ is the specific capacitance of the gate oxide, $V_G$ is the applied gate bias, and $V_{TH}$ is the threshold voltage. Accumulation layer mobility values ranging from of 100 to 200 $cm^2/V\text{-}s$ have been observed in lateral MOSFET structures [6]. The relatively high accumulation layer mobility, together with a smaller threshold voltage, reduces the channel resistance component in the shielded planar ACCUFET structure.

In the case of the 600 V shielded planar ACCUFET structure, the cell width will be assumed to be 8 μm with a channel length of 1 μm. The threshold voltage for the accumulation-mode structure is 2 V based up on using a doping concentration of $5 \times 10^{16}\ cm^{-3}$ for the N-base region. Using a gate oxide thickness is 500 Å and an inversion layer mobility of 180 $cm^2/V\text{-}s$ in the above equation, the specific resistance contributed by the channel at a gate bias of 10 V is found to be 0.407 $m\Omega\ cm^2$. This value is an order of magnitude smaller than that for the inversion-mode structure.

### 9.2.3.2 Total On-Resistance

The total specific on-resistance for the 600-V shielded planar 4H-SiC ACCUFET structure with a cell width of 8 μm and JFET width of 2 μm is obtained by adding all the components of the resistances within the device structure. For a gate bias of 10 V, the total specific on-resistance is found to be 0.767 $m\Omega\ cm^2$. The channel resistance constitutes only 53% of the total specific on-resistance due to the improved mobility for the electrons in the accumulation layer. The specific on-resistance for the 600-V shielded planar 4H-SiC ACCUFET structure is an order of magnitude smaller than that for 600-V shielded planar 4H-SiC MOSFET structure. Consequently, the specific on-resistance for the 600-V shielded planar

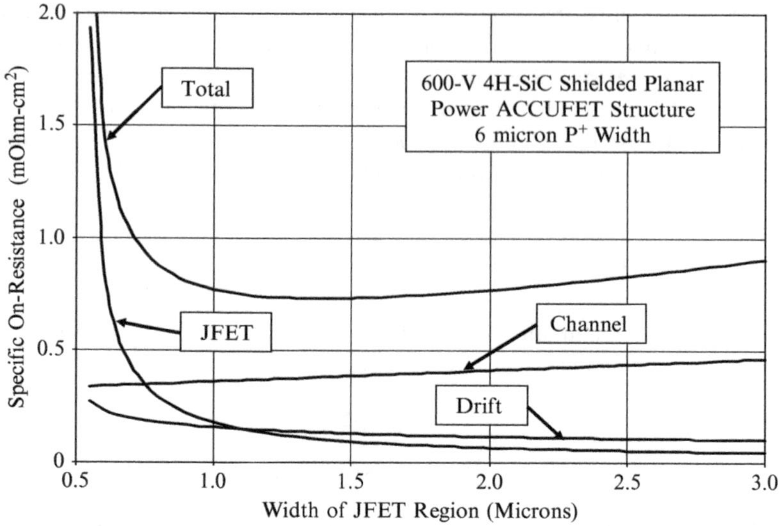


**Fig. 9.26** On-resistance for the 600-V 4H-SiC shielded planar ACCUFET structures

4H-SiC ACCUFET structure is two orders of magnitude smaller than that for the silicon 600-V power D-MOSFET and U-MOSFET structures. Even compared with the silicon 600-V GD-MOSFET and SJ-MOSFET structures, the specific on-resistance for the 600-V shielded planar 4H-SiC ACCUFET structure is about five-times smaller. Consequently, the 600-V shielded planar 4H-SiC ACCUFET structure is superior to the best silicon technology because of the improved channel resistance.

The impact of changing the width of the JFET region on the specific on-resistance of the shielded planar 4H-SiC ACCUFET structure can be determined by using the above analytical model. The results obtained for the case of a 600-V device structure with a $P^+$ shielding region width of 6 μm are provided in Fig. 9.26. For this analysis, an accumulation layer mobility of 180 $cm^2/V$-s was used. It can be observed that the specific resistance of the channel is now comparable to those of the other components in the structure. The specific on-resistance goes through a minimum at a larger width of the JFET region when compared with the inversion-mode structure due to the larger channel mobility. The JFET width can be optimized between 1 and 2 μm from the point of view of shielding the gate oxide and N-base region from the high electric fields in the drift region.

#### 9.2.3.3 Impact of Breakdown Voltage

The specific on-resistance for the shielded planar 4H-SiC MOSFET structure is plotted in Fig. 9.27 as a function of the breakdown voltage by using the analytical model. In performing the modeling, it is important to recognize that the thickness of

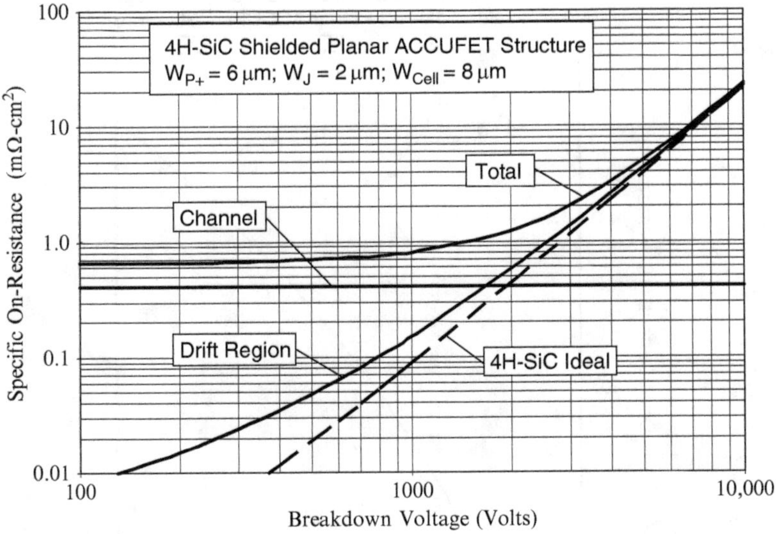


**Fig. 9.27** Specific on-resistance for the 4H-SiC shielded planar ACCUFET structure

the drift region (parameter 't' in Fig. 9.9) can become smaller than half the width of the $P^+$ shielding region at lower breakdown voltages. Under these conditions, the current does not distribute across the entire drift region under the $P^+$ shielding region. The device parameters used for the plot are: channel inversion mobility of 180 $cm^2$/V-s; a fixed JFET doping concentration of $5 \times 10^{16}$ $cm^{-3}$; a width of 6 μm for the $P^+$ shielding region; a width of 2 μm for the JFET region; a cell width of 8 μm; gate bias of 10 V; threshold voltage of 2 V; and a gate oxide thickness of 500 Å. From Fig. 9.27, it can be seen that specific on-resistance of 4H-SiC shielded planar ACCUFET structure is limited by the channel resistance for breakdown voltages below 1,000 V. However, the specific on-resistance obtained for the 4H-SiC shielded planar ACCUFET structure is an order of magnitude smaller than that for the 4H-SiC shielded planar MOSFET structure. The drift region resistance can be observed to be close to the ideal specific on-resistance when the breakdown voltage exceeds 1,000 V. The total specific on-resistance of the 4H-SiC shielded planar ACCUFET structure approaches that of the drift region only when the breakdown voltage exceeds 5,000 V.

#### 9.2.3.4 Simulation Results

The results of two-dimensional numerical simulations on the 600-V shielded 4H-SiC planar power MOSFET structure are described here to provide a more detailed understanding of the underlying device physics and operation. The device parameters for the structure used for the numerical simulations were provided in the Sect. 9.2.1.

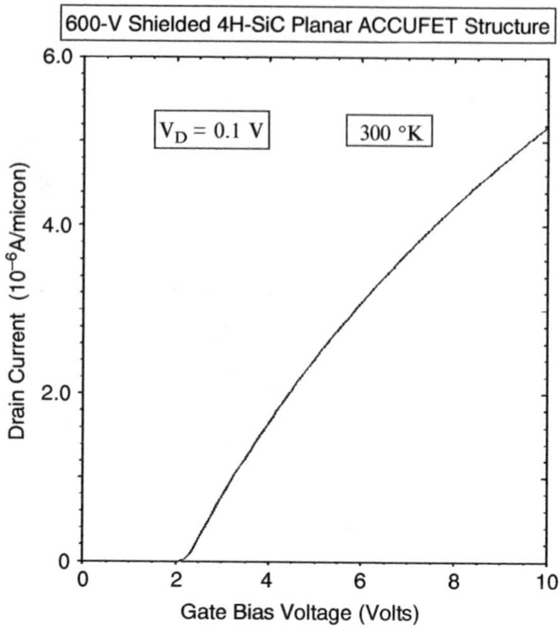


**Fig. 9.28** Transfer characteristic of the shielded 4H-SiC planar power ACCUFET structure

The transfer characteristic for the 600-V shielded 4H-SiC planar power ACCUFET structure was obtained using numerical simulations with a drain bias of 0.1 V at 300°K. The channel mobility was adjusted during the simulations to 180 cm$^2$/V-s. The resulting transfer characteristic is shown in Fig. 9.28. From this graph, a threshold voltage of 2 can be extracted at 300°K. This demonstrates that an optimum threshold voltage can be obtained in the accumulation-mode 4H-SiC power MOSFET structure by proper choice of the doping concentration for the N-base region ($5 \times 10^{16}$ cm$^{-3}$ in this case). The threshold voltage predicted by the analytical model (see Fig. 9.25) is in very good agreement with the value extracted from the numerical simulations. For the case of a gate bias of 10 V and 300°K, the specific in-resistance is found to be 0.772 mΩ cm$^2$ providing validation of the analytical model. The continuous increase in drain current with gate voltage indicates that the channel resistance is still dominant in the 600-V shielded 4H-SiC planar power ACCUFET structure even though the specific on-resistance has been greatly reduced.

The on-state current flow pattern within the 600-V shielded 4H-SiC planar power ACCUFET structure at a small drain bias of 0.1 V and a gate bias of 10 V is shown in Fig. 9.29. In the figure, the depletion layer boundary is shown by the dotted lines and the junction boundary is delineated by the dashed line. The depletion layer width ($W_0$) in the JFET region is about 0.25 μm in good agreement with the value computed using the analytical model. It can be observed from this

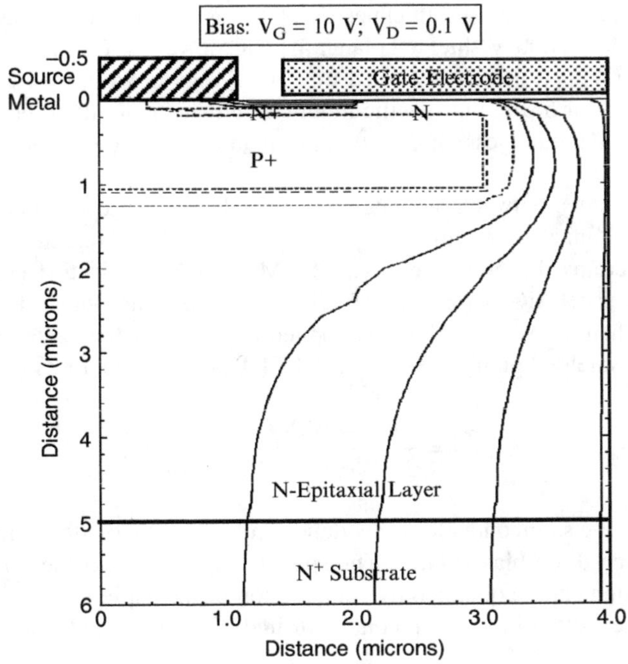


**Fig. 9.29** Current distribution in the shielded 4H-SiC planar power ACCUFET structure

figure that the channel current is confined to a very thin accumulation layer formed at the surface. No current flows through the bulk portion of the N-base region demonstrating that the ACCUFET structure is not a bulk channel device. It can also be observed that the current flows from the channel and distributes into the JFET region via the accumulation layer. Within the JFET region, the cross-sectional area is approximately constant with a width (a/2) of 0.75 μm. From the figure, it can be seen that the current spreads from the JFET region to the drift region at a 45° angle as assumed in the model and becomes uniform for the last 1 μm of the drift region.

### 9.2.4 Capacitances

The capacitances within the shielded planar 4H-SiC ACCUFET structure can be analytically modeled using the same approach as used for the shielded planar 4H-SiC MOSFET structure. The specific input (or gate) capacitance for the shielded planar 4H-SiC ACCUFET structure is given by

$$C_{IN,SP} = C_{N+} + C_P + C_{SM} = \frac{(W_G - W_J)}{W_{Cell}}\left(\frac{\varepsilon_{OX}}{t_{OX}}\right) + \frac{W_G}{W_{Cell}}\left(\frac{\varepsilon_{OX}}{t_{IEOX}}\right) \quad (9.48)$$

where $t_{OX}$ and $t_{IEOX}$ are the thicknesses of the gate and inter-electrode oxides, respectively. For a 600-V shielded planar 4H-SiC power ACCUFET structure with a cell width ($W_{CELL}$ in Fig. 9.9) of 8 μm, JFET region width of 2 μm, and gate electrode width of 5 μm, the specific input capacitance is found to be 30 nF cm$^{-2}$ for a gate oxide thickness of 500 Å and an inter-metal dielectric thickness of 5,000 Å.

The capacitance between the gate and drain electrodes (also called the reverse transfer capacitance) is determined by the width of the JFET region where the gate electrode overlaps the N-drift region. The MOS structure in this portion of the shielded planar 4H-SiC power ACCUFET structure operates under deep depletion conditions when a positive voltage is applied to the drain. The gate-drain capacitance for the shielded planar 4H-SiC ACCUFET power structure is given by

$$C_{GD,SP} = \frac{(W_G - W_J)}{W_{Cell}}\left(\frac{C_{OX}C_{S,M}}{C_{OX} + C_{S,M}}\right) \tag{9.49}$$

where $C_{S,M}$ is the semiconductor capacitance under the gate oxide, which decreases with increasing drain bias voltage. The specific capacitance of the semiconductor depletion region can be obtained by computation of the depletion layer width. The depletion layer width in the semiconductor under the gate oxide can be obtained using

$$W_{D,MOS} = \frac{\varepsilon_S}{C_{OX}}\left\{\sqrt{1 + \frac{2V_D C_{OX}^2}{q\ \varepsilon_S N_{DJ}}} - 1\right\} \tag{9.50}$$

where $N_{DJ}$ is the doping concentration of the JFET region. The specific capacitance for the semiconductor is then obtained using

$$C_{S,M} = \frac{\varepsilon_S}{W_{D,MOS}} \tag{9.51}$$

The gate drain (or reverse transfer) capacitance can be computed by using (9.49) with the above equations until the JFET region is completely depleted. It then abruptly reduces to nearly zero.

The output capacitance for the shielded planar 4H-SiC power ACCUFET structure is associated with the capacitance of the junction between the P$^+$ shielding region and the N-drift region. Due to pinch-off of the JFET region with increasing drain bias voltage, it is necessary to examine the change in the depletion region boundary with applied voltage. The depletion layer boundary inside the shielded planar 4H-SiC power ACCUFET structure prior to the pinch-off of the JFET region is similar to that previously shown by the dashed lines in Fig. 4.19 for the power SC-MOSFET structure. It can be observed that the depletion region has a vertical boundary inside the JFET region and a horizontal boundary below the P$^+$ shielding region. The capacitances associated with each of these regions are indicated in the figure as $C_{S1}$ and $C_{S2}$. The specific junction capacitance associated with the JFET region is given by

$$C_{S1,SP} = \frac{\varepsilon_S}{W_{DJ}}\left(\frac{2L_{P+}}{W_{Cell}}\right) \tag{9.52}$$

where the depletion region thickness ($W_{DJ}$) in the JFET region is related to the drain bias voltage

$$W_{DJ} = \sqrt{\frac{2\varepsilon_S(V_D + V_{bi})}{qN_{DJ}}} \tag{9.53}$$

The specific junction capacitance associated with the bottom of the $P^+$ shielding region is given by [1]

$$C_{S2,SP} = \frac{\varepsilon_S}{W_{DD}}\left(\frac{W_{P+}}{W_{Cell}}\right) \tag{9.54}$$

where the depletion region thickness ($W_{DD}$) in the drift region is related to the drain bias voltage

$$W_{DD} = \sqrt{\frac{2\varepsilon_S(V_D + V_{bi})}{qN_D}} \tag{9.55}$$

where $N_D$ is the doping concentration of the drift region. The specific output capacitance for the shielded planar 4H-SiC power ACCUFET structure can then be obtained by combining the above values

$$C_{O,SP} = C_{S1,SP} + C_{S2,SP} \tag{9.56}$$

### 9.2.5 *Gate Charge*

The gate charge components for the shielded planar 4H-SiC power ACCUFET structure are given by similar equations to those derived for the shielded planar 4H-SiC power MOSFET structure. If the screening effect is neglected, the gate transfer charge corresponding to the transition where the drain voltage changes from $V_{DS}$ to $V_{ON}$ is given by

$$Q_{GD} = \frac{2K_G q\varepsilon_S N_D}{C_{OX}}\left[\sqrt{1+\frac{2V_{DS}C_{OX}^2}{q\ \varepsilon_S N_D}} - \sqrt{1+\frac{2V_{ON}C_{GOX}^2}{q\ \varepsilon_S N_D}}\right] \tag{9.57}$$

where the parameter $K_G$ is given by

$$K_G = \frac{(W_G - W_J)}{W_{Cell}} \tag{9.58}$$

The other components of the gate charge are similar to those already provided in the textbook [3]

$$Q_{SW} = Q_{GS2} + Q_{GD} \tag{9.59}$$

$$Q_G = [C_{GS} + C_{GD}(V_{DS})]V_{GP} + Q_{GD} + [C_{GS} + C_{GD}(V_{ON})](V_G - V_{GP}) \tag{9.60}$$

Equations for the gate voltage, drain current, and drain voltage waveforms for the shielded planar 4H-SiC power ACCUFET structure are similar to those derived in Chap. 4 for the power SC-MOSFET structure. If the screening effect is neglected, the drain voltage is determined by the gate transfer capacitance given by (9.49)

$$v_D(t) = \frac{q\varepsilon_S N_D}{2C_{OX}^2}\left\{\left[\sqrt{1 + \frac{2V_{DS}C_{OX}^2}{q\,\varepsilon_S N_D}} - \frac{J_G C_{OX}(t - t_2)}{2K_G q\,\varepsilon_S N_D}\right]^2 - 1\right\} \tag{9.61}$$

from $t = t_2$ to $t = t_3$.

#### 9.2.5.1 Simulation Example

The gate charge for the 600-V shielded 4H-SiC planar power ACCUFET structure was extracted by using the results of two-dimensional numerical simulations of the cell described in the previous sections. The device was turned-on from blocking state with a drain bias of 400 V by driving it using a gate current of $1 \times 10^{-8}$ A/μm (equivalent to 0.25 A cm$^{-2}$ for the area of $4 \times 10^{-8}$ cm$^2$). Once the drain current density reached 320 A cm$^{-2}$, the drain current was held constant resulting in a reduction of the drain voltage. The gate plateau voltage for this drain current density was found to be 3.8 V due to the improved transconductance of the device as a result of the high accumulation layer mobility. Once the drain voltage reached the on-state value corresponding to the gate plateau voltage, the gate voltage increased to the steady-state value of 10 V.

The gate charge waveforms obtained by using an input gate current density of 0.25 A cm$^{-2}$ when turning on the 600-V shielded 4H-SiC planar power MOSFET structure from a blocking state with drain bias of 400 V are shown in Fig. 9.30. The on-state current density is 320 A cm$^{-2}$ at a DC gate bias of 10 V at the end of the turn-on transient. The gate voltage increases at a constant rate at the beginning of the turn-on process as predicted by the analytical model. When the gate voltage reaches the threshold voltage, the drain current begins to increase. The drain current

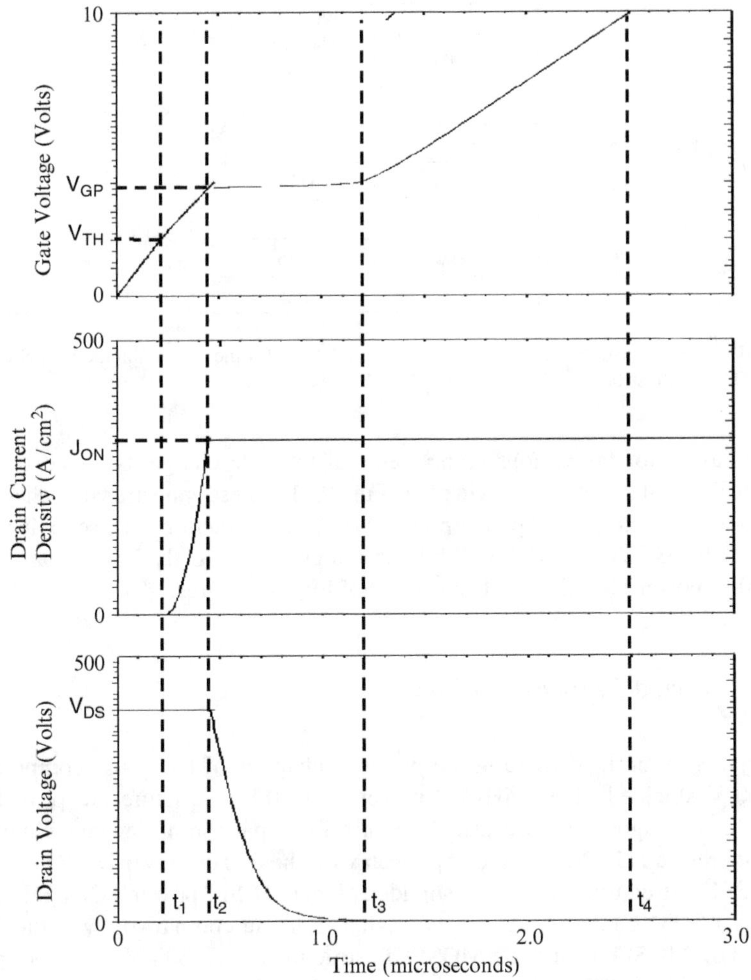


**Fig. 9.30** Turn-on waveforms for the 600-V shielded 4H-SiC planar power ACCUFET structure

increases more rapidly than in the case of the inversion-mode structure until it reaches the on-state current density of 320 A cm$^{-2}$ due to the improved transconductance of the device.

Once the drain current reaches the on-state value, the gate voltage remains approximately constant at the plateau voltage ($V_{GP}$). The plateau voltage for this structure is 3.8 V even for the larger drain current density of 320 A cm$^{-2}$ due to the improved transconductance of the device. The drain voltage drops rapidly from the supply voltage of 400 V of about 50 V (the JFET pinch-off voltage) as predicted by the analytical model. The drain voltage then decreases during the plateau phase in a non-linear manner. After the end of the plateau phase, the gate voltage again increases until it reaches the gate supply voltage.

| Specific Gate Charge | Numerical Simulation ($nC/cm^2$) |
|---|---|
| $Q_{GS1}$ | 53 |
| $Q_{GS2}$ | 60 |
| $Q_{GS}$ | 113 |
| $Q_{GD}$ | 188 |
| $Q_{SW}$ | 248 |
| $Q_G$ | 625 |

**Fig. 9.31** Gate charge extracted from numerical simulations for the 600-V shielded 4H-SiC planar power ACCUFET structure

The values for the various components of the gate charge extracted from the numerical simulations are provided in Fig. 9.31. These gate transfer charge for the 600-V shielded 4H-SiC planar power ACCUFET structure is close to that of the 600-V D-MOSFET and U-MOSFET structures, and slightly larger than that of the 600-V power GD-MOSFET and SJ-MOSFET structures.

### *9.2.6 Device Figures of Merit*

The figures of merit (defined in the previous chapters of the book) computed for the 600-V shielded planar 4H-SiC power ACCUFET structure are provided in Fig. 9.32. The figure of merit usually used for comparison of device technologies in the literature is FOM(C). In comparison with the 600-V power D-MOSFET and U-MOSFET structures, the 600-V shielded planar 4H-SiC power ACCUFET structure has a FOM(C) that is 150-times smaller. Even in comparison with the 600-V power GD-MOSFET and SJ-MOSFET structures, the 600-V shielded planar 4H-SiC power ACCUFET structure has a FOM(C) that is three-times smaller. Consequently, 600-V shielded planar 4H-SiC power MOSFET structure offers significant improvement in circuit performance compared with the conventional power MOSFET structure and is superior to the new silicon power MOSFET technology based up on the charge coupling concept.

| Figures of Merit | $V_G = 10$ V |
|---|---|
| FOM(A) ($\Omega^2 cm^4 s^{-1}$) | 25,567 |
| FOM(B)(ps) | 23 |
| FOM(C) (m$\Omega$*nC) | 144 |
| FOM(D) (m$\Omega$*nC) | 190 |

**Fig. 9.32** Figures of merit for the 600-V 4H-SiC shielded planar ACCUFET structures

### 9.2.7 *Inductive Load Turn-Off Characteristics*

As discussed in preceding chapters, high voltage power MOSFET devices are often used in adjustable speed motor drives which behave as inductive loads. The operation of the shielded planar 4H-SiC power ACCUFET structure in an inductive load circuit can be analyzed using the same approach as used for the shielded planar 4H-SiC power MOSFET structure. The gate plateau voltage for the shielded planar 4H-SiC power ACCUFET structure is given by

$$V_{GP} = V_{TH} + \sqrt{\frac{J_{D,ON} W_{Cell} L_{CH}}{\mu_{acc} C_{GOX}}} \tag{9.62}$$

where $C_{GOX}$ is the gate oxide capacitance. The time constant for discharging the gate of the shielded planar 4H-SiC power ACCUFET structure is $R_{G,SP}*[C_{GS} + C_{GD}(V_{ON})]$ and the gate voltage decreases exponentially with time as given by

$$v_G(t) = V_{GS} e^{-t/R_{G,SP}[C_{GS}+C_{GD}(V_{ON})]} \tag{9.63}$$

The time $t_4$ (using the notation from the textbook) for reaching the gate plateau voltage can be obtained by using this equation with (9.62) for the plateau voltage

$$t_4 = R_{G,SP}[C_{GS} + C_{GD}(V_{ON})] \ln\left[\frac{V_{GS}}{V_{GP}}\right] \tag{9.64}$$

This time can be considered to a *turn-off delay time* before the drain voltage begins to increase after the turn-off is initiated by the control circuit.

The drain voltage begins to increase at time $t_4$ but the drain current remains constant at the load current $I_L$ because the current cannot be transferred to the diode until the voltage at the drain of the MOSFET device exceeds the supply voltage $V_{DS}$ by one diode drop to forward bias the diode. Since the drain current density is constant, the gate voltage also remains constant at the gate plateau voltage. Consequently

$$J_{GP} = \frac{V_{GP}}{R_{G,SP}} \tag{9.65}$$

where $R_{G,SP}$ is the specific gate resistance. Since all the gate current is used to discharge the gate-drain capacitance during the plateau phase because there is no change in the voltage across the gate-source capacitance

$$J_{GP} = C_{GD,SP} \frac{dv_D}{dt} \tag{9.66}$$

where $C_{GD,SP}$ is the specific gate transfer capacitance of the power MOSFET structure which is a function of the drain voltage. This voltage dependence of the gate transfer capacitance was not taken into account in the derivation provided in the textbook but is important to include here to allow comparison of the behavior of various power MOSFET structures.

For simplicity of analysis, it will be assumed that any screening effect can be ignored. The gate transfer capacitance for the shielded planar 4H-SiC power ACCUFET structure is then given by

$$C_{GD,SP} = \frac{(W_G - W_J)}{W_{Cell}} \left( \frac{C_{OX} C_{S,M}}{C_{OX} + C_{S,M}} \right) \tag{9.67}$$

Using this expression in (9.66) yields the following differential equation for the voltage increase phase of the turn-off transient

$$dt = \left( \frac{W_G - W_J}{W_{Cell}} \right) \frac{1}{J_{GP}} \left[ \frac{C_{GOX}}{\sqrt{1 + \frac{2 v_D(t) C_{GOX}^2}{q\, \varepsilon_S N_D}}} \right] dv_D \tag{9.68}$$

Integration of this equation yields

$$(t - t_4) = \left( \frac{W_G - W_J}{W_{Cell}} \right) \frac{4 q \varepsilon_S N_D}{J_{GP} C_{GOX}} \left[ \sqrt{1 + \frac{2 v_D(t) C_{GOX}^2}{q\, \varepsilon_S N_D}} - \sqrt{1 + \frac{2 V_{ON} C_{GOX}^2}{q\, \varepsilon_S N_D}} \right] \tag{9.69}$$

In the case of the shielded planar 4H-SiC power ACCUFET structure, the drain voltage increases from the on-state voltage drop ($V_{ON}$) until it reaches the drain-supply voltage ($V_{DS}$). The voltage rise-time, i.e. the time taken for the voltage to increase from the on-state voltage drop ($V_{ON}$) to the drain supply voltage ($V_{DS}$)

$$t_{V,OFF}(t_5 - t_4) = \left( \frac{W_G - W_J}{W_{Cell}} \right) \frac{4 q \varepsilon_S N_D}{J_{GP} C_{GOX}} \left[ \sqrt{1 + \frac{2 V_{DS} C_{GOX}^2}{q\, \varepsilon_S N_D}} - \sqrt{1 + \frac{2 V_{ON} C_{GOX}^2}{q\, \varepsilon_S N_D}} \right] \tag{9.70}$$

A closed form solution for the rise in the drain voltage can be obtained from (9.69)

$$v_D(t) = \frac{q \varepsilon_S N_D}{2 C_{GOX}^2} \left\{ \left[ \frac{J_{GP} C_{GOX}}{4 q\, \varepsilon_S N_D} \left( \frac{W_{Cell}}{W_G - W_J} \right) (t - t_4) + \sqrt{1 + \frac{2 V_{ON} C_{GOX}^2}{q\, \varepsilon_S N_D}} \right]^2 - 1 \right\} \tag{9.71}$$

This equation describes the increase in the drain voltage from the on-state voltage drop until it reaches the drain supply voltage. The drain voltage has an approximately quadratic shape as a function of the time after the time $t_4$.

At the end of the plateau phase (at time $t_5$), the load current begins to transfer from the power MOSFET device to the free wheeling diode. Since the drain voltage remains constant, the gate-drain capacitance can also be assumed to remain constant during this phase. The current flowing through the gate resistance ($R_G$) discharges both the gate-drain and gate-source capacitances leading to an exponential fall in gate voltage from the plateau voltage

$$v_G(t) = V_{GP}e^{-(t-t_5)/R_{G,SP}C_{GS}} \tag{9.72}$$

The drain current follows the gate voltage as given by

$$J_D(t) = g_m[v_G(t) - V_{TH}] = \frac{\mu_{ni}C_{OX}}{L_{CH}W_{Cell}}[v_G(t) - V_{TH}]^2 \tag{9.73}$$

The drain current decreases rapidly with time due to the exponential reduction of the gate voltage, as given by (9.72), during the current fall phase. The drain current becomes equal to zero when the gate voltage reaches the threshold voltage. The current fall time can therefore be obtained from (9.72)

$$t_{I,OFF} = R_{G,SP}C_{GS}\ln\left(\frac{V_{GP}}{V_{TH}}\right) \tag{9.74}$$

Specific capacitances should be used in this expression for computation of the current fall time. Beyond this point in time, the gate voltage decreases exponentially until it reaches zero. The time constant for this exponential decay is different from the initial phase due to the smaller (zero) gate-drain capacitance.

The turn-off energy loss per cycle can be obtained using

$$E_{OFF} = \frac{1}{2}J_{ON}V_{DS}(t_{V,OFF} + t_{I,OFF}) \tag{9.75}$$

under the assumption that the drain current and voltage excursions are approximately linear with time. The energy loss during the voltage rise-time interval is comparable to the energy loss during the current fall-time interval for the shielded planar 4H-SiC power MOSFET structure.

#### 9.2.7.1 Simulation Results

The results of two-dimensional numerical simulations on the turn-off of the 600 V shielded 4H-SiC planar power MOSFET structure are described here. The drain supply voltage was chosen as 400 V for the turn-off analysis. During the turn-off simulations, the gate voltage was reduced to zero with a gate resistance of $1 \times 10^8\ \Omega\ \mu m$ for the 4 μm half-cell structure, which is equivalent to a specific gate resistance of $4\ \Omega\ cm^2$. The current density was initially held constant at an on-state current density of $360\ A\ cm^{-2}$ allowing the drain voltage to rise to the drain supply voltage. The drain supply voltage was then held constant allowing the drain current density to reduce to zero.

The turn-off waveforms obtained for the 600-V shielded 4H-SiC planar power ACCUFET structure by using the numerical simulations are shown in Fig. 9.33. The gate voltage initially reduces to the gate plateau voltage corresponding to the on-state current density. The drain voltage then increases quadratically from the on-

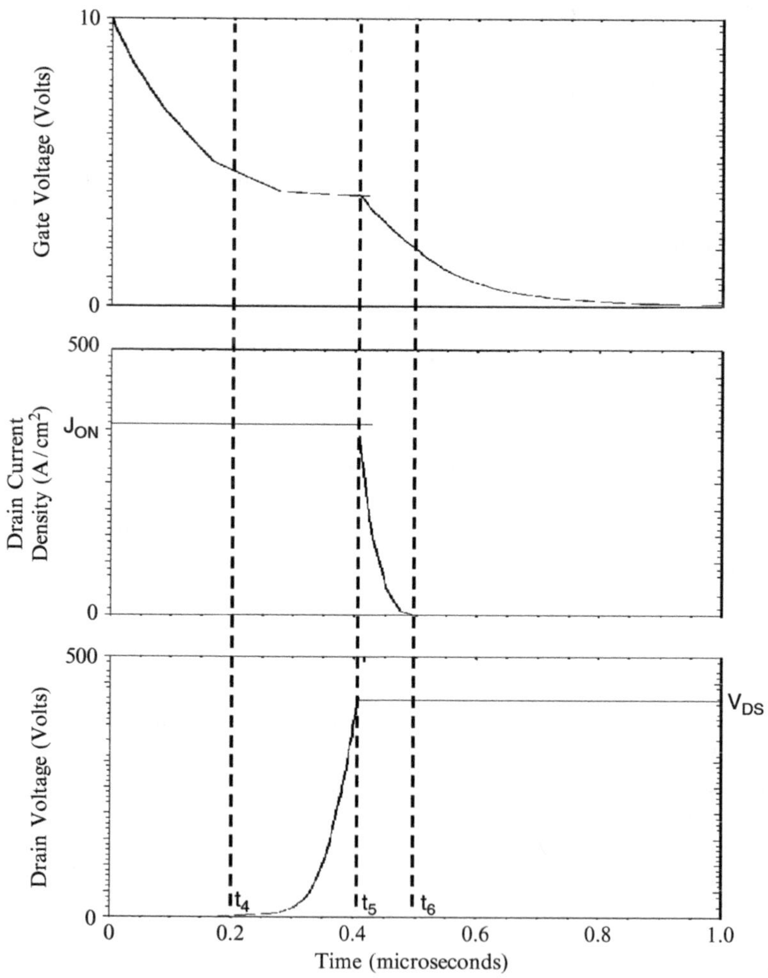


**Fig. 9.33** Turn-off waveforms for the 600-V shielded 4H-SiC planar power ACCUFET structure

state voltage drop to the drain supply voltage as predicted by the analytical model to the drain supply voltage. After this, the drain current reduces exponentially. The drain voltage rise-time ($t_5 - t_4$) and the drain current fall time ($t_6 - t_5$) are comparable for the shielded 4H-SiC planar power ACCUFET structure. The drain voltage rise-time obtained from the simulations of the 600-V shielded 4H-SiC planar power ACCUFET structure is 0.21 μs and the drain current fall-time obtained from the simulations is 0.09 μs. The energy loss per cycle for the shielded 4H-SiC planar power MOSFET structure computed by using the values from the simulations is 21.6 mJ $cm^{-2}$. This value is much larger than that observed for the 600-V silicon power SJ-MOSFET structure and the 600-V silicon power GD-MOSFET structure.

## 9.2.8 *Body-Diode Characteristics*

The shielded planar 4H-SiC power ACCUFET structure has an internal body-diode formed between the $P^+$ shielding region and the N-drift region. The on-state voltage drop for this body diode is about 3 V compared with 1 V for silicon devices. A large power loss would occur when this body diode is carrying current in the third quadrant for the shielded planar 4H-SiC power ACCUFET structure. However, as the forward bias for the junction between the $P^+$ shielding region and the N-drift region increases, the depletion width in the N-base region, at its junction with the $P^+$ shielding region, shrinks. The reduction of the depletion region opens a bulk current conduction path in the N-base region which bypasses the junction. The on-state voltage drop for the current flow in the third quadrant for the shielded planar 4H-SiC power ACCUFET structure can therefore become smaller than the typical on-state voltage drop of 3 V for the P-N junction. The suppression of minority carrier injection by the junction due to the alternate current path also improves the reverse recovery performance of the 'body-diode'.

### 9.2.8.1 Simulation Results

The results of two-dimensional numerical simulations on the operation of the 600-V shielded 4H-SiC planar power ACCUFET structure in the third quadrant are described here to provide a more detailed understanding of the underlying device physics and operation. The device parameters for the structure used for the numerical simulations were provided in the Sect. 9.2.1.

The forward *i–v* characteristic for the 'body-diode' in the 600-V shielded 4H-SiC planar power ACCUFET structure was obtained using numerical simulations with zero gate bias. A lifetime ($\tau_{n0}$ and $\tau_{p0}$) of 1 μs was used for the simulations as representative of the quality of recent epitaxial layers. The resulting characteristic is shown in Fig. 9.34 by the solid line. For comparison purposes, the *i–v* characteristic for the body-diode in the 600-V shielded 4H-SiC planar power MOSFET structure is also shown in the figure by the dashed line. It can be observed that the on-state

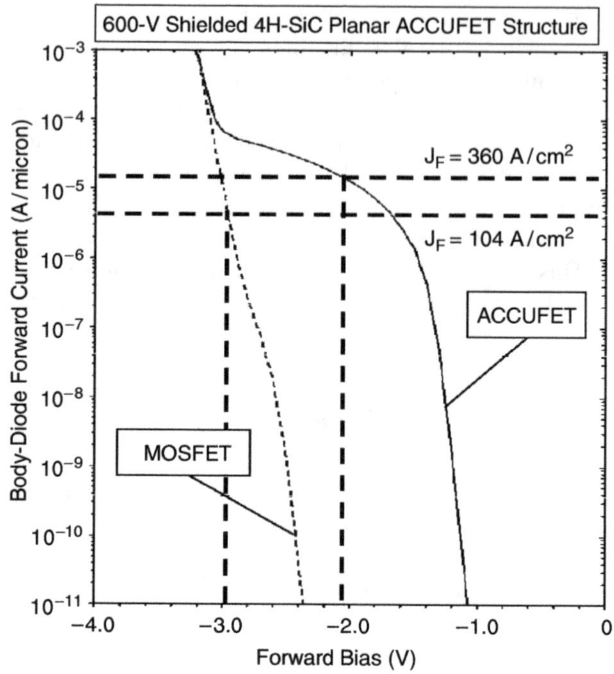


**Fig. 9.34** Forward characteristic of the body-diode within the shielded 4H-SiC planar power ACCUFET structure

voltage drop for the body-diode in this device is 3 V at its on-state current density of 104 A cm$^{-2}$. From this graph, it can be observed that the *i*–*v* characteristic for the 600-V shielded 4H-SiC planar power ACCUFET structure is shifted towards the right-hand-side. The on-state voltage drop at the on-state current density of 360 A cm$^{-2}$ for this device is only 2.04 V. An inflection in the *i*–*v* characteristics occurs when the drain bias exceed 3 V due to the on-set of current flow across the P-N junction in the 600-V shielded 4H-SiC planar power ACCUFET structure.

The hole concentration profile in the 600-V shielded 4H-SiC planar power ACCUFET structure is shown in Fig. 9.35 when the body diode is forward biased with a current density of 360 A cm$^{-2}$. A lifetime of 1 μs was used for these simulations. It can be observed that there are no holes injected into the N-type drift region under these conditions. In contrast, the injected hole concentration in the drift region for the 600-V shielded 4H-SiC planar power MOSFET structure was found to be approximately equal to the doping concentration. Based up on these results, it can be concluded that the injection of minority carriers is suppressed within the shielded 4H-SiC planar power ACCUFET structure. This is an unusual feature that is unique to the ACCUFET structure.

The current path for the 'body diode' in the 600-V shielded 4H-SiC planar power ACCUFET structure, when it is forward biased by a negative drain bias, is shown in Fig. 9.36. It can be observed that the current flow lines by-pass the junction between

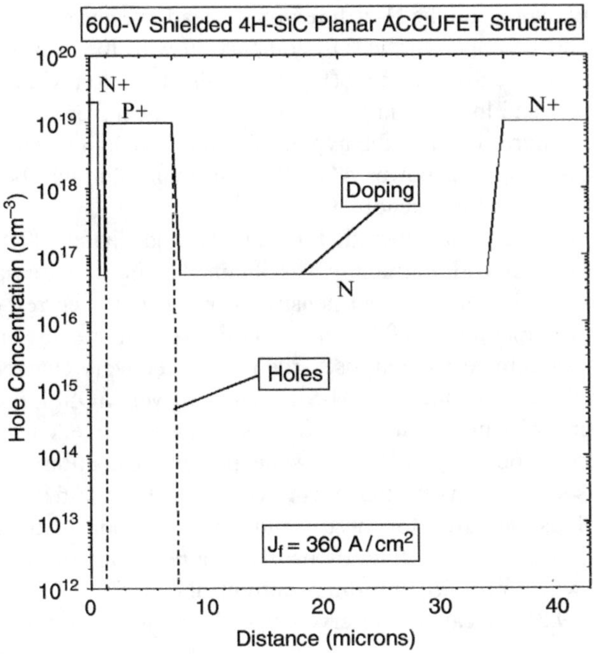


**Fig. 9.35** Hole distribution in the 600-V shielded 4H-SiC planar power ACCUFET structure

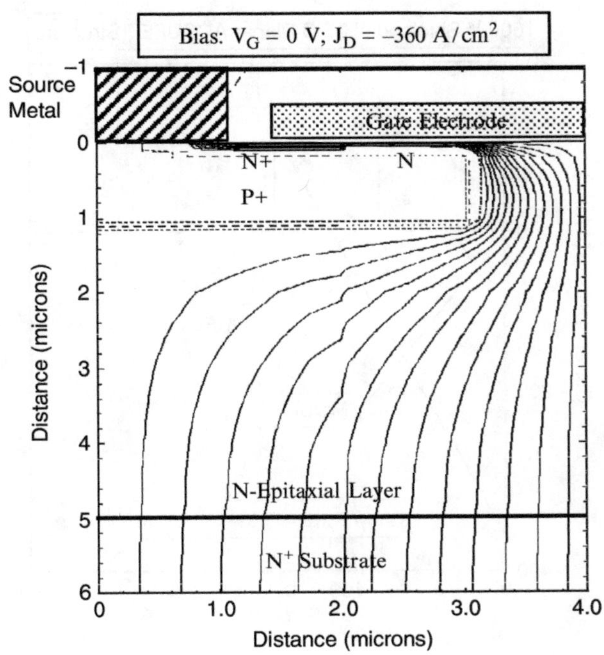


**Fig. 9.36** Body diode current path within the 600-V shielded 4H-SiC planar power ACCUFET structure

the $P^+$ shielding region and the N-drift region. Instead, the current flows through the N-base region at the upper surface. In comparison with the on-state current flow during operation in the first quadrant (Fig. 9.29), the depletion width at the junction is much narrower and the current flow-lines spread down from the semiconductor surface. These features confirm the hypothesis that unipolar current flow occurs in the shielded 4H-SiC planar power ACCUFET structure through its N-base region during operation in the third quadrant.

The reverse recovery characteristic for the body diode in the 600-V shielded 4H-SiC planar power ACCUFET structure was obtained by initially forward biasing the body diode with an on-state current density of 360 A cm$^{-2}$ corresponding with a on-state power dissipation of 100 W cm$^{-2}$ based up on the specific on-resistance of this structure. For comparison purposes, the reverse recovery characteristic for the body-diode in the 600-V shielded 4H-SiC planar power MOSFET structure, with initially forward biasing the body diode with an on-state current density of 104 A cm$^{-2}$ corresponding with a on-state power dissipation of 100 W cm$^{-2}$ based up on its specific on-resistance, is also provided in the figure (dashed line). A lifetime of 1 μs was used for these simulations. For both structures, the drain current was ramped from negative to positive at a ramp rate of 3,300 A cm$^{-2}$-μs. The resulting waveforms for the drain current and drain voltage are shown in Figs. 9.37 and 9.38. It can be observed that the peak reverse recovery current

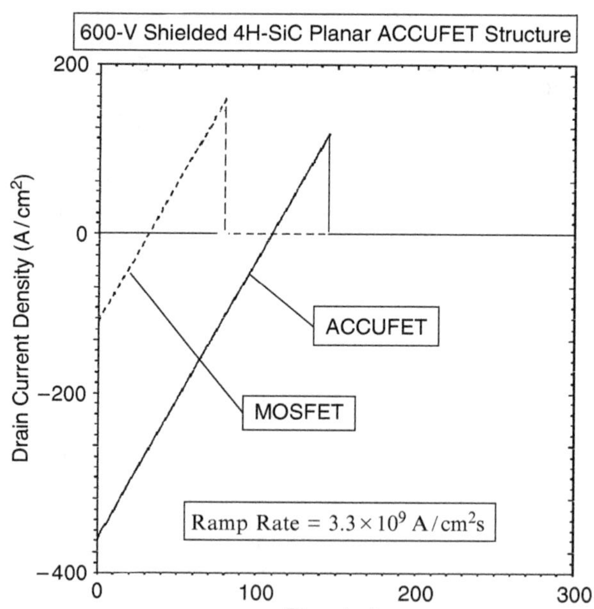


**Fig. 9.37** Reverse recovery current of the body diode within the 600-V shielded 4H-SiC planar power ACCUFET structure

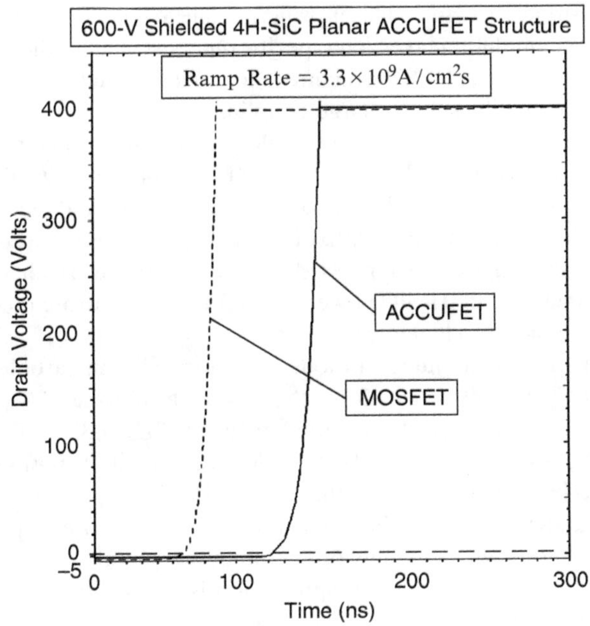


**Fig. 9.38** Reverse recovery voltage of the body diode within the 600-V shielded 4H-SiC planar power ACCUFET structure

density ($J_{PR}$) for the 600-V shielded 4H-SiC planar power ACCUFET structure is 117 A cm$^{-2}$. This value is only one-third of the on-state current density. In contrast, the peak reverse recovery current density ($J_{PR}$) for the 600-V shielded 4H-SiC planar power MOSFET structure is 156 A cm$^{-2}$. This value is 1.5-times the on-state current density. Based up on these results, it is clear that the shielded 4H-SiC planar power ACCUFET structure has a much superior (~ five-times) integral diode.

## 9.3 Discussion

The characteristics of the 600-V shielded planar 4H-SiC power MOSFET structure have been reviewed in this chapter. It is demonstrated that the specific on-resistance of the device is dominated by the channel resistance if a typical channel inversion layer mobility of 10–20 cm$^2$/V-s is assumed. The resulting specific on-resistance, although an order of magnitude smaller than the conventional silicon power MOSFET structures, is not competitive with the silicon power MOSFET structures that utilize charge-coupling. Moreover, the P-N body-diode in the structure has a relatively large on-state voltage drop and poor reverse recovery characteristics.

In contrast, it is demonstrated that the specific on-resistance of the 600-V shielded planar 4H-SiC power ACCUFET structure is an order of magnitude smaller than that of the inversion-mode structure. This is primarily due to the much larger mobility for electrons in an accumulation layer. The specific on-resistance for the 600-V shielded planar 4H-SiC power ACCUFET structure is found to be smaller than that obtained even for the silicon devices with charge-coupling.

A new mode of operation for current flow in the third quadrant is reported here for the first time for the shielded planar 4H-SiC power ACCUFET structure. This current flow occurs via the N-base region by-passing the P-N junction. Consequently, the on-state voltage drop of the 'body-diode' is reduced and minority carrier injection is completely suppressed. The reverse recovery characteristics for the 600-V shielded planar 4H-SiC power ACCUFET structure are much superior to those for the inversion-mode device.

The performance of the integral diode within the silicon carbide structures is summarized in Figs. 9.39 and 9.40 for the case of a lifetime of 1 μs in the drift region. The on-state voltage drop provided in the Fig. 9.39 is for the integral diode and in Fig. 9.40 for the MOSFET mode. From the information in the above figures, it can be concluded that the 600-V shielded planar 4H-SiC power ACCUFET structure offers the best performance. It has an exceptionally small specific on-resistance which allows operation at a high on-state current density. The chip area required to serve any application is proportionately reduced. The ratio of the peak reverse recovery current to the on-state current density for this device is much lower than for all of the 600-V power MOSFET structures discussed in the previous chapters. The figures-of-merit for the 600-V shielded planar 4H-SiC power ACCUFET structure are also an order of magnitude superior to those of the best silicon power MOSFET structures discussed in the previous chapters.

| Device Structure | $R_{DSON,SP}$ (mΩ-cm$^2$) | $J_{ON}$ (A/cm$^2$) | $V_{ON}$ (V) | $J_{RR}$ (A/cm$^2$) | $J_{RR}/J_{ON}$ | FOM(E) (V) |
|---|---|---|---|---|---|---|
| MOSFET | 9.20 | 104 | 2.96 | 156 | 1.50 | 1.44 |
| ACCUFET | 0.77 | 360 | 2.04 | 117 | 0.325 | 0.091 |

**Fig. 9.39** Integral diode parameters for the 600-V shielded 4H-SiC structures

| Device Structure | $R_{DSON,SP}$ (mΩ-cm$^2$) | $J_{ON}$ (A/cm$^2$) | $V_{ON}$ (V) | $Q_{RR}$ (μC/cm$^2$) | $Q_{RR}/J_{ON}$ (μs) | FOM(F) (nV-s) |
|---|---|---|---|---|---|---|
| MOSFET | 9.20 | 104 | 0.96 | 3.69 | 0.0355 | 33.95 |
| ACCUFET | 0.77 | 360 | 0.28 | 2.08 | 0.0058 | 1.60 |

**Fig. 9.40** Integral diode parameters for the 600-V shielded 4H-SiC structures

## References

1. B.J. Baliga, "Silicon Carbide Power Devices", World Scientific, Singapore, 2005.
2. B.J. Baliga, "Silicon Carbide Semiconductor Devices having Buried Silicon Carbide Conduction Barrier Layers Therein", U. S. Patent 5,543,637, Issued August 6, 1996.
3. B.J. Baliga, "Fundamentals of Power Semiconductor Devices", Springer-Science, New York, 2008.
4. S. Sridevan and B.J. Baliga, "Lateral N-channel Inversion Mode 4H-SiC MOSFETs", IEEE Electron Device Letters, Vol. 19, pp. 228–230, 1998.
5. S.H. Ryu, et al., "10-kV 5-A 4H-SiC Power DMOSFET", IEEE International Symposium on Power Semiconductor Devices, Abstract 7.2, pp. 265–268, 2006.
6. P.M. Shenoy and B.J. Baliga, "The Planar 6H-SiC ACCUFET", IEEE Electron Device Letters, Vol. 18, pp. 589–591, 1997.
7. N. Thapar and B.J. Baliga, "Analytical Model for the Threshold Voltage of Accumulation Channel MOS-Gated Devices", Solid State Electronics, Vol. 42, pp. 1975–1979, 1998.

# Chapter 10
# Synopsis

Power devices are required for systems that operate over a broad spectrum of power levels and frequencies as discussed in the textbook [1]. A useful classification for the applications that is based up on the operating voltage level is shown in Fig. 10.1.

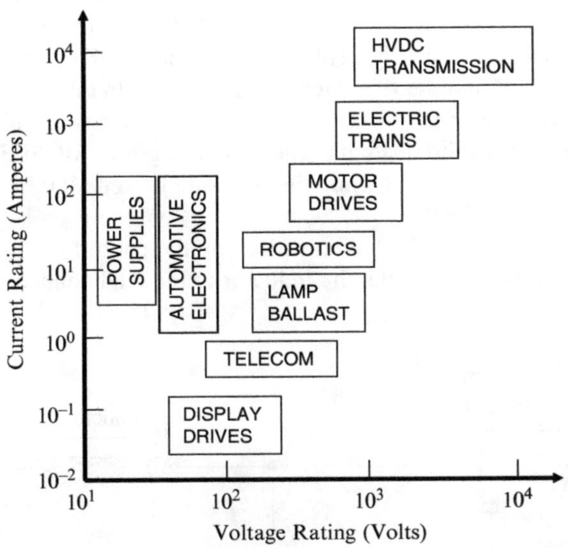


**Fig. 10.1** Applications for power MOSFET devices

At lower voltages (<50 V), a large number of power MOSFET devices are needed for use in computer power supplies and in automotive electronics. Discrete power MOSFET devices with larger voltage ratings are also required for motor control as shown in the figure. A variety of advanced power MOSFET structures have been discussed in previous chapters of this book for serving the applications

B.J. Baliga, *Advanced Power MOSFET Concepts*,
DOI 10.1007/978-1-4419-5917-1_10, © Springer Science+Business Media, LLC 2010

shown in Fig. 10.1. The choice of the optimum device suitable for each application depends upon the device voltage rating and the circuit switching frequency.

## 10.1 Computer Power Supplies

One of the major applications for silicon power MOSFET devices is in computer power supplies that provide power to microprocessors. The power delivery in a computer consists of conversion of the high voltage AC input power to a DC back-plane power source with a typical voltage of 17 V. The power delivery to the microprocessor is performed using a voltage regulator module (VRM) that converts the 17-V DC power to the 1-V level DC power required by modern microprocessors. Typical current levels required to operate the microprocessor can range above 100 A. This large current delivery is accomplished by using VRMs operated in parallel with each VRM delivering approximately 20 A.

The commonly used sync-buck topology used for the DC-to-DC voltage conversion using the VRM is shown in Fig. 10.2. Due to the relatively low operating voltage in this circuit, power MOSFET devices are typically used as the switch in the high-side and low-side locations as illustrated in the figure. The power MOSFET on the low-side can be replaced by a Schottky diode [2] but the power MOSFET is preferred in order to reduce on-state losses. In this case, the gate signal must be synchronized with the voltage across the low-side power MOSFET so that it is turned-on only in the third quadrant. The high-side power MOSFET is commonly referred to as the 'control-FET' while the low-side power MOSFET is commonly referred to as the 'sync-FET'.

When the high-side MOSFET is turned on by the control circuit, current flows from the DC input source through the inductor to the load connected at the output

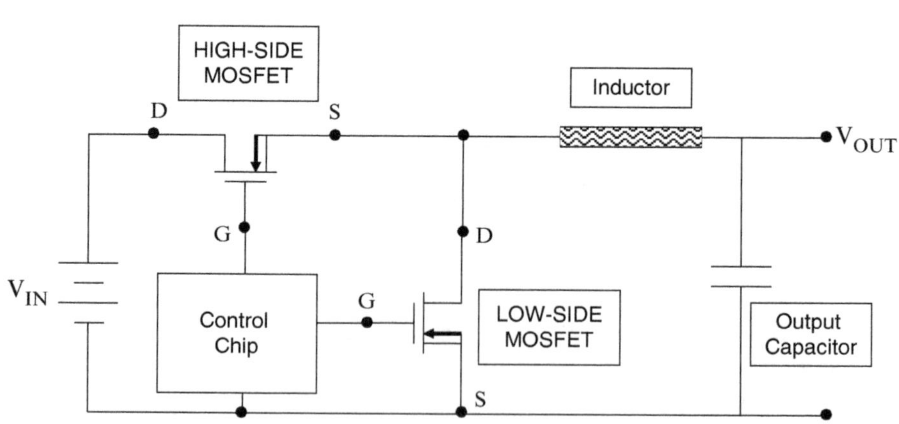


**Fig. 10.2** Sync-buck DC-DC converter circuit used for the VRM

terminals. When the transistor is switched off by the control circuit, the load current circulates through the MOSFET connected on the low-side and the inductor. The regulation of the DC output voltage can be achieved by adjusting the on-time of the transistor [3].

The low-side MOSFET is also referred to as a synchronous rectifier. The current flow in the low-side MOSFET is in the opposite direction to the normal operation of power MOSFET devices. If the low-side power MOSFET is turned on by the control circuit, the current flow occurs with a voltage drop determined by the on-resistance of the MOSFET structure. This voltage drop is usually much smaller than the built-in potential of the junction between the P-base region and the N-drift region (called the body-diode of the MOSFET). However, if the low-side MOSFET is not turned-on, the current will flow through the body-diode. This not only increases power losses due to the high on-state voltage drop of the P-N junction, it also severely slows down the switching time due to the long reverse recovery time for the body-diode as discussed in detail in Chap. 8.

The high-side and low-side MOSFET devices cannot be simultaneously turned-on during circuit operation to prevent short-circuiting the input power source. This requires delaying the turn-on of the low-side transistor after the gate bias to the high-side transistor has been turned-off. During this delay-time interval, the current in the low-side MOSFET flows through its body-diode. This upsets high frequency operation of the circuit and degrades the efficiency. One method to overcome this problem is to connect an external Schottky rectifier across the low-side switch. This has not been found to be an effective solution because of the inductance between the low-side MOSFET and the Schottky diode in the packages and circuit boards. An elegant solution to this problem is to integrate the Schottky rectifier into the power MOSFET structure as discussed in Chap. 8.

From the above discussion, the efficiency of the sync-buck converter is determined by the on-resistance and switching speed of the two transistors. As discussed in previous chapters the switching of the power MOSFET structures is determined by the gate transfer capacitance and the gate transfer charge.

### *10.1.1 Inadvertent Turn-On Suppression*

The voltage at the gate of the low-side transistor can inadvertently become larger than the threshold voltage when its drain voltage increases rapidly during VRM circuit operation. This allows the low-side transistor to turn-on even when no gate bias is applied by the control circuit. Since the high-side transistor is turned-on by the gate signal, this produces a significant current flow from the input power source which reduces the VRM efficiency and can even lead to destructive failure of the power MOSFET devices.

The criterion that governs the spike in the gate voltage to above the threshold voltage can be derived from the equivalent circuit for the power MOSFET shown in

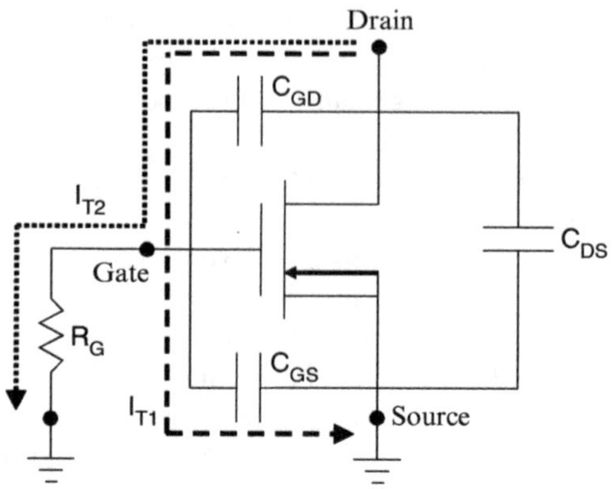


**Fig. 10.3** Transient [dV/dt] current paths in the power MOSFET structure

Fig. 10.3. When the drain voltage increases at a rapid rate [$dV_D/dt$], two transient currents are produced through the device gate transfer capacitance. If the magnitude of the gate circuit resistance is large, the transient current through the input capacitance will be dominant. The transient current through the gate transfer capacitance shown by the dashed line in Fig. 10.3 is then given by:

$$I_{T1} = \left(\frac{C_{GD}C_{GS}}{C_{GD}+C_{GS}}\right)\left[\frac{dV_D}{dt}\right] \tag{10.1}$$

The gate voltage produced by this transient current is given by:

$$V_{GS}(t) = \left(\frac{C_{GD}}{C_{GD}+C_{GS}}\right)\left[\frac{dV_D}{dt}\right]t \tag{10.2}$$

From this expression, it can be concluded that the inadvertent turn-on of the power MOSFET can be suppressed by reducing the ratio of gate transfer capacitance to the input capacitance. A useful figure-of-merit that allows comparison of power MOSFET structures is:

$$FOM(G) = \frac{C_{GD}}{C_{GS}} \tag{10.3}$$

If the magnitude of the gate circuit resistance is small, the transient current through the gate resistance will become dominant. The transient current through the gate transfer capacitance shown by the dotted line in Fig. 10.3 is then given by:

$$I_{T2} = C_{GD}\left[\frac{dV_D}{dt}\right] \tag{10.4}$$

The gate voltage produced by this transient current is given by:

$$V_{GS}(t) = C_{GD}R_G\left[\frac{dV_D}{dt}\right] \tag{10.5}$$

From this expression, it can be concluded that the inadvertent turn-on of the power MOSFET can be suppressed by reducing the magnitude of the gate transfer capacitance.

### *10.1.2 Device Active Area*

The size (or area) of any power device is important from the cost stand point. The active area for the power MOSFET devices is dictated by thermal considerations. The power dissipation incurred in the power MOSFET during circuit operation produces an increase in the device temperature. This temperature rise must be maintained below a typical junction temperature limit of 200°C to achieve the desired reliability targets.

The power dissipation in the power MOSFET structure occurs during the on-state, during the blocking state, and during the turn-on and turn-off transients. The power loss during the blocking state can be neglected due to the low leakage current for power MOSFET structures. For the comparison of power MOSFET structures, it will be assumed in this chapter that the active area is determined by limiting the power dissipation ($P_{D,ON}$) in the on-state to 100 W/cm$^2$. This approach does not account for the switching power loss. However, the switching power losses for the various power MOSFET structures can be separately compared.

The on-state current density for the power MOSFET structure can be obtained from the power dissipation and its specific on-resistance:

$$J_{ON} = \sqrt{\frac{P_{D,ON}}{R_{ON,SP}}} \tag{10.6}$$

The active area for the power MOSFET structure for any drain current rating ($I_D$) can be then obtained using:

$$A = \frac{I_D}{J_{ON}} = I_D\sqrt{\frac{R_{ON,SP}}{P_{D,ON}}} \tag{10.7}$$

From this expression, it can be seen that a reduced active area is possible for power MOSFET structures with smaller specific on-resistances.

### 10.1.3 Switching Power Losses

The power dissipated during the turn-on transient is dictated by the behavior of the fly-back rectifier in the circuit as discussed in the textbook [1]. For this reason, it is more appropriate to compare the turn-off power loss for the power MOSFET structures in terms of an energy loss per cycle. The various power MOSFET structures discussed in the previous chapters operate at different on-state current densities, which implies that they are handling different output power levels. It is only meaningful to compare the turn-off power losses for devices that are designed to operate with the same output power level. This can be achieved by comparing the turn-off power losses for power MOSFET structures with the same drain current rating.

Alternately, the turn-off energy loss can be normalized to the output power ($J_{ON}*V_{DS}$) controlled by the power MOSFET structure. In this case, the normalized energy loss per cycle is given by:

$$NE_{OFF} = \frac{1}{2}\left(t_{V,OFF} + t_{I,OFF}\right) \tag{10.8}$$

From this expression, it can be concluded that a shorter voltage rise-time ($t_{V,OFF}$) and current fall-time ($t_{I,OFF}$) are desirable for the power MOSFET structures. Analytical formulae for the voltage rise-time and the current fall-time were derived for each of the power MOSFET structures in the previous chapters together with quantitative values provided for each device.

### 10.1.4 Input Capacitance

The gate drive circuit for the power MOSFET structure must deliver enough current to rapidly charge and discharge the input capacitance for the device during each switching cycle. A larger gate drive current is required for power MOSFET structures with larger input capacitances to switch them on and off in the same time duration. Since a larger gate drive current requires the use of more expensive gate control circuits, it is preferable to utilize power MOSFET structures with smaller input capacitances. When comparing power MOSFET structures, it is not appropriate to use the specific input capacitance derived in the previous chapters because the devices suitable for a particular application will be of different sizes due to their different specific on-resistances. For this reason, in this chapter, the input capacitances for the various power MOSFET structures will be compared for a common drain current rating.

### 10.1.5 Device Comparison

Commercially available power D-MOSFET structures were originally used to develop switch mode power supplies. The operating frequency for the power supply was initially relatively low (~20 kHz). At this operating frequency, the specific on-resistance was the most important device parameter because conduction power loss was dominant in determining the efficiency. Consequently, the power U-MOSFET structure was developed to achieve a reduced specific on-resistance. In an effort to reduce the size and weight of the magnetic elements and to increase the efficiency, the operating frequency for the power supplies has been increased over the years to more than 200 kHz with some applications utilizing frequencies approaching 1 MHz. In this case, the switching power loss becomes as important as the on-state power loss. The power MOSFET devices required for these power supplies must have both low specific on-resistance and a small gate transfer charge (or capacitance). Recently, the power SC-MOSFET structure has been commercialized due to its combination of low specific on-resistance and small specific gate transfer charge.

The specific on-resistances at a gate bias of 10 V are provided in Fig. 10.4 for the various silicon power MOSFET structures discussed in the earlier chapters with blocking voltages of 30 V. These values are the same as those provided in the summary section of each chapter. The active area for the power MOSFET structures, computed by using (10.7), is also provided in Fig. 10.4 for the case of a drain current rating of 10 A based up on an on-state power dissipation of 100 W/cm$^2$. It can be observed that the active area for the power U-MOSFET structure is half that for the power D-MOSFET structure. A slightly smaller active area (47%) is obtained for the power SC-MOSFET structure making it an attractive option due to its planar device topology. A further reduction of the active area to just 20% of that for the power D-MOSFET structure is possible by utilizing either the power CC-MOSFET structure or the power GD-MOSFET structure. However, the active area for the power SJ-MOSFET structure is 46% of the active area for the power D-MOSFET structure. Consequently, it is not an attractive technology for this low blocking voltage rating of 30 V. This comparison demonstrates that the power CC-MOSFET and GD-MOSFET structures offer the most promise for improvement of power MOSFET technology.

The gate input capacitance and the gate reverse transfer capacitance are provided in Fig. 10.5 for the various silicon power MOSFET structures discussed in the

| **Power MOSFET Structure** | **Specific On-Resistance (mΩ-cm$^2$)** | **Active Area (cm$^{-3}$)** |
|---|---|---|
| D-MOSFET | 0.687 | 0.0262 |
| U-MOSFET | 0.171 | 0.0131 |
| SC-MOSFET | 0.149 | 0.0122 |
| CC-MOSFET | 0.0247 | 0.0050 |
| GD-MOSFET | 0.0337 | 0.0058 |
| SJ-MOSFET | 0.143 | 0.0120 |

**Fig. 10.4** Specific on-resistances and active areas for the 30-V power MOSFET structures with 10 A drain current rating

earlier chapters with blocking voltages of 30 V. The active area for the power MOSFET structures provided in Fig. 10.4 was used when computing these capacitances. It can be observed that the input capacitance for the power U-MOSFET structure is only slightly larger than that for the power D-MOSFET structure because its much larger specific input capacitance is compensated for by the much smaller active area. The input capacitance for the power SC-MOSFET structure is four-times smaller than that of the power D-MOSFET and U-MOSFET structure (consistent with data reported for commercial SSCFET devices) making it an attractive technology. The input capacitances for the power CC-MOSFET, power GD-MOSFET and SJ-MOSFET structures are all comparable and a factor of two-times smaller than that for the power D-MOSFET and U-MOSFET technologies. However, their input capacitance is twice as large as that for the power SC-MOSFET structure. This comparison indicates that the power SC-MOSFET structure is the best technology from the point of view of low input capacitance.

The gate transfer capacitances computed for the 10-A drain current rating of the 30-V silicon power MOSFET structures are also provided in Fig. 10.5 at a drain bias of 20 V. The gate transfer capacitance is usually determined at a drain bias ranging from 15 to 20 V in datasheets. The gate transfer capacitance for the power D-MOSFET and U-MOSFET structure are comparable. Much lower gate transfer capacitances are observed for the power SC-MOSFET, CC-MOSFET, GD-MOSFET, and SJ-MOSFET structures. These devices also have a very low FOM(G), i.e. the ($C_{GD}/C_{GS}$) ratio, making inadvertent turn-on under high [$dV_D/dt$] conditions virtually impossible.

It is usual practice to compare the gate transfer charge for 30-V power MOSFET structures when selecting them for the VRM applications. The gate transfer charge values are provided in Fig. 10.6 for a device with 10 A rated drain current. The gate transfer charge for the power D-MOSFET and U-MOSFET structure are comparable. Much lower gate transfer charges are observed for the power SC-MOSFET, CC-MOSFET, and GD-MOSFET, structures. The values for the figure-of-merit (C) for the 30-V power MOSFET structures are also provided in

| Power MOSFET Structure | Input Capacitance (pF) | Gate Transfer Capacitance (pF) | FOM(G) |
|---|---|---|---|
| D-MOSFET | 577 | 79 | 0.137 |
| U-MOSFET | 615 | 71 | 0.115 |
| SC-MOSFET | 159 | 1.5 | 0.009 |
| CC-MOSFET | 273 | 0 | 0 |
| GD-MOSFET | 250 | 2.5 | 0.01 |
| SJ-MOSFET | 311 | 0 | 0 |

**Fig. 10.5** Capacitances and figure-of-merit (G) for the 30-V power MOSFET structures with 10 A drain current rating

| Power MOSFET Structure | Gate Transfer Charge (nC) | FOM (C) (ps) |
|---|---|---|
| D-MOSFET | 5.21 | 136.7 |
| U-MOSFET | 4.66 | 60.88 |
| SC-MOSFET | 1.07 | 13.11 |
| CC-MOSFET | 1.48 | 7.29 |
| GD-MOSFET | 1.19 | 6.94 |
| SJ-MOSFET | 3.37 | 40.18 |

**Fig. 10.6** Gate transfer charge and figure-of-merit (C) for the 30-V power MOSFET structures with 10 A drain current rating

Fig. 10.6. It can be observed that the FOM(C) for the power U-MOSFET structure is two-times better than that for the power D-MOSFET structure. The FOM(C) for the power SC-MOSFET structure is ten-times better than that for the power D-MOSFET structure. The FOM(C) for the power CC-MOSFET and GD-MOSFET structures is about 20-times better than that for the power D-MOSFET structure. The FOM(C) for the power SJ-MOSFET structure is three-times better than that for the power D-MOSFET structure. The power SC-MOSFET structure was commercialized in 2004 based up on these favorable values [4]. In the future, the power CC-MOSFET and GD-MOSFET structures provide further enhancements in power MOSFET technology for devices with the 30-V rating.

## 10.2 High Voltage Motor Control

Power transistors are commonly used in motor control circuits. The most prevalent applications for commercial and industrial systems utilize an H-bridge configuration as shown in Fig. 10.7. The operating voltages for these applications typically range from 30 to 6,000 V. At voltage levels below 200 V, the best power switch is the power MOSFET as demonstrated in the textbook [1]. One important motor control application is in heating, ventilating, and air-conditioning (HVAC) systems. Another important motor control application is in electric and hybrid-electric vehicles. These applications utilize the H-bridge circuit with a 400-V DC bus which requires power switches with 600-V blocking capability. The power insulated gate bipolar transistor (IGBT) structure is now commonly used for these applications due to a favorable combination of on-state voltage drop, switching speed, and ruggedness. However, the advanced silicon and silicon carbide power MOSFET structures discussed in this book are potential alternative devices.

The specific on-resistances at a gate bias of 10 V are provided in Fig. 10.8 for the various power MOSFET structures discussed in the earlier chapters with blocking voltages of 600 V. These values are the same as those provided in the summary

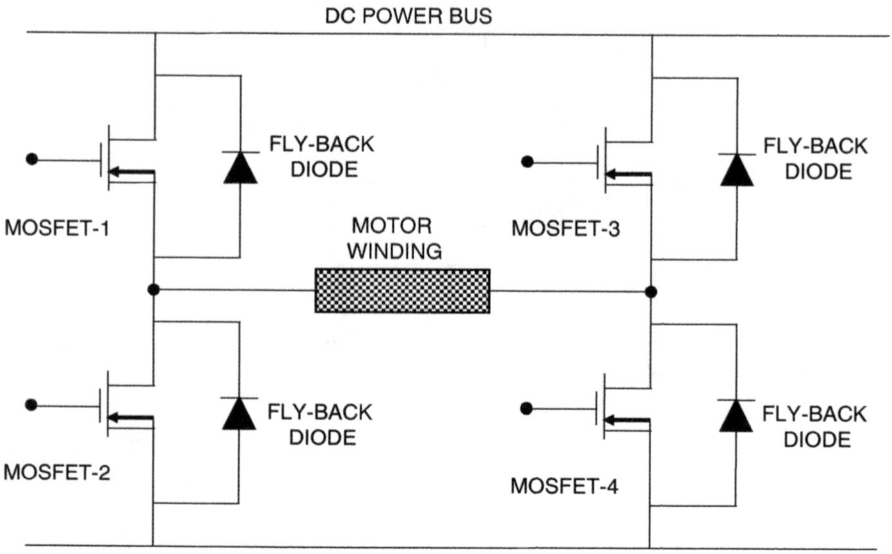


**Fig. 10.7** Typical H-bridge topology for motor control

| Power MOSFET Structure | Specific On-Resistance (mΩ-cm²) | On-State Current Density (A/cm²) | Active Area (cm⁻³) | On-State Voltage Drop (Volts) |
|---|---|---|---|---|
| D-MOSFET | 102 | 31.3 | 0.319 | 3.19 |
| U-MOSFET | 100 | 31.6 | 0.316 | 3.16 |
| SC-MOSFET | 99.6 | 31.6 | 0.315 | 3.16 |
| JBSFET | 95.3 | 32.4 | 0.309 | 3.09 |
| GD-MOSFET | 5.29 | 137.5 | 0.0727 | 0.73 |
| SJ-MOSFET | 5.42 | 136 | 0.0735 | 0.74 |
| SiC-MOSFET | 9.17 | 104 | 0.0962 | 0.95 |
| SiC-ACCUFET | 0.77 | 360 | 0.0278 | 0.28 |

**Fig. 10.8** Specific on-resistances and active areas for the 600-V power MOSFET structures with 10 A drain current rating

section of each chapter. The active areas for the power MOSFET structures are also provided in Fig. 10.8 for the case of a drain current rating of 10 A based up on an on-state power dissipation of 100 W/cm$^2$ (see (10.7)). It can be observed that the specific on-resistances for the power D-MOSFET, U-MOSFET, SC-MOSFET and JBSFET structures are essentially equal leading to the same active area. The power CC-MOSFET structure is not viable because of difficulty with scaling its breakdown voltage above 85 V. Among the silicon devices, the power GD-MOSFET and SJ-MOSFET structures offer significantly reduced specific on-resistances derived from the charge coupling phenomenon. In the ideal case with perfect charge balance, the specific on-resistance for the power SJ-MOSFET structure is smaller than that of the power GD-MOSFET structure. However, charge imbalance considerations result in an increase in specific on-resistance by a factor of five-times. Based up on this, it can be concluded that the power GD-MOSFET structure provides the best opportunity to reduce the size of the devices for the 600-V blocking capability. Further improvement in chip size is possible with the 4H-SiC shielded planar ACCUFET structure. However, the cost of this technology is far greater than that for silicon devices at this time.

The on-state current density and the on-state voltage drop for the various power MOSFET structures are provided in Fig. 10.8. In the case of the silicon power D-MOSFET, U-MOSFET, SC-MOSFET and JBSFET structures, the on-state voltage drop is above 3-V. This is larger than the typical on-state voltage drop of about 2-V for 600-V silicon power IGBT devices with fast switching speeds. Consequently, the 600-V silicon power MOSFET structures based up on one-dimensional voltage blocking physics are not competitive with the IGBT devices. However, the silicon power GD-MOSFET and SJ-MOSFET structures based up on two-dimensional charge coupling physics have an on-state voltage drop below 0.8 V making them superior to 600-V IGBT devices.

The gate input capacitance and the gate reverse transfer capacitance are provided in Fig. 10.9 for the various power MOSFET structures discussed in the earlier chapters with blocking voltages of 600 V. The active areas for the power MOSFET structures provided in Fig. 10.8 were used when computing these capacitances. It can be observed that the input capacitance for the power U-MOSFET structure is twice as large as that for the power D-MOSFET structure because of its much larger specific input capacitance. The input capacitances for the power SC-MOSFET and

| Power MOSFET Structure | Input Capacitance (pF) | Gate Transfer Capacitance (pF) | FOM(G) |
|---|---|---|---|
| D-MOSFET | 7,018 | 29 | 0.0041 |
| U-MOSFET | 14,852 | 50 | 0.0034 |
| SC-MOSFET | 4,095 | 7.2 | 0.0018 |
| JBSFET | 4,017 | 7.1 | 0.0018 |
| GD-MOSFET | 1,200 | 0 | 0 |
| SJ-MOSFET | 1,910 | 0 | 0 |
| SiC-MOSFET | 2,886 | 63 | 0.022 |
| SiC-ACCUFET | 834 | 18 | 0.022 |

**Fig. 10.9** Capacitances and figure-of-merit (G) for the 600-V power MOSFET structures with 10 A drain current rating

JBSFET structures are about 60% of that for the power D-MOSFET structure making them an attractive technology. The input capacitances for the power GD-MOSFET and SJ-MOSFET structures are comparable and a factor of five-times smaller than that for the power D-MOSFET structure. This comparison indicates that the power GD-MOSFET structure is the best silicon technology from the point of view of low input capacitance. The 600-V 4H-SiC shielded planar power ACCUFET structure has an even superior input capacitance.

The gate transfer capacitances computed for the 10-A drain current rating of the 600-V power MOSFET structures are also provided in Fig. 10.9. The gate transfer capacitance is calculated at a drain bias of 400 V. The gate transfer capacitance for the U-MOSFET structure is 1.7-times larger than that for the power D-MOSFET structure. In contrast, the gate transfer capacitances for the SC-MOSFET and JBSFET structures are four-times smaller than that for the power D-MOSFET structure. Very low gate transfer capacitances are observed for the GD-MOSFET and SJ-MOSFET structures. These devices also have a very low FOM(G), i.e. the ($C_{GD}/C_{GS}$) ratio, making inadvertent turn-on under high [$dV_D/dt$] conditions virtually impossible. The gate transfer capacitance for the silicon carbide devices are relatively large because the high drift region doping concentration produces small values for the depletion width. These devices also have relatively high values for the FOM(G).

It is usual practice to compare the gate transfer charge for 600-V power MOSFET structures when selecting them for the motor control applications. The gate transfer charge values are provided in Fig. 10.10. The gate transfer charge for the power D-MOSFET and U-MOSFET structure are comparable. Much lower gate transfer charges are observed for the power SC-MOSFET, JBSFET and GD-MOSFET structures. The gate transfer charge for the SJ-MOSFET structure is significantly larger than that for the power GD-MOSFET structure. The gate transfer charge for the 4H-SiC shielded planar power MOSFET structure is also relatively large. In contrast, gate transfer charge for the 4H-SiC shielded planar

| Power MOSFET Structure | Gate Transfer Charge (nC) | FOM (C) (ps) |
|---|---|---|
| D-MOSFET | 73.4 | 23,460 |
| U-MOSFET | 69.8 | 21,800 |
| SC-MOSFET | 12.0 | 3,785 |
| JBSFET | 11.7 | 3,618 |
| GD-MOSFET | 4.73 | 344 |
| SJ-MOSFET | 16.9 | 1247 |
| SiC-MOSFET | 22.1 | 2,116 |
| SiC-ACCUFET | 5.23 | 144 |

**Fig. 10.10** Gate transfer charge and figure-of-merit (C) for the 600-V power MOSFET structures with 10 A drain current rating

power ACCUFET structure is competitive with that for the silicon GD-MOSFET structure.

The figure-of-merit (C) values for the 600-V power MOSFET structures are also provided in Fig. 10.10. It can be observed that the FOM(C) for the power U-MOSFET structure is similar to that for the power D-MOSFET structure. In contrast, the FOM(C) for the power SC-MOSFET and JBSFET structures is six-times better than that for the power D-MOSFET and U-MOSFET structures. The FOM(C) for the GD-MOSFET structure is about 60-times better than that for the power D-MOSFET and U-MOSFET structures. The FOM(C) for the SJ-MOSFET structure is 3.4-times worse than that for the power GD-MOSFET structure. Based up on these values, the power GD-MOSFET structure offers the best performance among the silicon device structures. The 4H-SiC shielded planar ACCUFET structure has an even superior FOM(C).

It is worth comparing the performance of the body-diode within the high voltage power MOSFET structures. A power MOSFET structure with a good body-diode performance is preferable from the point of view of eliminating the external fly-back diode that is commonly used with silicon IGBT devices in H-bridge circuits. The performance of the integral diode within all the power MOSFET structures with 10 A drain current rating is provided in Figs. 10.11 and 10.12. Although the reverse recovery current for the power JBSFET structure is much smaller than for the power U-MOSFET structure, the on-state voltage drop for the body-diode is quite high. The body-diodes in the GD-MOSFET and GD-JBSFET structures have a low peak reverse recovery current and low on-state voltage drop. The power SJ-JBSFET structure has much superior reverse recovery characteristics compared with the power SJ-MOSFET structure. Its body-diode performance is also superior to that of the body-diode in the power GD-MOSFET and GD-JBSFET structures. The on-state voltage drop for the body diode in the 4H-SiC MOSFET structure is quite high. In contrast, the body-diode in the 4H-SiC ACCUFET structure has a lower on-state voltage drop and a small reverse recovery current.

| Device Structure | $R_{DSON,SP}$ (m$\Omega$-cm$^2$) | $J_{ON}$ (A/cm$^2$) | $V_{ON}$ (V) | $I_{PR}$ (A) | $I_{PR}/I_{ON}$ | FOM(E) (V) |
|---|---|---|---|---|---|---|
| U-MOSFET | 94.0 | 33 | 0.758 | 75 | 7.97 | 24.7 |
| JBSFET | 95.3 | 32.4 | 3.33 | 13.3 | 1.30 | 4.00 |
| GD-MOSFET | 6.42 | 125 | 1.21 | 14.5 | 1.60 | 1.28 |
| GD-JBSFET | 6.81 | 121 | 1.22 | 12.6 | 1.42 | 1.17 |
| SJ-MOSFET | 5.42 | 136 | 0.827 | 21.4 | 2.14 | 1.58 |
| SJ-JBSFET | 5.42 | 136 | 0.827 | 7.44 | 0.93 | 0.69 |
| SiC-MOSFET | 9.17 | 104 | 2.96 | 15.0 | 1.5 | 1.43 |
| SiC-ACCUFET | 0.77 | 360 | 2.04 | 3.25 | 0.325 | 0.09 |

**Fig. 10.11** Integral diode parameters for the 600-V power MOSFET structures with 10-A current rating

| Device Structure | $R_{DSON,SP}$ (mΩ-cm$^2$) | $J_{ON}$ (A / cm$^2$) | $V_{ON}$ (V) | $Q_{RR}$ (μC) | FOM(F) (nV-s) |
|---|---|---|---|---|---|
| U-MOSFET | 94.0 | 33 | 3.10 | 4.10 | 1220 |
| JBSFET | 95.3 | 32.4 | 3.08 | 0.084 | 26.0 |
| GD-MOSFET | 6.42 | 125 | 0.80 | 0.447 | 39.5 |
| GD-JBSFET | 6.81 | 121 | 0.82 | 0.365 | 30.1 |
| SJ-MOSFET | 5.42 | 136 | 0.74 | 0.928 | 68.5 |
| SJ-JBSFET | 5.42 | 136 | 0.74 | 0.179 | 13.2 |
| SiC-MOSFET | 9.17 | 104 | 0.96 | 0.355 | 33.8 |
| SiC-ACCUFET | 0.77 | 360 | 0.28 | 0.058 | 1.60 |

**Fig. 10.12** Integral diode parameters for the 600-V power MOSFET structures with 10-A current rating

The reverse recovery charge for power rectifiers is a good measure of the power losses during their reverse recovery transient. The reverse recovery charge for the internal diode for all the power MOSFET structures is provided in Fig. 10.12. The reverse recovery charge for the silicon power JBSFET structure is 50-times smaller than that for the silicon power U-MOSFET structure. The reverse recovery charge for the body diode in the silicon power GD-MOSFET and GD-JBSFET structures are an order of magnitude smaller than in the silicon power U-MOSFET structure. The reverse recovery charge for the body diode in the silicon power SJ-MOSFET structure is four-times smaller than in the power U-MOSFET structure. The reverse recovery charge for this structure can be greatly reduced by incorporation of the Schottky contact to create the power SJ-JBSFET structure. The reverse recovery charge in the 4H-SiC shielded power MOSFET structure is an order of magnitude smaller than in the silicon power U-MOSFET structure. The reverse recovery charge in the 4H-SiC shielded power ACCUFET structure is 70-times smaller than in the silicon power U-MOSFET structure.

Another comparison between the power MOSFET structures can be made on the basis of the power losses during inductive load turn-off. The energy loss per cycle for the power MOSFET structures with 10-A rating are provided in Fig. 10.13. It can be seen that the power GD-MOSFET structure has half the losses observed in the power U-MOSFET structure. The power losses in the power SJ-MOSFET are much better (by a factor six-times) due to the abrupt transition of the drain voltage. Relatively large power losses during turn-off are observed for the silicon carbide devices.

From the information in the above figures, it can be concluded that the best performance is observed with the 600-V power GD-JBSFET structure and the 600-V power SJ-JBSFET structure. They have an exceptionally small specific on-resistance which allows operation at a high on-state current density. The chip area required to serve any application is proportionately reduced. The peak reverse recovery current for the 600-V power GD-JBSFET and 600-V power SJ-JBSFET

| Device Structure | $t_{V,\,OFF}$ (ns) | $t_{I,\,OFF}$ (ns) | $E_{OFF}$ (mJ) |
|---|---|---|---|
| U-MOSFET | 167 | 11 | 0.370 |
| GD-MOSFET | 88 | 13 | 0.183 |
| SJ-MOSFET | 84 | 13 | 0.029 |
| SiC-MOSFET | 130 | 60 | 0.380 |
| SiC-ACCUFET | 210 | 90 | 0.600 |

**Fig. 10.13** Inductive load turn-off parameters for the 600-V power MOSFET structures with 10-A current rating

structures are only 142 and 93% of the on-state current density. The FOM(C) for the 600-V power GD-JBSFET and 600-V power SJ-JBSFET structures are also an order of magnitude superior to those of the conventional 600-V power U-MOSFET structure.

## 10.3 Device Comparison

In each of the preceding chapters devoted to individual power MOSFET structures, the discussion section provided a description of the capability of the structure over a broad range of blocking voltages. This section provides a comparison of all the different structures. The parameters used for the comparison are the specific on-resistance, the specific gate transfer charge, and the figure-of-merit (C).

The specific on-resistance is shown in Fig. 10.14 for all the different power MOSFET structures as a function of the breakdown voltage. The ideal specific on-resistance for silicon based up on Baliga's law for the impact ionization coefficients is included in the figure (dashed line). The values for the silicon carbide structures are indicated by the dotted lines. Among the silicon devices, it can be observed that the power D-MOSFET, U-MOSFET and SC-MOSFET structures have specific on-resistances that are always above the ideal limit because they rely up on the one-dimensional potential distribution. The specific on-resistances for the power GD-MOSFET and SJ-MOSFET structures fall below the ideal limit for silicon because they rely up on two-dimensional charge coupling. It can be seen from the figure that the power GD-MOSFET structure is superior to the power SJ-MOSFET structure at lower breakdown voltages. The power SJ-MOSFET structure becomes superior to the power GD-MOSFET structure at breakdown voltages above 300 V in this graph. However, the graph is based up on the case with ideal two-dimensional charge coupling. As discussed in Chap. 7, the power SJ-MOSFET structure is sensitive to charge imbalance. When this is taken into consideration, it specific on-resistance is three to five times larger than for the ideal case. Under these conditions, the power GD-MOSFET structure will be superior to

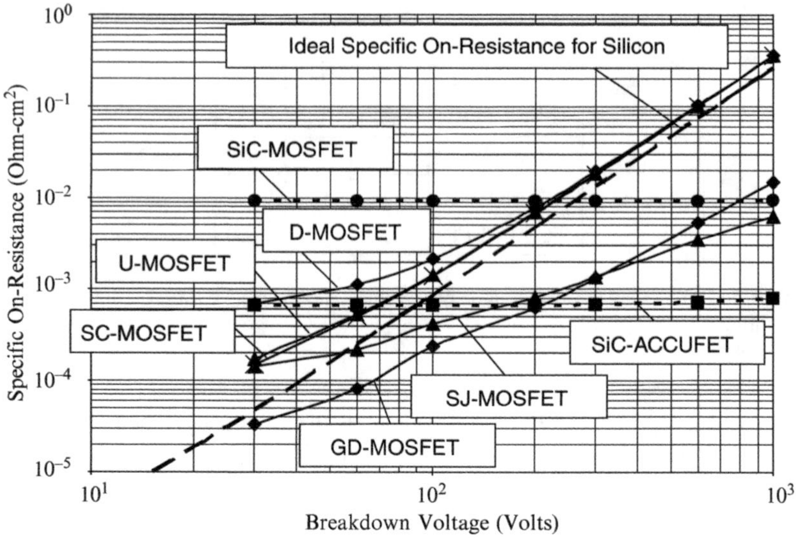


**Fig. 10.14** Comparison of the specific on-resistance for power MOSFET structures

the SJ-MOSFET structure for the entire range of breakdown voltages in Fig. 10.14. The 4H-SiC shielded planar power MOSFET structure is superior to the silicon power D-MOSFET, U-MOSFET and SC-MOSFET structures for breakdown voltages above 250 V. In contrast, the 4H-SiC shielded planar power ACCUFET structure is superior to the silicon power D-MOSFET, U-MOSFET and SC-MOSFET structures for breakdown voltages above 70 V.

The specific gate transfer charge is shown in Fig. 10.15 for all the different power MOSFET structures discussed in the previous chapters as a function of the breakdown voltage. The values for the silicon carbide structures are indicated by the dotted lines. Among the silicon devices, it can be observed that the power D-MOSFET, U-MOSFET and SJ-MOSFET structures have specific gate transfer charge values that are comparable. The specific gate transfer charge values for the power GD-MOSFET and SC-MOSFET structures are significantly smaller in magnitude. The 4H-SiC shielded planar power MOSFET and ACCUFET structures have specific gate transfer charge values similar in magnitude to the power D-MOSFET structure. Based up on Fig. 10.15, it can be concluded that the power SC-MOSFET structure is a very attractive device from the point of view of the gate transfer charge. However, the power GD-MOSFET structure is the most promising device because its specific on-resistance is far smaller than that for the power SC-MOSFET structure.

The Figure-of-Merit (C) for all the power MOSFET structures is compared in Fig. 10.16. This is a good technology figure-of-merit for selection of devices for high frequency switching applications because it takes into account the on-state and switching losses. Among the conventional silicon devices, the power SC-MOSFET

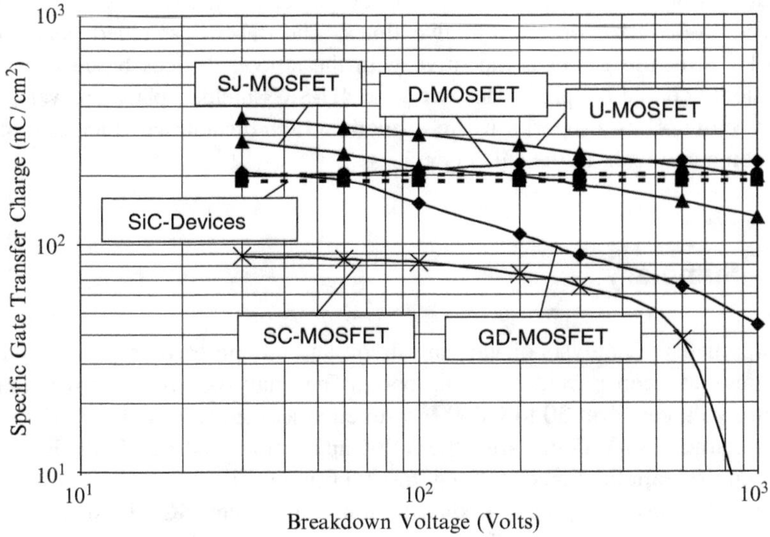


**Fig. 10.15** Comparison of the specific gate transfer charge for power MOSFET structures

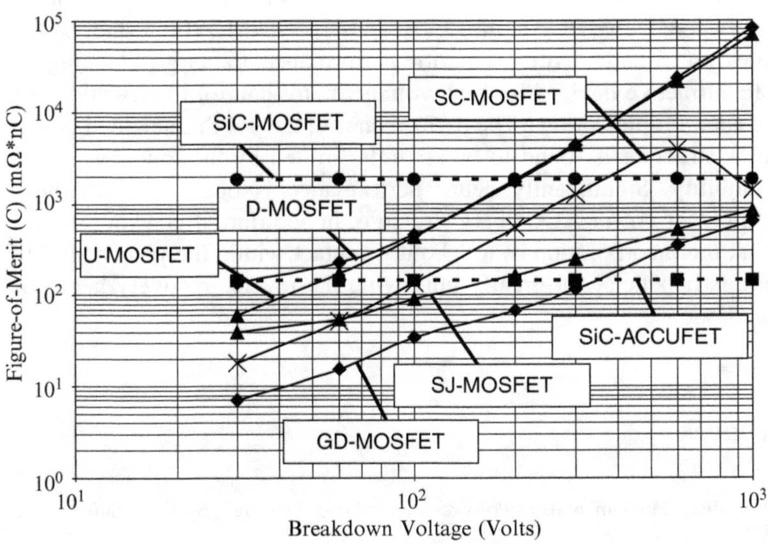


**Fig. 10.16** Comparison of the FOM(C) for power MOSFET structures

structure offers a significant improvement in performance due to its small gate transfer charge. Among the devices based on the two-dimensional charge coupling phenomenon, the power GD-MOSFET structure is superior to the power SJ-MOSFET structure despite the use of ideal charge balance conditions for the power SJ-MOSFET structure. In practice, the performance of the power SJ-MOSFET structure may be degraded by a factor of three to five times. The 4H-SiC shielded planar power MOSFET structure is not competitive with the silicon devices based up on two-dimensional charge coupling. However, the 4H-SiC shielded planar power ACCU-FET structure offers a superior figure-of-merit at high breakdown voltages (>400 V) when compared with all the silicon devices.

## 10.4 Summary

A comprehensive analysis of various silicon and silicon carbide power MOSFET structures has been provided in this book. The analysis cover a broad range of blocking voltages form 30 to 1,000 V with emphasis given to two specific voltage ratings: namely 30-V devices for the VRM application and 600-V devices for the motor control applications. Among the traditional silicon devices that operate with one-dimensional potential distribution, the power SC-MOSFET structure offers a unique combination of low specific on-resistance and small gate transfer charge with a planar device topology that is suitable for manufacturing at low cost. Significantly smaller specific on-resistance can be achieved in the low blocking voltage devices by utilizing the power CC-MOSFET structure which utilizes two-dimensional charge coupling. The power GD-MOSFET and SJ-MOSFET structures are found to offer superior performance for large blocking voltages. They are attractive devices for high voltage motor control in air-conditioning and electric vehicle applications. The performance of the 4H-SiC shielded planar power MOSFET structure is found to be severely limited by the low channel inversion layer mobility. Significantly better performance is observed with the 4H-SiC shielded planar power ACCUFET structure. In addition, it is demonstrated in this book that the incorporation of a Schottky contact within the power MOSFET cell structure is beneficial for greatly improving the reverse recovery behavior of the internal diode.

## References

1. B.J. Baliga, "Fundamentals of Power Semiconductor Devices", Springer-Science, New York, 2008.
2. B.J. Baliga, "Advanced Power Rectifier Concepts", Springer-Science, New York, 2009.
3. N. Mohan, T.M. Undeland, and W.P. Robbins, "Power Electronics", pp. 164–172, Wiley, New York, 1995.
4. Silicon Semiconductor Corporation Datasheets, www.siliconsemi.com.

# About the Author

Professor Baliga is internationally recognized for his leadership in the area of power semiconductor devices. In addition to over 500 publications in international journals and conference digests, he has authored and edited 16 books ("*Power Transistors*", IEEE Press 1984; "*Epitaxial Silicon Technology*", Academic Press 1986; "*Modern Power Devices*", John Wiley 1987; "*High Voltage Integrated Circuits*", IEEE Press 1988; "*Solution Manual: Modern Power Devices*", John Wiley 1988; "*Proceedings of the 3$^{rd}$ Int. Symposium on Power Devices and ICs*", IEEE Press 1991; "*Modern Power Devices*", Krieger Publishing Co. 1992; "*Proceedings of the 5$^{th}$ Int. Symposium on Power Devices and ICs*", IEEE Press 1993; "*Power Semiconductor Devices*"; PWS Publishing Company 1995; "*Solution Manual: Power Semiconductor Devices*"; PWS Publishing Company 1996; "*Cryogenic Operation of Power Devices*", Kluwer Press 1998; "*Silicon RF Power MOSFETs*", World Scientific Publishing Company 2005; "*Silicon Carbide Power Devices*", World Scientific Publishing Company 2006; "*Fundamentals of Power Semiconductor Devices*", Springer Science, 2008; "*Solution Manual: Fundamentals of Power Semiconductor Devices*", Springer Science, 2008; "*Advanced Power Rectifier Concepts*", Springer Science, 2009. In addition, he has contributed chapters to another twenty books. He holds 120 US Patents in the solid-state area. In 1995, one of his inventions was selected for the *B.F. Goodrich Collegiate Inventors Award* presented at the *Inventors Hall of Fame*.

Professor Baliga obtained his Bachelor of Technology degree in 1969 from the Indian Institute of Technology, Madras, India. He was the recipient of the *Philips India Medal* and the *Special Merit Medal (as Valedictorian)* at I.I.T, Madras. He obtained his Masters and Ph.D. degrees from Rensselaer Polytechnic Institute, Troy NY, in 1971 and 1974, respectively. His thesis work involved Gallium Arsenide diffusion mechanisms and pioneering work on the growth of InAs and GaInAs layers using Organometallic CVD techniques. At R.P.I., he was the recipient of the *IBM Fellowship* in 1972 and the *Allen B. Dumont Prize* in 1974.

From 1974 to 1988, Dr. Baliga performed research and directed a group of 40 scientists at the General Electric Research and Development Center in Schenectady, NY, in the area of Power Semiconductor Devices and High Voltage Integrated Circuits. During this time, he pioneered the concept of MOS-Bipolar functional integration to create a new family of discrete devices. He is the *inventor of the IGBT* which is now in production by many International Semiconductor companies. This invention is widely used around the globe for air-conditioning, home appliance (washing machines, refrigerators, mixers, etc) control, factory automation (robotics), medical systems (CAT scanners, uninterruptible power supplies), and electric street-cars/bullet-trains, as well as for the drive-train in electric and hybrid-electric cars under development for reducing urban pollution. The US Department of Energy has released a report that the variable speed motor drives enabled by IGBTs produce an energy savings of 2 quadrillion btus per year (equivalent to 70 Giga-Watts of power). The widespread adoption of Compact Fluorescent Lamps (CFLs) in place of incandescent lamps is producing an additional power savings of 30 Giga-Watts. The cumulative impact of these energy savings on the environment is a reduction in Carbon Dioxide emissions from Coal-Fired power plants by over One Trillion pounds per year. Most recently, the IGBT has enabled fabrication of very compact, light-weight, and inexpensive defibrillators used to resuscitate cardiac arrest victims. When installed in fire-trucks, paramedic vans, and on-board airlines, it is projected by the American Medical Association (AMA) to save 100,000 lives per year in the US. For this work, *Scientific American Magazine* named him one of the eight heroes of the semiconductor revolution in their 1997 special issue commemorating the Solid-State Century.

Dr. Baliga is also the originator of the concept of merging Schottky and p-n junction physics to create a new family of power rectifiers that are commercially available from various companies. In 1979, he theoretically demonstrated that the performance of power MOSFETs could be enhanced by several orders of magnitude by replacing silicon with other materials such as gallium arsenide and silicon carbide. This is forming the basis of a new generation of power devices in the twenty first Century.

In August 1988, Dr. Baliga joined the faculty of the Department of Electrical and Computer Engineering at North Carolina State University, Raleigh, North Carolina, as a Full Professor. At NCSU, in 1991 he established an international center called the *Power Semiconductor Research Center* (PSRC) for research in the area of power semiconductor devices and high voltage integrated circuits, and has served as its Founding Director. His research interests include the modeling of novel device concepts, device fabrication technology, and the investigation of the impact of new materials, such as GaAs and Silicon Carbide, on power devices. In 1997, in recognition of his contributions to NCSU, he was given the highest university faculty rank of *Distinguished University Professor of Electrical Engineering*.

In 2008, Professor Baliga was a key member of an NCSU team – partnered with four other universities - that was successful in being granted an Engineering Research Center from the National Science Foundation for the development of micro-grids that allow integration of renewable energy sources. Within this program, he is responsible for the fundamental sciences platform and the development of power devices from wide-band-gap semiconductors for utility applications.

Professor Baliga has received numerous awards in recognition for his contributions to semiconductor devices. These include two *IR 100 awards* (1983, 1984), the *Dushman* and *Coolidge Awards* at GE (1983), and being selected among the *100 Brightest Young Scientists in America* by Science Digest Magazine (1984). He was elected *Fellow of the IEEE* in 1983 at the age of 35 for his contributions to power semiconductor devices. In 1984, he was given the *Applied Sciences Award* by the world famous sitar maestro Ravi Shankar at the Third Convention of Asians in North America. He received the 1991 *IEEE William E. Newell Award*, the highest honor given by the Power Electronics Society, followed by the 1993 *IEEE Morris E. Liebman Award* for his contributions to the emerging *Smart Power Technology*. In 1992, he was the first recipient of the BSS Society's *Pride of India Award*. At the age of 45, he was elected as Foreign Affiliate to the prestigious *National Academy of Engineering*, and was one of only four citizens of India to have the honor at that time (converted to regular Member in 2000 after taking US Citizenship). In 1998, the University of North Carolina system selected him for the *O. Max Gardner Award*, which recognizes the faculty member among the 16 constituent universities who has made the greatest contribution to the welfare of the human race. In December 1998, he received the *J.J. Ebers Award*, the highest recognition given by the IEEE Electron Devices Society for his technical contributions to the Solid-State area. In June 1999, he was honored at the Whitehall Palace in London with the *IEEE Lamme Medal*, one of the highest forms of recognition given by the IEEE Board of Governors, for his contributions to development of an apparatus/technology of benefit to society. In April 2000, he was honored by his Alma Mater as a *Distinguished Alumnus*. In November 2000, he received the *R.J. Reynolds Tobacco Company Award for Excellence in Teaching, Research, and Extension* for his contributions to the College of Engineering at North Carolina State University.

In 1999, Prof. Baliga founded a company, *Giant Semiconductor Corporation*, with seed investment from Centennial Venture Partners, to acquire an exclusive license for his patented technology from North Carolina State University with the goal of bringing his NCSU inventions to the marketplace. A company, *Micro-Ohm Corporation*, subsequently formed by him in 1999, has been successful in licensing the GD-TMBS power rectifier technology to several major semiconductor companies for world-wide distribution. These devices have application in power supplies, battery chargers, and automotive electronics. In June 2000, Prof. Baliga founded another company, *Silicon Wireless Corporation*, to commercialize a novel super-linear silicon

RF transistor that he invented for application in cellular base-stations and grew it to 41 employees. This company (renamed *Silicon Semiconductor Corporation*) is located at Research Triangle Park, NC. It received an investment of $10 Million from *Fairchild Semiconductor Corporation* in December 2000 to co-develop and market this technology. Based upon his additional inventions, this company has also produced a new generation of Power MOSFETs for delivering power to microprocessors in notebooks and servers. This technology was licensed by his company to Linear Technologies Corporation with transfer of the know-how and manufacturing process. Voltage Regulator Modules (VRMs) using his transistors are currently available in the market for powering microprocessor and graphics chips in laptops and servers.

# Index


**A**

Abrupt junction, 6
Active area, 46, 55, 56, 63, 90, 91, 112, 151, 152, 160, 218, 219, 242, 283, 284, 333, 339, 372, 373, 375, 539, 541, 542, 544, 545
Adjustable speed motor drive, 106, 310, 385, 399, 499, 523
Ambipolar diffusion coefficient, 402
Ambipolar diffusion length, 402
Annealing, 63, 123, 160, 242, 413, 445
Anode, 400
Anti-parallel diode, 412, 424
Applications, 1, 2, 4, 5, 18–21, 23, 24, 26, 28, 37, 51, 55, 56, 61, 63, 64, 73, 86, 91, 106, 119, 121, 123, 124, 133, 139, 145, 147, 152, 153, 157, 161, 187, 188, 196, 208, 210, 212, 215, 217–220, 236, 243, 273, 278, 279, 282, 284, 310, 325, 358, 371–373, 385, 391, 400, 403, 409, 419, 424–475, 478, 484, 532, 535, 536, 540–543, 546, 548, 550, 552, 555, 556
Audio amplifiers, 5
Automotive electronics, 1, 424, 535, 555
Avalanche breakdown, 358

**B**

Baliga's figure of merit, 7
Baliga's power law, 13–17, 21, 58, 105, 114, 155, 237, 246, 254, 258, 260, 261, 294, 300, 304, 318, 347, 392
Band gap, 7, 8, 477, 555
Bipolar power devices, 1, 23, 34, 119, 479, 505
Blocking characteristics, 40, 75, 101, 102, 134, 174, 176, 182–184, 222, 224–226, 229–231, 253, 260, 262, 263, 286–288, 291, 292, 296, 297, 301, 302, 331, 332, 338, 339, 378, 379, 382, 417, 418, 434, 435, 447, 448, 454, 455, 481, 508, 510
Blocking voltage, 1, 6, 21, 26, 27, 33, 34, 37–44, 51, 56–58, 63, 70, 73–80, 88, 101, 113, 114, 123, 127, 133–138, 149, 153–155, 159, 161–185, 187, 189, 191, 210, 212, 220, 221, 224, 225, 227, 229, 230, 232, 233, 235–237, 241–263, 265, 269, 279, 285–287, 291, 296, 301, 307, 309, 317–319, 323, 325, 329, 350, 354, 356, 368, 374, 378, 381–383, 385, 391–393, 417, 477, 479, 505, 541–543, 545, 549, 552
Body diode, 19, 399–412, 415, 419–433, 437, 439–443, 445, 449–453, 457, 459–461, 463–466, 468–472, 503–504, 527–532, 537, 547, 548
Boltzmann's constant, 402, 485
Boron, 5, 63, 121, 123, 160, 161, 242, 243, 324, 325
Breakdown voltage, 3, 7, 9–13, 15–17, 21, 34, 37–39, 56, 57, 59, 60, 70, 73, 74, 101, 113, 114, 117, 127, 133, 153–156, 158, 161, 163–167, 169, 170, 172, 173, 182, 187, 191, 220, 221, 224, 226, 232, 234, 235, 237, 239, 242, 244–246, 248–250, 253, 254, 258, 260, 261, 265, 269, 285–288, 290–292, 296, 300, 303, 306–309, 317–321, 324–330, 333, 336, 338, 339, 343, 345–348, 350, 352, 354, 356, 373–375, 378, 381–384, 391–394, 396, 430, 491, 514–515, 545, 549, 550, 552

Buck converter, 18–19, 537
Built-in potential, 19, 30, 49, 84, 203, 352, 364, 401, 469, 488, 489, 504, 512, 537
Bulk mobility, 12, 30, 126, 489

**C**

Carrier distribution, 64
Cathode, 6, 9
Cell pitch, 9–12, 26, 33, 46–48, 51, 63, 64, 66, 68, 69, 81–83, 87, 93, 111, 114, 120, 121, 123–127, 140, 141, 143, 148, 153, 162, 187, 188, 190, 191, 197, 201, 202, 205, 210, 212, 236, 237, 243, 265–268, 273, 276, 279, 295, 300, 305, 309, 315, 318, 321, 326, 338, 350, 351, 353, 358, 360, 364, 368, 369, 388, 391, 416, 439, 446, 453, 463, 468
Channel aspect ratio, 338
Charge balance, 329, 331, 338, 345, 385, 545, 551
Charge coupled structures, 8–12, 173
Charge-coupling, 1, 159, 162–185, 191, 196, 224, 226, 241, 244–263, 269, 272, 285, 287, 291, 303, 317, 325–346, 353, 354, 375, 378, 531, 532
Chynoweth's law, 12
Computer power supply, 1, 400–424, 535–543
Conduction barrier, 477
Conduction loss, 541
Contact resistance, 25, 32, 67, 122, 186, 264, 349
Critical electric field, 7–12, 15, 16, 103, 163–169, 172, 244–248, 254, 258, 260, 261, 290, 294, 298, 304, 326–328, 336, 343, 346, 348, 360, 362, 363, 479, 505
Current constriction, 27
Current distribution, 24, 27, 31, 36, 37, 64, 73, 97, 121, 132, 192–195, 270, 271, 355, 418, 435, 447, 455, 493, 517
Current flow pattern, 27, 36, 67, 69, 123, 130, 192, 195, 270, 355, 417, 434, 446, 454, 486, 492, 512, 516
Current spreading, 29, 31, 36, 68, 69, 72, 96, 124, 126, 187–189, 264, 266, 351, 486, 487, 489, 512
Cylindrical junction, 38, 133

**D**

Damage, 63, 161, 242, 407, 428
DC-DC converter, 18–19, 401, 536
DC-DC transformer, 18
Depletion region, 8, 30, 38, 41, 46–49, 64, 73–75, 82–84, 93, 125, 134, 140–144, 161, 176, 198, 200–203, 227, 243, 256, 288, 292, 325, 333, 340, 348, 349, 352, 355, 359–364, 407, 408, 437, 461, 468, 479, 494, 495, 505, 518, 519, 527
Depletion width, 7, 30, 31, 33, 34, 38, 39, 70, 108, 125–127, 133, 134, 191, 200, 269, 312, 350, 354, 359, 360, 362, 363, 367, 383, 462, 488, 489, 527, 530, 546
Deposited oxide, 161, 162, 243
Destructive failure, 537
Dielectric, 46, 81, 140, 161, 162, 197, 243, 273, 358, 374, 375, 378, 493, 518
Dielectric constant, 9, 44, 78
Diffusion coefficients, 402
Diffusion length, 402, 407
Disk drives, 18
Displacement current, 61
Display drives, 1
Dopant ionization energy, 6, 120, 123
Doping concentration, 6, 7, 9–13, 16, 17, 24, 26, 27, 30–34, 37–39, 41, 44, 45, 49, 56, 57, 69, 70, 72–75, 78, 80, 84, 96, 101, 106, 111, 113, 120, 121, 123, 125–129, 133, 134, 138, 140, 141, 143, 153–155, 161–165, 168–171, 173, 174, 176, 181–183, 187, 189, 191, 195, 200, 202, 203, 224–230, 232–237, 241–243, 248, 250, 251, 256, 263, 265–269, 271, 276, 279, 285–287, 291, 296, 301, 303, 307, 309, 312, 323, 325–331, 333, 338–340, 343, 345–348, 350–354, 356, 357, 360, 362–364, 378, 379, 381–383, 385, 388, 391, 392, 394, 403, 404, 407–410, 415, 416, 426, 427, 430, 437, 440, 441, 446, 453, 461, 462, 464, 467–469, 477, 479–481, 485, 487–489, 491, 492, 494, 496, 504–506, 508, 512, 513, 515, 516, 518, 519, 528, 546
Doping profile, 27, 30, 34, 35, 38–44, 56, 70, 71, 73–80, 101–103, 113, 119, 120, 123, 126, 128–131, 140, 141, 153, 155, 160, 174, 175, 222, 225, 229, 230, 232, 236, 239, 241–244, 246–254, 256, 258, 267, 268, 286–288, 290–292, 294, 296–298, 300–304, 306, 317, 331, 332, 338, 415, 416, 437, 439, 440, 445, 446, 450, 453, 456, 457, 464, 469, 481, 482, 506, 507
Drift region, 5–8, 24, 64, 121, 159, 241, 323, 400, 477, 537
Drift region resistance, 5, 7, 8, 31–32, 68–69, 72, 96, 105, 121, 126–127, 186, 189–191, 241, 263, 266–269, 326, 346, 351–353, 356, 456, 491, 515

Duty cycle, 54, 90, 106, 151, 218, 283, 372

E
Edge, 5, 27, 29, 34, 68, 124, 128–130, 141, 188, 201, 221, 222, 266, 351, 374, 375, 478, 482, 487, 495, 504, 508
Edge termination, 27, 34, 38–40, 56, 70, 74, 75, 113, 123, 127, 133–138, 153, 176, 191, 220–224, 237, 269, 285, 317, 333, 354, 373–378, 392, 417
Electric field, 6–16, 30, 38, 39, 41, 43, 44, 64, 65, 74, 78–80, 98, 103, 104, 125, 133, 135, 137, 138, 159–161, 163–173, 178–182, 184, 185, 187, 221, 222, 224, 225, 227–232, 237, 239, 241, 243–248, 250, 253–256, 258–263, 265, 285, 286, 288–289, 292–294, 296, 298, 300, 303, 304, 317, 323, 325–328, 333, 336–338, 340, 342–346, 348, 350, 358, 360, 362, 363, 374, 375, 377–380, 414, 416, 419, 420, 436, 437, 445, 447, 449, 456, 457, 477–479, 482–484, 490, 505, 506, 508–511, 514
Electric field profile, 6, 9, 41, 44, 78, 80, 98, 103, 104, 135, 169–173, 178, 180–182, 227, 228, 230, 231, 241, 244, 246, 253, 255, 256, 258, 260–263, 288, 289, 292, 293, 298, 299, 303, 304, 333, 336, 337, 343–345, 379, 380, 419, 437, 449, 456, 482, 510, 511
Electric trains, 554
Electron injection, 464
Electron irradiation, 399, 402, 407, 425, 428
Electron mobility, 34, 70, 127, 191, 267, 269, 354
Energy band gap, 7
Environmental pollution, 554
Epitaxial growth, 324
Epitaxial layer, 63, 120, 160, 242, 324, 527

F
Fabrication, 23, 65, 74, 112, 153, 161, 162, 173, 174, 220, 225, 229, 236, 243, 250, 285, 286, 324, 374, 391, 554, 555
Field oxide, 221, 222
Fluorescent lamps, 554
Fly-back diode, 19
Forward characteristics, 528
Forward voltage drop, 3, 311, 386, 403, 409, 421, 427, 430, 440, 441, 450, 457, 464, 469, 527, 528
Fulop's power law, 12, 14–17
Fundamental properties, 6

G
Gauss's law, 247
Graphics chips, 18, 556

H
H-bridge, 399, 424–426, 434, 440, 464, 468, 543, 544, 547
High level injection, 403, 404, 427, 437, 461, 464
High level lifetime, 403, 404, 427, 430, 464
HVDC transmission, 2, 535
Hybrid electric cars, 554

I
Ideal drift region, 6–8, 31, 68, 70, 126, 189, 266, 351, 489
Ideal power rectifier, 548, 554, 555
Ideal specific on-resistance, 7–12, 16, 17, 24, 33, 34, 58, 64, 70, 105, 114, 121, 127, 128, 155, 159, 161, 191, 228, 236, 237, 243, 269, 294, 300, 304, 317–319, 323, 325, 346–348, 353, 354, 391–393, 487, 491, 515, 549
IGBT. *See* Insulated gate bipolar transistor
Impact ionization, 10, 12, 13, 17, 167, 246, 254, 255, 336, 338, 343, 344, 375
Impact ionization coefficients, 7, 12, 13, 16, 17, 21, 58, 105, 114, 155, 166, 237, 245, 254, 258, 260, 261, 294, 300, 304, 318, 327, 347, 393, 549
Inductive load, 106–112, 310–316, 385–391, 499–503, 523–527, 548, 549
Insulated gate bipolar transistor (IGBT), 1, 19, 20, 543, 545, 547, 554
Integral diode, 399–476, 531, 532, 547, 548
Intrinsic carrier concentration, 402
Ion implantation, 5, 6, 23, 26, 63, 67, 120, 121, 123, 129, 160, 161, 173, 225, 229, 242, 243, 250, 286, 324, 325, 330, 378, 478, 506
Ionization integral, 13, 166, 244

J
JBS rectifier. *See* Junction barrier Schottky rectifier
JTE. *See* Junction termination extension
Junction barrier Schottky (JBS) rectifier, 414, 443, 504
Junction depth, 26, 29, 31, 38, 47, 49, 56, 68, 73, 81, 83, 84, 93, 113, 123, 125, 140, 141, 143, 153, 187, 197, 201, 237, 265, 273, 317, 330, 331, 350, 392

**L**

Laptops, 1, 18, 119, 556
Leakage current, 2, 3, 182, 226, 227, 230, 232, 233, 287, 292, 296, 301, 307, 333, 339, 382, 399, 413, 414, 417, 419, 434, 437, 444–447, 449, 454, 456, 481, 508, 539
Lifetime, 399, 402–404, 407–410, 412, 419, 421, 422, 425, 427, 428, 430, 432, 434, 439–441, 449, 452, 457, 461, 463, 464, 467–470, 472–474, 527, 528, 530, 532
Lithography, 23, 26, 114, 123, 187, 265

**M**

Material properties, 7
Maximum depletion width, 7, 38, 39, 134
Maximum electric field, 7, 14, 41, 78, 80, 135, 163, 244, 419, 449, 456, 479, 482, 506, 510
Merged PN Schottky (MPS) rectifiers, 399, 430
Mesa, 63, 69, 70, 103, 109, 169–174, 176, 178, 189, 200–203, 205, 206, 209, 213, 217, 220, 224–230, 242–244, 246, 251, 254, 256, 271, 275, 278, 279, 282, 285–288, 291, 292, 296, 298, 301, 303, 311–313, 374, 379, 443–446, 450, 457, 459
Mesa width, 67, 111, 113, 114, 163–165, 167, 169, 172, 173, 182, 187, 198, 200, 202, 222, 225, 229, 237, 248, 250, 253, 260, 264, 265, 267, 268, 273, 276, 286, 309, 315, 317
Metal-semiconductor contacts, 417, 434, 447, 453, 454
Minority carrier lifetime, 399, 402, 403, 409, 419, 425, 427, 430, 440, 449, 457, 464
Minority carriers, 46, 410, 415, 421, 437, 461, 468, 470, 527, 528, 532
Mobility, 7, 8, 10–12, 17, 28–30, 33, 34, 36, 37, 56, 63, 70, 113, 124–127, 132, 153, 161, 188, 189, 191, 237, 242, 265–267, 269, 317, 347, 348, 350, 351, 354, 392, 477, 487–492, 497, 499, 505, 513–516, 520, 531, 532, 552
MOSFET, 1, 23, 63, 119, 159, 241, 323, 399, 477, 535
Motor control, 20, 391, 399, 424–472, 474, 478, 535, 543–549, 552
Motor current, 424, 449, 457
Motor windings, 19, 426, 437, 440, 461, 464, 468, 469
MPS rectifiers. *See* Merged PN Schottky rectifiers
Multiplication, 358

**O**

Ohmic contact, 32, 174, 225, 229, 250, 286, 413, 419, 437, 445, 449, 456
On-resistance, 1, 24, 63, 119, 159, 241, 323, 401, 477, 537
On-state characteristics, 403, 409, 420, 426, 431, 440, 450, 458, 465, 469
On-state voltage drop, 19, 52, 88, 111, 147, 149, 155, 158, 210, 213, 280, 313, 315, 352, 387, 388, 390, 402, 403, 407, 409, 413, 421, 427, 430, 440, 441, 444, 450, 457, 464, 469, 473, 475, 501, 503, 524, 525, 527, 528, 531, 532, 537, 543, 545, 547
Operating temperature, 539
Optimum spacing, 121
Oxide field, 221, 285, 374
Oxide passivation, 160

**P**

Paralleling devices, 103, 113, 133, 237, 378, 392
Parallel-plane junction, 13–16, 38, 133, 161, 164, 243, 317, 325, 391
Phosphorus, 5, 63, 123, 160, 242, 324, 506
Photoresist, 26, 123
Pinch-off voltage, 141, 147, 149, 155, 158, 282, 315, 360, 387, 388, 390, 394, 463, 467, 468, 494, 496, 498, 521
P-i-N rectifiers, 20, 402, 404, 407, 425, 427, 430
Planar edge termination, 220–224
Planar gate, 5, 23, 119, 128, 236, 399, 424
Planar junction, 38, 133
P-N junction, 8, 13, 14, 19, 38, 48, 83, 103, 159, 251, 258, 260, 261, 288, 294, 298, 304, 323, 400, 421, 430, 432, 437, 457, 459, 461, 469, 470, 472, 527, 528, 532, 537
Poisson's equation, 6, 163, 169, 244
Polysilicon, 5, 23, 26, 34, 47, 49, 51, 56, 120–123, 126, 129, 132, 141, 143, 153, 155, 161, 162, 237, 242, 243, 318, 358, 485
Potential barrier, 121, 133, 479, 505, 506
Potential contours, 40–43, 75–77, 103, 134–137, 176–179, 184, 185, 227, 254, 256, 257, 287–289, 292, 293, 296, 298, 303, 333–335, 339–342, 379, 380, 417, 419, 447, 448, 456, 481–483, 508
Potential crowding, 340
Potential distribution, 178, 184, 223, 224, 256, 333, 340, 375, 376, 456, 549, 552
Power dissipation, 2, 20, 54, 55, 59, 60, 90, 91, 106, 110, 111, 114, 117, 151, 152, 155, 156, 218, 239, 283, 310, 314, 315, 372,

385, 388, 394, 400, 402, 410, 422, 426–428, 430, 437, 440, 441, 452, 459, 461, 464, 467, 468, 470, 530, 539, 541, 544
Power loss, 2, 19, 20, 50, 54, 55, 86, 90, 91, 106, 145, 151, 152, 218, 283, 284, 310, 372, 385, 399, 412, 504, 527, 537, 539–541, 548
Power MOSFET, 1, 23, 63, 119, 159, 241, 323, 399, 477, 535
Pre-avalanche, 358
Pulse width modulation, 19
Punch-through, 379

**R**
Radius of curvature, 6
Reach-through, 133, 161, 187, 243, 265, 288, 294, 298, 304, 481
Reach-through breakdown, 41, 121, 477, 479, 505
Reactive ion etching, 63, 160, 242
Resistivity, 9, 10, 30–32, 126, 127, 189, 267, 346, 352, 477, 489
Reverse blocking, 382
Reverse characteristics, 399
Reverse recovery, 19, 20, 399–402, 404–408, 410–412, 421–425, 428–430, 432, 433, 437–439, 441–443, 452, 453, 459–461, 463, 464, 466–468, 470–475, 527, 530–532, 537, 547, 548, 552
Rupture, 222, 477, 479, 482, 506, 510

**S**
Schottky barrier lowering, 414, 419, 447
Schottky contact, 8, 413–417, 419, 421, 422, 424, 425, 430, 432, 434, 437, 443–459, 467–470, 472, 504, 548, 552
Schottky rectifier, 6, 8, 19, 20, 399, 413–415, 443, 444, 504, 537
Shielded Schottky rectifiers, 504
Shielding, 119, 196, 272, 437, 478
Shielding region, 120, 121, 123, 125, 126, 128, 129, 133, 135, 138, 416, 419, 478–482, 488–491, 494, 495, 503–506, 508, 510, 514, 515, 518, 519, 527, 530
Short-circuiting, 19, 401, 479, 505, 537
Silicon carbide, 1, 7, 8, 20, 21, 477–479, 484–486, 505, 506, 510, 512, 532, 543, 546, 548–550, 552
Silicon dioxide, 222
Silicon P-i-N rectifier, 20
Spacing, 3–5, 120, 256
Specific capacitance, 50, 85, 99, 146, 207, 208, 277, 366
Specific on-resistance, 8, 12, 17, 57, 58, 115, 154, 156, 233–235, 238, 308, 319, 347, 384, 393, 490, 491, 514, 515, 550
Spreading angle, 31, 126, 187, 264
Spreading resistance, 188, 266
Stored charge, 400, 402, 404, 407, 410, 421, 425, 430, 437, 461, 467, 468
Sub-micron, 23
Substrate resistance, 32
Sub-surface region, 478, 504
Switching characteristics, 3, 4, 44, 80, 138, 195, 271, 357
Switching energy, 540
Switching loci, 44, 80, 138, 195, 271, 357
Switching losses, 54, 55, 90, 91, 138, 147, 151, 152, 195, 212, 217, 218, 272, 279, 283, 284, 371, 372, 550
Switching speed, 1, 19, 23, 46, 50, 86, 119, 139, 145, 196, 273, 358, 399, 537, 543, 545
Switching transient, 310, 385
Switch mode power supplies, 55, 90, 152, 218, 372, 540
Synchronous rectifier, 19, 401, 537
Synopsis, 535–552

**T**
Technology, 11, 23, 32, 55, 61, 63, 64, 90, 91, 151, 152, 191, 218, 269, 283, 286, 317, 319, 353, 372, 474, 475, 490, 499, 514, 522, 541–543, 545, 546, 550
Telecommunications, 2, 535
Temperature coefficient, 402
Thermal oxidation, 161, 162, 235, 242, 309
TMBS rectifier. *See* Trench MOS barrier Schottky rectifier
Transformer, 18
Transient current, 538, 539
Transition time, 52, 54, 55, 88, 91, 209, 279
Trench, 6, 24, 63, 119, 159, 241, 323, 399
Trench bottom, 65, 67, 92–101, 113, 153, 198, 201, 219, 220, 284, 296
Trench corner, 79
Trench depth, 9, 68, 83, 93, 95, 111, 174, 226, 230, 245, 251, 253, 291, 296, 300, 315, 317
Trench MOS barrier Schottky (TMBS) rectifier, 414
Trench refill, 376, 377
Trench sidewalls, 70, 96, 161, 162, 174, 187, 188, 197, 198, 207, 221, 222, 226, 235, 242, 243, 251, 253, 254, 264, 266, 273, 285, 323

Trench width, 69, 82, 84, 87, 93, 113, 114, 173, 188–191, 197, 201, 205, 210, 212, 225, 229, 237, 250, 266, 279, 286, 318
TSBS rectifier. *See* Trench Schottky barrier Schottky rectifier
Two-dimensional numerical simulations, 23, 34, 40, 45, 49, 52, 70, 75, 80, 84, 88, 96, 99, 101, 105, 111, 128, 134, 139, 144, 149, 173, 195, 206, 214, 222, 224, 229, 232, 250, 272, 276, 280, 286, 290, 296, 300, 307, 315, 329, 345, 364, 369, 374, 378, 382, 389, 403, 409, 415, 426, 430, 439, 445, 452, 464, 468, 478, 480, 497, 502, 506, 515, 520, 525, 527
Typical power rectifier, 536

**U**

Unipolar power devices, 6–8

**V**

Variable frequency motor drives, 19–21

**W**

Wide band-gap semiconductors, 8
Wireless base-station, 4
Work function, 417, 434, 447, 454, 468, 485, 512

**Z**

Zero-bias depletion width, 30, 31, 125, 126, 488, 489

www.ingramcontent.com/pod-product-compliance
Lightning Source LLC
LaVergne TN
LVHW020018181026
850497LV00003B/114
* 9 7 8 1 4 4 1 9 5 9 1 6 4 *